Lecture Notes in Computer Science 8587

Commenced Publication in 1973
Founding and Former Series Editors:
Gerhard Goos, Juris Hartmanis, and Jan van Leeuwen

Markus Holzer   Martin Kutrib (Eds.)

Implementation and Application of Automata

19th International Conference, CIAA 2014
Giessen, Germany, July 30 – August 2, 2014
Proceedings

 Springer

Volume Editors

Markus Holzer
Martin Kutrib

Universität Giessen, Institut für Informatik
Arndtstrasse 2, 35392 Giessen, Germany
E-mail: {holzer, kutrib}@informatik.uni-giessen.de

ISSN 0302-9743   e-ISSN 1611-3349
ISBN 978-3-319-08845-7   e-ISBN 978-3-319-08846-4
DOI 10.1007/978-3-319-08846-4
Springer Cham Heidelberg New York Dordrecht London

Library of Congress Control Number: 2014942440

LNCS Sublibrary: SL 1 – Theoretical Computer Science and General Issues

Typesetting: Camera-ready by author, data conversion by Scientific Publishing Services, Chennai, India

Printed on acid-free paper

Springer is part of Springer Science+Business Media (www.springer.com)

Preface

The 19th International Conference on Implementation and Application of Automata (CIAA 2014) was organized by the Institut für Informatik of the Universität Giessen and took place at the campus of natural sciences. It was a four-day conference starting July 30 and ending August 2, 2014. The Universität Giessen is one of the older universities in the German-speaking part of Europe. It was founded in 1607.

The CIAA conference series is a major international venue for the dissemination of new results in the implementation, application, and theory of automata. The previous 18 conferences were held in the following locations: Halifax (2013), Porto (2012), Blois (2011), Winnipeg (2010), Sydney (2009), San Francisco (2008), Prague (2007), Taipei (2006), Nice (2005), Kingston (2004), Santa Barbara (2003), Tours (2002), Pretoria (2001), London Ontario (2000), Potsdam (WIA 1999), Rouen (WIA 1998), London Ontario (WIA 1997 and WIA 1996).

This volume contains the invited contributions and the accepted papers presented at CIAA 2014. The submission and refereeing process was supported by the EasyChair conference management system. In all, 36 papers were submitted by authors in 23 different countries, including Argentina, Brazil, Canada, China, Czech Republic, France, Germany, Greece, Hungary, Italy, Japan, Korea Republic, Latvia, The Netherlands, Poland, Portugal, Romania, Russian Federation, Slovakia, Spain, Switzerland, Ukraine, and the USA. Each submission was reviewed by at least three referees and discussed by the Program Committee. A total of 21 full papers were selected for presentation at the conference. There were four invited talks presented by Javier Esparza, Friedrich Otto, Giovanni Pighizzini, and Georgios Ch. Sirakoulis. We warmly thank the invited speakers and all authors of the submitted papers and the members of the Program Committee for their excellent work in making this selection. We also thank the additional external reviewers for their careful evaluation. All these efforts were the basis for the success of the conference. The collaboration with Springer for preparing this volume was very efficient and pleasant. We like to thank in particular Alfred Hofmann and Anna Kramer from Springer for their help. We are grateful to the additional members of the Organizing Committee consisting of Susanne Gretschel, Sebastian Jakobi, Andreas Malcher, Katja Meckel, Heinz Rübeling, Bianca Truthe, and Matthias Wendlandt for their support of the sessions and the accompanying events.

Finally, we are indebted to all participants for attending the conference. We hope that this conference will be a successful and fruitful meeting, will bear new ideas for investigations, and will bring together people for new scientific collaborations. Looking forward to CIAA 2015 in Umeå, Sweden.

August 2014   Markus Holzer
                                                                     Martin Kutrib

Organization

CIAA 2014 was organized by the Institut für Informatik of the Universität Giessen, Germany. The conference took place at the campus of natural sciences.

Invited Speakers

Javier Esparza	Technische Universität München, Germany
Friedrich Otto	Universität Kassel, Germany
Giovanni Pighizzini	Università degli studi di Milano, Italy
Georgios Ch. Sirakoulis	Democritus University of Thrace, Greece

Program Committee

Holger Bock Axelsen	University of Copenhagen, Denmark
Cezar Câmpeanu	University of Prince Edward Island, Canada
Jean-Marc Champarnaud	Université de Rouen, France
Stefano Crespi-Reghizzi	Politechnico di Milano, Italy
Jürgen Dassow	Otto-von-Guericke-Universität Magdeburg, Germany
Frank Drewes	Umeå University, Sweden
Manfred Droste	Universität Leipzig, Germany
Markus Holzer	Universität Giessen, Germany
Oscar Ibarra	University of California, USA
Katsunobu Imai	Hiroshima University, Japan
Galina Jirásková	Slovak Academy of Sciences, Slovak Republic
Christos Kapoutsis	Carnegie Mellon University in Qatar, State of Qatar
Juhani Karhumäki	University of Turku, Finland
Ondřej Klíma	Masaryk University, Czech Republic
Stavros Konstantinidis	St. Mary's University, Canada
Martin Kutrib	Universität Giessen, Germany
Andreas Maletti	Universität Stuttgart, Germany
Denis Maurel	Université Francois Rabelais, France
Carlo Mereghetti	Università degli Studi di Milano, Italy
Brink van der Merwe	University of Stellenbosch, South Africa
Nelma Moreira	Universidade do Porto, Portugal
František Mráz	Univerzita Karlova, Czech Republic
Cyril Nicaud	Université Paris-Est, France

Daniel Reidenbach   Loughborough University, UK
Kai Salomaa   Queen's University, Canada
Klaus Sutner   Carnegie Mellon University, USA
György Vaszil   University of Debrecen, Hungary
Hsu-Chun Yen   National Taiwan University, Taiwan

External Referees

Cyril Allauzen   Ludovic Mignot
Henning Bordihn   Mark-Jan Nederhof
Alessandra Cherubini   Timothy Ng
Joel Day   Florent Nicart
Nathalie Friburger   Alexander Okhotin
Zoltan Fülöp   Bruno Patrou
Daniel Goc   Vincent Penelle
Jozef Gruska   Vitaly Perevoshchikov
Petr Jančar   Giovanni Pighizzini
Nataša Jonoska   Daniel Quernheim
Andrzej Kisielewicz   Bala Ravikumar
Sylvain Lombardy   Sylvain Schmitz
Jean-Gabriel Luque   Shinnosuke Seki
António Machiavelo   Olivier Serre
Eva Maia   Nicolas Thiéry
Dino Mandrioli   Nicholas Tran
Tomás Masopust   Stéphane Vialette
Jonathan May   Janis Voigtländer

Organizing Committee

Susanne Gretschel   Katja Meckel
Markus Holzer   Heinz Rübeling
Sebastian Jakobi   Bianca Truthe
Martin Kutrib   Matthias Wendlandt
Andreas Malcher

Sponsoring Institutions

Universität Giessen

Table of Contents

Invited Papers

FPSOLVE: A Generic Solver for Fixpoint Equations over Semirings 1
    Javier Esparza, Michael Luttenberger, and Maximilian Schlund

Restarting Automata for Picture Languages: A Survey on Recent
Developments 16
    Friedrich Otto

Investigations on Automata and Languages over a Unary Alphabet 42
    Giovanni Pighizzini

Cellular Automata for Crowd Dynamics 58
    Georgios Ch. Sirakoulis

Regular Papers

Counting Equivalent Linear Finite Transducers Using a Canonical
Form 70
    Ivone Amorim, António Machiavelo, and Rogério Reis

On the Power of One-Way Automata with Quantum and Classical
States 84
    Maria Paola Bianchi, Carlo Mereghetti, and Beatrice Palano

On Comparing Deterministic Finite Automata and the Shuffle
of Words 98
    Franziska Biegler and Ian McQuillan

Minimal Partial Languages and Automata 110
    Francine Blanchet-Sadri, Kira Goldner, and Aidan Shackleton

Large Aperiodic Semigroups 124
    Janusz Brzozowski and Marek Szykuła

On the Square of Regular Languages 136
    Kristína Čevorová, Galina Jirásková, and Ivana Krajňáková

Unary Languages Recognized by Two-Way One-Counter Automata 148
    Marzio De Biasi and Abuzer Yakaryılmaz

A Type System for Weighted Automata and Rational Expressions...... 162
    *Akim Demaille, Alexandre Duret-Lutz, Sylvain Lombardy,
    Luca Saiu, and Jacques Sakarovitch*

Bounded Prefix-Suffix Duplication 176
    Marius Dumitran, Javier Gil, Florin Manea, and Victor Mitrana

Recognition of Labeled Multidigraphs by Spanning Tree Automata 188
    Akio Fujiyoshi

Reset Thresholds of Automata with Two Cycle Lengths............... 200
    Vladimir V. Gusev and Elena V. Pribavkina

On the Ambiguity, Finite-Valuedness, and Lossiness Problems in
Acceptors and Transducers.. 211
    Oscar H. Ibarra

Kleene Closure on Regular and Prefix-Free Languages 226
    Galina Jirásková, Matúš Palmovský, and Juraj Šebej

Left is Better than Right for Reducing Nondeterminism of NFAs 238
    Sang-Ki Ko and Yo-Sub Han

Analytic Functions Computable by Finite State Transducers........... 252
    Petr Kůrka and Tomáš Vávra

Partial Derivative and Position Bisimilarity Automata 264
    Eva Maia, Nelma Moreira, and Rogério Reis

The Power of Regularity-Preserving Multi Bottom-up Tree
Transducers 278
    Andreas Maletti

Pushdown Machines for Weighted Context-Free Tree Translation 290
    Johannes Osterholzer

Weighted Variable Automata over Infinite Alphabets 304
    Maria Pittou and George Rahonis

Implications of Quantum Automata for Contextuality 318
    Jibran Rashid and Abuzer Yakaryılmaz

Pairwise Rational Kernels Obtained by Automaton Operations 332
    Abiel Roche-Lima, Michael Domaratzki, and Brian Fristensky

Author Index... 347

FPSOLVE: A Generic Solver for Fixpoint Equations over Semirings*

Javier Esparza, Michael Luttenberger, and Maximilian Schlund

Technische Universität München, Munich, Germany
{esparza,luttenbe,schlund}@in.tum.de

Abstract. We introduce FPSOLVE, an implementation of generic algorithms for solving fixpoint equations over semirings. We first illustrate
the interest of generic solvers by means of a scenario. We then succinctly
describe some of the algorithms implemented in the tool, and provide
some implementation details.

1 Introduction

We present FPSOLVE[1], a solver for *algebraic systems of equations* first introduced
in [17]. These are systems of equations of the form

$$X_1 = f_1(X_1, X_2, \ldots, X_n) \quad \cdots \quad X_n = f_n(X_1, X_2, \ldots, X_n)$$

where $f_1, \ldots, f_n$ are polynomials in the variables $X_1, X_2, \ldots, X_n$. The coefficients of the polynomials can be elements of any semiring satisfying some weak
conditions, which ensure that there exists a unique smallest solution. FPSOLVE
implements a number of *generic* algorithms, i.e. algorithms parametric in the
semiring operations of addition and multiplication, plus possibly the Kleene star
operation.

Algebraic systems naturally arise in various settings:

- The language of a context-free grammar like $X \to aXX \mid b$ is the least
  solution of the equation $X = aXX + b$ over the semiring whose elements are
  languages, with union and concatenation of languages as sum and product.
- Shortest-paths problems on finite graphs and on some infinite graphs, like
  those generated by weighted pushdown automata, can be reduced to solving
  fixed-point equations over a semiring having the possible edge weights as
  elements [7,16].
- Data-flow equations associated to many intra- and interprocedural dataflow
  analyses are fixed-point equations over complete lattices [13], which can often
  be recast as equations on semirings [16,6].

---

* This work was funded by the DFG project "Polynomial Systems on Semirings: Foundations, Algorithms, Applications".

[1] Freely available from https://github.com/mschlund/FPsolve

M. Holzer and M. Kutrib (Eds.): CIAA 2014, LNCS 8587, pp. 1–15, 2014.

– Authorization problems (like, for instance, the authorization problem for the SPKI/SDSI authorization system), can be recast as a reachability problem in weighted pushdown automata [12], and thus to algebraic systems [6].
– Computing the reputation of a principal in a reputation system (a system in which principals can recommend other principals, and rules are used to compute reputation out of a set of direct recommendations) reduces to solving an algebraic system [5].
– Evaluating a Datalog query can be reformulated as the problem of deciding whether a non-terminal of a context-free grammar is productive or not, and so it also amounts to solving a system of equations. Moreover, several problems concerning the computation of *provenance information*, an important research topic in database theory, reduces to solving an associated algebraic system over different semirings. [11]

The paper is structured as follows. Section 2 motivates by means of a scenario the interest of generic solvers for algebraic systems. Section 3 describes the basic algorithms and data structures used in FPSOLVE . Finally, Section 4 briefly describes the implementation.

2   Scenario: A Recommendation System

We succinctly describe SDSIREC, a recommendation system inspired by the SDSI authorization system [12], and very close to the reputation system described in [5].

SDSIREC distinguishes *customers* (denoted by x, y, z) and *products* (denoted by p). Given a collection of individual recommendations of products by customers, SDSIREC computes an aggregated customer rating for each product. Individual recommendations are described in SDSIREC by means of *rules* of the form:

$$x.\mathsf{Rec} \xrightarrow{w} p \tag{1}$$

$$x.\mathsf{Trust} \xrightarrow{w} y \tag{2}$$

The term $x.\mathsf{Rec}$ denotes the fuzzy set of all products recommended by customer x. The rule $x.\mathsf{Rec} \xrightarrow{w} p$ denotes that p belongs to $x.\mathsf{Rec}$ with weight w, i.e., that x recommends p with "rating" w. Analogously, $x.\mathsf{Trust}$ denotes the fuzzy set of all customers (whose recommendations are) trusted by x. The set of all weights, denoted by S, contains the special weight $\mathbb{0}$, which explicitly states that p resp. y does not belong to $x.\mathsf{Rec}$ resp. $x.\mathsf{Trust}$; assigning a rule the weight $\mathbb{0}$ is equivalent to removing the rule from the input.

Besides direct recommendation and direct trust, SDSIREC also takes into account indirect recommendation of products via trust in other customers. For instance, consider the following scenario:

$$\mathrm{JESSE}.\mathsf{Trust} \xrightarrow{w_1} \mathrm{WALT}$$

$$\mathrm{WALT}.\mathsf{Rec} \xrightarrow{w_2} \mathrm{FPSOLVE}$$

Since WALT recommends FPSOLVE with weight w_2, and JESSE trusts the recommendations of WALT (his former high-school teacher) with weight w_1, SDSIREC infers that JESSE indirectly recommends FPSOLVE with some weight $w_1 \odot w_2$, where $\odot$ abstracts from the concrete way how the weights should be combined into a new weight. The operator must satisfy $0 \odot w = 0 = w \odot 0$, so that the interpretation of 0 as a non-existing rule is preserved. The inference is modeled be the following (hard coded) rules:

$$x.\mathsf{Trust} \xrightarrow{\lambda} x.\mathsf{Trust}.\mathsf{Trust} \qquad (3)$$

$$x.\mathsf{Rec} \xrightarrow{\mu} x.\mathsf{Trust}.\mathsf{Rec} \qquad (4)$$

Rule (3) states that the set of customers trusted by x contains the set of customers trusted by customers trusted by x. Analogously, rule (4) states that the set of products recommended by x contains all products recommended by customers trusted by x. As these rules may lead to cycles, i.e. x might trust herself, thereby recommending to herself the products recommended by her, SDSIREC allows one to specify discount factors λ and μ to dampen resp. penalize these effects. The special weight 1 (required to satisfy $w \odot 1 = w = 1 \odot w$) can be used to disable this discounting. SDSIREC then treats the rules (1) to (4) as rewrite rules in the sense of a pushdown system [16]. For instance, we get

$$\mathsf{JESSE}.\mathsf{Rec} \xrightarrow{\mu} \mathsf{JESSE}.\mathsf{Trust}.\mathsf{Rec} \xrightarrow{w_1} \mathsf{WALT}.\mathsf{Rec} \xrightarrow{w_2} \mathsf{FPSOLVE}$$

$$\mathsf{JESSE}.\mathsf{Rec} \xrightarrow{\mu} \mathsf{JESSE}.\mathsf{Trust}.\mathsf{Rec} \xrightarrow{\lambda} \mathsf{JESSE}.\mathsf{Trust}.\mathsf{Trust}.\mathsf{Rec} \xrightarrow{w_1} \mathsf{WALT}.\mathsf{Trust}.\mathsf{Rec}$$

The first "path" with weight $\mu \odot w_1 \odot w_2$ captures that JESSE indirectly recommends FPSOLVE. The second path is an example of a path that cannot be extended to a recommendation of p: Since WALT trusts nobody (as specified by the input system), SDSIREC can never rewrite WALT.Trust to WALT.

In order to compute to what extent p belongs to $x.\mathsf{Rec}$ SDSIREC finally aggregates the weight of the (possibly infinitely many) paths leading from $x.\mathsf{Rec}$ to p. We use $\oplus$ to denote the operator that is used to aggregate the weights of different paths. It is well-known that if $\langle S, \oplus, \odot, 0, 1 \rangle$ forms an ω-continuous semiring, then the problem of aggregating over all possible paths can be recast as computing the least solution of an algebraic system (see below) [9,16,6]. Recall that $\langle S, \oplus, \odot, 0, 1 \rangle$ is a semiring if $\oplus$ and $\odot$ are associative and have neutral elements 0 and 1, respectively, $\oplus$ is commutative, $\odot$ distributes over $\oplus$, and any product with 0 as factor evaluates to 0. Given $a, b \in S$, we say $a \sqsubseteq b$ if there is $c \in S$ such that $a + c = b$. A semiring is *naturally ordered* if the relation $\sqsubseteq$ is a partial order. An ω-*continuous* semiring is a naturally ordered semiring extended by an infinite summation-operator $\sum$ that satisfies some natural properties. In particular, for every sequence $(a_i)_{i \geq 0}$ the supremum $\sup\{\sum_{0 \leq i \leq k} a_i \mid k \in \mathbb{N}\}$ w.r.t. $\sqsubseteq$ exists, and is equal to $\sum_{i \in \mathbb{N}} a_i$ [14].

Let R_{xp} and T_{xy} be variables standing for the total weights with which x recommends p or trusts y, and let r_{xp}, t_{xy} denote the weights of the direct recommendation, or the direct trust of x in p and y, respectively (i.e. $x.\mathsf{Rec} \xrightarrow{r_{xp}} p$ resp. $x.\mathsf{Trust} \xrightarrow{t_{xy}} y$).

If $\langle S, \oplus, \odot, \mathbb{0}, \mathbb{1} \rangle$ is an ω-continuous semiring, then the total weights are the unique smallest solution w.r.t. $\sqsubseteq$ of the following algebraic system (cf. [9,16,6])

$$R_{xp} = r_{xp} \oplus \bigoplus_y \mu \odot T_{xy} \odot R_{yp} \qquad \text{for all consumers } x, \text{ products } p$$

$$T_{xy} = t_{xy} \oplus \bigoplus_z \lambda \odot T_{xz} \odot T_{zy} \qquad \text{for all consumers } x, y$$

The key point of our argumentation is that in an application like the above we are interested in solving the same set of equations over many different semirings. Even further, users of the system may be interested in first defining their own semiring, and then solving the system. To illustrate this, let us examine several different interpretations of "weight", all of them very natural:

Weights as Scores. The most natural interpretation of weights is perhaps as scores. A consumer x gives a product p or another consumer Y a score, corresponding to its degree of satisfaction with p, or its degree of trust in the recommendations of Y. If we assume that scores are real numbers in the interval $[0, 1]$, and choose $\oplus$ and $\otimes$ as sum and product of real numbers, we obtain the probabilistic semiring. Then R_{xp} represents the total weight of all recommendation paths leading from x to p. If we choose the Viterbi semiring $\langle [0, 1], \max, \cdot, 0, 1 \rangle$ instead, then R_{xp} returns the weight of the strongest recommendation path.

Weights as Expire Times. Direct recommendations and trust, represented by rules of types (1) and (2,) can (and should) have an expire time. If we choose $\oplus$ to be the maximum and $\odot$ the minimum over the reals, then R_{xp} returns the earliest time at which all recommendation paths from x to p will have expired.

Weights as Provenance Information. If a system user does not trust some consumers, she may wish to compute, for each recommendation path from x to p, the set of consumers in the path. Or she may want to know the set of consumers visited along the recommendation path of maximal weight. Such provenance information can be computed within the semiring framework. For this it is convenient to treat all non-zero parameters $r_{xp}, t_{xy}, \lambda, \mu$ as formal parameters (free variables).

- To compute for each path the set of consumers involved, one can use the Why-semiring, well-known in provenance theory. Semiring elements are sets of sets of consumers. We set $r_{xp} = \{\{x\}\}$ and $t_{xy} = \{\{x, y\}\}$ (and treat λ as $\{\{\lambda\}\}$ and μ as $\{\{\mu\}\}$), and define:
  - $\{X_1, \ldots, X_n\} \odot \{Y_1, \ldots, Y_n\} := \{X_1 \cup Y_1, x_1 \cup Y_2, \ldots, X_n \cup Y_m\}$, and
  - $\{X_1, \ldots, X_n\} \oplus \{Y_1, \ldots, Y_n\} := \{X_1, \ldots, X_n, Y_1, \ldots, Y_m\}$.
- If we wish to compute the provenance of the recommendation of maximal weight, we can use the following semiring: as semiring elements we choose the pairs (α, X), where $\alpha \in [0, 1]$ and X is a set of consumers. We set $t_{xy} = (w, \{x, y\})$ with $w \in (0, 1]$, and analogously for r_{xp}. The abstract operators are instantiated as follows:

- $(\alpha_1, X_1) \odot (\alpha_2, X_2) := (\alpha_1 \alpha_2, X_1 \cup X_2)$, and
- $(\alpha_1, X_1) \oplus (\alpha_2, X_2) := (\max\{\alpha_1, \alpha_2\}, \text{if } \alpha_1 \geq \alpha_2 \text{ then } X_1 \text{ else } X_2)$

These examples show that, instead of creating new tools for each new semiring, it can be better to implement a generic tool, with generic algorithms applicable to any semiring, or at least to any semiring in a broad class.

3 Algorithms and Data Structures

The two main generic schemes implemented in FPSOLVE for the approximation (and sometimes exact computation) of the least solution of an algebraic system are classical fixpoint iteration and Newton's method. Following [15,10], we introduce them as procedures that "unfold" the algebraic system up to a certain depth which allows both to unify and at the same time simplify their presentation.

Classical Fixed-Point Iteration. Given an algebraic system of equations $X = F(X)$ over an ω-continuous semiring, Kleene's theorem states that the system has a unique least solution μF with respect to the natural order $\sqsubseteq$, and that μF is the supremum of the sequence $F(0), F^2(0), \ldots, F^i(0)$ [14]. So μF can be approximated by computing successive elements of the sequence. If the semiring further satisfies the *ascending chain property* (for every ω-chain $a_1 \sqsubseteq a_2 \sqsubseteq \ldots$ eventually $a_k = a_{k+1} = a_{k+2} = \ldots$) then $\mu F = F^i(0)$ for some $i \geq 0$, and so μF can be effectively computed.

As explained in e.g. [15], an algebraic system can be associated a context-free grammar. For instance, for the system

$$X = a \odot X \odot X \oplus b \tag{5}$$

we obtain the grammar

$$X \to aXX \mid b \tag{6}$$

Conversely, we assign to a derivation tree of the grammar a value in the semiring, given by the product of its leaves (the ordered product if the semiring is not commutative); further, we assign to a set of derivation trees the sum of the values of its elements. The following result can be proved by a simple induction on k:

$F^k(0)$ is the sum over the set of all derivation trees of the grammar of height less than k.

Building on this observation, for every k we "unfold" (5) into an acyclic system over variables $X^{<h}$ and $X^{=h}$ for every $h \leq k$, such that the solutions of $X^{<h}$ and $X^{=h}$ are the values of the derivation trees of height *less* than h and *equal* to h, respectively. For this, we obviously have to set

$$X^{<0} = 0 \quad X^{=0} = b \quad \text{and} \quad X^{<h+1} = X^{<h} \oplus X^{=h} \text{ for all } h \in \mathbb{N} \tag{7}$$

In order to obtain the defining equation of $X^{=h}$ we observe that the trees of height h can be partitioned into those whose left subtree has height $h-1$, and those whose left subtree has height strictly smaller than $h-1$ *and* whose right subtree has height $h-1$, i.e. we partition by means of the first position from the right at which a subtree of height exactly $h-1$ is rooted. This leads to

$$X^{=h} = a \odot X^{<h} \odot X^{=h-1} \oplus a \odot X^{=h-1} \odot X^{<h-1}. \tag{8}$$

We can see the unfolding up to depth k as a symbolic representation of $F^k(0)$, which implicitly uses subterm sharing (*arithmetic circuit*). When the coefficients of the algebraic system (a, b in our example) are formal parameters, we can efficiently compute $F^k(0)$ for different values a, b, by just plugging them into the unfolded system.

Newton's Method. Newton's method for arbitrary ω-continuous semirings, as described in [9], can be much faster than Kleene iteration. It is shown in [10] that the method can also be presented as an unfolding of the algebraic system: This time, the system is unfolded w.r.t. the *Strahler number* or *dimension* of its associated derivation trees (see [10,15]). The dimension of a rooted tree t is defined as the height of the largest perfect binary tree that is a minor of t.

Consider again equation (5). We split X into a family of variables $X^{<d}$ and $X^{=d}$ for $d \in \mathbb{N}$. The solutions of the unfolded system for $X^{<d}$ and $X^{=d}$ will now be the value of the derivation trees of dimension *less* than d, and *equal* to d, respectively. Just as before, we have

$$X^{<0} = 0 \quad X^{=0} = b \quad \text{and} \quad X^{<d+1} = X^{<d} \oplus X^{=d} \text{ for all } d \in \mathbb{N}. \tag{9}$$

In order to derive the defining equation of $X^{=d}$, observe that there are three possible cases for a tree of dimension d: either the left subtree has dimension d, and the right subtree has dimension at most $d-1$; or vice versa; or both subtrees have dimension exactly $d-1$ (this is the case in which the root of the minor coincides with the root of the tree). So we get

$$X^{=d} = a \odot X^{=d} \odot X^{<d} \oplus a \odot X^{<d} \odot X^{=d} \oplus a \odot X^{=d-1} \odot X^{=d-1}. \tag{10}$$

However, this unfolding does not represent an arithmetic circuit as it is not yet acyclic: $X^{=d}$ appears on both sides of the equation. But equation (10) is linear in $X^{=d}$, and so, if multiplication is commutative, we can replace it by (with $1 \oplus 1 = 2$)

$$X^{=d} = 2 \odot a \odot X^{=d} \odot X^{<d} \oplus a \odot X^{=d-1} \odot X^{=d-1} \tag{11}$$

and use Kleene's theorem [14] to replace it by

$$X^{=d} = \left(2 \odot a X^{<d}\right)^* \odot a \odot \left(X^{=d-1}\right)^2 \tag{12}$$

where the Kleene star is defined, as usual, by $x^* := \sum_{k \in \mathbb{N}} x^k$ (and is well defined for any ω-continuous semiring).[2] The new system is acyclic, i.e. an arithmetic

---

[2] In the noncommutative case, one may resort to an instance of the semiring of contexts in order to obtain a rational tree expression.

circuit w.r.t. $\oplus$, $\odot$, and $*$, and as in the previous case, can be used as a compact symbolic representation of the d-th Newton approximation, useful when a, b are formal parameters (see Fig. 1 for an example). In particular, every Newton approximation can be represented by means of a rational expression.

To actually compute the solution for particular values of a and b, we can then use straight-forward constant propagation going from bottom $(X^{=0}, X^{<0})$ to top $(X^{<d})$. However, for this the Kleene star x^* must be effectively computable in the given semiring representation. This is indeed the case for several important semirings. The simplest example is the probability semiring, where for every rational number $x \in [0, 1)$ we have $x^* = 1/(1 - x)$. Tropical semirings are another example. For instance, over the integers extended by least $(-\infty)$ and greatest element $(+\infty)$, with addition given by min and multiplication given by $+$ (on $\mathbb{Z}$), we have $x^* = 0$ if $x \geq 0$ and $x = -\infty$ otherwise. A third example is the semiring of semilinear sets of vectors with components in $\mathbb{N} \cup \{\infty\}$, with $X \oplus Y := X \cup Y$, and $X \odot Y := \{x + y \mid x \in X, y \in Y\}$.

Connection to Newton's Method over the Reals. Applying Newton's method to $g(X) := f(X) - X = aX^2 - X + b$ (interpreted over the reals) starting form the initial approximation $X = 0$ we obtain the sequence:

$$X_0 := 0 \quad X_{d+1} = X_d - \frac{g(X_d)}{g'(X_d)} = X_d - \frac{aX_d^2 - X_d + b}{2aX_d - 1} = X_d + \frac{aX_d^2 - X_d + b}{1 - 2aX_d}.$$

Setting $Y_d := X_{d+1} - X_d$ this can be written as

$$Y_d = 2aX_dY_d + (aX_d^2 - X_d + b).$$

Straight-forward induction now shows that over the nonnegative reals the values of X_d and $X^{<d}$ resp. Y_d and $X^{=d}$ coincide [10]. In particular, the defining equation of $X^{=d}$ can be seen as the generalization of the derivative of aX^2 in the noncommutative case.

Multivariate Case. Both unfoldings immediately generalize to the setting of multiple variables $X, Y, \ldots$. As mentioned above, in the univariate case we use the fact that the solution of an equation $X = aX + b$ is given by a^*b. In the multivariate case, when the semiring is commutative, we have to deal with systems of linear equations $X = AX + B$ for a matrix A and a vector B over the semiring. It is well known that the solution is given by A^*B, where $A^* = \sum_{i=0}^{\infty} A^i$, and matrix multiplication is defined as for the natural or the real numbers, but replacing sum and product by the operations of the semiring being considered. In the next section, we describe the two algorithms for computing A^* implemented in FPSOLVE.

3.1   Solving Linear Equations

As mentioned above, solving a linear equation amounts to computing A^* for a given square matrix A over a semiring. Given a matrix A, the algorithms returns a matrix whose elements are semiring expressions over the semiring operations and the Kleene star. So, intuitively, the algorithms reduce the problem of computing the star of a matrix to computing the star of semiring elements.

FPSOLVE implements both the well-known Floyd-Warshall algorithm, and the recursive divide-and-conquer approach.

Generalized Floyd-Warshall. The Floyd-Warshall algorithm for solving the all-pairs-shortest-path problem in weighted (finite) graphs carries directly over to the setting of generic ω-continuous semirings, even if addition $\oplus$ is not idempotent (cf. [7]). (In fact, it suffices when the semiring $\langle S, \oplus, \odot, \mathbb{0}, \mathbb{1} \rangle$ is closed but not necessarily ω-continuous.) The following description is an optimized variant of the algorithm in [7] which reduces the number of semiring operations required.

> **input** : Matrix $\mathbf{A} \in S^{n \times n}$ over a semiring S.
> **output**: Reflexive-transitive closure $\mathbf{A}^*$.

$B := A$
for $k = 1 \ldots n$ **do**
  $B_{k,k} := B_{k,k}^*$
  for $i = 1 \ldots n,\, i \neq k$ **do**
    $B_{i,k} = B_{i,k} \odot B_{k,k}$
    for $j = 1 \ldots n,\, j \neq k$ **do**
      | $B_{i,j} := B_{i,j} \oplus B_{i,k} \odot B_{k,j}$
    end
  end
  for $j = 1 \ldots n, j \neq k$ **do**
    | $B_{k,j} = B_{k,k} \odot B_{k,j}$
  end
end
return B

Algorithm 1. Generalized Floyd-Warshall algorithm over semirings

From the description of the algorithm it is easy to count that the total number of semiring operations (i.e. $+, \cdot, ^*$) needed is $T(n) = 2n^3 - 2n^2 + n \in \Theta(n^3)$.

Divide-and-Conquer. This algorithm recursively applies the formula for computing the Kleene star of a 2×2-matrix:

$$\mathbf{M} = \begin{bmatrix} \mathbf{A} \ \mathbf{B} \\ \mathbf{C} \ \mathbf{D} \end{bmatrix} \qquad \mathbf{M}^* = \begin{bmatrix} \mathbf{F} & \alpha \mathbf{G}^* \\ \mathbf{G}^* \beta & \mathbf{G}^* \end{bmatrix} \qquad \text{with} \quad \begin{aligned} \alpha &= \mathbf{A}^* \mathbf{B} \\ \beta &= \mathbf{C} \mathbf{A}^* \\ \mathbf{G} &= \mathbf{D} + \mathbf{C}\alpha \\ \mathbf{F} &= \alpha \mathbf{G}^* \beta + \mathbf{A}^* \end{aligned}$$

Given a $n \times n$-matrix $\mathbf{M}$ ($n > 2$), the entries $\mathbf{A}, \mathbf{B}, \mathbf{C}, \mathbf{D}$ become submatrices of $\mathbf{M}$ to which the algorithm is then applied recursively; the recursion stops when either $n = 2$ or $n = 1$. A formal proof of correctness (for any *Conway* semiring) goes back to Ésik and Kuich (cf. [7]).

Altogether we need two recursive calls, six matrix multiplications (the term αG^* appears twice and thus needs to be evaluated only once), and two matrix additions. Hence, the number of operations needed by this algorithm can be expressed by the recurrence relation[3]

$$T(n) = 2T\left(\frac{n}{2}\right) + 6\left[2\left(\frac{n}{2}\right)^3 - \left(\frac{n}{2}\right)^2\right] + 2\left(\frac{n}{2}\right)^2$$

$$= 2T\left(\frac{n}{2}\right) + \frac{3}{2}n^3 - n^2$$

Which can be solved exactly (setting $T(n) = 1$ and $n = 2^l$) resulting in $T(n) = 2n^3 - 2n^2 + n \in \Theta(n^3)$. Hence this algorithm uses the same number of operations as Floyd-Warshall. Both algorithms need $n^3 - 2n^2 + n$ additions, $n^3 - n$ multiplications, and n Kleene stars.

Symbolic Solving. Recall our initial example

$$X = a \odot X \odot X \oplus b$$

and its unfolding w.r.t. dimension for an arbitrary $d \in \mathbb{N}$ (assuming $\odot$ is commutative)

$$X^{=d} = 2 \odot a \odot X^{=d} \odot X^{<d} \oplus a \odot X^{=d-1} \odot X^{=d-1}$$

As $X^{<d} = X^{<d-1} \oplus X^{=d-1}$, every iteration of Newton's method essentially consists of solving this linear equation after substituting for the variables $X^{<d}$ and $X^{=d-1}$ the already computed solution. Analogously, in the multivariate setting essentially the same linear equation system has to be solved over and over again. FPSOLVE thus allows to first compute a symbolic solution of the linear system by treating $X^{<d}$ and $X^{=d-1}$ as formal parameters which allows to share common subexpressions and thus obtain a succinct symbolic representation of a Newton approximation. This allows to efficiently evaluate a Newton approximation of an algebraic system for several different semiring interpretations.

Consider the generic linear equation system

$$\begin{pmatrix} x \\ y \end{pmatrix} = \begin{pmatrix} a & b \\ c & d \end{pmatrix} \cdot \begin{pmatrix} x \\ y \end{pmatrix} + \begin{pmatrix} e \\ f \end{pmatrix}$$

Treating $a, \ldots, f$ as formal parameters over some semiring, the (symbolic) solution of this system is given by

$$\begin{pmatrix} x \\ y \end{pmatrix} = \begin{pmatrix} a^*b(ca^*b \oplus d)^*ca^*e \oplus a^*b(ca^*b \oplus d)^*f \oplus a^*e \\ (ca^*b \oplus d)^*ca^*e \oplus (ca^*b \oplus d)^*f \end{pmatrix}$$

where   we   have   omitted   the $\odot$ for   readability.

---

[3] Note that multiplying two $n \times n$ matrices requires $n^3 - n^2$ operations (via the school-book method – we cannot use e.g. Strassen's algorithm as we lack a difference operator!).

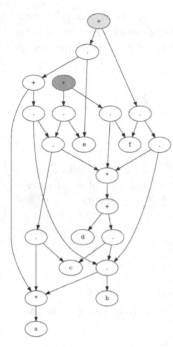

Fig. 1. Succinctly representing all terms of the
product $A^* \cdot (e, f)^T$ via a
BDD-like sharing of subexpressions. By reversing the
direction of all edges this
can be read as an arithmetic
circuit with output gates
colored in grey.

The internal representation of these terms is shown
in Fig. 1: FPSOLVE stores the expressions as part
of an "abstract syntax DAG" (reversing the direction of the edges we obtain an arithmetic circuit
with gates for addition, multiplication, and Kleene
star) similar to BDD libraries like CUDD, where
we have colored the x- resp. y-component in light
resp. dark grey; this representation allows to reduce both the memory consumption, and the re-
evaluation of identical subterms.

In this simple 2×2 case the concrete recursive
approach (as stated above) computes 10 semiring
operations, the same if the symbolic solution is computed (using the same recursive algorithm). This
holds in general if all elements of the input matrix
are different. However, in general input matrices
can have the same element in many different positions, then even the recursive algorithm will compute some identical subexpressions multiple times
(that occur in different execution branches) since it
cannot guess them a priori. In this case, symbolic
solving allows for a *global* subexpression detection
after the whole matrix-star has been computed.

Although the symbolic approach significantly reduces the number of semiring operations needed,
the overhead from computing and storing the symbolic solution is not always negligible. This is particularly true for numeric semirings (like the semiring
of positive reals) that are implemented using machine precision floating point numbers – for these
the semiring operations are so fast that the overhead outweighs the benefits of symbolic solving.

We therefore give the user the freedom of choice whether to use the concrete
(i.e. in every iteration) or symbolic (i.e. solve once then plug in in every iteration)
method of solving linear equations.

3.2 Decomposition into Strongly-Connected Components

To efficiently process large algebraic systems, FPSOLVE supports a decomposition of the system into strongly connected components (SCCs). To make this
precise recall the definition of *dependency graph*: Its nodes are the variables occurring in the algebraic system; its edges are induced by the defining equations:
we have an edge from variable X to variable Y if Y occurs in the defining equation of X. X *depends* on Y if there is a path from X to Y in the dependency
graph. To determine the value of variable X it then suffices to determine the values of all variables on which X depends. Using Tarjan's algorithm we therefore

partition the dependency graph into SCCs, and process these SCCs in reverse topological ordering ("bottom up"). In particular when using Newton's method this can lead to a noticeable speed-up in the computation of the Kleene star.

4 Implementation

Currently, FPSOLVE comprises roughly $8,000$ lines of C++. The code can be obtained freely from `https://github.com/mschlund/FPsolve`. We use several existing frameworks and libraries:

- CPPUNIT for writing unit-tests.
- BOOST for IO-tasks (parsing, command-line arguments).
- GENEPI, MONA, and LASH for representing semilinear sets via NDDs.
- LIBFA for representing elements of "lossy" semirings (i.e. semirings satisfying $1 \sqsubseteq a$ for any semiring element $a \neq 0$ – this generalizes the downward closure of languages) as finite automata.

FPSOLVE features data structures for commutative as well as non-commutative polynomials, different solvers (semi-naive fixpoint iteration, Newton's method), and several predefined semirings (semilinear sets, real numbers, tropical and boolean semiring) as well as some generic constructions (via C++ templates) to build new semirings from existing ones like the direct product of two semirings or the semiring of matrices over some semiring.

The focus of our library is to provide generic algorithms and to be easily extensible. One of our goals was to make it easy for users to write their own semiring-constructions or tailor the generic solving algorithms to their needs.

The library consists of three main parts:

- Data structures (polynomials, matrices, BDD-like DAG-structure to support subterm sharing)
- Semirings (semilinear sets, positive real numbers, why semiring, generic product semiring, ...)
- Solvers (Kleene solver, Newton solver)

Figure 2 shows a simplified view of the main structure of our library. Observe that many classes are templated which produces efficient code due to compile-time polymorphism.

4.1 Invocation of the Standalone Solver

FPSOLVE also includes a callable solver and a parser for equation systems that demonstrates the use of the library.

To apply the standalone solver, one has to describe the algebraic system as a BNF-style context-free grammar. Variables of the system are enclosed in angle brackets, multiplication is not explicitly written, the addition $x + y$ is written as $x \mid y$. To solve the following system over the reals

$$X = 0.5XY + 0.5 \qquad Y = 0.3Y + 0.7X$$

we would create a text file `test.g` containing:

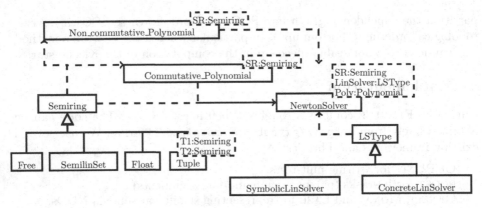

Fig. 2. A (simplified) part of our architecture

```
<X> ::= 0.5 <X> <Y>  | 0.5;
<Y> ::= 0.3 <Y> | 0.7 <X>;
```

The simplest invocation of the tool is then

```
$ ./fpsolve -f test.g --float
```

This minimal set of parameters specifies

1. the input file containing the algebraic system (`-f test.g`).
2. the semiring over which the system and its constants (like `0.3`) are to be interpreted (here `--float`).

The tool outputs:

```
$ ./fpsolve -f test.g --float
Newton Concrete
Iterations: 3
Solving time:   0 ms (196 us)
X == 0.875
Y == 0.875
```

By default, the number of Newton iterations for a system of n equations is $n + 1$ – for commutative, idempotent semirings this suffices to compute the exact solution [8].

A more sophisticated use of the tool's options would be the following:

```
$ ./fpsolve -f test.g --float -i 10 -s newtonSymb
Newton Symbolic
Iterations: 10
Solving time: 0 ms (536 us)
X == 0.999023
Y == 0.999023
```

Here, we select

1. the number of iterations (-i 10)
2. the solving algorithm to use (switch -s), possible choices are newtonSymb, newtonConc, kleene.

For larger equation systems there is the possibility to decompose the system into SCCs and solve them bottom-up (switch --scc).

4.2 Custom Semirings and Extensions

It is very easy and straightforward to extend our library with new semirings, it merely requires three steps:

- Implement all semiring operations (addition $\oplus$, multiplication $\odot$, star $*$)
- Define a constructor that takes a string-argument (effectively a small parser)
- Add a new command-line switch to the main-method together with a call to the solving function.

The second point delegates the IO/parsing task for semiring-elements to the implementer. This enables us to parse equation systems into the most general intermediate format (non-commutative polynomials over the free semiring) and then to map these to the user-defined semiring. Since our input-parser takes quite some time to compile (due to boost::spirit and templates), by this approach we avoid to touch the parser and the need for recompilation.

The semiring operations $\odot$, $\oplus$ (+ and *) are implemented in the abstract base-class Semiring using += and *=. Any new semiring should be derived from the abstract class StarableSemiring and has to implement the three operations *=, +=, star(). Take for instance the "MaxProvenance" semiring from the end of Section 2 consisting of pairs (α, X) of real numbers and sets of variables with

$$(\alpha_1, X_1) \odot (\alpha_2, X_2) := (\alpha_1\alpha_2, X_1 \cup X_2)$$
$$(\alpha_1, X_1) \oplus (\alpha_2, X_2) := (\max\{\alpha_1, \alpha_2\}, \text{if } \alpha_1 \geq \alpha_2 \text{ then } X_1 \text{ else } X_2)$$
$$(\alpha, X)^* := (1, \emptyset)$$

To implement this simple semiring, we derive from StarableSemiring the new class MaxProvSR with members weight and prov storing α (e.g. as a float) and X (e.g. as a set<>), respectively. What remains is then to implement the three operators *=, +=, star(). For instance, the addition-assignment operator could be implemented as

```
MaxProvSR MaxProvSR::operator+=(const MaxProvSR& elem)
{
  if(this->weight < elem.weight) {
    this->weight = elem.weight;
    this->prov = elem.prov;
  }
  return *this;
}
```

Inheritance then takes care of the implementation of the addition operator. Implementing the remaining two operators is just as straight-forward. To make the semiring available in the command line tool, a corresponding switch and a parser for reading semiring elements from the input have to implemented in addition.

To check our claim of "easy extendability", we made a rather naive implementation of the Why-semiring for this paper which took about two hours (until all bugs were eliminated[4]). Once a new semiring is defined and the main-method is adapted, all solvers just work out-of-the-box to solve algebraic systems like the following (file `test/grammars/bintrees.g`):

```
<X> ::= a<X><X> | c;

$ ./fpsolve --why -f ../test/grammars/bintrees.g
Newton Concrete
Iterations: 2
Solving time: 0 ms (214 us)
X == {{a,c},{c}}

$ ./fpsolve --why -f ../test/grammars/bintrees.g -s kleene -i 2
Kleene solver
Iterations: 2
Solving time: 0 ms (281 us)
X == {{a,c},{c}}
```

5   Conclusions and Related Tools

We have introduced FPSOLVE, an implementation of generic algorithms for solving fixpoint equations on semirings. The algorithms are parametric on the semiring. New semirings can be easily added by defining implementations of the sum, product and (possibly) Kleene star operations.

As mentioned in the introduction, many program analysis problems can be reduced to solving fixpoint equations on semirings. This has lead to a number of implementations and tools. An early effort is the Fixpoint-Analysis Machine for solving systems of boolean fixpoint equations [18]. The tool can deal with hierarchical and alternating fixpoints, but is not parametric on the equation domain. The Weighted Pushdown Systems Library and Weighted Automata Library (see [16,3,2]), and GOBLINT (see [4,1] implement many sophisticated algorithms for semirings satisfying the ascending chain condition.

While FPSOLVE is currently an academic tool, we have illustrated its potential interest outside theoretical computer science by means of an application scenario, namely a recommendation system. Genericity allows the users of the system to aggregate the information given by individual recommendations in different, personalized ways, by defining their own semiring.

---

[4] We developed a small collection of unit-tests (also generic tests that can be instantiated with any semiring) and encourage any user who implements new semirings to use and adapt them during development.

Acknowledgments. We thank Michael Kerscher and Michał Terepeta for their help in developing and extending FPSOLVE .

References

1. Goblint, http://goblint.in.tum.de/
2. WALi: The Weighted Automata Library, https://research.cs.wisc.edu/wpis/wpds/
3. Weighted Pushdown Systems Library, http://www2.informatik.uni-stuttgart.de/fmi/szs/tools/wpds/
4. Apinis, K., Seidl, H., Vojdani, V.: How to combine widening and narrowing for non-monotonic systems of equations. In: Boehm, H.J., Flanagan, C. (eds.) PLDI, pp. 377–386. ACM (2013)
5. Bouajjani, A., Esparza, J., Schwoon, S., Suwimonteerabuth, D.: SDSIrep: A Reputation System Based on SDSI. In: Ramakrishnan, C.R., Rehof, J. (eds.) TACAS 2008. LNCS, vol. 4963, pp. 501–516. Springer, Heidelberg (2008)
6. Bouajjani, A., Esparza, J., Touili, T.: A generic approach to the static analysis of concurrent programs with procedures. Int. J. Found. Comput. Sci. 14(4), 551 (2003)
7. Droste, M., Kuich, W., Vogler, H.: Handbook of Weighted Automata. Springer (2009)
8. Esparza, J., Kiefer, S., Luttenberger, M.: On Fixed Point Equations over Commutative Semirings. In: Thomas, W., Weil, P. (eds.) STACS 2007. LNCS, vol. 4393, pp. 296–307. Springer, Heidelberg (2007)
9. Esparza, J., Kiefer, S., Luttenberger, M.: Newtonian Program Analysis. J. ACM 57(6), 33 (2010)
10. Esparza, J., Luttenberger, M.: Solving fixed-point equations by derivation tree analysis. In: Corradini, A., Klin, B., Cîrstea, C. (eds.) CALCO 2011. LNCS, vol. 6859, pp. 19–35. Springer, Heidelberg (2011)
11. Green, T.J., Karvounarakis, G., Tannen, V.: Provenance semirings. In: PODS, pp. 31–40 (2007)
12. Jha, S., Reps, T.W.: Model checking SPKI/SDSI. Journal of Computer Security 12(3-4), 317–353 (2004)
13. Knoop, J., Steffen, B.: The interprocedural coincidence theorem. In: Pfahler, P., Kastens, U. (eds.) CC 1992. LNCS, vol. 641, pp. 125–140. Springer, Heidelberg (1992)
14. Kuich, W.: Semirings and Formal Power Series: Their Relevance to Formal Languages and Automata. In: Handbook of Formal Languages, ch. 9, vol. 1, pp. 609–677. Springer (1997)
15. Luttenberger, M., Schlund, M.: Convergence of Newton's Method over Commutative Semirings. In: Dediu, A.-H., Martín-Vide, C., Truthe, B. (eds.) LATA 2013. LNCS, vol. 7810, pp. 407–418. Springer, Heidelberg (2013)
16. Reps, T., Schwoon, S., Jha, S., Melski, D.: Weighted pushdown systems and their application to interprocedural dataflow analysis. Science of Computer Programming 58(1-2), 206–263 (2003); Special Issue on the Static Analysis Symposium 2003
17. Schlund, M., Terepeta, M., Luttenberger, M.: Putting Newton into Practice: A Solver for Polynomial Equations over Semirings. In: McMillan, K., Middeldorp, A., Voronkov, A. (eds.) LPAR-19 2013. LNCS, vol. 8312, pp. 727–734. Springer, Heidelberg (2013)
18. Steffen, B., Claßen, A., Klein, M., Knoop, J., Margaria, T.: The fixpoint-analysis machine. In: Lee, I., Smolka, S.A. (eds.) CONCUR 1995. LNCS, vol. 962, pp. 72–87. Springer, Heidelberg (1995)

Restarting Automata for Picture Languages: A Survey on Recent Developments

Friedrich Otto

Fachbereich Elektrotechnik/Informatik, Universität Kassel
34109 Kassel, Germany
`otto@theory.informatik.uni-kassel.de`

Abstract. Much work has been done to obtain classes of picture languages that would correspond to the classes of the Chomsky hierarchy for string languages, and finally the class REC of recognizable picture languages has been agreed on as the class that corresponds to the 'regular string languages.' This class has several nice characterizations in terms of regular expressions, tiling automata, and on-line tesselation automata, and it has nice closure properties, but it also has two main drawbacks: all its characterizations are highly nondeterministic in nature, and it contains languages that are NP-complete. Consequentially, various deterministic subclasses of REC have been defined. Mainly, however, these definitions are quite complex, and it is not clear which of the resulting classes should be considered as 'the' class of deterministic recognizable picture languages. Here we present some recent developments obtained in a research project that aims at finding a deterministic model of a two-dimensional automaton that has the following desirable properties:

- the automaton should be conceptually simple,
- the class of languages accepted should be as large as possible,
- it should have nice closure properties,
- the membership problem for each of these languages should be solvable in polynomial time,
- but when restricted to one-row pictures (that is, strings), only the regular languages should be accepted.

In the course of the project, several types of two-dimensional automata have been defined and investigated. Here these types of automata and the classes of picture languages accepted by them are compared to each other and to the classes REC and DREC, and their closure properties and algorithmic properties are considered.

Keywords: picture language, two-dimensional automaton, Sgraffito automaton, restarting automaton.

1 Introduction

The theory of automata and formal languages is one of the classical subjects of theoretical computer science. One of its most celebrated achievements is certainly the hierarchy of language classes known as the 'Chomsky hierarchy.' Particularly the two lower classes, that is, the class REG of *regular languages* and

M. Holzer and M. Kutrib (Eds.): CIAA 2014, LNCS 8587, pp. 16–41, 2014.

the class CFL of *context-free languages* and its subclass DCFL of *deterministic context-free languages*, have found many applications in theoretical research as well as in practical work. It may, however, be less well-known that already in the 1960's work started on extending these notions from strings to pictures, that is, from one-dimensional objects to two-dimensional objects. Here at least three approaches are to be mentioned:

- M. Blum and C. Hewitt extended the notion of finite-state acceptor to two dimensions, obtaining the *four-way finite automaton* (4FA) and the *four-way marker automaton* [3],
- A. Rosenfeld introduced the *isometric array grammars* [28] which generate pictures by performing rewrite sequences in the plane,
- G. Siromoney, R. Siromoney, and K. Krithivasan presented the *matrix grammars* [29] which generate pictures by first deriving a horizontal string of intermediate symbols and by then performing vertical rewrites in parallel on that string.

The 4FAs and the *right-linear matrix grammars* of [29] are quite weak with respect to their expressive power, while the isometric array grammars are very expressive. Accordingly, in the years since, many more analytical and generative methods have been proposed and investigated for describing picture languages. One of the major concerns in these studies has been the quest for a class of picture languages that could rightfully be considered as the two-dimensional equivalent of the class REG of regular (string) languages. Here we mention only some of them:

- the *two-dimensional on-line tessellation automaton* (2OTA) of K. Inoue and A. Nakamura [10] is a two-dimensional cellular automaton that processes a given picture of size m by n by performing $m + n - 1$ global steps that sweep across the picture from the top-left corner to the bottom-right corner,
- D. Giammarresi and A. Restivo presented *tiling systems* in [6], which consist of a finite number pictures of size 2 by 2 (the so-called *tiles*) over a finite alphabet Γ and a projection $\pi : \Gamma \to \Sigma$, and which define the set of all pictures P over Σ for which there exists a picture P' of the same size over Γ that can be covered with the given tiles such that $P = \pi(P')$ holds,
- in [7] D. Giammarresi and A. Restivo also presented *regular expressions* for picture languages, defining several different classes of picture languages.

As it turned out, the class of picture languages that are accepted by 2OTAs coincides with the class of picture languages that are defined by tiling systems and with the class of languages that can be defined by complementation-free regular expressions with projection.

This class is now known as the class REC of *recognizable two-dimensional languages*. It has many nice closure properties and various different characterizations that can be interpreted as two-dimensional analogs to characterizations of the regular (string) languages, and in addition, the string languages contained in REC are just the regular languages. Accordingly, this class is generally considered as the appropriate generalization of the regular languages to two dimensions.

However, there is a serious drawback: the class REC contains languages that are NP-complete [16]. Thus, from a complexity theoretical point of view, the languages are much more complicated than the regular (string) languages. Consequentially, much work has been spent on deriving deterministic variants of the different nondeterministic characterizations of REC. These include the following:

- Already in [3], M. Blum and C. Hewitt considered *deterministic 4-way finite automata* (4DFA), showing that these automata are strictly weaker than the nondeterministic variants.
- In [10] K. Inoue and A. Nakamura also introduced *deterministic two-dimensional on-line tessellation automata* (2DOTA), and they proved that the class of picture languages accepted by these automata is incomparable to the class of languages accepted by (deterministic) 4-way finite automata with respect to inclusion.
- The class DREC of *deterministic recognizable two-dimensional languages* of M. Anselmo, D. Giammarresi and M. Madonia [1], which are defined by deterministic tiling systems. DREC is a proper subclass of REC, and the membership problem for a language L in DREC is solvable deterministically in linear time (in the size of the given picture).
- A different notion of *deterministically recognizable two-dimensional languages* was obtained by K. Reinhardt [27], who uses tiles of size 2 by 1 and of size 1 by 2, and who describes a rewriting process that is to produce a unique tiling for a given input picture.
- As an extension of the previous notion, B. Borchert and K. Reinhardt also introduced the notion of *Sudoku-deterministically recognizable picture languages* [4], where a transformation process similar to the way in which a Sudoku puzzle is being solved is used to transform a given input picture into a unique tiling.
- *Deterministic tiling automata* have been studied by M. Anselmo, D. Giammarresi, and M. Madonia in [2]. Given an input picture P, such an automaton scans P based on a fixed scanning strategy using a given tiling system and a particular data structure. It turned out that the class of picture languages that are accepted by these automata coincides with the class DREC mentioned above.

Thus, quite a large number of different deterministic subclasses of the recognizable picture languages have been considered, and it is not clear which of them is the 'right' one. Some of them are too small, while for others the accepting devices are not very intuitive. This led to a research program that aims at deriving a class of two-dimensional automata that satisfy all of the following conditions:

- the automata should be conceptually simple, that is, it should be 'easy' to design automata of this type for interesting example languages;
- they should be more powerful than the class DREC of deterministic recognizable languages of [1], but when restricted to the one-dimensional case (that is, strings), this model should only accept the regular languages;
- the membership problem for the accepted languages should be decidable in polynomial time;
- and the class of accepted picture languages should have nice closure properties.

In the present survey we present some recent models of two-dimensional automata that were developed in the course of this program:

- the *deterministic Sgraffito automaton* of F. Mráz and D. Průša [23],
- the *restarting tiling automaton* of the same authors [22],
- the *deterministic two-dimensional three-way ordered restarting automaton* [19], and
- the *deterministic two-dimensional extended two-way ordered restarting automaton* [18].

We will present these automata in turn, providing simple examples and stating their main properties. As we will see, these types of deterministic two-dimensional automata are quite expressive, and the corresponding membership problems are solvable in polynomial time, but unfortunately, none of them meets all the conditions listed above. Accordingly, the quest for such a type of two-dimensional automaton is continuing.

This paper is structured as follows. In the next section we introduce basic notions and notation on picture languages, and we restate the definitions of some types of two-dimensional automata from the literature in short. Then we consider the Sgraffito automaton from [23] and its main properties, and we study the *deterministic Sgraffito automaton*, presenting the main results from [25] and [26]. In the next section we turn to two-dimensional variants of the restarting automaton by looking at the restarting tiling automaton, before we study deterministic 2-dimensional 4-way ordered restarting automata in Section 6, deterministic 2-dimensional 3-way ordered restarting automaton in Section 7, and deterministic 2-dimensional (extended) 2-way ordered restarting automata in Section 8. The paper closes with a short summary and some open problems.

2   Pictures and Picture Languages

We use the common notation and terms on pictures and picture languages (see, e.g., [7]). For a finite alphabet Σ, $\Sigma^{*,*}$ denotes the set of rectangular pictures over Σ, that is, if $P \in \Sigma^{*,*}$, then P is a two-dimensional array of symbols from Σ. We denote the number of rows and columns of a picture P by $\mathrm{rows}(P)$ and $\mathrm{cols}(P)$, respectively. The pair $(\mathrm{rows}(P), \mathrm{cols}(P))$ is called the *size* of P. The *empty picture* Λ is defined as the only picture of size $(0,0)$, and for all $m, n \geq 1$, $\Sigma^{m,n}$ denotes the set of pictures of size (m,n) over Σ. A *picture language* over Σ is a subset of $\Sigma^{*,*}$. For $1 \leq i \leq \mathrm{rows}(P)$ and $1 \leq j \leq \mathrm{cols}(P)$, $P(i,j)$ (or shortly $P_{i,j}$) identifies the symbol located in row i and column j of P.

Two (partial) binary operations are used to concatenate pictures. Let P and Q be pictures over Σ of sizes (k,l) and (m,n), respectively. The *column concatenation* $P \oplus Q$ is defined if and only if $k = m$, while the *row concatenation* $P \ominus Q$ is defined if and only if $l = n$. These products are depicted below:

$$P \oplus Q = \begin{bmatrix} P_{1,1} \cdots P_{1,l} & Q_{1,1} \cdots Q_{1,n} \\ \vdots \ddots \vdots & \vdots \ddots \vdots \\ P_{k,1} \cdots P_{k,l} & Q_{m,1} \cdots Q_{m,n} \end{bmatrix} \text{ and } P \ominus Q = \begin{bmatrix} P_{1,1} \cdots P_{1,l} \\ \vdots \ddots \vdots \\ P_{k,1} \cdots P_{k,l} \\ Q_{1,1} \cdots Q_{1,n} \\ \vdots \ddots \vdots \\ Q_{m,1} \cdots Q_{m,n} \end{bmatrix}.$$

We also define $\Lambda \ominus P = P \ominus \Lambda = \Lambda \oplus P = P \oplus \Lambda = P$ for any picture P. These operations are extended to languages by taking $L_1 \oplus L_2 = \{ P_1 \oplus P_2 \mid P_i \in L_i \}$ and $L_1 \ominus L_2 = \{ P_1 \ominus P_2 \mid P_i \in L_i \}$.

The *column closure* $L^{*\oplus}$ of a picture language L is defined as $L^{*\oplus} = \bigcup_{i \geq 0} L^{i\oplus}$, where $L^{0\oplus} = \{\Lambda\}$ and $L^{i\oplus} = L \oplus L^{(i-1)\oplus}$ for all $i \geq 1$. Similarly, the *row closure* $L^{*\ominus}$ is defined as $L^{*\ominus} = \bigcup_{i \geq 0} L^{i\ominus}$, where $L^{0\ominus} = \{\Lambda\}$ and $L^{i\ominus} = L \ominus L^{(i-1)\ominus}$ for all $i \geq 1$. These two operations can be seen as the extensions of the Kleene star from string languages to picture languages.

Also we consider the *clockwise rotation* P^R and the *transposition* P^T of a picture $P \in \Sigma^{m,n}$:

$$P^R = \begin{bmatrix} P_{m,1} \cdots P_{1,1} \\ \vdots \ddots \vdots \\ P_{m,n} \cdots P_{1,n} \end{bmatrix}, \text{ and } P^T = \begin{bmatrix} P_{1,1} \cdots P_{m,1} \\ \vdots \ddots \vdots \\ P_{1,n} \cdots P_{m,n} \end{bmatrix}.$$

Let $\pi : \Gamma \to \Sigma$ be a mapping, where Γ is an alphabet. Then π induces a mapping from $\Gamma^{*,*}$ to $\Sigma^{*,*}$ by sending $P \in \Gamma^{m,n}$ to $P' \in \Sigma^{m,n}$ such that $P'(i,j) = \pi(P(i,j))$ for each $1 \leq i \leq m$ and $1 \leq j \leq n$. This mapping is called a *projection* from $\Gamma^{*,*}$ to $\Sigma^{*,*}$, and P' is simply written as $\pi(P)$. Note that each of these operations naturally extends to languages.

Let $\mathcal{S} = \{\vdash, \dashv, \top, \bot, \#\}$ be a set of five special markers, called *sentinels*. In what follows, we will always assume implicitly that $\Sigma \cap \mathcal{S} = \emptyset$ for any alphabet Σ considered. In order to enable our automata to be defined below to detect the border of a picture $P \in \Sigma^{m,n}$ easily, we define the *boundary picture* $\widehat{P}$ over $\Sigma \cup \mathcal{S}$ of size $(m+2, n+2)$, which is illustrated in Figure 1. Here the symbols $\vdash, \dashv, \top$ and $\bot$ uniquely identify the corresponding borders (left, right, top, bottom) of $\widehat{P}$, while the symbol $\#$ marks the corners of $\widehat{P}$.

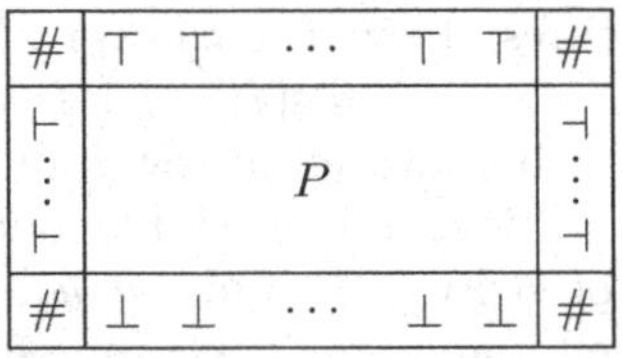

#	⊤ ⊤	⋯	⊤ ⊤	#
⊢				⊣
⋮		P		⋮
⊢				⊣
#	⊥ ⊥	⋯	⊥ ⊥	#

Fig. 1. The boundary picture $\widehat{P}$

As mentioned in the introduction, a large variety of accepting devices has been studied in the literature. Here we restate some of them in short.

The *4-way finite automaton* (4FA) is the two-dimensional variant of the finite-state acceptor. Such an automaton is defined by a tuple $\mathcal{A} = (Q, \Sigma, q_0, q_a, q_r, \delta)$, where Q is a finite set of states, $q_0, q_a, q_r \in Q$ are the initial state, the accepting state, and the rejecting state, respectively, and $\delta : ((Q \smallsetminus \{q_a, q_r\}) \times \Sigma) \rightarrow 2^{(Q \times \{L,R,U,D\})}$ is the transition relation [3]. Here L, R, U, and D denote directions (left, right, up, down). The 4FA $\mathcal{A}$ accepts a picture $P \in \Sigma^{*,*}$ if it has an accepting computation on P starting in its initial state q_0 from position $(1,1)$. If δ is a (partial) function, then $\mathcal{A}$ is a *deterministic 4-way finite automaton* (4DFA). It is known that, in contrast to the one-dimensional case, $\mathcal{L}(4DFA)$ is a proper subclass of $\mathcal{L}(4FA)$.

The *two-dimensional on-line tessellation automaton* (2OTA) is a restricted type of cellular automaton [10]. For an input $P \in \Sigma^{*,*}$, a cell is placed at each position (i, j) of P. Then a computation is performed that consists in $\text{rows}(P) + \text{cols}(P) - 1$ many parallel steps. During the k-th step, each cell at coordinates (i, j), where $i + j - 1 = k$, performs a state-transition that depends on $P(i, j)$ and on the states of its left and top neighbour cells. If a neighbour lies at the border of P, it is a fictive cell the state of which is defined as the corresponding symbol of $\widehat{P}$. The result of the computation is determined by the final state of the cell at the bottom-right corner of P.

Also for the 2OTA, the deterministic variant, the 2DOTA, is strictly weaker than the nondeterministic variant, and $\mathcal{L}(4FA)$ is properly contained in $\mathcal{L}(2OTA)$, while $\mathcal{L}(2DOTA)$ is incomparable to both, $\mathcal{L}(4DFA)$ and $\mathcal{L}(4FA)$, with respect to inclusion [10].

Finally, we turn to tiling systems. A *tile* is a square picture of size $(2,2)$, and for a picture P, $B_{2,2}(\widehat{P})$ denotes the set of all tiles that are subpictures of $\widehat{P}$. Now a picture language $L \subseteq \Sigma^{*,*}$ is called a *local language*, if there exists a finite set of tiles Θ such that $L = \{ P \in \Sigma^{*,*} \mid B_{2,2}(\widehat{P}) \subseteq \Theta \}$, that is, $P \in L$ if and only if all tiles that are subpictures of $\widehat{P}$ belong to Θ. This fact is expressed as $L = L(\Theta)$.

A *tiling system* (TS) is given through a tuple $\mathcal{T} = (\Sigma, \Gamma, \Theta, \pi)$, where Σ and Γ are two finite alphabets, Θ is a finite set of tiles over $\Gamma \cup \mathcal{S}$, and $\pi : \Gamma \rightarrow \Sigma$ is a projection. The language $L = L(\mathcal{T}) \subseteq \Sigma^{*,*}$ that is defined by $\mathcal{T}$ is the projection $\pi(L(\Theta))$ of the local language specified by Θ. $\mathcal{L}(TS)$ is called the class of *tiling recognizable languages* [6]. Since $\mathcal{L}(TS) = \mathcal{L}(2OTA)$ [11], and since this class has many nice closure properties, it is simply being referred to as REC. The class DREC is obtained by the restriction to d-deterministic tiling systems, where d is a corner-to-corner direction, that is, based on the chosen direction d, the tiling of a given picture $P \in L$ is unique [1]. The class DREC coincides with the closure of $\mathcal{L}(2DOTA)$ under rotation, and $\mathcal{L}(4DFA) \not\subseteq$ DREC.

3   Sgraffito Automata

Our first model is the so-called *Sgraffito automaton* of D. Průša and F. Mráz introduced in [23]. It is a two-dimensional extension of the one-dimensional *constant-visit machine* studied by Hennie [8].

Let $\mathcal{H} = \{R, L, D, U, Z\}$ be the set of possible *head movements*, where the first four elements denote directions (right, left, down, up), while Z stands for zero (no) movement. Furthermore, let $\nu : (\mathcal{S} \setminus \{\#\}) \to \mathcal{H}$ denote the mapping that is defined by $\nu(\vdash) = R$, $\nu(\dashv) = L$, $\nu(\top) = D$, and $\nu(\bot) = U$, that is, for each occurrence of a sentinel from $\{\vdash, \dashv, \top, \bot\}$ within a boundary picture $\widehat{P}$, ν yields the direction to the nearest field of the proper picture P.

Definition 1. *A two-dimensional Sgraffito automaton (2SA) is given by a tuple $\mathcal{A} = (Q, \Sigma, \Gamma, \delta, q_0, Q_F, \mu)$, where*

- *Q is a finite, nonempty set of states,*
- *Σ is an input alphabet,*
- *Γ is a working alphabet such that $\Sigma \subseteq \Gamma$,*
- *$q_0 \in Q$ is the initial state,*
- *$Q_F \subseteq Q$ is the set of final states,*
- *$\delta : (Q \setminus Q_F) \times (\Gamma \cup \mathcal{S}) \to 2^{Q \times (\Gamma \cup \mathcal{S}) \times \mathcal{H}}$ is a transition relation such that the following properties are satisfied for each pair $(q, a) \in (Q \setminus Q_F) \times (\Gamma \cup \mathcal{S})$ and each transition $(q', a', d) \in \delta(q, a)$:*
  - *if $a \in \mathcal{S}$, then $d = \nu(a)$ and $a' = a$, and*
  - *if $a \notin \mathcal{S}$, then $a' \notin \mathcal{S}$,*
- *and $\mu : \Gamma \to \mathbb{N}$ is a weight function such that the condition $\mu(a') < \mu(a)$ holds for all transitions $(q', a', d) \in \delta(q, a)$ satisfying $a \in \Gamma$.*

The Sgraffito automaton $\mathcal{A}$ is deterministic, *that is, a* 2DSA, *if $|\delta(q, a)| \leq 1$ for all $q \in Q$ and $a \in \Gamma \cup \mathcal{S}$.*

The notions of configuration and computation of the 2SA $\mathcal{A}$ are defined as usual. Let $P \in \Sigma^{*,*}$ be a given input picture for $\mathcal{A}$. In the initial configuration of $\mathcal{A}$ on input P, the working tape contains the boundary picture $\widehat{P}$, $\mathcal{A}$ is in its initial state q_0, and its head scans the top-left corner of P, that is, cell $(2, 2)$ of $\widehat{P}$. Now $\mathcal{A}$ walks across this boundary picture, and it cannot leave the space covered by $\widehat{P}$. Furthermore, when executing a transition step at a position (i, j) of P, then it must replace the current symbol, say a, at that position by a symbol, say a', that has strictly less weight than a. It follows that $\mathcal{A}$ can visit each position of P, and therewith of $\widehat{P}$, at most $|\Gamma|$ many times. The machine $\mathcal{A}$ is said to *accept* P if there is a computation of $\mathcal{A}$ that starts in the initial configuration on input P and that finishes in a state from Q_F. By $L(\mathcal{A})$ we denote the picture language that consists of all pictures that are accepted by $\mathcal{A}$, and $\mathcal{L}(2SA)$ ($\mathcal{L}(2DSA)$) denotes the class of all picture languages that are accepted by (deterministic) Sgraffito automata.

The advantage of the Sgraffito automaton over the constant-visit machine is its constructiveness. While it is undecidable in general whether a given Turing machine is a Hennie machine (see, e.g., [21]), the property of being weight-reducing can easily be checked algorithmically.

The first result on Sgraffito automata shows that, when restricted to one-row pictures, that is, strings, the 2SAs accept exactly the class of regular languages. Actually, this is just a slight generalization of a theorem by Hennie [8].

Theorem 2. [23] *Let $\mathcal{A}$ be a 2SA that accepts a one-dimensional picture language $L(\mathcal{A}) \subseteq \Sigma^{1,*} = \Sigma^*$. Then $L(\mathcal{A})$ is a regular (string) language.*

In order to formulate the next result, we consider the two-dimensional variants L_{copy} of the *copy language* and $\hat{L}_{\mathrm{copy}}$ of the *marked copy language*. The former contains those pictures over $\Sigma = \{\Box, \blacksquare\}$ that are of the form $U \oplus U$, where U is a square picture over Σ, and the latter is the picture language over $\hat{\Sigma} = \{\Box, \boxtimes, \blacksquare\}$ that consists of all pictures $U \oplus C \oplus U$, where U is a square picture over Σ, and C is a column of symbols $\boxtimes$ (see Figure 2). L_{copy} and $\hat{L}_{\mathrm{copy}}$ are widely used examples of rather complicated picture languages [7].

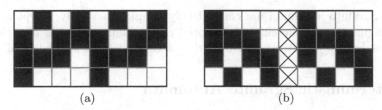

(a)   (b)

Fig. 2. Sample pictures from (a) L_{copy} and (b) $\hat{L}_{\mathrm{copy}}$

The following technical result can be proved based on a notion of *horizontal crossing sequence* [23] that is a rather straightforward generalization of the notion of crossing sequence as, for example, presented in [9].

Lemma 3. (a) *Let $L \subseteq L_{\mathrm{copy}}$ be a picture language over Σ accepted by a 2SA, and let $f : 2^{\Sigma^{*,*}} \times \mathbb{N} \to \mathbb{N}$ be the function that is defined by taking $f(L,n)$ to be the number of pictures in L of size $(n, 2n)$. Then $f(L,n) \in 2^{\mathcal{O}(n \log n)}$.*
(b) *Let $L \subseteq \hat{L}_{\mathrm{copy}}$ be a picture language over $\hat{\Sigma}$ accepted by a 2SA, and let $f : 2^{\Sigma^{*,*}} \times \mathbb{N} \to \mathbb{N}$ be the function that is defined by taking $f(L,n)$ to be the number of pictures in L of size $(n, 2n+1)$. Then $f(L,n) \in 2^{\mathcal{O}(n \log n)}$.*

As an immediate application we obtain the following negative results.

Corollary 4. [23] *L_{copy} and $\hat{L}_{\mathrm{copy}}$ are not accepted by any 2SA.*

Concerning the various operations on languages, the following closure properties have been established for Sgraffito automata.

Theorem 5. [23] *The language class $\mathcal{L}(2SA)$ is closed under union, intersection, rotation, transposition, row and column concatenation, row and column closure, and projection, but it is not closed under complement.*

The language of 'permutations' L_{perm} is the subset of L_{copy} that consists of those pictures $Q \oplus Q$ for which each row and each column of Q contains the symbol $\blacksquare$ exactly once. An example is shown in Figure 3. It is known that L_{copy} and L_{perm} are not in REC, while their complements belong to REC [7,15]. It is easily seen that a 2OTA can be simulated by a 2SA, and it can be shown that L_{perm} is accepted by a 2DSA. This yields the following proper inclusion.

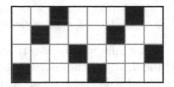

Fig. 3. A sample picture from L_perm

Theorem 6. [23] REC $\subsetneq$ $\mathcal{L}$(2SA).

Thus, Sgraffito automata are conceptually simple, they only accept regular string languages, and the class of accepted picture languages has nice closure properties, but they are too powerful as they accept a proper superclass of REC.

4   Deterministic Sgraffito Automata

The language $L_\mathrm{perm} \notin$ REC is accepted by a deterministic Sgraffito automaton. Further, the following closure and non-closure properties hold.

Theorem 7. [23] *The language class* $\mathcal{L}$(2DSA) *is closed under union, intersection, rotation, transposition, and complement, but it is neither closed under row or column concatenation nor under projection.*

As DREC coincides with the closure of $\mathcal{L}$(2DOTA) under rotation [2], and as each 2DOTA can be simulated by a 2DSA, we obtain the following.

Theorem 8. [23] DREC $\subsetneq$ $\mathcal{L}$(2DSA).

In fact, the 2DSA is very expressive. This follows from the observation that a depth-first search (DFS) on certain graphs that are represented by two-dimensional arrays (pictures) can be implemented on 2DSAs.

Let $G = (V, E)$ be a directed graph that satisfies the following conditions:

1. $V \subseteq \{1, \ldots, m\} \times \{1, \ldots, n\} \times U$ for some integers m, n and a finite set U.
2. For every edge $((i_1, j_1, u_1), (i_2, j_2, u_2))$ in E, $|i_1 - i_2| + |j_1 - j_2| \leq 1$.

Then G can be represented by a picture P of size (m, n), where the field at a position (i, j) records the vertices of the form (i, j, u) $(u \in U)$ in V and the outgoing edges of these vertices. Since the edges only go to the vertices represented in the field itself and in its neighbouring fields, it is only necessary to represent $\mathcal{O}(|U|)$ many vertices and edges in each tape field.

Assume that a 2DSA $\mathcal{A}$ is given this representation of G as input. To traverse G, it assigns a status to vertices as well as to edges. Initially, each vertex has status *unexplored*. When a vertex v is visited during the DFS for the first time, its status is changed to *open*, and when the DFS backtracks to v, then its status is set to *explored*. Analogously, each edge e has initially the status *unexplored*. This changes when e is being traversed. If it leads to an *unexplored* vertex, its status is set to *discovery*, otherwise its status is set to *back*. The edges with status *discovery* will form DFS-trees at the end of the search. This search can now be realized as follows:

1. If there is no vertex with status *unexplored* at the current field, go to 3.
2. While there is a vertex v with status *unexplored* (possibly fulfilling some additional requirement), mark it as *open* and start the DFS, which will return to v at the end.
3. If not all fields have been scanned yet, move the head to the next field in the current row or to the first field of the next row when the right border is reached. Continue with 1.

The whole process visits and rewrites each tape field $\mathcal{O}(|U|)$ many times, and hence, it can be realized by a 2DSA.

For example, a simple application of the DFS could be used to check whether the black pixels within a black and white picture form just a single connected component. Further, this strategy can be used to obtain simulations of other two-dimensional types of automata by deterministic Sgraffito automata.

Let $\mathcal{B} = (Q, \Sigma, q_0, q_a, q_r, \delta)$ be a 4FA (see Section 2), and let $P \in \Sigma^{*,*}$ be an input picture. With $\mathcal{B}$ and P we associate a directed graph $G = (V, E)$:

- $V = \{ (i, j, q) \mid 1 \leq i \leq \text{rows}(P), 1 \leq j \leq \text{cols}(P), q \in Q \}$,
- $E = \{ ((i_1, j_1, q_1), (i_2, j_2, q_2)) \mid (q_2, d) \in \delta(q_1, P(i_1, j_1)),$ and (i_2, j_2) is reached from (i_1, j_1) by a move according to $d \}$.

Then $\mathcal{B}$ accepts on input P iff there is a vertex of the form (i, j, q_a) that is reachable from $(1, 1, q_0)$ in G. Now a 2DSA $\mathcal{A}$ can be designed that, on input P, simulates the DFS on the graph G, and that accepts iff $\mathcal{B}$ accepts, which yields $\mathcal{L}(4FA) \subseteq \mathcal{L}(2DSA)$.

Using the same approach the following results have been derived, where 4AFA denotes the *four-way alternating automaton* [15], SDREC denotes the *Sudoku-deterministically recognizable picture languages* [4], and 2DM$_1$A denotes the class of *deterministic four-way one-marker automata*, that is, deterministic four-way finite automata that can use an additional *marker* [3].

Theorem 9. [23,25,26] $\mathcal{L}(4AFA) \cup \text{SDREC} \cup \mathcal{L}(2DM_1A) \subseteq \mathcal{L}(2DSA)$.

D. Giammaresi and A. Restivo studied the problem of which functions can be represented by recognizable picture languages. Let $\Sigma = \{\Box\}$ denote a one-letter alphabet. A function $f : \mathbb{N} \to \mathbb{N}$ is called *representable* if the language $L_f = \{ \Box^{n, f(n)} \mid n \in \mathbb{N} \}$ belongs to REC. A representable function cannot grow faster than an exponential function [7]. However, using 2DSAs, functions can be represented that grow faster than any exponential function.

Theorem 10. [26] *The language* $L_! = \{ \Box^{n, n!} \mid n \in \mathbb{N} \}$ *is accepted by a 2DSA.*

Since the number of different crossing sequences of a 2DSA between two neighbouring columns of height n is $2^{\mathcal{O}(n \log n)}$, the above is a function with the fastest possible growth that is accepted by a 2DSA. Also the following result is an immediate consequence.

Corollary 11. $\mathsf{DREC} \cap \{\Box\}^{*,*} \subsetneq \mathcal{L}(\mathsf{2DSA}) \cap \{\Box\}^{*,*}$.

Thus, the deterministic Sgraffito automaton is a type of two-dimensional automaton that is very intuitive, that is quite expressive, and that only accepts regular string languages. In addition, it is easily seen that the membership problem for the language $L(\mathcal{A})$ is decidable in linear time for each 2DSA $\mathcal{A}$. Unfortunately, $\mathcal{L}(\mathsf{2DSA})$ is not closed under some important operations.

5   Restarting Tiling Automata

The restarting automaton was proposed by P. Jančar, F. Mráz, M. Plátek, and J. Vogel in [13] as a formal way of modelling the linguistic technique of 'analysis by reduction.' On input $w \in \Sigma^*$, a restarting automaton M scans w from left to right until it detects a small factor u that can be simplified, that is, u is rewritten into a shorter factor v, say. Then M starts the whole process again, this time on the simplified string obtained by the rewrite process. This continues until either an error is detected, and M rejects, or until a simple string is obtained that is then accepted by M. Since the 1990's many different variants and extensions of the basic model of the restarting automaton have been proposed and investigated, and many well-known language classes have been characterized by certain types of restarting automata (see, e.g., [20] for a survey).

As the restarting automaton is a fairly intuitive model, it is only natural to look for two-dimensional extensions of it. However, it is required in general that each rewrite step of a restarting automaton is strictly length-reducing, a requirement that is not easily carried over to the two-dimensional setting. H. Messerschmidt and M. Stommel have studied a corresponding model in [17], in which each rewrite step consists in an application of a replacement rule that is then followed by a consolidation step that transforms the result of the replacement into a proper picture. Their main result states that each *Church-Rosser picture language* is accepted by a deterministic two-dimensional restarting automaton of an appropriate type. Here we will not go into any details concerning these automata as the model is technically quite involved and not very intuitive. However, it may have practical applications (see [17]).

Instead we turn to the restarting tiling automaton, which was introduced by F. Mráz and D. Průša in [22] (see [24] for an extended presentation). The restarting tiling automaton combines the rewrite/restart capability of the restarting automaton with the acceptance condition of the tiling automaton. To describe this model, we introduce the notion of a 'scanning strategy.'

A *scanning strategy* determines the way in which an automaton scans a given input picture. It is given through a pair $s = (c_s, f)$, where $c_s \in \{1, 2, 3, 4\}$ is a *starting position*, and $f : \mathbb{N}^4 \to \mathbb{N}^2$ is a computable partial function. Here $c_s = 1$ denotes the top-left corner, 2 denotes the top-right corner, 3 stands for the bottom-right corner, and 4 stands for the bottom-left corner, and $f(i, j, m, n) = (i', j')$ states that in a picture of size (m, n), position (i', j') is scanned immediately after position (i, j). It is required that, if (i_0, j_0) is the starting position in a picture P of size (m, n), and if $(i_r, j_r) = f(i_{r-1}, j_{r-1}, m, n)$ for all $r \geq 1$, then the sequence $(i_0, j_0), (i_1, j_1), \ldots, (i_{(m+1)\cdot(n+1)-1}, j_{(m+1)\cdot(n+1)-1})$ is a permutation of the set of all positions of tiles of size $(2, 2)$ in the boundary picture $\hat{P}$. In [2] and in [5] some additional restrictions are studied for scanning strategies, but no such additional restrictions are used by F. Mráz and D. Průša in [22]. An example is the scanning strategy $\nu_{\text{row}} = (1, f_{\text{row}})$, where

$$f_{\text{row}}(i, j, m, n) = \begin{cases} (i, j+1), \text{ if } j < n+1, \\ (i+1, 1), \text{ if } j = n+1 \text{ and } i < m+1, \end{cases}$$

which scans a picture row by row from left to right starting at the top-left corner.

Definition 12. *A two-dimensional restarting tiling automaton, a 2RTA for short, is given through a sixtuple* $\mathcal{M} = (\Sigma, \Gamma, \Theta_F, \delta, \nu, \mu)$, *where* Σ *is a finite input alphabet,* Γ *is a finite working alphabet containing* Σ, $\Theta_F \subseteq (\Gamma \cup \mathcal{S})^{2,2}$ *is a set of accepting tiles,* $\nu = (c_\nu, f_\nu)$ *is a scanning strategy,* $\mu : \Gamma \to \mathbb{N}$ *is a weight function, and* $\delta \subseteq \{ (U \to V) \mid U, V \in (\Gamma \cup \mathcal{S})^{2,2} \}$ *is a set of rewrite rules such that, in every rule* $(U \to V) \in \delta$, *there is a single position* (i, j) *of* U *that is changed, and moreover, if* $U(i, j) = a \in \Gamma$ *is rewritten into* $V(i, j) = b \in \Gamma$, *then* $\mu(a) > \mu(b)$ *holds.*

The symbols from $\Gamma \setminus \Sigma$ are called *auxiliary symbols*, as they cannot occur in any input picture. Given a picture $P \in \Sigma^{m,n}$ as input, the automaton $\mathcal{M}$ works in cycles, proceeding as follows:

- at the beginning of each cycle, the scanning window of $\mathcal{M}$ of size $(2, 2)$ is placed on the corner of the boundary picture $\hat{P}$ that corresponds to the starting position c_ν of the scanning strategy ν;
- now $\mathcal{M}$ scans $\hat{P}$ step by step, moving its window according to the function f_ν;
- when $\mathcal{M}$ reaches a position such that the tile being scanned is of the form U for some rule $(U \to V) \in \delta$, then $\mathcal{M}$ replaces this tile by the corresponding tile V and restarts, that is, its scanning window is repositioned on the initial position corresponding to c_ν, completing the current cycle;
- when $\mathcal{M}$ succeeds to scan $\hat{P}$ completely without encountering a possible rewrite, then it halts, finishing the current computation. $\mathcal{M}$ is said to *accept* if at this point the boundary picture obtained only contains tiles of size $(2, 2)$ that belong to the set Θ_F.

By $L(\mathcal{M})$ we denote the language accepted by $\mathcal{M}$, which is the set of all pictures $P \in \Sigma^{*,*}$ for which $\mathcal{M}$ has an accepting computation.

Observe that a 2RTA $\mathcal{M}$ may contain several different rewrite rules with the same left-hand side U. Hence, $\mathcal{M}$ is in general nondeterministic. If, however, δ contains at most one rule with left-hand side U for every tile $U \in (\Gamma \cup \mathcal{S})^{2,2}$, then $\mathcal{M}$ is a *deterministic two-dimensional restarting tiling automaton* (a 2DRTA). We illustrate the workings of a 2RTA by a simple example taken from [22].

Example 13. Let $\mathcal{M} = (\Sigma, \Gamma, \Theta_F, \delta, \nu_{\text{row}}, \mu)$ be defined as follows:

- $\Sigma = \{a\}$ and $\Gamma = \{a, 1\}$,

$$- \Theta_F = \left\{ \begin{array}{c}\#\top\\\vdash 1\end{array}, \begin{array}{c}\top\top\\1\,a\end{array}, \begin{array}{c}\top\top\\a\,a\end{array}, \begin{array}{c}\top\#\\a\,\dashv\end{array}, \begin{array}{c}a\,\dashv\\a\,\dashv\end{array}, \begin{array}{c}a\,\dashv\\a\,\dashv\end{array}, \begin{array}{c}1\,a\\1\,\dashv\end{array}, \begin{array}{c}a\,1\\a\,1\end{array}, \begin{array}{c}a\,a\\a\,a\end{array}, \begin{array}{c}a\,a\\a\,a\end{array}, \right.$$

$$\left. \begin{array}{c}\#\#\\\#\#\end{array}, \begin{array}{c}\top\#\\1\,\dashv\end{array}, \begin{array}{c}\vdash 1\\\#\bot\end{array}, \begin{array}{c}1\,\dashv\\\bot\#\end{array}, \begin{array}{c}a\,1\\\bot\bot\end{array}, \begin{array}{c}a\,a\\\bot\bot\end{array}, \begin{array}{c}\vdash a\\\#\bot\end{array}, \begin{array}{c}\vdash a\\\vdash a\end{array}, \begin{array}{c}\vdash 1\\\vdash a\end{array} \right\},$$

$$- \delta = \left\{ \begin{array}{c}\#\top\\\vdash a\end{array} \to \begin{array}{c}\#\top\\\vdash 1\end{array}, \begin{array}{c}1\,a\\a\,a\end{array} \to \begin{array}{c}1\,a\\a\,1\end{array} \right\}, \text{ and}$$

- $\mu : \Gamma \to \mathbb{N}$ is defined by $\mu(a) = 2$ and $\mu(1) = 1$.

Then $\mathcal{M}$ is a 2DRTA that accepts the unary picture language $L(\mathcal{M}) = \{ P \in \Sigma^{*,*} \mid \text{rows}(P) = \text{cols}(P) \}$ of quadratic pictures over Σ. In fact, given a picture $P \in \Sigma^{m,n}$ as input, $\mathcal{M}$ rewrites each letter a on the main diagonal of P into 1, proceeding from the top to the bottom. The picture is accepted, if the bottom-right corner is hit by this process, which happens if and only if $m = n$.

We denote by ν-2RTA the class of two-dimensional restarting tiling automata that use the scanning strategy ν, and $\mathcal{L}(\nu\text{-2RTA})$ denotes the corresponding class of picture languages. A picture language L is called *strategy independent* if, for each scanning strategy ν, there exists a ν-2RTA $\mathcal{M}_\nu$ such that $L(\mathcal{M}_\nu) = L$ holds. The class of all strategy independent languages is denoted by si-2RTL. Analogously, one can define the class si-2DRTL of strategy independent languages that are accepted by 2DRTAs. For these classes the following closure properties have been obtained.

Theorem 14. [22] *The classes* si-2RTL *and* si-2DRTL *are closed under union, intersection, projection, transposition, and rotation.*

For 2DRTAs also the following result has been obtained.

Theorem 15. [22] *If ν is a scanning strategy for which the last position scanned is always in the same corner for any input picture of any positive size, then the class $\mathcal{L}(\nu\text{-2DRTA})$ is closed under complement.*

It has been noted that $\mathcal{L}(\nu_{\text{row}}\text{-2RTA})$ is also closed under row and column concatenation, but it is open whether this result holds for the class si-2RTL. Concerning the expressive power of the 2RTA, the following is known.

Theorem 16. [24]
(a) $\mathcal{L}(\text{2SA})$ *is contained in* si-2RTL, *and* $\mathcal{L}(\text{2DSA})$ *is contained in* si-2DRTL.
(b) *There exists a non-regular string language that is accepted by a 2DRTA.*

The example language in (b) is based on a very particular non-continuous scanning strategy. With the scanning strategy ν_{row}, 2RTA can only accept regular string languages, which implies the following.

Theorem 17. [22] *When restricted to languages of one-row pictures, then the class* si-2RTL *coincides with the class of regular (string) languages.*

6   Ordered Restarting Automata

Here we consider a model of a restarting automaton for picture languages that is closer to the original idea of the restarting automaton, the *deterministic two-dimensional four-way ordered restarting automaton*. It has a finite set of states, and it has a scanning window of size $(3,3)$. Based on its current state and the current contents of the window, it can move in either of the four possible directions and change its state, or it can perform a combined rewrite/restart step similar to the restarting tiling automaton. It accepts by executing a specific accept instruction. Formally this automaton is defined as follows, where $\mathcal{H} = \{R, L, D, U\}$ is the set of possible *window movements*.

Definition 18. *A deterministic two-dimensional four-way ordered restarting automaton, a* det-2D-4W-ORWW-*automaton for short, is given through a 7-tuple* $\mathcal{M} = (Q, \Sigma, \Gamma, \mathcal{S}, q_0, \delta, >)$, *where*

- *Q is a finite set of states containing the initial state q_0,*
- *Σ is a finite input alphabet, Γ is a finite tape alphabet containing Σ such that $\Gamma \cap \mathcal{S} = \emptyset$, $>$ is a partial ordering on Γ, and*
- *$\delta : Q \times (\Gamma \cup \mathcal{S})^{3,3} \to (Q \times \mathcal{H}) \cup \Gamma \cup \{\mathsf{Accept}\}$ is the transition function that satisfies the following five restrictions for all $q \in Q$ and all $C \in (\Gamma \cup \mathcal{S})^{3,3}$:*
    1. *if $C_{1,2} = \top$, then $\delta(q, C) \neq (q', U)$ for all $q' \in Q$,*
    2. *if $C_{2,3} = \dashv$, then $\delta(q, C) \neq (q', R)$ for all $q' \in Q$,*
    3. *if $C_{2,1} = \vdash$, then $\delta(q, C) \neq (q', L)$ for all $q' \in Q$,*
    4. *if $C_{3,2} = \bot$, then $\delta(q, C) \neq (q', D)$ for all $q' \in Q$,*
    5. *if $\delta(q, C) = b \in \Gamma$, then $C(2,2) > b$ with respect to the ordering $>$.*

To simplify the presentation we say that the window of $\mathcal{M}$ is *at position* (i, j) to mean that the field in the center of the window is at row i and column j of P. Given a picture $P \in \Sigma^{m,n}$ as input, $\mathcal{M}$ begins its computation in state q_0 with its read/write window reading the subpicture of size $(3,3)$ of $\widehat{P}$ at the upper left corner, that is, the window is at position $(1,1)$. Thus, $\mathcal{M}$ sees the subpicture $\begin{pmatrix} \# & \top & \top \\ \vdash & P_{1,1} & P_{1,2} \\ \vdash & P_{2,1} & P_{2,2} \end{pmatrix}$. Applying its transition function, $\mathcal{M}$ now moves through $\widehat{P}$ until it reaches a state q and a position with current contents C of the read/write window such that

- either $\delta(p, C)$ is undefined, or
- $\delta(p, C) = \mathsf{Accept}$, or
- $\delta(p, C) = b$ for some letter $b \in \Gamma$ such that $C_{2,2} > b$.

In the first case, $\mathcal{M}$ gets stuck, and so the current computation ends without accepting, in the second case, $\mathcal{M}$ halts and accepts, and in the third case, $\mathcal{M}$ replaces the symbol $C_{2,2}$ by the symbol b, moves its read/write window back to the upper left corner, and reenters its initial state q_0. This latter step is called a *combined rewrite/restart step*. A picture $P \in \Sigma^{*,*}$ is *accepted* by $\mathcal{M}$, if the computation of $\mathcal{M}$ on input P ends with an Accept instruction. By $L(\mathcal{M})$ we denote the language consisting of all pictures over Σ that $\mathcal{M}$ accepts.

Next we illustrate the way in which the det-2D-4W-ORWW-automaton works by an example.

Example 19. L_{copy} is not accepted by any Sgraffito automaton (Corollary 4). Here we present a det-2D-4W-ORWW-automaton M_{4c} for this language. On input $P \in \Sigma^{m,n}$, M_{4c} proceeds as follows, where $\Sigma = \{\Box, \blacksquare\}$, $\Box_1, \blacksquare_1, \Box_2, \blacksquare_2$ are four new symbols, and $\Box > \blacksquare > \Box_1 > \blacksquare_1 > \Box_2 > \blacksquare_2$ holds:

1. M_{4c} first checks that P satisfies the condition $n = 2m$. If $n \neq 2m$, then M_{4c} halts without acceptance, otherwise, it moves to position $(1, m+1)$, marks the letter $P(1, m+1) = a \in \Sigma$ by replacing it by the symbol a_1, and restarts.
2. In the next cycles, M_{4c} checks that each row of P contains a string of the form $uu \in \Sigma^{2m}$. For doing so M_{4c} proceeds row by row, from the top to the bottom. In the first of these cycles, it will eventually reach the symbol a_1. At that point, it stores the symbol a in its finite-state control, and it moves to the left end of the first row, where it compares the symbol a to the symbol $b = P(1, 1) \in \Sigma$. If $a \neq b$, then M_{4c} halts without acceptance; otherwise, it replaces $P(1, 1) = a$ by a_1 and restarts.
3. In the next cycle, M_{4c} will encounter the symbol a_1 at position $(1, 1)$. It will then move right until it gets to the symbol a_1 at position $(1, m+1)$, which it will replace by the symbol a_2.
4. In the next cycle, M_{4c} will encounter the symbol a_2 at position $(1, m+1)$. It will then move to the left and replace a_1 at position $(1, 1)$ by a_2 as well.
5. In the next cycles the contents of the first row has the form $u_2 v u_2 w$, where $u_2 \in \{\Box_2, \blacksquare_2\}^+$ and $v, w \in \Sigma^{m-|u_2|}$. M_{4c} now compares the first letter c of w to the first letter d of v by first rewriting c into c_1, by then rewriting d into d_1, if $c = d$, by rewriting c_1 into c_2, and finally by rewriting d_1 into d_2.
6. After all rows have been checked successfully, M_{4c} halts on reaching the bottom-right corner and accepts.

This example shows that det-2D-4W-ORWW-automata are quite expressive. In [26] it is shown that 2DSAs are strictly weaker than the *two-dimensional deterministic forgetting automata* (det-FA-automata) of Jiřička and Král [14], which are bounded two-dimensional Turing machines that are allowed to rewrite by only using a special symbol @. In comparison to the 2DSA, there is no bound on the number of visits to any tape field. However, the det-2D-4W-ORWW-automata are at least as expressive as these automata.

Theorem 20. $\mathcal{L}(\text{det-FA}) \subseteq \mathcal{L}(\text{det-2D-4W-ORWW})$.

Proof. Let $\mathcal{A}$ be a det-FA, and let $P \in \Sigma^{*,*}$ be an input picture. We describe a det-2D-4W-ORWW-automaton $\mathcal{M}$ that simulates $\mathcal{A}$. Obviously, $\mathcal{M}$ can simulate $\mathcal{A}$ step by step, performing the same move operations and the same rewrite operations. However, there is the problem that $\mathcal{M}$ restarts after executing a rewrite operation. Hence, we must devise a way in which $\mathcal{M}$ can rediscover the position of the latest rewrite operation.

For that, we introduce auxiliary letters of the form $[a, q, i]$, where $a \in \Sigma$ is an input symbol, q is a state of $\mathcal{A}$, and $i \in \{1, 2\}$ is an additional index. The pair (a, q) will be used to encode the information that $\mathcal{A}$ replaced the symbol a by @ while in state q. The index i will be used to help $\mathcal{M}$ to find the correct position. This is done as follows, where we distinguish four cases.

1. If the current picture does not contain any of the above auxiliary symbols, no rewrite has been executed yet. Thus, in this situation $\mathcal{M}$ simulates $\mathcal{A}$ step by step until it reaches the position at which $\mathcal{A}$ would execute its first rewrite step. If this position contains the symbol $a \in \Sigma$, and if $\mathcal{A}$ is in state q, then $\mathcal{M}$ replaces a by the symbol $[a, q, 2]$ and restarts.
2. If the current picture contains a single auxiliary symbol which is of the form $[a, q, 2]$, then $\mathcal{M}$ starts simulating $\mathcal{A}$ from that position, but without actually performing the replacement of a by @. This continues until the next rewrite operation of $\mathcal{A}$ is detected, which would replace a symbol b by @, while $\mathcal{A}$ is in some state p. The automaton $\mathcal{M}$ will then replace the symbol b by the auxiliary symbol $[b, p, 1]$ and restart.
3. If the current picture contains an auxiliary symbol of the form $[a, q, 2]$ and an auxiliary symbol of the form $[b, p, 1]$, then the former marks the last but one rewrite operation of $\mathcal{A}$, while the latter marks the latest rewrite operation of $\mathcal{A}$. Now $\mathcal{M}$ simply rewrites $[a, q, 2]$ into @ and restarts.
4. If the current picture contains a single auxiliary symbol which is of the form $[b, p, 1]$, then $\mathcal{M}$ simply rewrites $[b, p, 1]$ into $[b, p, 2]$ and restarts.

Thus, $\mathcal{M}$ uses a single cycle to simulate the computation of $\mathcal{A}$ up to the first rewrite step, and then it uses three cycles each time it simulates a part of the computation of $\mathcal{A}$ that leads from a rewrite step to the next one.   □

It remains open whether the inclusion in Theorem 20 is proper. Further, it is known that det-FA working over strings accept all deterministic context-free languages [12]. This implies that det-2D-4W-ORWW-automata are too expressive for our purposes. To overcome this problem we restrict the potential head movements of the two-dimensional ordered restarting automata.

7 Deterministic Three-Way ORWW-Automata

Here we consider the *deterministic two-dimensional three-way ordered restarting automaton*.

Definition 21. *A deterministic two-dimensional three-way ordered restarting automaton, a* det-2D-3W-ORWW-*automaton for short, is given through a 7-tuple* $\mathcal{M} = (Q, \Sigma, \Gamma, \mathcal{S}, q_0, \delta, >)$, *where all components are defined as for a* det-2D-4W-ORWW-*automaton with the restriction that* $\mathcal{H} = \{R, D, U\}$ *is taken in the definition of the transition function* δ, *that is, no move-left steps are possible.*

In principle, it could happen that a det-2D-3W-ORWW-automaton $\mathcal{M}$ does not terminate on some input picture, as it may get stuck on a column, moving up and down. To avoid this, we *require explicitly* that $\mathcal{M}$ halts on all input pictures! This can be realized by either providing a simple pattern, e.g., up* − down* − up* − down*, such that on each column, the sequence of up and down movements must fit this pattern, or one could use an external counter that, for each column entered in the course of a computation, counts the number of uninterrupted up and down movements, making sure that the computation fails as soon as more than $(m \cdot |Q|)$- many such steps are encountered on a column of height m. Actually, in most cases termination follows from the fact that within a column, an automaton is just looking for an occurrence of a specific symbol, and if that is not found, then the computation fails anyway.

Given a picture $P \in \Sigma^{m,n}$ as input, a det-2D-3W-ORWW-automaton $\mathcal{M} = (Q, \Sigma, \Gamma, \mathcal{S}, q_0, \delta, >)$ can execute at most $m \cdot n \cdot (|\Gamma| - 1)$ many cycles. As each cycle takes at most $m \cdot n \cdot |Q|$ many steps, we see that for accepting P, $\mathcal{M}$ executes at most $m^2 \cdot n^2 \cdot (|\Gamma| - 1) \cdot |Q|$ many steps. A multi-tape Turing machine that stores P column by column needs m steps to simulate a single move-right step of $\mathcal{M}$. As $\mathcal{M}$ can execute at most $n - 1$ move-right steps in any cycle, we obtain the following result.

Theorem 22. [19] $\mathcal{L}(\text{det-2D-3W-ORWW}) \subseteq \text{DTIME}((\text{size}(P))^2)$.

When restricted to one-row pictures $P \in \Sigma^{1,*}$, then the det-2D-3W-ORWW-automaton coincides with the deterministic ordered restarting automaton introduced in [19]. Accordingly, the following result holds.

Theorem 23. [19] *When restricted to one-row inputs, the* det-2D-3DW-ORWW-*automaton just accepts the regular (string) languages.*

It is known that deterministic Sgraffito automata can be simulated by det-2D-3W-ORWW-automata [19]. In fact, we even have the following inclusion, as a 2DRTA that is scanning its input column by column from left to right is easily simulated by a det-2D-3W-ORWW-automaton.

Theorem 24. si-2DRTL $\subsetneq \mathcal{L}(\text{det-2D-3W-ORWW})$.

To see that this inclusion is proper, consider the following picture language $L_{1\text{col}}$ over $\Sigma = \{a, b\}$:

$$L_{1\text{col}} = \{ P \in \Sigma^{2n,1} \mid n \geq 1, P(1,1) \ldots P(n,1) = (P(n+1,1) \ldots P(2n,1))^R \},$$

that is, $L_{1\text{col}}$ consists of all pictures with a single column of even length such that the content of this column read from top to bottom is a palindrome. Based on the fact that a det-2D-3W-ORWW-automaton can freely move up and down a column, it can be shown that this language is accepted by some det-2D-3W-ORWW-automaton. By Theorem 14, the class si-2DRTL is closed under the operation of *rotation*. This operation turns the language $L_{1\text{col}}$ into the string language $L_{\text{pal}} = \{ w \in \Sigma^+ \mid |w| = 0 \mod 2, w = w^R \}$ of palindromes of even length, which is a non-regular language. As si-2DRTL only contains regular string languages (Theorem 17), it follows that $L_{1\text{col}}$ is not contained in si-2DRTL. This shows that the inclusion in Theorem 24 is proper. It also shows the following negative result.

Corollary 25. $\mathcal{L}$(det-2D-3W-ORWW) *is neither closed under rotation nor under transposition.*

In addition, the following closure and non-closure properties can be shown.

Theorem 26. *The language class* $\mathcal{L}$(det-2D-3W-ORWW) *is closed under union, intersection and complement, but it is neither closed under projection nor under column concatenation.*

To conclude this section we turn to another example that illustrates the limitations of the det-2D-3W-ORWW-automaton. Let $L_{\text{pal},2}$ be the following picture language over $\Sigma = \{a, b, \square\}$:

$$L_{\text{pal},2} = \{\, P \in \Sigma^{2,2n} \mid n \geq 1, P(1,1) \ldots P(1,n) = (P(1,n+1) \ldots P(1,2n))^R,$$
$$P(1,i) \in \{a,b\} \text{ and } P(2,i) = \square \text{ for all } 1 \leq i \leq 2n \,\},$$

that is, $L_{\text{pal},2}$ consists of all two-row pictures such that the first row contains a palindrome of even length over $\{a,b\}$, and the second row contains $\square$-symbols.

Proposition 27. $L_{\text{pal},2} \notin \mathcal{L}$(det-2D-3W-ORWW).

Proof. Assume to the contrary that there exists a det-2D-3W-ORWW-automaton $\mathcal{M} = (Q, \Sigma, \Gamma, \mathcal{S}, q_0, \delta, >)$ on $\Sigma = \{a, b, \square\}$ such that $L(\mathcal{M}) = L_{\text{pal},2}$. We analyze the accepting computations of $\mathcal{M}$ on pictures of the form $P_w \odot P_w^R \in L_{\text{pal},2}$, where

$$P_w = \begin{bmatrix} a_1 & a_2 & \cdots & a_n & a \\ \square & \square & \cdots & \square & \square \end{bmatrix}, P_w^R = \begin{bmatrix} a & a_n & \cdots & a_2 & a_1 \\ \square & \square & \cdots & \square & \square \end{bmatrix}, \text{ and } w = a_1 \ldots a_n \in \{a,b\}^n.$$

We say that $\mathcal{M}$ is *on* P_w (P_w^R) if the *active position* of its read/write window is on a tape field that belongs to P_w (P_w^R). As $\mathcal{M}$ cannot make any move-left steps, we see that each cycle of $\mathcal{M}$ consists of an initial part during which $\mathcal{M}$ is on P_w, which is then possibly followed by a part during which $\mathcal{M}$ is on P_w^R. In particular, when $\mathcal{M}$ performs a rewrite step on P_w, then it did not visit P_w^R at all during the corresponding cycle. Accordingly, the computation of $\mathcal{M}$ on input $P_w \odot P_w^R$ can be divided into two types of phases:

- a *left-phase* consists of a sequence of cycles during which $\mathcal{M}$ executes rewrite steps on P_w;
- a *right-phase* consists of a sequence of cycles during which $\mathcal{M}$ executes rewrite steps on P_w^R.

Obviously, an accepting computation of $\mathcal{M}$ on an input of the form $P_w \odot P_w^R$ cannot consist of a single left-phase only. Thus, after a (possibly empty) left-phase, a right-phase follows. As $\mathcal{M}$ is deterministic, this right-phase cannot end until at least one element in the first column of P_w^R has been rewritten, and the same is true also for all later right-phases. In fact, we say that a right-phase ends as soon as an element of the first column of P_w^R is being rewritten, and we associate with each right-phase a triple (r, m, X) that indicates that this right-phase ended by placing the symbol $X \in \Gamma$ into the tape field at position

$(m, n + 2)$. It follows either a left-phase or already the next right-phase. In addition, we see that altogether there are at most $2 \cdot (|\Gamma| - 1) + 1$ many right-phases and at most as many left-phases.

We are interested in the configurations of $\mathcal{M}$ in which it enters the tape fields of P_w^R during a right-phase, which are called *enter configurations*. An enter configuration is described by a pair (q, m), where $q \in Q$ is the current state of $\mathcal{M}$, and $m \in \{1, 2\}$ is the row of P on which the *active field* of the window is located.

During the execution of a right-phase, the behaviour of $\mathcal{M}$ is influenced by the current contents of the last column of P_w. This column has two entries, each of which can be rewritten at most $|\Gamma| - 1$ times. Hence, in order to keep track of the contents of this column, we use the triple (l, m, X) to denote that $\mathcal{M}$ executes a rewrite that places the symbol $X \in \Gamma$ into the tape field at position $(m, n + 1)$.

Now we can associate a *generalized crossing sequence* $\mathrm{GCS}(\mathcal{M}, P_w \oplus P_w^R)$ with the accepting computation of $\mathcal{M}$ on input $P_w \oplus P_w^R$ by appending the triples of the form (l, m, X), the enter configurations of the form (q, m), and the triples of the form (r, m, X) in the sequence in which the corresponding operations occur in the computation. Based on the observations above, it can be shown that there is only a constant number of such generalized crossing sequences.

However, there are 2^n many pictures of the form $P_w \oplus P_w^R$. Hence, if n is sufficiently large, then there are more pictures of this form than there are generalized crossing sequences. Hence, there exist two strings $w, x \in \{a, b\}^n$, $w \neq x$, such that $\mathrm{GCS}(\mathcal{M}, P_w \oplus P_w^R) = \mathrm{GCS}(\mathcal{M}, P_x \oplus P_x^R)$. It can now be shown that together with these pictures, $\mathcal{M}$ also accepts the picture $P_w \oplus P_x^R \notin L_{\mathrm{pal},2}$.   □

Using the same kind of reasoning it can be shown that the language L_{copy} is not accepted by any det-2D-3W-ORWW-automaton, either. As $\mathcal{L}(\text{det-2D-3W-ORWW})$ is closed under complement, this implies that $(L_{\mathrm{copy}})^c \in \mathcal{L}(2\mathrm{SA}) \smallsetminus \mathcal{L}(\text{det-2D-3W-ORWW})$. Together with the fact that $L_{1\mathrm{col}} \in \mathcal{L}(\text{det-2D-3W-ORWW}) \smallsetminus \mathcal{L}(2\mathrm{SA})$, this yields the following incomparability results.

Corollary 28. *The class of picture languages $\mathcal{L}(\text{det-2D-3W-ORWW})$ is incomparable to the classes $\mathcal{L}(2\mathrm{SA})$ and REC with respect to inclusion.*

8   Deterministic Two-Way ORWW-Automata

To get rid of the termination problem for two-dimensional ORWW-automata, we restrict the possible window movements even further.

Definition 29. *A deterministic two-dimensional two-way ordered restarting automaton, a det-2D-2W-ORWW-automaton for short, is given through a 7-tuple $\mathcal{M} = (Q, \Sigma, \Gamma, \mathcal{S}, q_0, \delta, >)$, where all components are defined as for a det-2D-4W-ORWW-automaton with the restriction that $\mathcal{H} = \{\mathrm{R}, \mathrm{D}\}$ is taken in the definition of the transition function δ, that is, no move-left or move-up steps are possible.*

By interchanging move-right steps and move-down steps, it is immediate that the language class $\mathcal{L}(\text{det-2D-2W-ORWW})$ is closed under transposition. In fact, the following closure and non-closure results can be derived.

Theorem 30. *The language class $\mathcal{L}$(det-2D-2W-ORWW) is closed under transposition, union, intersection, and complement, but it is neither closed under projection nor under column or row concatenation.*

When restricted to one-row pictures, then the det-2D-2W-ORWW-automaton coincides with the det-2D-3W-ORWW-automaton, and so it only accept regular string languages. Concerning the expressive power of the det-2D-2W-ORWW-automaton, we have the following result.

Theorem 31. DREC $\subsetneq \mathcal{L}$(det-2D-2W-ORWW) $\subseteq \mathcal{L}$(2DSA).

Proof. First it can be shown that each 2DOTA can be simulated by a det-2D-2W-ORWW-automaton. Actually, this simulation works for each possible corner-to-corner direction, and hence, as DREC coincides with the closure of $\mathcal{L}$(2DOTA) under rotation, the first inclusion follows. In [2] it is shown that the language L_{frames} over $\{0, 1\}$ that consists of all square pictures P for which (i) the first row equals the first column, (ii) the second row equals the reverse of the second-last column, (iii) the second-last row equals the reverse of the second column, and (iv) the last row equals the last column, is not in DREC, but as $\mathcal{L}$(det-2D-2W-ORWW) is closed under intersection, and as each of the defining conditions for the elements of L_{frames} can be verified by a det-2D-2W-ORWW-automaton, it follows that this language is accepted by a det-2D-2W-ORWW-automaton.

Finally, let $\mathcal{M}$ be a det-2D-2W-ORWW-automaton on Σ. On input $P \in \Sigma^{*,*}$, $\mathcal{M}$ scans this picture, starting at the upper left corner, until it executes a rewrite step at some position (i, j). Now a 2DSA $\mathcal{A}$ can simulate the computation of $\mathcal{M}$ step by step until it reaches position (i, j). As $\mathcal{A}$ must perform a rewrite in each step, we can assume that $\mathcal{A}$ encodes the corresponding state of $\mathcal{M}$ at each position visited together with the information on the last two steps that $\mathcal{M}$ performed when it moved to the current position. After performing its rewrite step, $\mathcal{M}$ restarts, but as it is deterministic, it will follow the same path again, entering the same states at the same positions, until it reaches a position $(k, l) \in \{(i-1, j), (i, j-1), (i-1, j-1)\}$ such that the new symbol at (i, j) is contained in its window of size $(3, 3)$. From the information stored at position (i, j), $\mathcal{A}$ can determine this position. It then moves to the position (k, l), extracts the corresponding state of $\mathcal{M}$ from the symbol written at that position, and continues to simulate $\mathcal{M}$ from that point on. It follows that $\mathcal{A}$ can enter a particular tape field at most a constant number of times (once at the first time this position is reached by $\mathcal{M}$, once each time a field immediately to the right or down is rewritten, and once again each time the contents of the field itself has been rewritten). $\square$

It is, however, still open whether the second inclusion in Theorem 31 is proper. The problem lies in the fact that, in order to simulate the computation of a 2DSA, a det-2D-2W-ORWW-automaton must be able to rediscover the actual head position of the 2DSA after each restart step. In fact, it is not known whether the language L_{perm} (see Section 3) is accepted by any det-2D-2W-ORWW-automaton. We therefore consider an extension of this model in which the move-right and move-down steps are slightly more general.

Definition 32. *A deterministic two-dimensional extended two-way ordered restarting automaton, a* det-2D-x2W-ORWW-*automaton for short, is given through a 7-tuple* $\mathcal{M} = (Q, \Sigma, \Gamma, \mathcal{S}, q_0, \delta, >)$, *where all components are defined as for a* det-2D-2W-ORWW-*automaton, but the move-right and move-down steps are extended as follows:*

1. *When the window contains the right border marker* ⊣, *but not the bottom marker, then an* extended move-right step *shifts the window to the beginning of the next row, that is, if the central position of the window is on the last field of row* i *for some* $i < \mathrm{rows}(P)$, *then it is moved to the first field of row* $i + 1$.

2. *When the window contains the bottom marker* ⊥, *but not the right border marker, then an* extended move-down step *shifts the window to the top of the next column, that is, if the central position of the window is on the bottom-most field of column* j *for some* $j < \mathrm{cols}(P)$, *then it is moved to the top-most field of column* $j + 1$.

3. *In any cycle, as soon as* $\mathcal{M}$ *executes an extended move-right (move-down) step, then for the rest of this cycle, it cannot execute any extended move-down (move-right) steps.*

Finally, $\mathcal{M}$ *is called a* stateless det-2D-x2W-ORWW-*automaton (or a* stl-det-2D-x2W-ORWW-*automaton) if it has just a single state. The components* Q *and* q_0 *refering to states will be suppressed within the description of such an automaton.*

When restricted to one-row pictures $P \in \Sigma^{1,*}$, then the det-2D-x2W-ORWW-automaton coincides with the det-2D-3W-ORWW-automaton. Thus, we obtain the following result, where the part on stateless variants is an easy extension.

Corollary 33. *When restricted to one-row inputs, then the* (stl-)det-2D-x2W-ORWW-*automaton just accepts the regular (string) languages.*

Given a picture $P \in \Sigma^{m,n}$ as input, a det-2D-x2W-ORWW-automaton $\mathcal{M}$ can execute at most $m \cdot n \cdot (|\Gamma| - 1)$ many cycles. In each cycle $\mathcal{M}$ can either execute up to n move-right steps, $n \cdot (m - 1)$ move-down steps, and $(n - 1)$ extended move-down steps, or m move-down steps, $m \cdot (n - 1)$ move-right steps, and $(m - 1)$ extended move-right steps. Thus, $\mathcal{M}$ executes at most $m^2 \cdot n^2 \cdot (|\Gamma| - 1)$ many steps. Hence, a two-dimensional Turing machine can simulate $\mathcal{M}$ in time $O(m^2 \cdot n^2)$. A multi-tape Turing machine $\mathcal{T}$ that stores P column by column needs m steps to simulate a single move-right step of $\mathcal{M}$, and it needs $m \cdot n$ steps to simulate an extended move-right step, that is, it may need up to $O(m^3 \cdot n^2)$ many steps to simulate $\mathcal{M}$. Hence, we obtain the following upper bound.

Theorem 34. [18] $\mathcal{L}(\text{det-2D-x2W-ORWW}) \subseteq \mathrm{DTIME}((\mathrm{size}(P))^3)$.

As a 2DRTA is stateless, it is easily seen that a 2DRTA that is scanning its input row by row from top to bottom can be simulated by a stateless det-2D-x2W-ORWW-automaton. This yields the following inclusion.

Theorem 35. si-2DRTL $\subseteq \mathcal{L}(\text{stl-det-2D-x2W-ORWW})$.

It is still open whether or not this is a proper inclusion. From the definition the following closure properties are immediate.

Proposition 36. *The classes of picture languages* $\mathcal{L}$(det-2D-x2W-ORWW) *and* $\mathcal{L}$(stl-det-2D-x2W-ORWW) *are closed under transposition and complement.*

Thus, the language L_{1col} is not accepted by any det-2D-x2W-ORWW-automaton. On the other hand, the following result holds.

Proposition 37. [18] $L_{pal,2} \in \mathcal{L}$(det-2D-x2W-ORWW).

Proof. Let $M_{pal,2} = (Q, \Sigma, \Gamma, \mathcal{S}, q_0, \delta, >)$ be defined by $\Gamma = \Sigma \cup \{a_1, a_2, b_1, b_2, \uparrow_1, \uparrow_2\}$, where $a > b > a_1 > b_1 > a_2 > b_2 > \square > \uparrow_1 > \uparrow_2$, and by defining δ in such a way that $M_{pal,2}$ proceeds as follows. As the det-2D-4W-ORWW-automaton M_{4c} of Example 19, $M_{pal,2}$ marks the letters in the first row with indices 1 and 2, alternating between marking the first unmarked letter from the left and the first unmarked letter from the right. To distinguish these two cases, the second row is used. First M scans the first row completely from left to right. If during this sweep it realizes that the first unmarked letter from the right must be marked, then it simply does this and restarts. If, however, it realizes at the end of this sweep that the first unmarked letter from the left should have been marked, then it executes an extended move-right step, and then it replaces the letter in row 2 that is below the first unmarked letter from the left in row 1 by the symbol $\uparrow_1$ or $\uparrow_2$, depending on the current situation. In this way it indicates that the corresponding letter in row 1 must be marked in the next cycle by index 1 or 2, respectively. $\square$

From the results on L_{1col} and $L_{pal,2}$ we obtain the following.

Corollary 38. *The class of picture languages* $\mathcal{L}$(det-2D-x2W-ORWW) *is incomparable under inclusion to the class of picture languages* $\mathcal{L}$(det-2D-3W-ORWW).

Actually, $L_{pal,2}$ also separates the det-2D-x2W-ORWW-automata from their stateless variants.

Proposition 39. [18] $L_{pal,2} \notin \mathcal{L}$(stl-det-2D-x2W-ORWW).

Proof. Assume that $\mathcal{M} = (\Sigma, \Gamma, \mathcal{S}, \delta, >)$ is a stl-det-2D-x2W-ORWW-automaton over $\Sigma = \{a, b, \square\}$ such that $L(\mathcal{M}) = L_{pal,2}$. Given $P_w \oplus P_w^R$ (see the proof of Prop. 27) as input, $\mathcal{M}$ will perform an accepting computation, which consists of a finite sequence of cycles that is followed by an accepting tail computation. We now split this computation into a finite number of *phases*, where we distinguish between four types of phases:

1. A *left-only phase* (O) consists of a sequence of cycles in which the window of M stays on the left half of the picture.
2. An *upper-right phase* (U) consists of a sequence of cycles in which all rewrite steps are performed on the right half of the picture, and in addition, in the first of these cycles, $\mathcal{M}$ enters the right half of the picture through move-right steps in row 1.

3. A *lower-left phase* (L) is a sequence of cycles in which all rewrite steps are performed in the left half of the picture, and in addition, the first of these cycles contains an extended move-right step.

4. A *lower-right phase* (R) is a sequence of cycles in which all rewrite steps are performed in the right half of the picture, and in addition, in the first of these cycles, $\mathcal{M}$ enters the right half of the picture through a move-right step in row 2 or after executing an extended move-down step.

Obviously, the sequence of cycles of the computation of $\mathcal{M}$ on input $P_w \oplus P_w^R$ can uniquely be split into a sequence of phases if we require that each phase is of maximum length. Thus, this computation can be described in a unique way by a string α over the alphabet $\Omega = \{O, U, L, R\}$.

Concerning the possible changes from one phase to the next there are some restrictions based on the fact that $\mathcal{M}$ is stateless. For example, while $\mathcal{M}$ is in a lower-right phase (R), it just moves through the left half of the current picture after each rewrite/restart step. Thus, $\mathcal{M}$ cannot get into another phase until it performs a rewrite step that replaces a symbol in the first column of the right half of the picture. Only then may follow a left-only phase (O) or a lower-left phase (L). However, in a fixed column, less than $2 \cdot |\Gamma|$ many rewrite steps can be performed, and so $|\alpha|_R \leq 1 + 2 \cdot |\Gamma|$. By analyzing all cases, it can be shown that $|\alpha| \leq 13 + 24 \cdot |\Gamma|$.

Now we associate a *generalized crossing sequence* $\mathrm{GCS}(P_w \oplus P_w^R)$ with the computation of $\mathcal{M}$ on $P_w \oplus P_w^R$ by inserting a 2-by-2 picture $\begin{pmatrix} c & d \\ e & f \end{pmatrix}$ after each letter X of α, where $\binom{c}{e}$ is the contents of the rightmost column of the left half and $\binom{d}{f}$ is the contents of the leftmost column of the right half of the picture at the end of the phase represented by the letter X. Thus, $\mathrm{GCS}(P_w \oplus P_w^R)$ is a string of length at most $26 + 48 \cdot |\Gamma|$ over the finite alphabet $\Omega \cup \Gamma^{2,2}$ of size $4 + |\Gamma|^4$, that is, there are only finitely many different such crossing sequences. If n is sufficiently large, then there are two strings $w, x \in \{a, b\}^n$, $w \neq x$, such that $\mathrm{GCS}(P_w \oplus P_w^R) = \mathrm{GCS}(P_x \oplus P_x^R)$. As $\mathcal{M}$ accepts both $P_w \oplus P_w^R$ and $P_x \oplus P_x^R$, it follows that $\mathcal{M}$ will also accept on input $P_w \oplus P_x^R \notin L_{\mathrm{pal},2}$, a contradiction. $\square$

Together with Proposition 37 this yields the following separation result.

Theorem 40. $\mathcal{L}(\text{stl-det-2D-x2W-ORWW}) \subsetneq \mathcal{L}(\text{det-2D-x2W-ORWW})$.

By using the technique from the proof of Proposition 37, also the following can be shown.

Proposition 41. $L_{\mathrm{copy}} \in \mathcal{L}(\text{det-2D-x2W-ORWW})$.

Thus, we see that $\mathcal{L}(\text{det-2D-x2W-ORWW})$ is not contained in $\mathcal{L}(\text{2SA})$, but it remains the question of whether $\mathcal{L}(\text{2SA}) \subset \mathcal{L}(\text{det-2D-x2W-ORWW})$ holds. We complete this section with some more closure properties.

Theorem 42. [18] *The language classes* $\mathcal{L}(\text{stl-det-2D-x2W-ORWW})$ *and* $\mathcal{L}(\text{det-2D-x2W-ORWW})$ *are closed under union and intersection.*

9 Conclusion

We have studied six different models of deterministic two-dimensional restarting automata, comparing them to the (deterministic) Sgraffito automata and the well-known classes REC and DREC. The taxonomy of classes of picture languages obtained is summarized by the diagram in Figure 4.

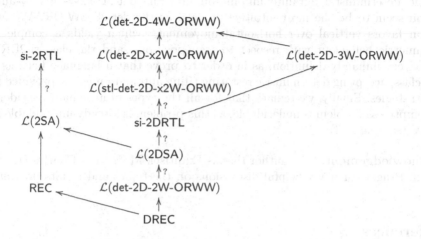

Fig. 4. Hierarchy of classes of picture languages accepted by various types of two-dimensional automata. Question marks indicate inclusions not known to be proper.

For all these deterministic types of automata, the membership problem is solvable in polynomial time. When restricted to one-row pictures, that is, string languages, then all these models (and also the non-deterministic restarting tiling automaton and the Sgraffito automaton) only accept the regular languages with the sole exception of the **det-2D-4W-ORWW**-automaton, which accepts some string languages that are not even growing context-sensitive. Accordingly, the latter model is far too expressive for our purposes.

Class of Picture Languages	$\cup$	$\cap$	c	R	T	π	$\oplus$	$\ominus$	$*\oplus$	$*\ominus$
REC	+	+	−	+	+	+	+	+	+	+
DREC	−	−	+	+	+	?	?	?	?	?
$\mathcal{L}$(2DSA)	+	+	+	+	+	−	−	−	?	?
si-2DRTL	+	+	?	+	+	+	?	?	?	?
$\mathcal{L}$(det-2D-2W-ORWW)	+	+	+	?	+	−	−	−	?	?
$\mathcal{L}$(stl-det-2D-x2W-ORWW)	+	+	+	?	+	?	?	?	?	?
$\mathcal{L}$(det-2D-x2W-ORWW)	+	+	+	?	+	?	?	?	?	?
$\mathcal{L}$(det-2D-3W-ORWW)	+	+	+	−	−	−	−	?	?	?

Fig. 5. The known closure properties for some deterministic classes of picture languages. An entry '+' stands for a known closure property, '−' marks a known negative result, and '?' indicates that the corresponding result is still open.

For the deterministic Sgraffito automaton and the deterministic tiling automaton the most closure properties have been established, but compared to the class REC, all other classes have less nice closure properties. A summary of the known closure properties is presented in the table in Figure 5, where c denotes the operation of complement, R denotes rotation, T stands for transposition, and π denotes projections.

The deterministic Sgraffito automaton and the det-2D-x2W-ORWW-automaton seem to be the most intuitive models, as the det-2D-3W-ORWW-automaton favors vertical over horizontal movements, which yields a completely asymmetric behaviour with respect to transpositions, and the class si-2DRTL has a very unhandy definition, as in order to prove that a language belongs to this class, accepting deterministic restating tiling automata must be provided for *all* strategies. Finally, we remark that for all the types of automata considered the emptiness problem is undecidable, as this problem is already undecidable for DREC (see, e.g., [7]).

Acknowledgement. The author thanks František Mráz from Charles University in Prague for many helpful discussions on the topics and results presented here.

References

1. Anselmo, M., Giammarresi, D., Madonia, M.: From determinism to non-determinism in recognizable two-dimensional languages. In: Harju, T., Karhumäki, J., Lepistö, A. (eds.) DLT 2007. LNCS, vol. 4588, pp. 36–47. Springer, Heidelberg (2007)
2. Anselmo, A., Giammarresi, D., Madonia, M.: A computational model for tiling recognizable two-dimensional languages. Theor. Comput. Sci. 410, 3520–3529 (2009)
3. Blum, M., Hewitt, C.: Automata on a 2-dimensional tape. In: SWAT 1967, pp. 155–160. IEEE Computer Society, Washington, DC (1967)
4. Borchert, B., Reinhardt, K.: Deterministically and sudoku-deterministically recognizable picture languages. In: Loos, R., Fazekas, S., Martín-Vide, C. (eds.) LATA 2007, Preproc. Report 35/07, pp. 175–186. Research Group on Mathematical Linguistics, Universitat Rovira i Virgili, Tarragona (2007)
5. Cherubini, A., Pradella, M.: Picture Languages: From Wang Tiles to 2D Grammars. In: Bozapalidis, S., Rahonis, G. (eds.) CAI 2009. LNCS, vol. 5725, pp. 13–46. Springer, Heidelberg (2009)
6. Giammarresi, D., Restivo, A.: Recognizable picture languages. Intern. J. Pattern Recognition and Artificial Intelligence 6(2-3), 241–256 (1992)
7. Giammarresi, D., Restivo, A.: Two-dimensional languages. In: Rozenberg, G., Salomaa, A. (eds.) Handbook of Formal Languages, vol. 3, pp. 215–267. Springer, New York (1997)
8. Hennie, F.: One-tape, off-line Turing machine computations. Informat. Control 8(6), 553–578 (1965)
9. Hopcroft, J., Ullman, J.: Introduction to Automata Theory, Languages, and Computation. Addison-Wesley, Reading (1979)
10. Inoue, K., Nakamura, A.: Some properties of two-dimensional on-line tessellation acceptors. Inform. Sci. 13, 95–121 (1977)

11. Inoue, K., Takanami, I.: A characterization of recognizable picture languages. In: Nakamura, A., Saoudi, A., Inoue, K., Wang, P.S.P., Nivat, M. (eds.) ICPIA 1992. LNCS, vol. 654, pp. 133–143. Springer, Heidelberg (1992)

12. Jančar, P., Mráz, F., Plátek, M.: Characterization of context-free languages by erasing automata. In: Havel, I.M., Koubek, V. (eds.) MFCS 1992. LNCS, vol. 629, pp. 307–314. Springer, Heidelberg (1992)

13. Jančar, P., Mráz, F., Plátek, M., Vogel, J.: Restarting automata. In: Reichel, H. (ed.) FCT 1995. LNCS, vol. 965, pp. 283–292. Springer, Heidelberg (1995)

14. Jiřička, P., Král, J.: Deterministic forgetting planar automata are more powerful than non-deterministic finite-state planar automata. In: Rozenberg, G., Thomas, W. (eds.) DLT 1999, pp. 71–80. World Scientific, Singapore (2000)

15. Kari, J., Moore, C.: New results on alternating and non-deterministic two-dimensional finite-state automata. In: Ferreira, A., Reichel, H. (eds.) STACS 2001. LNCS, vol. 2010, pp. 396–406. Springer, Heidelberg (2001)

16. Lindgren, K., Moore, C., Nordahl, M.: Complexity of two-dimensional patterns. J. Stat. Phys. 91(5-6), 909–951 (1998)

17. Messerschmidt, H., Stommel, M.: Church-Rosser picture languages and their applications in picture recognition. J. Autom. Lang. Comb. 16, 165–194 (2011)

18. Otto, F., Mráz, F.: Extended two-way ordered restarting automata for picture languages. In: Dediu, A.-H., Martín-Vide, C., Sierra-Rodríguez, J.-L., Truthe, B. (eds.) LATA 2014. LNCS, vol. 8370, pp. 541–552. Springer, Heidelberg (2014)

19. Mráz, F., Otto, F.: Ordered restarting automata for picture languages. In: Geffert, V., Preneel, B., Rovan, B., Štuller, J., Tjoa, A.M. (eds.) SOFSEM 2014. LNCS, vol. 8327, pp. 431–442. Springer, Heidelberg (2014)

20. Otto, F.: Restarting automata. In: Ésik, Z., Martín-Vide, C., Mitrana, V. (eds.) Recent Advances in Formal Languages and Applications. SCI, vol. 25, pp. 269–303. Springer, Heidelberg (2006)

21. Průša, D.: Weight-reducing Hennie machines and their descriptional complexity. In: Dediu, A.-H., Martín-Vide, C., Sierra-Rodríguez, J.-L., Truthe, B. (eds.) LATA 2014. LNCS, vol. 8370, pp. 553–564. Springer, Heidelberg (2014)

22. Průša, D., Mráz, F.: Restarting tiling automata. In: Moreira, N., Reis, R. (eds.) CIAA 2012. LNCS, vol. 7381, pp. 289–300. Springer, Heidelberg (2012)

23. Průša, D., Mráz, F.: Two-dimensional sgraffito automata. In: Yen, H.-C., Ibarra, O.H. (eds.) DLT 2012. LNCS, vol. 7410, pp. 251–262. Springer, Heidelberg (2012)

24. Průša, D., Mráz, F.: Restarting tiling automata. Intern. J. Found. Comput. Sci. 24(6), 863–878 (2013)

25. Průša, D., Mráz, F., Otto, F.: Comparing two-dimensional one-marker automata to sgraffito automata. In: Konstantinidis, S. (ed.) CIAA 2013. LNCS, vol. 7982, pp. 268–279. Springer, Heidelberg (2013)

26. Průša, D., Mráz, F., Otto, F.: New results on deterministic sgraffito automata. In: Béal, M.-P., Carton, O. (eds.) DLT 2013. LNCS, vol. 7907, pp. 409–419. Springer, Heidelberg (2013)

27. Reinhardt, K.: On some recognizable picture-languages. In: Brim, L., Gruska, J., Zlatuška, J. (eds.) MFCS 1998. LNCS, vol. 1450, pp. 760–770. Springer, Heidelberg (1998)

28. Rosenfeld, A.: Isotonic grammars, parallel grammars, and picture grammars. In: Meltzer, B., Michie, D. (eds.) Machine Intelligence, vol. 6, pp. 281–294. Edinburgh University Press (1971)

29. Siromoney, G., Siromoney, R., Krithivasan, K.: Abstract families of matrices and picture languages. Computer Graphics and Image Processing 1(3), 284–307 (1972)

Investigations on Automata and Languages over a Unary Alphabet

Giovanni Pighizzini*

Dipartimento di Informatica
Università degli Studi di Milano, Italy
pighizzini@di.unimi.it

Abstract. The investigation of automata and languages defined over a one letter alphabet shows interesting differences with respect to the case of alphabets with at least two letters. Probably, the oldest example emphasizing one of these differences is the collapse of the classes of regular and context-free languages in the unary case (Ginsburg and Rice, 1962). Many differences have been proved concerning the state costs of the simulations between different variants of unary finite state automata (Chrobak, 1986, Mereghetti and Pighizzini, 2001). We present an overview of those results. Because important connections with fundamental questions in space complexity, we give emphasis to unary two-way automata. Furthermore, we discuss unary versions of other computational models, as one-way and two-way pushdown automata, even extended with auxiliary workspace, and multi-head automata.

In Memory of Alberto Bertoni,
who taught me how beautiful theoretical computer science is.

1 Introduction

In 1962, Ginsburg and Rice discovered that in the case of languages defined oven a one letter alphabet, called *unary languages*, the classes of regular and context-free languages collapse [13]. Probably, this is the oldest example which emphasizes a difference between the unary and the general case. Many other results in formal language and complexity theory have been proved under the hypothesis of unrestricted or at least two letter alphabets, while different properties have been obtained in the one letter case, showing that unary languages (also called *tally sets*) have interesting and sometimes surprising properties.

In this paper we present an overview of some of these results. A large part of the paper is devoted to devices accepting unary regular languages and to their descriptional complexity. In particular, we discuss the state costs of the optimal simulations between different types of unary automata. We will emphasize the role of normal forms for unary one-way and two-way automata in these

---

* Partially supported by MIUR under the project PRIN "Automi e Linguaggi Formali: Aspetti Matematici e Applicativi", code H41J12000190001.

M. Holzer and M. Kutrib (Eds.): CIAA 2014, LNCS 8587, pp. 42–57, 2014.

studies. We also consider pushdown automata in both the deterministic and the nondeterministic cases.

In the last part of the paper, we shortly discuss some extensions of these models that are able to recognize unary nonregular languages, as one-way auxiliary pushdown automata, two-way pushdown automata, and multi-head automata. An interesting topic related to these extensions is the study of minimal amounts of extra resources (e.g., workspace on auxiliary tapes, number of head reversals) which make these devices more powerful than finite automata in the recognition of unary languages. Many problems in this area are still open.

The unary case could seem a kind of "separate world": special properties, different methods (many of them deriving from number theory), *etc.* Actually the investigation of the unary case is also related to some general and relevant questions. For instance, as we will discuss in the paper, the question of the optimal state cost of removing nondeterminism from two-way automata in the unary case is strictly related to the question of the power of nondeterminism in space bounded machines over general alphabets. In fact, proving that polynomially many states are not sufficient to simulate unary two-way nondeterministic automata by equivalent two-way deterministic automata will separate deterministic and nondeterministic logarithmic space [11].

We will keep the presentation at an informal level, trying to avoid, as much as possible, technical details, that can be found in the references. Due to many results on the unary case, the paper does not pretend to be exhaustive. We partially cover some topics that in our opinion are relevant and that are related to our research experience.

Throughout the paper, we use the symbol Σ to denote an *alphabet*, namely a non empty finite set of symbols. Sometimes we write *general alphabet* to emphasize that there are no restrictions on the cardinality of Σ, in contrast with *unary alphabet*, to indicate that the cardinality of Σ is 1, in which case we stipulate $\Sigma = \{a\}$. We also write *nonunary alphabet* when we want to restrict to the case $\#\Sigma \geq 2$. A similar meaning is associated with *general*, *unary*, and *nonunary case*. We use standard notations from formal language theory as in [15,46]. Since in the unary case each string is represented by an integer, there is a natural bijection between unary languages and subsets of $\mathbb{N}$, namely unary languages can be identified with sets of nonnegative integers.

2 Unary Finite Automata: Optimal Simulations

In this section we discuss some properties of finite automata over a unary alphabet. The focus is mainly on the costs, in term of states, of the simulations between different variant of automata.

A *one-way finite state automaton* is denoted as a tuple $A = (Q, \Sigma, \delta, q_0, F)$, where, as usual, Q is the finite set of states, Σ is the input alphabet, δ is the *transition function*, $q_0 \in Q$ is the *initial state*, and $F \subseteq Q$ is the set of final states. The language accepted by A is denoted as $L(A)$. As usual, we consider *deterministic* and *nondeterministic* versions of these devices indicated, for brevity, as

1DFAs and 1NFAs, respectively. We emphasize that those devices are *one-way*, namely at each transition they move the input head one position to the right, in contrast with *two-way* deterministic and nondeterministic variants, indicated as 2DFAs and 2NFAs, respectively. The transition function of a 2DFA associates with each pair $(q, a) \in Q \times \Sigma$ a next state p and a direction $d \in \{-1, 0, +1\}$, where $d = -1$ and $d = +1$ mean that the head is moved one position to the left and to the right, respectively, while $d = 0$ means that the head is kept on the same tape cell. In the case of 2NFAs, the transition function can associate to each pair (q, a) a set of pairs (p, d), representing different nondeterministic choices. Furthermore, in two-way automata the input is assumed to be surrounded by two special symbols, the *left end-marker* and the *right end-marker*. Hence, the transition function is defined even on those symbols, with the restriction that from the left end-marker the head cannot move to the left, while from the right end-marker it cannot move to the right. In the literature, slightly different acceptance conditions for two-way automata have been used (e.g., just reaching a final state or reaching a final state with the head on the right end-marker). Since these small technical differences do not affect the power of the models and slightly change the number of states, here we can avoid to enter into these details. However, it is useful to emphasize that computations of two-way automata can enter into infinite loops. All those computations are rejecting.

Since the beginning of automata theory, it is well-known that all the just mentioned variants of finite automata have the same computational power, in fact all of them characterize the class of *regular languages* [41]. The state costs of the simulations between these variants of finite automata have been investigated, in the general case, in several papers (e.g., [47,35,43]). Here, we discuss these costs in the unary case. Before doing that, we present some fundamental properties of unary automata.

It can be easily observed that the transition graph of a 1DFA A consists of a path (frequently called *tail*) of $\mu \geq 0$ states which starts from the initial state and of a loop of $\lambda \geq 1$ states, which is reachable by an edge from the last state of the initial path (see Figure 1). From this observation, it follows that $L(A) \subseteq \{a^0, a^1, \ldots, a^{\mu-1}\} \cup \{a^\mu\} \cdot \{a^0, a^1, \ldots, a^{\lambda-1}\} \cdot \{a^\lambda\}^*$. In other words, for each $K \geq \mu$, $a^K \in L(A)$ if and only if $a^{K+\lambda} \in L(A)$. Hence, to each unary regular language corresponds an ultimately periodic set of integers. It is trivial to observe that the converse also holds.

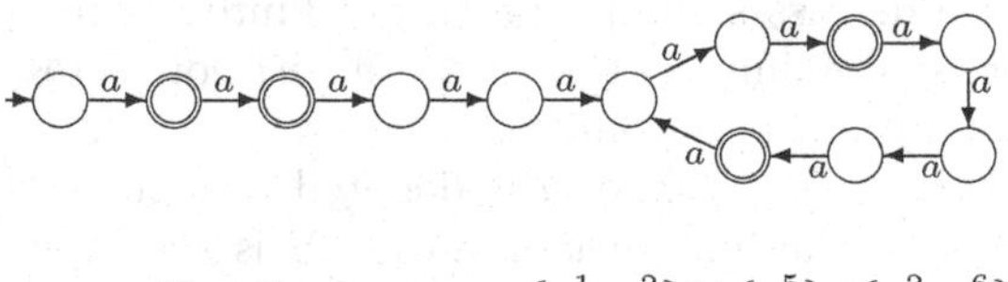

Fig. 1. A 1DFA accepting the language $\{a^1, a^2\} \cup \{a^5\} \cdot \{a^2, a^6\} \cdot \{a^7\}^*$

There are two special cases.

- If $\mu = 0$, namely the initial path is empty and the initial state belongs to the loop, then the set of integers corresponding to $L(A)$ is periodic. In this case $L(A)$ is said to be a λ-*cyclic language*, or just a *cyclic language*.
- If we allow the transition function to be partial, then the graph could contain only the initial path, thus implying that the accepted language is a subset of $\{a^0, a^1, \ldots, a^{\mu-1}\}$. Of course, the transition function can be made complete by introducing a loop consisting of just one (rejecting) state.

Now, let us turn our attention to the nondeterministic case. Each directed graph with a selected vertex marked as initial state and some vertices appointed as final states can be interpreted as a 1NFA. This form does not look so simple as the one of unary 1DFAs. In spite of that, as shown by Chrobak, *with only a polynomial increasing in the number of states* each 1NFA can be turned into an equivalent 1NFA whose transition graph has a very simple structure, which we are now going to describe [4].

A 1NFA is in *Chrobak Normal form* if its transition graph consists of an initial deterministic path and $s \geq 0$ disjoint loops. The last state in the path is connected via s outgoing edges to each of the loops. This is the only place where a nondeterministic decision can be taken by the automaton. (See Figure 2.) Hence, during each computation, the automaton can make at most one nondeterministic choice. (Notice that for $s = 1$ we obtain 1DFAs.) Chrobak proved that each unary n-state 1NFA can be converted into a 1NFA with an initial path of $O(n^2)$ states and a total number of the states in the loops $\leq n$.

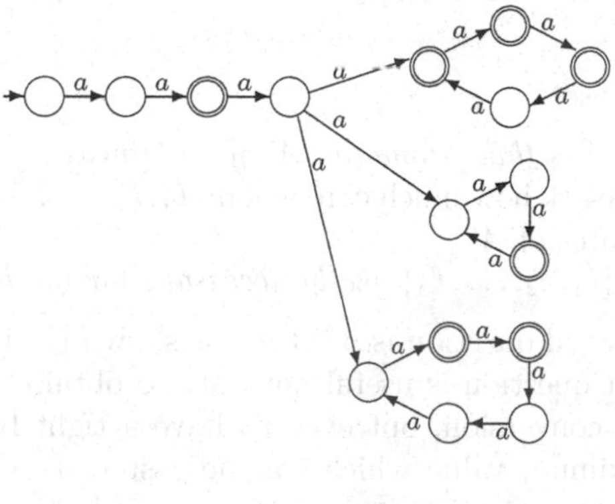

Fig. 2. A 1NFA in Chrobak normal form

Some remarks on this fundamental result and on related works are suitable. The original proof by Chrobak contains a subtle error which has been discovered and fixed by To [51]. A polynomial time algorithm for converting unary 1NFAs into Chrobak normal form has been obtained by Martinez and independently improved by Gawrychowski and Sawa [31,7,45]. Geffert gave a completely different and very detailed proof of the Chrobak normal form, reducing the state bound: except for the case of the trivial loop of length n, each unary n-state 1NFA can be converted into an equivalent 1NFA in Chrobak normal form with at most $n^2 - 2$ states on the initial path and a total number of states in the loops $\leq n - 1$ [8]. The length of the initial path has been further reduced to $n^2 - n$ [7]. Summing up:

Theorem 1. *For each unary n-state 1NFA different from the trivial loop of length n there exists an equivalent 1NFA in Chrobak normal form with an initial path consisting of at most $n^2 - n$ states and a set of loops whose total number of states is at most $n - 1$.*

Once a 1NFA A is in Chrobak normal form, it quite easy to convert it into an equivalent 1DFA A', by replacing the loops by a single loop which simulates "in parallel" all of them. Suppose that A contains $s \geq 1$ loops of length $\ell_1, \ell_2, \ldots, \ell_s$, respectively. Then a loop of length $\mathrm{lcm}\{\ell_1, \ell_2, \ldots, \ell_s\}$ (the least common multiple of loop lengths) is enough for A'. (See Figure 3.)

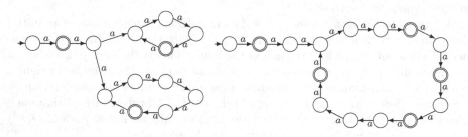

Fig. 3. A 1NFA in Chrobak normal form and an equivalent 1DFA

Now two questions arise:

1. *How many states has the automaton A' so obtained?*
   In other terms, we ask how much can be $\mathrm{lcm}\{\ell_1, \ell_2, \ldots, \ell_s\}$ with respect to n, the number of states of A'.
2. *Is the length $\mathrm{lcm}\{\ell_1, \ell_2, \ldots, \ell_s\}$ really necessary for the loop?*

The answer to the second question is positive, as shown in [4, Thm. 4-5]. Hence, answering to the first question is useful not only to obtain an upper bound for the state cost of the conversion, but even to have a tight bound. To this aim, we consider the maximum value which can be assumed by the least common multiple of positive integers with a given sum, namely the following function, studied by Landau at the beginning of the last century [24,25]:

$$F(n) = \{\mathrm{lcm}\{\ell_1, \ell_2, \ldots, \ell_s\} \mid s \geq 1 \text{ and } \ell_1 + \ell_2 + \cdots + \ell_s = n\}.$$

The best known approximation of $F(n)$ is due to Szalay [49]. For our purposes, the estimation $F(n) = e^{\Theta(\sqrt{n \cdot \ln n})}$ is enough [5].

Combining the state bounds for the conversion of 1DFAs into Chrobak normal form, with the above estimation of $F(n)$, the following result follows:

Theorem 2 ([28,4]). *Each unary n-state 1NFA can be simulated by an equivalent 1DFA with $e^{\Theta(\sqrt{n \cdot \ln n})}$ states. Furthemore, this bound is tight.*

In the general case, it is well-known that using the subset construction each n-state 1NFA can be converted into an equivalent 1DFA with 2^n states. Furthermore, this bound cannot be reduced in the worst case, as witnessed by examples over a two letter alphabet [29,35,37]. Hence, the cost of the conversion of 1NFAs into equivalent 1DFAs in the unary case is strictly lower than in the general case, even if it is superpolynomial.

Chrobak normal form suggests a different way to deterministically check the membership to the language accepted by a unary 1NFA, when it is possible to rescan the input tape, as in *two-way automata*.

A 1NFA A_c in Chrobak normal form with an initial path of μ states and s loops of length $\ell_1, \ell_2, \ldots, \ell_s$, respectively, can be simulated by a 2DFA A' which completely traverses the input at most $s + 1$ times as follows. A first traversal is made to check whether or not the input length is less than μ. If so, the input is accepted or rejected according to the initial path of A_c. Otherwise, in a second traversal, it is tested if the input is accepted on the first loop. This can be implemented using ℓ_1 states to simulate a counter modulo ℓ_1. If the input is not accepted, then a third traversal is made to check, using a counter modulo ℓ_2, if the input is accepted by the second loop of A_c, and so on. The 2DFA A' can be implemented using $\mu + \ell_1 + \cdots + \ell_s + 2$ states ($\mu + 1$ for the first traversal, ℓ_i for each of the next traversals, $i = 1, \ldots, s$, plus one accepting state). Considering the state costs of the conversion into Chrobak normal form, it turns out that each unary n-state 1NFA can be converted into an equivalent 2DFA with $O(n^2)$ states. Furthermore, the bound is tight [4]. Actually, when the language is cyclic, the first scan can be removed and an equivalent 2DFA with no more than n states can be obtained [33].

Using similar ideas, the tight costs of the simulation of 2DFAs by 1DFAs and 1NFAs has been proved to be $e^{\Theta(\sqrt{n \cdot \ln n})}$ [4]. These researches have been deepened by Mereghetti and Pighizzini, considering unary 2NFAs [34]. They proved that the state cost of the simulation of unary n-state 2NFAs by 1DFAs is also $e^{\Theta(\sqrt{n \cdot \ln n})}$. This means that removing two-way motion or nondeterministic choices has the same cost as removing both of them.

These simulations with their costs are summarized in Figure 4. Notice that the state cost of the simulation of 2NFAs by 2DFAs is unknown. This question is open even in the general case, where it has been formulated by Sakoda and Sipser in 1978, together with the question of the state cost of the optimal conversion of 1NFAs into equivalent 1DFAs, which in the general case is also open [43]. Since the question of Sakoda and Sipser is quite relevant in automata and complexity theory (for recent discussions and overviews see [19,39]), and even in the unary case important results related to this question have been obtained, the next section will be devoted to present some interesting properties of unary 2NFAs, and some results related to the Sakoda and Sipser question in the unary case.

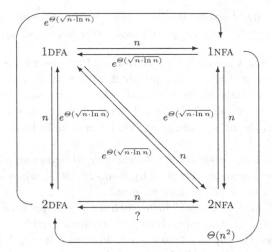

Fig. 4. Costs of the *optimal* simulations between different kinds of *unary* automata. An arc labeled $f(n)$ from a vertex x to a vertex y means that a unary n-state automaton in the class x can be simulated by an $f(n)$-state automaton in the class y. The $e^{\Theta(\sqrt{n \cdot \ln n})}$ costs for the simulations of 1NFAs and 2DFAs by 1DFAs as well as the cost $\Theta(n^2)$ for the simulation of 1NFAs by 2DFAs have been proved in [4]. The $e^{\Theta(\sqrt{n \cdot \ln n})}$ cost for the simulation of 2NFAs by 1DFAs has been proved in [34]. The other $e^{\Theta(\sqrt{n \cdot \ln n})}$ costs are easy consequences. All the n costs are trivial. The arc labeled "?" represents the open question of Sakoda and Sipser. For an upper bound see Theorem 4.

3 Unary Two-Way Automata

Due to the possibility of nondeterministic choices and head reversals, even in the unary case the computations of 2NFAs can have very complicated structures. However, as shown by Mereghetti and Pighizzini studying the simulation of unary 2NFAs by 1DFAs, for each computation of a unary 2NFA it is always possible to find another computation which has a simple pattern and which is in some sense equivalent [34]. This analysis was further refined in [9], obtaining a kind of normal form which can be seen as a generalization of Chrobak normal form to 2NFAs.

Let us go back for a while to unary 1NFAs. Chrobak normal form essentially states that a unary 1NFA on each sufficiently long string has to compute the input length modulo *one* integer which is nondeterministically selected in a given set. In the normal form for unary 2NFAs, the acceptance of each sufficiently long string is decided by computing the input length modulo *some* integers from a given set. The input tape is completely traversed several times to compute, in each traversal, the input length modulo one on these integers. We now present and discuss this form and its main consequences.

We say that two automata A and A' are *almost equivalent* if their accepted languages $L(A)$ and $L(A')$ coincide with the possible exception of a finite number of strings. For example, from the discussion in Section 2 we can observe that each 1DFA is almost equivalent to a 1DFA whose transition graph is a simple loop (i.e, a 1DFA accepting a cyclic language).

The normal form for unary 2NFAs is given in following result, proved in [9] (for further details see also [11]).

Theorem 3. *For each unary n-state 2NFA A there exists an* almost equivalent 2NFA M *such that:*

- *M can makes nondeterministic choices and can reverse the head direction only when the input head is visiting the end-markers,*
- *M has at most $2n + 2$ states.*

More into details, the set of states of M is the union of $s+1$ disjoint sets $\{q_I, q_F\}$, $Q_1, \ldots, Q_s$, such that:

- *the set $\{q_I, q_F\}$ consists of the initial and final states only, with $q_I \neq q_F$,*
- *each Q_i, $i = 1, \ldots, s$, represents a deterministic loop of $\ell_i \leq n$ states, used to traverse the entire input tape either from the left to the right end-marker or in the opposite direction,*
- *the computation of M starts on the left end-marker in the state q_I,*
- *the accepting state q_F can be reached only on the left end-marker; at that moment M stops and accepts the input.*

Furthermore, $L(A)$ and $L(M)$ can differ only on strings of length at most $5n^2$.

We observe that the automaton M obtained in Theorem 3 uses the nondeterminism and head reversals in a very restricted way, i.e., only at the end-markers. Hence, on "real" input symbols the behavior of M is deterministic, without the possibility of changing the head direction. Looking at the details given in the second part of the statement, we can observe that a computation of M is a sequence of complete traversals of the input (alternating the direction from left to right and from right to left), where in each traversal M counts the length of the input modulo an integer in a given set $\{\ell_1, \ldots, \ell_s\}$, where $\ell_i = \#Q_i$, $i = 1, \ldots, s$.

We also point out that the 2NFA M can be easily turned into an automaton "fully" equivalent to the original 2NFA A, by adding $O(n^2)$ states, used to fix the "errors", in a preliminary scan of the input.

Theorem 3 was the starting point to prove several relevant results on unary 2NFAs, that we now briefly mention.

Theorem 4 ([9]). *Each unary n-state 2NFA can be simulated by an equivalent 2DFA with $e^{O(\ln^2 n)}$ states.*

The proof of this result has been given using a divide-and-conquere procedure, similar to the famous one used by Savitch to remove nondeterminism from space bounded machines (by squaring the space) [44]. The key point is the observation that if the 2NFA M in Theorem 3 has an accepting computation on an input a^m, then it should also have an accepting computation which visits the left end-marker at most N times, where N is the number of states of M. The procedure tests state reachability at the left end-marker using a recursion on the number of visits of this end-marker.

At the moment we do not have a matching lower bound for this simulation. The best known lower bound is quadratic and derives from the cost of the simulation of 1NFAs by 2DFAs (cf. Fig. 4). Indeed, this is the already mentioned open question of Sakoda and Sipser restricted to the unary case.

Theorem 5 ([10]). *Each unary n-state* 2NFA *accepting a language L can be transformed into a* 2NFA *with* $O(n^8)$ *states accepting the complement of L.*

Even the proof of this result starts from the above observation which gives a bound on the number of visits to the left end-marked in shortest accepting computations. In this case, an inductive counting procedure (as in the proof of the famous Immerman-Szelepcsényi result [17,50]) is used to generate all the states that can be reached at the left end-marker in a given number of visits.

A further result that has been obtained using the form of 2NFAs given in Theorem 3 states an important relationship between the question of Sakoda and Sipser and the problem of the power of the nondeterminism in logarithmic space bounded computations. Let us denoted by L the class of languages accepted by deterministic Turing machines which work in logarithmic space and by NL the corresponding nondeterministic class. The question L $\overset{?}{=}$ NL is still open. The *Graph Accessibility Problem* (GAP, for short: given a direct graph $G = (V, E)$ with two fixed vertices s, t, decide whether or not G contains a path from the *source s* to the *destination t*) is complete for NL [18]. As a consequence, GAP $\in$ L if and only if L = NL.

Given a 2NFA M in the form given in Theorem 3, the membership problem for $L(M)$ can be reduced to GAP, as proved by Geffert and Pighizzini [11]. The reduction associates with each input string a^m a graph $G_m = (Q, E_m)$ whose set of vertices coincides with the set Q of the states of M and whose set of edges E_m contains the pair (p, q) if and only if M can traverse the input a^m starting at one end-marker in the state p and reaching the opposite end-marker in the state q. Since an accepting computation should start in the state q_I with the head at the left end-marker and end in the state q_F at the same end-marker, deciding whether or not a^m is accepted by M is equivalent to decide if the answer to GAP for the graph G_m (with source q_I and destination q_F) is positive. Due to the properties of M, the above outlined reduction can be computed by a deterministic transducer having a number of states polynomial with respect to the number of states of M. Adapting techniques used in space complexity to two-way automata, the following result has been proved:

Theorem 6 ([11]). *If* L = NL *then each unary n-state* 2NFA *can be simulated by an equivalent* 2DFA *with a number of states polynomial in n.*

The best known upper bound for the simulation of unary 2NFAs by equivalent 2DFAs is the superpolynomial bound presented in Theorem 4. As a consequence of Theorem 6 proving the optimality of such a bound, or proving a smaller but still superpolynomial lower bound for the simulation of unary 2NFAs by equivalent 2DFAs would separate L and NL *in the general case,* thus solving a longstanding open problem in complexity theory. Hence, this result gives evidence of the fact that the unary case is not a "separate world": its investigation has relevant implications in the general case.

Concerning Theorem 6, it is quite natural to wonder if the converse also holds. Trying to prove that, the main problem is related to uniformity. Adding uniformity conditions to the conversion of 2NFAs into equivalent 2DFAs even a stronger result has been obtained [11]. More interestingly, by replacing L with its nonuniform variant L/poly, defined in terms of machines *with polynomial advice* [21], Kapoutis and Pighizzini proved the following:

Theorem 7 ([20]). L/poly $\supseteq$ NL *if and only if each unary n-state 2NFA can be simulated by an equivalent 2DFA with a number of states polynomial in n.*

The *only-if* direction in Theorem 7 derives immediately from Theorem 6. The proof of the converse implication also uses GAP. In particular this problem, restricted to graphs with n vertices, is reduced to the membership problem for the unary language accepted by a 2NFA A_n with a number of state polynomial in n. The details can be found in [20] together with other results proving the equivalence of L/poly $\supseteq$ NL with other statements.

Using a result from nonuniform complexity [42] and adapting the technique used to prove Theorem 6, the following result has also been obtained:

Theorem 8 ([11]). *Each unary n-state 2NFA can be simulated by an equivalent unambiguous 2NFA with a number of states polynomial in n.*

4   Beyond Finite Automata

As mentioned in the introduction, each unary context-free language is regular. So the analysis presented in Section 2 on the costs of the optimal simulations between finite automata in the unary case (summarized in Figure 4) can be extended to include pushdown automata (PDAs, for short) and context-free grammars (CFGs).

While in the case of finite automata the number of states is a reasonable measure for the size of the description, in the case of PDAs we need to consider some other parameters (we remind the reader that each context-free language can be accepted by a PDA with just one state). We can measure the size of a PDA by counting the number of symbols needed to write down its transition table. In a similar way, for a context-free grammar, we can consider the number of symbols used to write down the productions. In the case of grammars in Chomsky normal form, the number of variables is considered a reasonable parameter, because it is polynomially related to the total size of the description. For more details we address the reader to [14].

In the general case, in 1971 Meyer and Fischer proved that for any given recursive function f and for arbitrarily large integers n there exists a context-free grammar G whose description uses n symbols, such that G generates a regular language and each 1DFA accepting L requires at least $f(n)$ states, thus implying that there is no a recursive bound between the size of context-free grammars (and of pushdown automata) generating regular languages and the number of states of equivalent finite automata [35]. The construction of Meyer and Fischer uses a binary alphabet, leaving open the unary case, which has been studied by Pighizzini, Shallit, and Wang proving, in contrast, *recursive bounds* [40].

Theorem 9. *For any unary* CFG *in Chomsky normal form with h variables there exists an equivalent* 1NFA *with at most* $2^{2h-1} + 1$ *states and an equivalent* 1DFA *with less than* 2^{h^2} *states.*

The bounds stated in Theorem 9 cannot be significantly improved. In fact in the same paper, for each integer h a very simple Chomsky normal form grammar generating the language $(a^{2^{h-1}})^+$, which requires $2^{h-1} + 1$ states to be accepted by a 1NFA, was presented. A more complicated example was also given, showing for infinitely many integers h a unary grammar in Chomsky normal form with h variables such that each equivalent 1DFA needs at least 2^{ch^2} states, for a suitable constant c.

Since the conversion of a context-free grammar G into Chomsky normal form produces a grammar with a polynomial number of variables with respect to the total size of the description of the original grammar, and since the conversion of a pushdown automaton into an equivalent context-free grammar is also polynomial in the size, we can conclude that the conversions of unary context-free grammars and pushdown automata into equivalent 1DFAs or 1NFAs are polynomial in size.

We more closely look to the case of pushdown automata. Each PDA can be turned into a normal form where each push operation add exactly one symbol on the stack (this conversion is also polynomial in size). As a corollary of Theorem 9, it has been proved that each unary PDA in this form, with n states and m stack symbols can be transformed into an equivalent 1DFA with $2^{O(n^4 m^2)}$ states [40]. In the case of *deterministic pushdown automata* the upper bound has been reduced to 2^{nm} [38]. Since the size of a pushdown automaton in this normal form is polynomial in n and m, it turns out that this bound is exponential in the size of the given DPDA. Furthermore, this bound cannot be significantly improved. (An alternative proof and other results relating unary DPDA with straight-line programs have been recently obtained by Chistikov and Majumdar [2]).

In the general case, in 1966 Ginsburg and Greibach proved that each DPDA of size s accepting a regular language can be simulated by a finite automaton with a number of states bounded by a function which is triply exponential in s [12]. The bound was reduced to a double exponential in 1975 by Valiant and it cannot be further reduced because a matching lower bound [52,35]. In the unary case, however, an exponential upper bound in s has been obtained. By summarizing:

Theorem 10. *Let M be a unary* PDA *with n states and a pushdown alphabet of m symbols such that each operation can push at most one symbol on the stack. Then M can be simulated by an equivalent* 1DFA *A with $2^{O(n^4 m^2)}$ states. The number of states of A reduces to 2^{nm} if M is deterministic.*

5   Beyond Regular Languages

In the previous sections we discussed descriptional complexity aspects of unary finite and pushdown automata. Now, we focus on devices that, even in the unary case, are more powerful. We are not going to consider general devices as Turing machines, but some computational models obtained adding a few extra features to the ones considered in previous sections. A tool that will be used in this section is the set of powers of 2 written in unary notation, namely the language $\mathcal{L}_p = \{a^{2^m} \mid m \geq 0\}$. Being a nonregular language, $\mathcal{L}_p$ cannot be accepted by any of the devices considered in the previous sections.

Let us start by considering two extensions of PDAs.

The first extension is obtained by adding a "small" worktape to those devices. The model so obtained is called *one-way auxiliary pushdown automata* (1AUX-PDAs, for short) [1]. The *space* used by those devices is measured by taking into account only the worktape. Hence, 1AUX-PDAs working in space $O(1)$ are equivalent to PDAs.

The language $\mathcal{L}_p$ can be recognized by a unary 1AUX-PDA M that in each configuration keeps in the pushdown store and in the auxiliary tape an encoding of the number of input symbols read so far. Given an integer m, let us consider its binary representation. Let $0 \leq k_1 < k_2 < \cdots < k_s$ be the positions of the digit 1 (from the less significant position), namely, $m = 2^{k_1} + 2^{k_2} + \cdots + 2^{k_s}$. After reading a^m the auxiliary tape contains k_1 written in binary, while the pushdown store contains (from the top) an encoding of the sequence $k_2, \ldots, k_s$. It is easy to write down the operations that M has to perform, while moving one position to the right, to get from the representation of m that of $m + 1$. On input a^n, the maximum number that can be written on the worktape is the position of the most significant digit in the representation of n, which is written in binary. Hence, M uses $O(\log \log n)$ space. Furthermore, M is *deterministic*. This suggested the following questions:

– Is it possible to use a smaller amount of space to recognize $\mathcal{L}_p$ on these devices (possibly making use also of nondeterminism)?
– Does there exist a language which is accepted in $o(\log \log n)$ space but not in $O(1)$ space on those devices?

To answer the last question, we need to be more precise about space notions. In fact, different space notions can be formulated (for an overview see, e.g., [32]). Among them, *strong space* is defined by requiring that the space bound is satisfied by *all computations* on each input, while *weak space* is defined by requiring that the bound is satisfied by *at least one accepting computation* on each string in the language. Brandenburg has shown that each language accepted by a 1AUX-PDA in strong $o(\log \log n)$ space is context-free, i.e., it is accepted in $O(1)$ space [1]. Hence, in the strong case the answer to both questions is negative.

For weak space, the situation is completely different. Chytil proved that for each $k \geq 1$ there exists a noncontext-free language accepted by a 1AUX-PDA in weak $O(\underbrace{\log \log \cdots \log}_{k \text{ times}} n)$ space [6]. Hence, the answer to the second question is positive. However, the witness language is defined over a nonunary alphabet. Hence, this does not give any answer to the first question and, in general, on the unary case. Indeed, the result by Chytil cannot be proved with unary witnesses: using Theorem 10, it has been shown that in the unary case 1AUX-PDAs working in weak $o(\log \log n)$ space are no more powerful that PDAs, hence they recognize only regular languages [40]. As a consequence, $\mathcal{L}_p$ is an example of nonregular language accepted by a 1AUX-PDA using a minimal amount of extra space. We believe that this language requires a minimal amount of resources even on other computational models, as we are now going to discuss.

Another extension of PDAs is obtained by allowing to move the head on both directions on the input tape. The resulting models, *two-way pushdown automata* (2PDAs), are more powerful than PDAs, even when restricted to a unary alphabet. For instance the language $\mathcal{L}_p$ is accepted using iterated divisions by 2. To this aim, we can define a 2PDA A which, while moving the input head from left to right, for each two input positions pushes one symbol on the stack, and while popping symbols off the stack, for each two symbols that are removed, A moves the input head one position to the left. At the beginning, A scans all the input from left to right, leaving $N/2$ symbols on the pushdown store, where N is the input length. If N is odd then the input is rejected. Otherwise, A makes its pushdown store empty, while moving the input head to the left, so reaching the cell at distance $N/4$ from the right end-marker and rejecting if $N/2$ is odd. At this point, A starts to move the input head to the right to leave $N/8$ symbols on the pushdown, and so on, representing in this way $N/2^k$ for increasing values of k. When $N/2^k = 1$, A stops and accepts.

The language $\mathcal{L}_p$ is just a very simple example of nonregular language accepted by unary 2PDAs. Actually, these devices are very powerful: as shown in 1984 by Monien, the unary encoding of each language in P, the class of languages accepted in polynomial time by deterministic Turing machines, is accepted by a 2PDA [36].

We point out that, in the previous example, a string of length n is accepted by reversing the direction of the input head $O(\log n)$ times. As a consequence of a result by Liu and Weiner, it follows that if a 2PDA accepts a unary language L by reversing its input head a *constant* number of times, then L is regular [27]. It should be interesting to know whether or not there is a unary nonregular language accepted by a 2PDA using $o(\log n)$ reversals. In our knowledge, at the moment this problem is open. Furthermore, unary 2PDAs which can make only a constant number of "turns" of the pushdown accept only regular languages [3]. Hence a similar problem can be formulated with respect to the number of turns of the pushdown store. For recent results on 2PDAs with restrictions on head reversals and number of turns we address the reader to [30].

We finally consider *multi-head finite automata*. The language L_p can be recognized by deterministic finite automaton with 2 two-way heads. The algorithm is very similar to the one above outlined for accepting the same language on 2PDAs. We can also observe that the total number of reversals used on inputs of length n is $O(\log n)$. As a consequence of a result by Sudborough, multi-head automata making a constant number of head reversals accept only regular languages [48]. Even in this case, we can ask about the existence of a unary nonregular language accepted by a multi-head automaton using $o(\log n)$ reversals. Descriptional complexity aspects of unary one-way multi-head automata have been recently studied in [23]. We point out that even adding multiple input heads to 2PDAs, if the number of reversals is constant then in the unary case only regular languages are accepted [16].

6 Final Remarks

We presented an overview on some results related to unary automata and languages. Some of the results we mentioned (e.g., those from [27,16,48]) are more general, taking into account semilinear sets or bounded languages. Other results originally proved for the unary case have been generalized. We shortly mention a few of those generalizations.

First of all, the *Chrobak normal form* for unary automata has been generalized by Kopczynski and To, providing a normal form for the representation of semilinear subsets of $\mathbb{N}^m$ [22]. Lavado, Pighizzini, and Seki studied automata determinization under Parikh equivalence, proving that for each n-state 1NFA accepting a language L there exists a 1DFA with $e^{\Theta(\sqrt{n \cdot \ln n})}$ states accepting a language L' with the same Parikh image as L [26]. The proof makes use of the above mentioned extension of Chrobak normal form to semilinear sets. They also proved that for each n-variable context-free grammar in Chomsky normal form generating a language L, there exists a 1DFA with $2^{O(h^2)}$ states accepting a Parikh equivalent language L', so extending the results discussed in Section 4.

References

1. Brandenburg, F.J.: On one-way auxiliary pushdown automata. In: Tzschach, H., Walter, H.K.-G., Waldschmidt, H. (eds.) GI-TCS 1977. LNCS, vol. 48, pp. 132–144. Springer, Heidelberg (1977)
2. Chistikov, D., Majumdar, R.: Unary pushdown automata and straight-line programs. In: Esparza, J., Fraigniaud, P., Husfeldt, T., Koutsoupias, E. (eds.) ICALP 2014, Part II. LNCS, vol. 8573, pp. 146–157. Springer, Heidelberg (2014)
3. Chrobak, M.: A note on bounded-reversal multipushdown machines. Inf. Process. Lett. 19(4), 179–180 (1984)
4. Chrobak, M.: Finite automata and unary languages. Theor. Comput. Sci. 47(3), 149–158 (1986); errata: [5]
5. Chrobak, M.: Errata to: Finite automata and unary languages. Theor. Comput. Sci. 302(1-3), 497–498 (2003)
6. Chytil, M.: Almost context-free languages. Fundamenta Informaticae IX, 283–322 (1986)
7. Gawrychowski, P.: Chrobak normal form revisited, with applications. In: Bouchou-Markhoff, B., Caron, P., Champarnaud, J.-M., Maurel, D. (eds.) CIAA 2011. LNCS, vol. 6807, pp. 142–153. Springer, Heidelberg (2011)
8. Geffert, V.: Magic numbers in the state hierarchy of finite automata. Inf. Comput. 205(11), 1652–1670 (2007)
9. Geffert, V., Mereghetti, C., Pighizzini, G.: Converting two-way nondeterministic unary automata into simpler automata. Theor. Comput. Sci. 295, 189–203 (2003)

10. Geffert, V., Mereghetti, C., Pighizzini, G.: Complementing two-way finite automata. Inf. Comput. 205(8), 1173–1187 (2007)
11. Geffert, V., Pighizzini, G.: Two-way unary automata versus logarithmic space. Inf. Comput. 209(7), 1016–1025 (2011)
12. Ginsburg, S., Greibach, S.A.: Deterministic context free languages. Information and Control 9(6), 620–648 (1966)
13. Ginsburg, S., Rice, H.G.: Two families of languages related to ALGOL. J. ACM 9(3), 350–371 (1962)
14. Harrison, M.A.: Introduction to Formal Language Theory. Addison-Wesley Longman Publishing Co., Inc., Boston (1978)
15. Hopcroft, J.E., Ullman, J.D.: Introduction to Automata Theory, Languages and Computation. Addison-Wesley (1979)
16. Ibarra, O.H.: A note on semilinear sets and bounded-reversal multihead pushdown automata. Inf. Process. Lett. 3(1), 25–28 (1974)
17. Immerman, N.: Nondeterministic space is closed under complementation. SIAM J. Comput. 17(5), 935–938 (1988)
18. Jones, N.D.: Space-bounded reducibility among combinatorial problems. J. Comput. Syst. Sci. 11(1), 68–85 (1975)
19. Kapoutsis, C.A.: Minicomplexity. In: Kutrib, M., Moreira, N., Reis, R. (eds.) DCFS 2012. LNCS, vol. 7386, pp. 20–42. Springer, Heidelberg (2012)
20. Kapoutsis, C.A., Pighizzini, G.: Two-way automata characterizations of l/poly versus nl. In: Hirsch, E.A., Karhumäki, J., Lepistö, A., Prilutskii, M. (eds.) CSR 2012. LNCS, vol. 7353, pp. 217–228. Springer, Heidelberg (2012)
21. Karp, R., Lipton, R.: Turing machines that take advice. In: Engeler, E., et al. (eds.) Logic and Algorithmic, pp. 191–209. L'Enseignement Mathématique, Genève (1982)
22. Kopczynski, E., To, A.W.: Parikh images of grammars: Complexity and applications. In: LICS, pp. 80–89. IEEE Computer Society (2010)
23. Kutrib, M., Malcher, A., Wendlandt, M.: Size of unary one-way multi-head finite automata. In: Jurgensen, H., Reis, R. (eds.) DCFS 2013. LNCS, vol. 8031, pp. 148–159. Springer, Heidelberg (2013)
24. Landau, E.: Über die maximalordnung der permutation gegebenen grades. Archiv der Mathematik und Physik 3, 92–103 (1903)
25. Landau, E.: Handbuch der Lehre von der Verteilung der Primzahlen I. Teubner, Leipzig (1909)
26. Lavado, G.J., Pighizzini, G., Seki, S.: Converting nondeterministic automata and context-free grammars into Parikh equivalent one-way and two-way deterministic automata. Inf. Comput. 228, 1–15 (2013)
27. Liu, L., Weiner, P.: Finite-reversal pushdown automata and semi-linear sets. In: Proc. of Sec. Ann. Princeton Conf. on Inf. Sciences and Systems, pp. 334–338 (1968)
28. Ljubič, J.: Bounds for the optimal determinization of nondeterministic autonomous automata. Sibirskij Matematičeskij Žurnal 2, 337–355 (1964) (in Russian)
29. Lupanov, O.: A comparison of two types of finite automata. Problemy Kibernet 9, 321–326 (1963) (in Russian); German translation: Über den Vergleich zweier Typen endlicher Quellen. Probleme der Kybernetik 6, 329–335 (1966)
30. Malcher, A., Mereghetti, C., Palano, B.: Descriptional complexity of two-way pushdown automata with restricted head reversals. Theor. Comput. Sci. 449, 119–133 (2012)

31. Martinez, A.: Efficient computation of regular expressions from unary nfas. In: Dassow, J., Hoeberechts, M., Jürgensen, H., Wotschke, D. (eds.) DCFS, vol. Report No. 586, pp. 174–187. Department of Computer Science, The University of Western Ontario, Canada (2002)
32. Mereghetti, C.: Testing the descriptional power of small Turing machines on non-regular language acceptance. Int. J. Found. Comput. Sci. 19(4), 827–843 (2008)
33. Mereghetti, C., Pighizzini, G.: Two-way automata simulations and unary languages. Journal of Automata, Languages and Combinatorics 5(3), 287–300 (2000)
34. Mereghetti, C., Pighizzini, G.: Optimal simulations between unary automata. SIAM J. Comput. 30(6), 1976–1992 (2001)
35. Meyer, A.R., Fischer, M.J.: Economy of description by automata, grammars, and formal systems. In: FOCS. pp. 188–191. IEEE (1971)
36. Monien, B.: Deterministic two-way one-head pushdown automata are very powerful. Inf. Process. Lett. 18(5), 239–242 (1984)
37. Moore, F.: On the bounds for state-set size in the proofs of equivalence between deterministic, nondeterministic, and two-way finite automata. IEEE Transactions on Computers C-20(10), 1211–1214 (1971)
38. Pighizzini, G.: Deterministic pushdown automata and unary languages. Int. J. Found. Comput. Sci. 20(4), 629–645 (2009)
39. Pighizzini, G.: Two-way finite automata: Old and recent results. Fundam. Inform. 126(2-3), 225–246 (2013)
40. Pighizzini, G., Shallit, J., Wang, M.: Unary context-free grammars and pushdown automata, descriptional complexity and auxiliary space lower bounds. J. Comput. Syst. Sci. 65(2), 393–414 (2002)
41. Rabin, M.O., Scott, D.: Finite automata and their decision problems. IBM J. Res. Dev. 3(2), 114–125 (1959)
42. Reinhardt, K., Allender, E.: Making nondeterminism unambiguous. SIAM Journal on Computing 29(4), 1118–1131 (2000)
43. Sakoda, W.J., Sipser, M.: Nondeterminism and the size of two way finite automata. In: Lipton, R.J., Burkhard, W.A., Savitch, W.J., Friedman, E.P., Aho, A.V. (eds.) STOC, pp. 275–286. ACM (1978)
44. Savitch, W.J.: Relationships between nondeterministic and deterministic tape complexities. J. Comput. Syst. Sci. 4(2), 177–192 (1970)
45. Sawa, Z.: Efficient construction of semilinear representations of languages accepted by unary nondeterministic finite automata. Fundam. Inform. 123(1), 97–106 (2013)
46. Shallit, J.O.: A Second Course in Formal Languages and Automata Theory. Cambridge University Press (2008)
47. Shepherdson, J.C.: The reduction of two-way automata to one-way automata. IBM J. Res. Dev. 3(2), 198–200 (1959)
48. Sudborough, I.H.: Bounded-reversal multihead finite automata languages. Information and Control 25(4), 317–328 (1974)
49. Szalay, M.: On the maximal order in S_n and S_n^*. Acta Arithmetica 37, 321–331 (1980)
50. Szelepcsényi, R.: The method of forced enumeration for nondeterministic automata. Acta Inf. 26(3), 279–284 (1988)
51. To, A.W.: Unary finite automata vs. arithmetic progressions. Inf. Process. Lett. 109(17), 1010–1014 (2009)
52. Valiant, L.G.: Regularity and related problems for deterministic pushdown automata. J. ACM 22(1), 1–10 (1975)

Cellular Automata for Crowd Dynamics

Georgios Ch. Sirakoulis

Department of Electrical and Computer Engineering
Democritus University of Thrace
67100 Xanthi, Greece
gsirak@ee.duth.gr
http://gsirak.ee.duth.gr

Abstract. Cellular Automata (CA) as bio-inspired parallel computational models of self-reproducing organisms can capture the essential features of systems where global behavior arises from the collective effect of simple components which interact locally. In this aspect, CAs have been considered as a fine candidate to model pedestrian behavior and crowd dynamics in a fine manner. In specific, for crowd modeling, the CA models show evidence of a macroscopic nature with microscopic extensions, i.e. they provide adequate details in the description of human behavior and interaction, whilst they retain the computational cost at low levels. In this paper several CA models for crowd evacuation taking into consideration different modeling principles, like potential fields techniques, obstacle avoidance, follow-the-leader principles, grouping theory, etc. will be presented in an attempt to accomplish efficient crowd evacuation simulation. Moreover, an integrated system based on CAs that operates as an anticipative crowd management tool in cases of medium density crowd evacuation for indoor and outdoor environments is also shown, and its results different real world cases and different environments prove its efficiency. Finally, robot guided evacuation with the help of CAs is also presented. Quite recently, an evacuation system was proposed, based on an accurate CA model capable of assessing the human behavior during emergency situations takes advantage of the simulation output to provide sufficient information to a mobile robotic guide, which in turn guides people towards a less congestive exit at a time.

Keywords: Cellular Automata, Crowd Dynamics, Modeling, Simulation, Hardware Implementation.

1 Introduction

When we are at a major sporting event or traveling on public transport or shopping around in shopping precincts, our safety and comfort depend crucially on our fellow crowd members and on the design and operation of the facility we are in. Consequently, the need for realistic and efficient in case of emergency crowd dynamics modeling approaches is of utter importance. As a result, pedestrian dynamics have been reported following a great variety of approaching methods, such as Cellular Automata (CA)-based [1], lattice-gas and social

M. Holzer and M. Kutrib (Eds.): CIAA 2014, LNCS 8587, pp. 58–69, 2014.
© Springer International Publishing Switzerland 2014

force models [2], fluid-dynamic [3] and agent-based [4], and methods related to game theory [5]. All approaches can be qualitatively distinguished, focusing on different characteristics that each of them dominantly display. In general, crowd movement simulation models can also be categorized into macroscopic and microscopic ones. In some models, pedestrians are ideally considered as homogeneous individuals, whereas in others, they are treated as heterogeneous groups with different features (e.g., gender, age, psychology). There are methods, where collective phenomena emerge from the complex interactions among individuals (self-organizing effects), thus describing pedestrian dynamics in a microscopic scale. Other methods treat crowd as a whole, modeling pedestrian dynamics on a macroscopic scale. There are models discrete in space and time and others spatial-temporally continuous.

Moreover, crowd movement could be defined as a non linear problem with many factors affecting it. A system of Partial Differential Equations (PDEs) could effectively approach it. The result is a system of PDEs very difficult to handle, which would also be demanding in terms of computer power, complexity and computation time. CA can act as an alternative to PDEs [12]. In CA based approach, the space under study is presented as a unified grid of cells with local attributes, which are generated by a set of rules that describe the behavior of the individuals. The state of each cell is defined according to the rules, the state of the cell at the previous time step and the current state of neighboring cells. Consequently, literature reports a variety of CA-based models investigating crowd behavior under different circumstances [6,7].

In this paper, some CA models for the simulation of crowd dynamics are going to be presented. In specific, a CA model based on electrostatic-induced potential, inspired by the Coulomb force as motion-driving mechanism, calculates the Euclidean distance between the destination (source) and the pedestrian (test charge), allowing smoother change of direction. Introducing an electric field approach, charges of different magnitude represent main or internal exits as well as obstacles and walls [8,9,11]. A somehow different CA model also applies an efficient method to overcome obstacles. Based on the generation of a virtual field along obstacles, a pedestrian moves along the axis of the obstacle towards the direction that the field increases its values, leading her/him to avoid the obstacle effectively [13]. Moreover, a bio–inspired CA-based model for crowd dynamics where the driving mechanism emanates from the characteristic collective behavior of biological organisms (e.g. school of fishes, flock of birds, etc.) is also presented [14,15]. The adoption of a CA approach enhanced with memory capacity allows the development of a microscopically–induced model with macroscopical characteristics. Due to the fact that in terms of simplicity, regularity, ease of mask generation, silicon-area utilization, and locality of interconnections [10,11], CA are the most promising computational architecture, the CA models can be easily implemented in Field Programmable Gate Array (FPGA) Circuits leading to support decision system for monitoring crowd dynamics in real-time, providing valuable near optimum management of crowd services. In this direction, an anticipative crowd management system preventing clogging in exits during pedestrian evacuation processes based on some of the proposed CA models is also presented. Finally, robot guided evacuation is presented [16]; namely, an evacuation system based on an accurate CA model and capable of assessing the human behavior during emergency situations takes advantage of the simulation output to provide sufficient information to a mobile robotic guide, which in turn guides people towards a less congestive exit at a time.

2   CA Models for Crowd Dynamics

CA Basics: A CA consists of a regular uniform n-dimensional lattice (or array), usually of infinite extent. At each site of the lattice (cell), a physical quantity takes on values. The value of this physical quantity over all the cells is the global state of the CA, whereas the value of this quantity at each site is its local state. A CA is characterized by five properties:

1. the number of spatial dimensions (n);
2. the width of each side of the array (w). w_j is the width of the jth side of the array, where $j = 1, 2, 3, ..., n$;
3. the width of the neighborhood of the cell (r);
4. the states of the CA cells;
5. the CA rule, which is an arbitrary function F.

The state of a cell, at time step $(t + 1)$, is computed according to F, a function of the state of this cell at time step (t) and the states of the cells in its neighborhood at time step (t). For a 2-d CA, two neighborhoods are often considered: Von Neumann, which consists of a central cell and its four geographical neighbors north, west, south and east; and the Moore neighborhood contains, in addition, second nearest neighbors northeast, northwest, southeast and southwest, i.e. nine cells. In most practical applications, when simulating a CA rule, it is impossible to deal with an infinite lattice. The system must be finite and have boundaries, resulting to various types of boundary conditions such as periodic (or cyclic), fixed, adiabatic or reflection.

CA model based on electrostatic-induced potential fields: A 2-d CA model based on electrostatic-induced potential fields was introduced to simulate efficiently crowd dynamics in the process of developing an active guiding system of crowd during evacuation. Certain attributes of crowd behavior, such as collective effects, blockings in front of exits as well as random to coherent motion due to a common purpose have been successfully incorporated in the model. Motion mechanism is based on an virtual potential field generated by electric charges at selected positions that attract pedestrians towards exit point or repel them from obstacles and walls. Assuming that each bounded area that includes an exit corresponds to an independent level then coupling among different fields is avoided. Efficient updating rules demonstrate global behavioral patterns that distinctly characterize mass egress. Each pedestrian is represented by a test charge, with such a small magnitude that when placing it at a point has a negligible affect on the field around the point (Fig. 1). Furthermore, the direction towards each of pedestrians should move is precisely determined. The model calculates the exact Euclidean distance between the destination (source) and the pedestrian (test charge); hence it achieves advanced estimation of crowd behavior. It is a field similar to an electrostatic one, described by the equation below, though it has some differences in order to be applicable in pedestrian motion.

$$\vec{F}(\vec{r}) = \frac{q}{4\pi\epsilon_o} \sum_{i=1}^{N} \frac{Q_i(\vec{r} - \vec{r_i})}{|\vec{r} - \vec{r_i}|^3} = \frac{q}{4\pi\epsilon_o} \sum_{i=1}^{N} \frac{Q_i}{R_i^2} \hat{R}_i = q\vec{E}(\vec{r}) \tag{1}$$

In Eq. 1, Q_i and $\vec{r_i}$ are the magnitude and the position of the ith charge, respectively, $\hat{R}_i$ is the unit vector in the direction of $\vec{R}_i = \vec{r} - \vec{r_i}$, i.e. a vector pointing from charge Q_i to charge q and R_i is the magnitude of $\vec{R}_i$, i.e. the distance between charges Q_i and q and $\vec{E}$ the corresponding electric field. The unit vector in space is expressed, in Cartesian notation, as a linear combination of $i = [1; 0]$ and $j = [0; 1]$ and the values of its scalar components are equal to the cosine of the angle formed by the unit vector with the respective basis vector [8]. The force is attracting when generated by charges located at exits and repulsive when generated by charges that represent obstacles or walls [9]. The distance calculated is the exact Euclidean distance, thus introducing increased precision in the model as far as the direction of pedestrian concerns [11].

In the CA grid every cell covers an extent of approximately $40 \times 40 \ cm^2$ [17]. Each cell corresponds to the fixed area that a person could occupy [1]. CA cells obtain discrete values, thus indicating their status; either free or occupied. During each time step, the algorithm aims at the definition of the direction that an individual should move towards to reach the closest exit trying to occupy one of the eight possible states of its closest neighborhood (Moore). More explicitly, according to the attracting force from the exit of the room and the repelling forces from obstacles and walls, the coordinates of the next cell-target are calculated. In case that the cell-target is free or it has not been defined as target from another pedestrian, then the initial pedestrian moves towards. Otherwise, the pedestrian searches for a neighboring cell equidistant from the exit with the initial target-cell. In case that the target-cell is an exit, it is checked whether this is free or not. Only if the exit is free does the pedestrian move towards it. The convergence of the resultant force plays significant role on the driving mechanism of the model. In fact, the convergence of the resultant force upon a test charge towards the point that the closest source is located defines the direction of movement of each pedestrian, which is towards the closest exit (as shown in Fig. 1(c)). More details about the mathematical calculations of the convergence of the resultant force can be found in [11].

CA model for automated obstacle avoidance: A distinct feature of the model is an automated, computationally fast and efficient method to enables obstacle avoidance based on the effect of a virtual field generated near obstacles [13]. Inside the field, a pedestrian moves towards the direction of greater field values. Following that direction a pedestrian is enabled to overcome efficiently

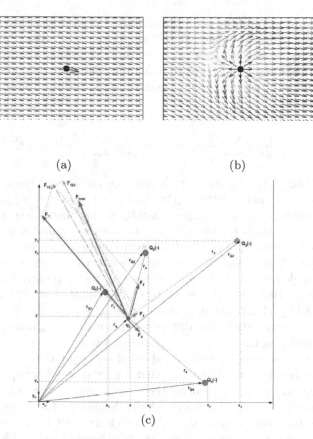

(a)   (b)

(c)

Fig. 1. (a) The effect of a test charge, which represents a pedestrian, on the field around the point and (b) the corresponding effect of an ordinary charge (right). (c) Graphical solution of the case of the convergence of the resultant force upon a test charge.

even complex obstacles. Specifically, in the general case, obstacle field values are increasing forming a parabola, which is described by the following equation:

$$F(x) = \frac{1}{2p}(x - x_o)^2, \ p > 0 \forall \in \{[x_{wl}, x_{wr}] \cap (x_A - x_{wl} \neq 0 \cup x_{wr} - x_B \neq 0)\}$$

(2)

$$x_o = \frac{x_B - x_A}{2}$$

(3)

In eq. 2, p corresponds to the parameter of the parabola, which also defines the distance between the two branches of the graphical representation of the function. In fact, as $\frac{1}{2p} \to 0$, then the width of the parabola increases. Moreover, (x_A, y_A), (x_B, y_B) represent the coordinates of the edges of the obstacle, whereas, x_{wl}, x_{wr} correspond respectively to the very left and very right x-axis coordinate of the walls. Equation 3 defines x_o, which corresponds to the x-axis coordinate of the middle point of the obstacle. In case that the obstacle is bonded to a wall, then the field is generated according to the common coordinate of the obstacle and the wall, as described by eqs. 4 and 5:

$$x_A - x_{wl} = 0 \Rightarrow F(x) = \frac{1}{2p}(x - x_{wl})^2 \tag{4}$$

$$x_{wl} - x_B = 0 \Rightarrow F(x) = \frac{1}{2p}(x - x_{wr})^2 \tag{5}$$

The length of the obstacle is given by:

$$L_{obstacle} = x_B - x_A \tag{6}$$

whereas the length of the area between walls is given by:

$$m = x_{wr} - x_{wl} \tag{7}$$

In the case of complex obstacles the corresponding field is generated by the superposition of fields that correspond to fundamental obstacles. Fig. 2 clarifies the effect of the auto-defined obstacle field to the direction of pedestrian movement, in correspondence to the location and the shape of the obstacle. It should be mentioned that above mathematical presentation takes into account geometrically shaped obstacles, however with slight modification can be successfully applied to arbitrary shaped obstacles as well. More details can be found in [13].

Hardware Implementation of CA models: It should be also noticed that the aforementioned CA model is orientated as a real-time processing module of an embedded system that could prevent clogging in exits under emergency conditions. More specifically, the initialization process could be originated along with a detecting and tracking algorithm supported by cameras and the automatic response of the processor provides the location of pedestrians around escape points. Consequently, the realization of the model becomes a rational additional step. Moreover, in terms of circuit design and layout, ease of mask generation, silicon-area utilization and maximization of achievable clock speed CA are perhaps the computational structures best suited for a fully parallel hardware realization [10,13]. In contrast to the serial computers, the implementation of the model is motivated by parallelism, an inherent feature of CA that contributes to further acceleration of the model's operation. The hardware implementation of the presented model is based on FPGA logic (for example the schematic design of the closest exit tracker can be seen in Fig 3. The dedicated processor could be used as a real-time processing module of an embedded system, dedicated to surveillance that responds fast under crowd evacuation emergency conditions. More techical details about CA design and the target FPGA in [11].

Follow-the-Leader CA Model: Furthermore, a CA-based computational model has been developed that simulates the movement of a crowd formed by individuals, following some of the basic principles of flocking. The driving mechanism of the model is based on the acceptance that each member of the crowd moves independently. Whenever possible, a group of individuals approaches the

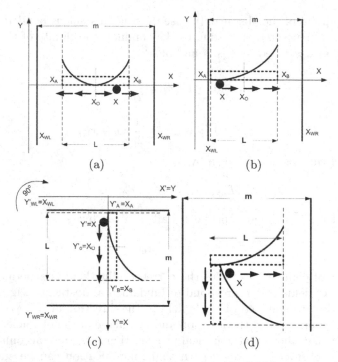

Fig. 2. (a) The graphical representation of the obstacle field in case that the obstacle lays between walls. The values of the field increase in the direction from left to right for half the length of the obstacle and the vice versa for the other half. Thus, the pedestrian is enabled to move following the one direction or the other, as indicated by arrows. (b) The response of the obstacle field, in case that the exit is closer to the left edge of the obstacle ($X_A = X_{wl}$). (c) The case of a vertical obstacle and the corresponding field. (d) The case of a complex obstacle. The final field is generated by the superposition of field cases (b) and (c).

closest exit following the shortest route. Thus, each member is supposed to have a complete knowledge of the space topology and acts completely rational. The model has been developed to simulate crowd movement both in 2-d and 3-d, according to flocking principles and incorporating the Follow-the-Leader technique [14]. Following nature's practice, the model allows dynamical transitions of the role of the leader among the members of the group. Particularly, in case that a member of the group appears in front of the leader also following the same direction of movement, then a leader's role transition occurs. The member in front becomes the leader, whereas the leader turns into a simple member. An individual follows the leader until it reaches the target, e.g. the exit. Furthermore, the model enables the creation of different groups in the crowd, assigning to each group a leader and the corresponding members. It favors the dynamic grouping rather than the static one.

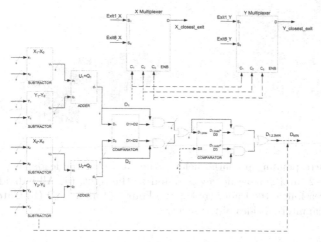

Fig. 3. Structure of a cell of the CA grid for Euclidean based calculations

As far as the members of the group concern, their movement is defined by the same set of rules as the one that defines the movement of individuals towards an exit. From a mathematical point of view, the direction of movement of the individuals relies on a potential field. It derives from the negative gradient of a function that involves the distance (Manhattan) of each point of the area from the position of the leader. In the case of two–dimensions, the function $f(x,y)$ is defined as follows:

$$f(x,y) = abs\,(x - x_o) + abs\,(y - y_o) \tag{8}$$

where (x_o, y_o) correspond to the coordinates of the leader.

The corresponding gradient of the function $f(x,y)$ is defined as:

$$-\nabla f(x,y) = -\left(\frac{\partial f(x,y)}{\partial x}\vec{i} + \frac{\partial f(x,y)}{\partial y}\vec{j} \right) \tag{9a}$$

$$-\nabla f(x,y) = -\left(\frac{\partial abs(x - x_o)}{\partial x}\vec{i} + \frac{\partial abs(y - y_o)}{\partial y}\vec{j} \right) \tag{9b}$$

It can be thought as a collection of vectors pointing in the direction of decreasing values of $f(x,y)$.

Spatially, the whole process is divided in eight subsections, i.e. for the case that the leader moves i) downwards $(y > y_o)$, ii) upwards $(y < y_o)$, iii) to the left $(x > x_o)$, iv) to the right $(x < x_o)$, v) downwards and to the right $(y > y_o)$ and $(x < x_o)$, vi) downwards and to the left $(y > y_o)$ and $(x > x_o)$, vii) upwards and to the left $(y < y_o)$ and $(x > x_o)$ and viii) upwards and to the right $(y < y_o)$ and $(x < x_o)$, respectively. For instance, in the case that the leader moves downwards and to the right (i.e. case (v)), the corresponding potential field that is derived from equations 9, is depicted in Figure 4. The adoption of sectors is based on the simple fact that even in real life bounded areas are divided to

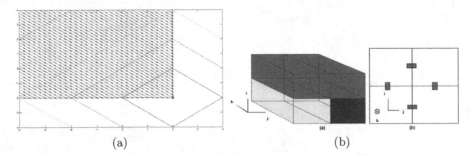

(a)   (b)

Fig. 4. The corresponding potential field in the case that the leader is positioned at $(x_o = 2, y_o = -2)$ and moving downwards and to the right. (a) All individuals within blue–arrows area follow the leader (red spot). Four sectors in 3–D (b) Gates placed at the center of the internal sides of the sectors.

multiple sub-areas with their own formation and their own exits. Each sub-area shares the same properties with the total area, thus enabling the use of the property of the superposition. Hence, the scalability of the method is reassured, allowing its application in more complicated areas. The movement of the leaders through sections takes place as follows: depending on the direction of the leaders motion, each section is supplied with two gates that the leaders use to enter the section. The leaders move towards the exits following the same rules that the members of the flock use to follow the leader.

The 3-d CA is defined in a cubic space, the dimensions of which are variable, taking into consideration that each cell needs three coordinates (i, j, k) to be properly defined. The neighborhood of each cell is shaped by its 26 closest cells, whereas there are four (4) sectors that divide the space in four rectangular parallelepipeds (Fig. 4a). In case that we wish to test the behavior of the model in 3-d dimensions, the following scheme takes place; the leaders pass through the sectors following the 1-2-3-4-1 sequence for the clockwise direction or 4-3-2-1-4 for the anti–clockwise one. Adopting similar logic as in two-dimensions, the gate that influences one sector lies inside the following sector. The gates are placed at the center of the internal sides of the sectors (Fig. 4b).

Different simulation processes were taken into account in order to verify the response of the model and investigate its efficiency [14]. Particularly, these various simulation scenarios demonstrated distinct features of crowd movement such as flocking, increasing crowd density in turnings and crowd movement deceleration as self-organized groups try to pass obstacles, transition from a random to a coordinated motion, etc. Please also check [14],[15] for further analytical presentation of the under study simulation scenarios and the corresponding results. *Anticipative crowd management tool based on CA model*: An integrated system that operates as an anticipative crowd management tool in cases of medium density crowd evacuation was also developed based on CA models [12]. Preliminary real data evaluation processes indicate that it responds fast in order to prevent clogging in exits under emergency conditions. The system consists of three modules; the detecting and tracking algorithm, the CA model of possible route estimation and the sound and optical signals. The initialization process is originated from the detecting and tracking algorithm, which is supported by cameras. The automatic response of the algorithm provides the location of pedestrians around escape points at any time, thus providing instant initialization data to the model of possible route estimation. However, its role is not confined exclusively for initialization purposes. Instead, it also operates as a control and rectifying mechanism, by checking and correcting periodically the response of the CA dynamic model originated from electrostatic-induced potential fields. The response of the route estimation model is compared to the output of the tracking algorithm. In cases of large differences, the model is re-initialized according to the current conditions of the tracking algorithm. Finally, sound and optical signals enable the system to redirect pedestrians, enhancing its effectiveness and efficiency [12]. System operation is developed in four successive stages, setting out with the detection and tracking of pedestrians that enable dynamic initialization and continuing with the estimation of their possible route for the very near future. Then, among all possible exit points, the most suitable is proposed as an alternative, triggering the activation of appropriate guiding signals, sound and optical. The criterion of suitability is the distance of the congested exit from

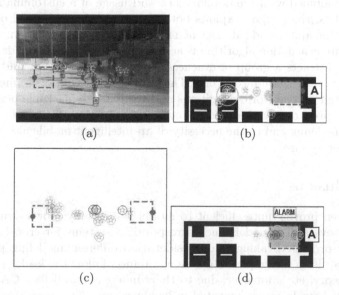

(a)   (b)

(c)   (d)

Fig. 5. (a) Initialization of the pedestrian movement model outdoor. (c) The transition from the first stage of the anticipative system, i.e. the detection algorithm to the second one, i.e. the crowd movement model. Red-dotted areas correspond to areas of interest in front of exits. (b) Two successive frames displaying response of individuals during alarm activation in a teaching room. In frame (b), people move towards exit A, not having reached the area of surveillance yet. Alarm is activated in frame (d).

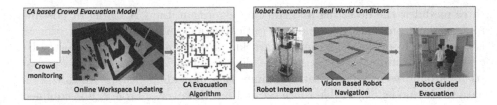

Fig. 6. The proposed robot guided evacuation system's architecture

an alternative one. Hence, the closest free exit is preferred. A few paradigms of the system process for outdoor and indoor study cases are depicted in Fig. 5.

CA based Robot Crowd Evacuation: Recently, a robot guided evacuation was proposed, to the best of our knowledge, for the first time in literature [16]. The proposed framework relies on the well established CA simulation models, while it employs a real world evacuation implementation assisted by a mobile robot. More specifically, the implementation of a CA model capable of assessing the humans'behavior during evacuation occasions has been presented. Then, an evacuation framework based on an assistant robot that deploys in emergency situations is exhibited. The main attribute of the introduced method is the coexistence of a discrete CA simulation model and a real wold continuous implementation combined with the development and usage of a custommade robotic platform. Thus, the method exploits both the computational speed of the discrete simulation and the added value of a real robotic implementation. Additionally, the entire evacuation algorithm is accompanied by a custommade assistant robot which attracts a group of evacuees from a congestive exit and redirects them towards to a less crowded one. The proposed evacuation framework has been evaluated on real world conditions and exhibited remarkable performance in terms of speed during the evacuation proving: a) the credibility of the CA simulation modeling and b) the necessity of an intelligent mobile aid during the evacuation procedure.

3 Conclusions

CA have been proven quite efficient to model successfully crowd dynamics. In this paper, several CA models and corresponding systems for crowd dynamics were briefly presented taking into consideration different modeling principles, like potential fields techniques, obstacle avoidance, follow the leader principles, grouping theory, etc. Moreover, due to their inherent parallelism CA, some of these models have been implemented in hardware and have been considered as basis of an anticipation crowd management system which is able of preventing clogging in exits during crowd evacuation processes. Finally, robot guided evacuation was presented, based on an accurate CA model capable of assessing the human behavior during emergency situations takes advantage of the simulation output to provide sufficient information to a mobile robotic guide, which in turn guides people towards a less congestive exit at a time.

References

1. Burstedde, C., Klauck, K., Schadschneider, A., Zittartz, J.: Simulation of pedestrian dynamics using a two-dimensional cellular automaton. Physica A 295, 507–525 (2001)
2. Helbing, D., Farkas, I., Vicsek, T.: Simulating dynamical features of escape panic. Nature 407, 487–490 (2000)
3. Goldstone, R.L., Janssen, M.A.: Computational models of collective behavior. Trends in Cognitive Sciences 9(9), 424–430 (2005)
4. Bonabeau, E.: Agent-based modeling: Methods and techniques for simulating human systems. PNAS 99(3), 7280–7287 (2002)
5. Lo, S.M., Huang, H.C., Wang, P., Yuen, K.K.: A game theory based exit selection model for evacuation. Fire Safety Journal 41, 364–369 (2006)
6. Nishinari, K., Sugawara, K., Kazama, T., Schadschneider, A., Chowdhury, D.: Modelling of self-driven particles: foraging ants and pedestrians. Physica A 372, 132–141 (2006)
7. Georgoudas, I.G., Sirakoulis, G.C., Andreadis, I.: A simulation tool for modelling pedestrian dynamics during evacuation of large areas. In: Maglogiannis, I., Karpouzis, K., Bramer, M. (eds.) AIAI 2006. IFIP AICT, vol. 204, pp. 618–626. Springer, Heidelberg (2006)
8. Georgoudas, I.G., Kyriakos, P., Sirakoulis, G.C., Andreadis, I.: A Cellular Automaton Evacuation Model Based on Electric and Potential Fields Technique. In: First International Conference on Evacuation Modeling (RISE 2008), September 23-25, Delft, The Netherlands (2009)
9. Georgoudas, I.G., Sirakoulis, G.C., Andreadis, I.T.: Hardware implementation of a Crowd Evacuation Model based on Cellular Automata. In: PED 2008, pp. 451–463 (2008)
10. Mardiris, V., Sirakoulis, G.C., Mizas, C., Karafyllidis, I., Thanailakis, A.: A CAD system for modeling and Simulation of Computer Networks using Cellular Automata. IEEE Transactions on Systems, Man and Cybernetics, Part C 38(2), 253–264 (2008)
11. Georgoudas, I.G., Kyriakos, P., Sirakoulis, G.C., Andreadis, I.: An FPGA Implemented Cellular Automaton Crowd Evacuation Model Inspired by the Electrostatic-Induced Potential Fields. Microprocessors and Microsystems 34(7-8), 285–300 (2010)
12. Georgoudas, I.G., Sirakoulis, G.C., Andreadis, I.: An Anticipative Crowd Management System Preventing Clogging in Exits during Pedestrian Evacuation Processes. IEEE Systems 5(1), 129–141 (2011)
13. Georgoudas, I.G., Koltsidas, G., Sirakoulis, G.C., Andreadis, I.T.: A Cellular Automaton Model for Crowd Evacuation and its Auto-Defined Obstacle Avoidance Attribute. In: Bandini, S., Manzoni, S., Umeo, H., Vizzari, G. (eds.) ACRI 2010. LNCS, vol. 6350, pp. 455–464. Springer, Heidelberg (2010)
14. Vihas, C., Georgoudas, I.G., Sirakoulis, G.C.: Follow-the-Leader Cellular Automata based Model Directing Crowd Movement. In: Sirakoulis, G.C., Bandini, S. (eds.) ACRI 2012. LNCS, vol. 7495, pp. 752–762. Springer, Heidelberg (2012)
15. Vihas, C., Georgoudas, I., Sirakoulis, G.C.: Cellular Automata incorporating Follow the Leader Principles to Model Crowd Dynamics. Journal of Cellular Automata 8(5-6), 333–346 (2013)
16. Boukas, E., Kostavelis, I., Gasteratos, A., Sirakoulis, G.C.: Robot Guided Crowd Evacuation. Accepted for publication in IEEE Transactions on Automation Science and Engineering (2014)

Counting Equivalent Linear Finite Transducers Using a Canonical Form

Ivone Amorim, António Machiavelo, and Rogério Reis

CMUP, Faculdade de Ciências da Universidade do Porto
Rua do Campo Alegre, 4169-007 Porto, Portugal
{ivone.amorim,rvr}@dcc.fc.up.pt, ajmachia@fc.up.pt

Abstract. The notion of linear finite transducer (LFT) plays a crucial role in a family of cryptosystems introduced in the 80's and 90's. However, as far as we know, no study was ever conducted to count and enumerate these transducers, which is essential to verify if the size of the key space, of the aforementioned systems, is large enough to prevent an exhaustive search attack. In this paper, we determine the cardinal of the equivalence classes on the set of the LFTs with a given size. This result is sufficient to get an approximate value, by random sampling, for the number of non-equivalent injective LFTs, and subsequently for the size of the key space. We introduce a notion of canonical LFT, give a method to verify if two LFTs are equivalent, and prove that every LFT has exactly one equivalent canonical LFT. We then show how this canonical LFT allows us to calculate the size of each equivalence class on the set of the LFTs with the same number of states.

1 Introduction

Transducers, in the most used sense in automata theory, are automata with output that realise rational functions. They are widely studied in the literature, having numerous applications to real world problems. They are essential, for example, in language and speech processing [4].

In this work we deal only with transducers as defined by Renji Tao [7], and our motivation comes from their application to Cryptography. According to that definition, a transducer is a finite state sequential machine given by a quintuple $\langle \mathcal{X}, \mathcal{Y}, S, \delta, \lambda \rangle$, where: $\mathcal{X}$, $\mathcal{Y}$ are the nonempty input and output alphabets, respectively; S is the nonempty finite set of states; $\delta : S \times \mathcal{X} \to S$, $\lambda : S \times \mathcal{X} \to \mathcal{Y}$, are the state transition and output functions, respectively. These transducers are deterministic and can be seen as having all the states as final. Every state in S can be used as initial, and this gives rise to a transducer in the usual sense, *i.e.*, one that realises a rational function. Therefore, in what follows, a transducer is a family of classical transducers that share the same underlying digraph.

A transducer is called linear if its transition and output functions are linear maps. These transducers play a core role in a family of cryptosystems, named FAPKCs, introduced in a series of papers by Tao [8,11,9,10]. Those schemes seem to be a good alternative to the classical ones, being computationally attractive

M. Holzer and M. Kutrib (Eds.): CIAA 2014, LNCS 8587, pp. 70–83, 2014.

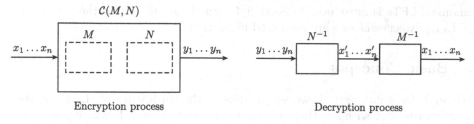

Fig. 1. Schematic representation of FAPKC working principle

and thus suitable for application on devices with very limited computational resources, such as satellites, cellular phones, sensor networks, and smart cards [9]. Roughly speaking, in these systems, the private key consists of two injective transducers, denoted by M and N in Figure 1, where M is a linear finite transducer (LFT), and N is a non-linear finite transducer (non-LFT) of a special kind, whose left inverses can be easily computed. The public key is the result of applying a special product for transducers, $\mathcal{C}$, to the original pair, thus obtaining a non-LFT, denoted by $\mathcal{C}(M, N)$ in Figure 1. The crucial point is that it is easy to obtain an inverse of $\mathcal{C}(M, N)$ from the inverses of its factors, M^{-1} and N^{-1}, while it is believed to be hard to find that inverse without knowing those factors. On the other hand, the factorization of a transducer seems to be hard by itself [12].

The LFTs in the FAPKC systems are of core importance in the invertibility theory of finite transducers, on which part of the security of these systems relies on [1]. They also play a crucial role in the key generation process, since in these systems a pair (public key, private key) is formed using a LFT and two non-LFTs, as explained above. Consequently, for these cryptosystems to be feasible, injective LFTs have to be easy to generate, and the set of non-equivalent injective LFTs has to be large enough to make an exhaustive search intractable.

Several studies were made on the invertibility of LFTs [5,6,13,12,3,1], and some attacks to the FAPKC systems were presented [2,13,7]. However, as far as we know, no study was conducted to determine the size of the key space of these systems. To evaluate that size, one first needs to determine the number of non-equivalent injective LFTs, the exact value of which seems to be quite hard to obtain. In order to be able to get an approximate value, one needs to know the different sizes of the equivalence classes. This is crucial to construct a LFT's uniform random generator.

In this work we describe a method to determine the sizes of those equivalence classes. To accomplish that, a notion of canonical LFT is introduced, being proved that each equivalence class has exactly one of these canonical LFTs. It is also shown how to construct the equivalent canonical LFT to any LFT in its matricial form, and, by introducing a new equivalence test for LFTs, to enumerate and count the equivalent transducers with the same number of states.

The paper is organized as follows. In Section 2 we introduce the basic definitions. Section 3 is devoted to the equivalence test on LFTs. The concept of

canonical LFTs is introduced in Section 4, and the results about the size of the
LFTs equivalence classes are presented in Section 5.

2   Basic Concepts

As usual, for a finite set A, we let $|A|$ denote the cardinality of A, A^n be the
set of words of A with length n, where $n \in \mathbb{N}$, and $A^0 = \{\varepsilon\}$, where ε denotes
the empty word. We put $A^\star = \cup_{n \geq 0} A^n$, the set of all finite words, and $A^\omega =
\{a_0 a_1 \cdots a_n \cdots \mid a_i \in A\}$ is the set of infinite words. Finally, $|\alpha|$ denotes the
length of $\alpha \in A^\star$.

In what follows, a *finite transducer* (FT) is a finite state sequential machine
which, in any given state, reads a symbol from a set $\mathcal{X}$, and produces a symbol
from a set $\mathcal{Y}$, and switches to another state. Thus, given an initial state and
a finite input sequence, a transducer produces an output sequence of the same
length. The formal definition of a finite transducer is the following.

Definition 1. *A finite transducer is a quintuple* $\langle \mathcal{X}, \mathcal{Y}, S, \delta, \lambda \rangle$, *where:* $\mathcal{X}$ *is a
nonempty finite set, called the* input alphabet; $\mathcal{Y}$ *is a nonempty finite set, called
the* output alphabet; S *is a nonempty finite set called the* set of states; δ :
$S \times \mathcal{X} \to S$, *called the* state transition function; *and* $\lambda : S \times \mathcal{X} \to \mathcal{Y}$, *called the*
output function.

Let $M = \langle \mathcal{X}, \mathcal{Y}, S, \delta, \lambda \rangle$ be a finite transducer. The state transition function δ
and the output function λ can be extended to finite words, i.e., elements of $\mathcal{X}^\star$,
recursively, as follows:

$$\delta(s, \varepsilon) = s \qquad\qquad \delta(s, x\alpha) = \delta(\delta(s, x), \alpha)$$
$$\lambda(s, \varepsilon) = \varepsilon \qquad\qquad \lambda(s, x\alpha) = \lambda(s, x)\, \lambda(\delta(s, x), \alpha),$$

where $s \in S$, $x \in \mathcal{X}$, and $\alpha \in \mathcal{X}^\star$. In an analogous way, λ may be extended to
$\mathcal{X}^\omega$.

From these definitions it follows that, for all $s \in S, \alpha \in \mathcal{X}^\star$, and for all
$\beta \in \mathcal{X}^\star \cup \mathcal{X}^\omega$,

$$\lambda(s, \alpha\beta) = \lambda(s, \alpha)\, \lambda(\delta(s, \alpha), \beta).$$

The notions of equivalent states and minimal transducer considered here are
the classical ones.

Definition 2. *Let* $M_1 = \langle \mathcal{X}, \mathcal{Y}_1, S_1, \delta_1, \lambda_1 \rangle$ *and* $M_2 = \langle \mathcal{X}, \mathcal{Y}_2, S_2, \delta_2, \lambda_2 \rangle$ *be two
FTs. Let* $s_1 \in S_1$, *and* $s_2 \in S_2$. *One says that* s_1 *and* s_2 *are* equivalent, *and
denote this relation by* $s_1 \sim s_2$, *if*

$$\forall \alpha \in \mathcal{X}^\star, \ \lambda_1(s_1, \alpha) = \lambda_2(s_2, \alpha).$$

Definition 3. *A finite tranducer is called* minimal *if it has no pair of equivalent
states.*

We now introduce the notion of equivalent transducers used in this context.

Definition 4. M_1 and M_2 are said to be equivalent, denoted by $M_1 \sim M_2$, if the following two conditions are satisfied:

$$\forall s_1 \in S_1, \exists s_2 \in S_2 : s_1 \sim s_2 \ \text{and} \ \forall s_2 \in S_2, \exists s_1 \in S_1 : s_1 \sim s_2.$$

This relation $\sim$ defines an equivalence relation on the set of FTs.

Definition 5. Let $M_1 = \langle \mathcal{X}, \mathcal{Y}, S_1, \delta_1, \lambda_1 \rangle$ and $M_2 = \langle \mathcal{X}, \mathcal{Y}, S_2, \delta_2, \lambda_2 \rangle$ be two FTs. M_1 and M_2 are said to be isomorphic if there exists a bijective map ψ from S_1 onto S_2 such that

$$\psi(\delta_1(s_1, x)) = \delta_2(\psi(s_1), x)$$
$$\lambda_1(s_1, x) = \lambda_2(\psi(s_1), x)$$

for all $s_1 \in S_1$, and for all $x \in X$. The map ψ is called an isomorphism from M_1 to M_2.

Finally, we give the definition of linear finite transducer (LFT).

Definition 6. If $\mathcal{X}, \mathcal{Y}$ and S are vector spaces over a field $\mathbb{F}$, and both $\delta : S \times \mathcal{X} \to S$ and $\lambda : S \times \mathcal{X} \to \mathcal{Y}$ are linear maps, then $M = \langle \mathcal{X}, \mathcal{Y}, S, \delta, \lambda \rangle$ is called linear over $\mathbb{F}$, and we say that $dim(S)$ is the size of M.

Let $\mathcal{L}$ be the set of LFTs over $\mathbb{F}$, and $\mathcal{L}_n$ the set of the transducers in $\mathcal{L}$ with size n. The restriction of $\sim$ to $\mathcal{L}$ is also represented by $\sim$, and the restriction to $\mathcal{L}_n$ is denoted by $\sim_n$.

Definition 7. Let M_1 and M_2 be two LFTs. M_1 and M_2 are said to be similar if there is a linear isomorphism from M_1 to M_2.

Let $M = \langle \mathcal{X}, \mathcal{Y}, S, \delta, \lambda \rangle$ be a LFT over a field $\mathbb{F}$. If $\mathcal{X}, \mathcal{Y}$, and S have dimensions l, m and n, respectively, then there exist matrices $A \in \mathcal{M}_{n,n}(\mathbb{F})$, $B \in \mathcal{M}_{n,l}(\mathbb{F})$, $C \in \mathcal{M}_{m,n}(\mathbb{F})$, and $D \in \mathcal{M}_{m,l}(\mathbb{F})$, such that

$$\delta(s, x) = As + Bx,$$
$$\lambda(s, x) = Cs + Dx,$$

for all $s \in S$, $x \in \mathcal{X}$. The matrices A, B, C, D are called the *structural matrices* of M, and l, m, n are called its *structural parameters*. Notice that if M_1 and M_2 are two equivalent LFTs with structural parameters l_1, m_1, n_1 and l_2, m_2, n_2, respectively, then, from the definition of equivalent transducers, one has $l_1 = l_2$ and $m_1 = m_2$.

A LFT such that C is the null matrix (with the adequate dimensions) is called *trivial*.

One can associate to a LFT, M, with structural matrices A, B, C, D, a family of matrices which are very important in the study of its equivalence class, as will be clear throughout this paper.

Definition 8. *Let $M \in \mathcal{L}_n$ with structural matrices A, B, C, D. The matrix*

$$\Delta_M^{(k)} = \begin{bmatrix} C \\ CA \\ \vdots \\ CA^{k-1} \end{bmatrix}$$

is called the k-diagnostic matrix of M, where $k \in \mathbb{N} \cup \{\infty\}$.

The matrix $\Delta_M^{(n)}$ will be simply denoted by Δ_M and will be referred to as the *diagnostic matrix* of M. The matrix $\Delta_M^{(2n)}$ will be denoted by $\hat{\Delta}_M$ and called the *augmented diagnostic matrix* of M.

Definition 9. *Let V be a k-dimensional vector subspace of $\mathbb{F}^n$, where $\mathbb{F}$ is a field. The unique basis $\{b_1, b_2, \ldots, b_k\}$ of V such that the matrix $[b_1 \ b_2 \ \cdots \ b_k]^T$ is in row echelon form will be here referred to as the* standard basis *of V.*

3 Testing the Equivalence of LFTs

Let $M = \langle \mathcal{X}, \mathcal{Y}, S, \delta, \lambda \rangle$ be a LFT over a field $\mathbb{F}$ with structural matrices A, B, C, D. Starting at a state s_0 and reading an input sequence $x_0 x_1 x_2 \ldots$, one gets a sequence of states $s_0 s_1 s_2 \ldots$ and a sequence of outputs $y_0 y_1 y_2 \ldots$ satisfying the relations

$$s_{t+1} = \delta(s_t, x_t) = As_t + Bx_t,$$
$$y_t = \lambda(s_t, x_t) = Cs_t + Dx_t,$$

for all $t \geq 0$. The following result is then easily proven by induction [7, Theorem 1.3.1].

Theorem 1. *For a LFT as above, $s_{i+1} = A^i s_0 + \sum_{j=0}^{i-1} A^{i-j-1} Bx_j$, and $y_i = CA^i s_0 + \sum_{j=0}^{i} H_{i-j} x_j$, for $i \in \{0, 1, \ldots\}$, where $H_0 = D$, and $H_j = CA^{j-1}B$, $j > 0$.*

Tao, in his book, presents the following necessary and sufficient condition, the only one known so far, for the equivalence of two states of LFTs [7, Theorem 1.3.3]:

Theorem 2. *Let $M_1 = \langle \mathcal{X}, \mathcal{Y}_1, S_1, \delta_1, \lambda_1 \rangle$ and $M_2 = \langle \mathcal{X}, \mathcal{Y}_2, S_2, \delta_2, \lambda_2 \rangle$ be two LFTs. Let $s_1 \in S_1$, and $s_2 \in S_2$. Then, $s_1 \sim s_2$ if and only if the null states of M_1 and M_2 are equivalent, and $\lambda_1(s_1, 0^\omega) = \lambda_2(s_2, 0^\omega)$.*

And, as a consequence, he also presents a necessary and sufficient condition for the equivalence of two LFTs [7, Theorem 1.3.3]:

Corollary 1. *Let M_1 and M_2 be two LFTs. Then, $M_1 \sim M_2$ if and only if their null states are equivalent, and $\{\lambda_1(s_1, 0^\omega) \mid s_1 \in S_1\} = \{\lambda_2(s_2, 0^\omega) \mid s_2 \in S_2\}$.*

However, both conditions cannot be checked efficiently, since they involve working with infinite words. In this section, we explain how they can be reduced to a couple of conditions that can effectively be verified. These new results will be essential in Section 5 to compute the sizes of the equivalence classes in $\mathcal{L}_n/\sim_n$.

The following two Lemmas, which play an important role in the proofs of the subsequent results, are immediate consequences of the basic fact that right multiplication performs linear combinations on the columns of a matrix.

Lemma 1. *Let $A \in \mathcal{M}_{m\times k}$, and $B \in \mathcal{M}_{m\times l}$. Then, $\mathrm{rank}([A|B]) = \mathrm{rank}(A)$ if and only if there $X \in \mathcal{M}_{k\times l}$ such that $B = AX$.*

Lemma 2. *Let $A, B \in \mathcal{M}_{m\times k}$. Then, $\mathrm{rank}(A) = \mathrm{rank}([A|B]) = \mathrm{rank}(B)$ if and only if there is an invertible matrix $X \in \mathcal{M}_{k\times k}$ such that $B = AX$.*

For the remainder of this Section, let M_1, M_2 be two LFTs with structural matrices A_1, B_1, C_1, D_1, and A_2, B_2, C_2, D_2 respectively. Let l_1, m_1, n_1 be the structural parameters of M_1, and l_2, m_2, n_2 be the structural parameters of M_2. To simplify the notation, take $\tilde{\Delta}_1 = \Delta_{M_1}^{(n_1+n_2)}$ and $\tilde{\Delta}_2 = \Delta_{M_2}^{(n_1+n_2)}$.

Lemma 3. *Let $s_1 \in S_1$ and $s_2 \in S_2$. Then, $\lambda_1(s_1, 0^\omega) = \lambda_2(s_2, 0^\omega)$ if and only if $\tilde{\Delta}_1 s_1 = \tilde{\Delta}_2 s_2$.*

Proof. From Theorem 1, one has that $\lambda_1(s_1, 0^\omega) = \lambda_2(s_2, 0^\omega)$ if and only if $C_1 A_1^i s_1 = C_2 A_2^i s_2$, for $i \geq 0$. Let p_1 be the characteristic polynomial of A_1, and p_2 the characteristic polynomial of A_2. Then, p_1 and p_2 are monic polynomials of order n_1 and n_2, respectively. Moreover, by the Cayley-Hamilton theorem, $p_1(A_1) = p_2(A_2) = 0$. Thus, $p = p_1 p_2$ is a monic polynomial of order $n_1 + n_2$ such that $p(A_1) = p(A_2) = 0$. Therefore $A_1^{n_1+n_2+k}$ and $A_2^{n_1+n_2+k}$, with $k \geq 0$, are linear combinations of lower powers of A_1 and A_2, respectively, with the same coefficients. Consequently, $C_1 A_1^i s_1 = C_2 A_2^i s_2$ for $i \geq 0$ is equivalent to $C_1 A_1^i s_1 = C_2 A_2^i s_2$ for $i = 0, 1, \ldots, n_1 + n_2 - 1$, and the result follows.   □

The next result states that the $(n_1 + n_2)$-diagnostic matrices of two equivalent LFTs, of sizes n_1 and n_2, can be used to verify if two of their states are equivalent. It follows from the previous Lemma, and from the fact that if $M_1 \sim M_2$ then, by Theorem 2, $s_1 \sim s_2$ if and only if $\lambda_1(s_1, 0^\omega) = \lambda_2(s_2, 0^\omega)$.

Theorem 3. *Let $s_1 \in S_1$ and $s_2 \in S_2$. If $M_1 \sim M_2$, then $s_1 \sim s_2$ if and only if $\tilde{\Delta}_1 s_1 = \tilde{\Delta}_2 s_2$.*

Corollary 2. *Let M be a LFT, and $s_1, s_2 \in M$. Then, $s_1 \sim s_2$ if and only if $\Delta_M s_1 = \Delta_M s_2$.*

Proof. From the last Theorem, $s_1 \sim s_2$ if and only if $\hat{\Delta}_M s_1 = \hat{\Delta}_M s_2$, that is, if and only if $C A^i s_1 = C A^i s_2$, for $i = 0, 1, \ldots, 2n-1$. Since the minimal polynomial of A has, at most, degree n, this latter condition is equivalent to $C A^i s_1 = C A^i s_2$, for $i = 0, 1, \ldots, n - 1$. Thus, $s_1 \sim s_2$ if and only if $\Delta_M s_1 = \Delta_M s_2$.   □

Corollary 3. *Let M be a LFT over a field $\mathbb{F}$. Then M is minimal if and only if* $\operatorname{rank}(\Delta_M) = \operatorname{size}(M)$.

Proof. It is enough to notice that the linear application $\varphi : S/\sim \to \mathbb{F}^{nm}$ defined by $\varphi([\,s\,]_\sim) = \Delta_M s$ is well-defined and injective, by the previous Corollary. $\square$

The following theorem gives a pair of conditions that have to be satisfied for two LFTs to be equivalent.

Theorem 4. *For LFTs M_1 and M_2 as above, $M_1 \sim M_2$ if and only if the following two conditions are simultaneously verified:*

1. $\operatorname{rank}(\tilde{\Delta}_1) = \operatorname{rank}([\tilde{\Delta}_1 \mid \tilde{\Delta}_2]) = \operatorname{rank}(\tilde{\Delta}_2)$;
2. $D_1 = D_2$ *and* $\tilde{\Delta}_1 B_1 = \tilde{\Delta}_2 B_2$.

Proof. From Corollary 1 one has that $M_1 \sim M_2$ if and only if the null states of M_1 and M_2 are equivalent, and $\{\lambda_1(s_1, 0^\omega) \mid s_1 \in S_1\} = \{\lambda_2(s_2, 0^\omega) \mid s_2 \in S_2\}$.

The null states of M_1 and M_2 are equivalent if and only if $\forall \alpha \in \mathcal{X}^*$, $\lambda_1(0, \alpha) = \lambda_2(0, \alpha)$. By Theorem 1, this is equivalent to: $\sum_{j=0}^{i} H_{i-j} x_j = \sum_{j=0}^{i} H'_{i-j} x_j$, $i = 0, 1, \ldots, |\alpha|$, where $\alpha = x_0 x_1 \cdots x_{|\alpha|} \in \mathcal{X}^*$, $H_0 = D_1$, $H'_0 = D_2$ and $H_j = C_1 A_1^{j-1} B_1$, $H'_j = C_2 A_2^{j-1} B_2$, for $j > 0$. That is, $\forall x_0, x_1, \cdots, x_{|\alpha|} \in \mathcal{X}$ the following equations are simultaneously satisfied:

$$D_1 x_0 = D_2 x_0$$
$$D_1 x_1 + C_1 B_1 x_0 = D_2 x_1 + C_2 B_2 x_0$$
$$D_1 x_2 + C_1 B_1 x_1 + C_1 A_1 B_1 x_0 = D_2 x_2 + C_2 B_2 x_1 + C_2 A_2 B_2 x_0$$
$$\vdots$$
$$D_1 x_{|\alpha|} + \cdots + C_1 A_1^{(|\alpha|-1)} B_1 x_0 = D_2 x_{|\alpha|} + \cdots + C_2 A_2^{(|\alpha|-1)} B_2 x_0.$$

Using the characteristic polynomials of A_1 and A_2, as in the proof of Lemma 3, one sees that when $|\alpha| \geq u$ the equations after the first u of them are implied by the previous ones. From the arbitrariness of α, it then follows that system is satisfied if and only if $D_1 = D_2$ and $\tilde{\Delta}_1 B_1 = \tilde{\Delta}_2 B_2$.

From Lemma 3, one has that $\{\lambda_1(s_1, 0^\omega) \mid s_1 \in S_1\} = \{\lambda_2(s_2, 0^\omega) \mid s_2 \in S_2\}$ if and only if $\{\tilde{\Delta}_1 s_1 \mid s_1 \in S_1\} = \{\tilde{\Delta}_2 s_2 \mid s_2 \in S_2\}$. This means that the column space of $\tilde{\Delta}_1$ is equal to the column space of $\tilde{\Delta}_2$, which is true if and only if there exist matrices X, Y such that $\tilde{\Delta}_2 = \tilde{\Delta}_1 X$ and $\tilde{\Delta}_1 = \tilde{\Delta}_2 Y$. But, from Lemma 1, this happens if and only if $\operatorname{rank}(\tilde{\Delta}_1) = \operatorname{rank}([\tilde{\Delta}_1 \mid \tilde{\Delta}_2])$ and $\operatorname{rank}(\tilde{\Delta}_2) = \operatorname{rank}([\tilde{\Delta}_1 \mid \tilde{\Delta}_2])$. $\square$

Using the conditions in the previous result, it is not hard to write an algorithm to test the equivalence of two LFTs. The running time of such an algorithm will be of the same order as the running time of well known algorithms to compute the rank of a matrix.

Corollary 4. $M_1 \sim M_2$ *implies* $D_1 = D_2$.

It is important to recall, at this moment, that the size of an LFT is the only structural parameter that can vary between transducers of the same equivalence class in $\mathcal{L}/\sim$. Moreover, the size of an LFT of an equivalence class $[M]_\sim$, can never be smaller than $\mathrm{rank}(\Delta_{M'})$, where M' is a minimal transducer in $[M]_\sim$. These facts will be important in Section 5.

The following Corollary is a direct consequence of Lemma 2 and of the first point of Theorem 4.

Corollary 5. *If $n = n_1 = n_2$, $S_1 = S_2$, and $M_1 \sim M_2$, then there is an invertible matrix $X \in \mathcal{M}_{n \times n}$ such that $\hat{\Delta}_{M_2} = \hat{\Delta}_{M_1} X$.*

4   Canonical LFTs

In this section we prove that every equivalence class in $\mathcal{L}/\sim$ has one and only one LFT that satisfies a certain condition[1]. We also prove that, given the structural matrices of a LFT, M, one can identify and construct the transducer in $[M]_\sim$ that satisfies that aforesaid condition. LFTs that satisfy that condition are what we call *canonical* LFTs.

Lemma 4. *Let $M \in \mathcal{L}_n$ with structural matrices A, B, C, D. Then,*

$$\mathrm{rank}(\Delta_M^{(k)}) = \mathrm{rank}(\Delta_M), \ \forall k \geq n.$$

Proof. The degree of the minimal polynomial of A is at most n, and so the matrices CA^k, for $k \geq n$, are linear combinations of $C, CA^1, \cdots, CA^{n-1}$.   □

The following result shows that if two minimal LFTs, with the same set of states, are equivalent, then the two vector spaces generated by the columns of their diagnostic matrices are equal.

Corollary 6. *Let $M_1 = \langle \mathcal{X}, \mathcal{Y}, S, \delta_1, \lambda_1 \rangle$ and $M_2 = \langle \mathcal{X}, \mathcal{Y}, S, \delta_2, \lambda_2 \rangle$ be two minimal LFT such that $M_1 \sim M_2$. Then, $\{\Delta_{M_1} s \mid s \in S\} = \{\Delta_{M_2} s \mid s \in S\}$.*

Proof. If $M_1 \sim M_2$, then $\{\lambda_1(s, 0^\omega) \mid s \in S\} = \{\lambda_2(s, 0^\omega) \mid s \in S\}$, by Corollary 1. That is, $\{\Delta_{M_1}^{(\infty)} s \mid s \in S\} = \{\Delta_{M_2}^{(\infty)} s \mid s \in S\}$. Since M_1 and M_2 are minimal, from Lemma 4 and Corollary 3 one concludes that $\{\Delta_{M_1} s \mid s \in S\} = \{\Delta_{M_2} s \mid s \in S\}$.   □

If M is a minimal LFT, then the columns of Δ_M form a basis of the space $\{\Delta_M s \mid s \in S\}$. Therefore, if M_1 and M_2 are minimal and equivalent, there is an invertible matrix X (with adequate dimensions) such that $\Delta_{M_1} X = \Delta_{M_2}$. Note that this condition, here obtained for minimal transducers, is less demanding than the one we have in Corollary 5.

The next result, together with its proof, gives a way to generate LFTs in $[M]_\sim$, where M is a LFT defined by its structural matrices.

---

[1] The equivalence classes formed by trivial LFTs are excluded.

Lemma 5. *Let $M_1 = \langle \mathcal{X}, \mathcal{Y}, S, \delta_1, \lambda_1 \rangle$ be a non-trivial LFT. Let $\psi : S \to S$ be a vector space isomorphism. Then, there is exactly one LFT $M_2 = \langle \mathcal{X}, \mathcal{Y}, S, \delta_2, \lambda_2 \rangle$ such that ψ is a linear isomorphism from M_1 to M_2. Moreover, M_1 is minimal if and only if M_2 is minimal.*

Proof. Let P be the matrix of ψ relative to the standard basis. From its definition ψ is an isomorphism between M_1 and M_2 if and only the conditions mentioned in Section 2 are satisfied. Let $x = 0$ and $s_1 \in S$. From the first condition, one gets

$$\psi(\delta_1(s_1, 0)) = \delta_2(\psi(s_1), 0) \Leftrightarrow PA_1 s_1 = A_2 P s_1 \Leftrightarrow (PA_1 - A_2 P)s_1 = 0.$$

From the arbitrariness of s_1, this is equivalent to $PA_1 - A_2 P = 0$. Since P is invertible, one gets $A_2 = PA_1 P^{-1}$. The second condition yields

$$\lambda_1(s_1, 0) = \lambda_2(\psi(s_1), 0) \Leftrightarrow C_1 s_1 = C_2 P s_1 \Leftrightarrow (C_1 - C_2 P)s_1 = 0.$$

Again, from the arbitrariness of s_1, this is equivalent to $C_1 - C_2 P = 0$. Thus, $C_2 = C_1 P^{-1}$.

Now, let $s_1 = 0$ and $x \in X$. Using a similar method, one gets $B_2 = PB_1$ and $D_1 = D_2$. Hence, the transducer M_2 satisfying the conditions of the theorem is uniquely determined by ψ. It is then easy to see that the transducer given by the structural matrices $A_2 = PA_1 P^{-1}$, $B_2 = PB_1$, $C_2 = C_1 P^{-1}$, and $D_2 = D_1$ is such that ψ is a linear isomorphism from M_1 to M_2.

Since M_1 and M_2 are isomorphic, they are equivalent. Therefore, M_1 is minimal if and only if M_2 is minimal. □

Recalling that $GL_n(\mathbb{F})$ denotes the set of $n \times n$ invertible matrices over the field $\mathbb{F}$, one has:

Corollary 7. *Let $M \in \mathcal{L}_n$ be a non-trivial minimal LFT over a finite field $\mathbb{F}$. Then, the number of minimal LFTs in $[M]_\sim$ is $|GL_n(\mathbb{F})|$.*

Moreover, from the proof of Lemma 5, one gets that, given an invertible matrix X, there is exactly one minimal transducer in $[M]_\sim$ which has $\Delta_M X$ as diagnostic matrix. The same is not true if M is not minimal, as it will be shown in the next section. The aforementioned proof also gives an explicit way to obtain that transducer from the structural matrices of M.

Proposition 1. *Let $M_1 = \langle \mathcal{X}, \mathcal{Y}, S, \delta_1, \lambda_1 \rangle$ be a LFT. Let $\psi : S \to S$ be a vector space isomorphism. Let M_2 be the LFT constructed from M_1 and $\psi(s) = Ps$ as described in the proof of the last Theorem. Then, $\Delta_{M_1} s = \Delta_{M_2} \psi(s)$.*

Proof. Let $s \in S$, then

$$\Delta_{M_2} \psi(s) = \begin{bmatrix} C_1 P^{-1} \\ C_1 A_1 P^{-1} \\ \vdots \\ C_1 A_1^{n-1} P^{-1} \end{bmatrix} Ps = \begin{bmatrix} C_1 \\ C_1 A_1 \\ \vdots \\ C_1 A_1^{n-1} \end{bmatrix} s = \Delta_{M_1} s.$$

□

The next Theorem gives the condition that was promised at the beginning of this section.

Theorem 5. *Every non-trivial equivalence class in $\mathcal{L}/\sim$ has exactly one LFT $M = \langle \mathcal{X}, \mathcal{Y}, S, \delta, \lambda \rangle$ which satisfies the condition that $\{ \Delta_M e_1, \Delta_M e_2, \cdots, \Delta_M e_n \}$ is the standard basis of $\{ \Delta_M s \mid s \in S \}$, where $\{ e_1, e_2, \cdots, e_n \}$ is the standard basis of S.*

Proof. Given the structural matrices of a LFT, Tao shows [7, Theorem 1.3.4] how to compute an equivalent minimal LFT. This implies, in particular, that every LFT is equivalent to a minimal LFT. Thus, to get the result here claimed, it is enough to prove that if $M_1 = \langle \mathcal{X}, \mathcal{Y}, S, \delta_1, \lambda_1 \rangle$ is a non-trivial minimal LFT, then M_1 is equivalent to exactly one finite transducer $M_2 = \langle \mathcal{X}, \mathcal{Y}, S, \delta_2, \lambda_2 \rangle$ such that $\{ \Delta_{M_2} e_1, \Delta_{M_2} e_2, \ldots, \Delta_{M_2} e_n \}$ is the standard basis of $\{ \Delta_{M_1} s \mid s \in S \}$. First, let us notice that, since M_1 is minimal, Δ_{M_1} is left invertible, and consequently s is uniquely determined by $\Delta_{M_1} s$. Let $\mathcal{B} = \{ b_1, b_2, \cdots, b_n \}$ be the standard basis of $\{ \Delta_{M_1} s \mid s \in S \}$. Let s_i be the unique vector in S such that $b_i = \Delta_{M_1} s_i$, for $i = 1, 2, \ldots, n$. Let $\psi : S \to S$ be defined by $\psi(s_i) = e_i$. Then ψ is a vector space isomorphism. Let M_2 be the LFT constructed from M_1 and ψ as described in the proof of Lemma 5. Then, $M_2 \sim M_1$ and M_2 is minimal, which, by Corollary 6, implies $\{ \Delta_{M_2} s \mid s \in S \} = \{ \Delta_{M_1} s \mid s \in S \}$. From Proposition 1 one also has $\Delta_{M_2} e_i = \Delta_{M_2} \psi(s_i) = \Delta_{M_1} s_i = b_i$, for $i = 1, 2, \cdots, n$. Therefore, $\{ \Delta_{M_2} e_1, \Delta_{M_2} e_2, \ldots, \Delta_{M_2} e_n \}$ is the standard basis of $\{ \Delta_{M_1} s \mid s \in S \}$. The uniqueness easily follows from the fact that all choices made are unique.   □

Finally we can state the definition of canonical LFT here considered.

Definition 10. *Let $M = \langle \mathcal{X}, \mathcal{Y}, S, \delta, \lambda \rangle$ be a linear finite transducer. One says that M is a canonical LFT if $\{ \Delta_M e_1, \Delta_M e_2, \cdots, \Delta_M e_n \}$ is the standard basis of $\{ \Delta_M s \mid s \in S \}$, where $\{ e_1, e_2, \cdots, e_n \}$ is the standard basis of S.*

The proofs of Theorem 5 and Lemma 5 show that given the structural matrices of a LFT, M, one can identify and construct the canonical transducer in $[M]_\sim$.

5   On the Size of Equivalence Classes of LFTs

In what follows we only consider LFTs defined over finite fields with q elements, $\mathbb{F}_q$, because these are the ones commonly used in Cryptography.

In this section we explore how the size of the equivalence classes in $\mathcal{L}_n/\sim_n$ varies with the size n. Given a minimal LFT M_1 in $\mathcal{L}_{n_1}$, our aim is to count the number of transducers in $\mathcal{L}_{n_2}$, with $n_2 \geq n_1$, that are equivalent to M_1.

The following result shows that given $M_1 \in \mathcal{L}_{n_1}$, one can easily construct an equivalent transducer in $\mathcal{L}_{n_2}$, for any $n_2 \geq n_1$, which can then be used to count the number of transducers in $\mathcal{L}_{n_2}$ that are equivalent to M_1, as well as the size of the equivalence classes in S.

Proposition 2. *Let M_1 be the LFT over $\mathbb{F}_q$ with structural matrices A_1, B_1, C_1, D_1, and structural parameters l, m, n_1. Let $n' \in \mathbb{N}$, and M_2 be the LFT with structural matrices*

$$A_2 = \left[\begin{array}{c|c} A_1 & 0_{n_1 \times n'} \\ \hline 0_{n' \times n_1} & 0_{n' \times n'} \end{array}\right], \quad B_2 = \left[\begin{array}{c} B_1 \\ 0_{n' \times l} \end{array}\right], \quad C_2 = \left[\begin{array}{cc} C_1 & 0_{m \times n'} \end{array}\right], \quad and \quad D_2 = D_1.$$

Then, $M_1 \sim M_2$. The structural parameters of M_2 are l, m, n_2, where $n_2 = n_1 + n'$.

Proof. Take $u = n_1 + n_2$. Notice that $C_2 A_2^i = [C_1 A_1^i \ \ 0_{m \times n'}]$, for $i = 0, 1, \ldots, u-1$. That is, $\Delta_{M_2}^{(u)} = [\Delta_{M_1}^{(u)} \ \ 0_{um \times n'}]$. The result is then trivial by Theorem 4. $\quad\square$

The next result counts the number of LFTs in $\mathcal{L}_{n_2}$ that are equivalent to M_2, where M_2 is the LFT defined from M_1 as described in Proposition 2. Because $M_1 \sim M_2$, this yields the number of LFTs in $\mathcal{L}_{n_2}$ that are equivalent to M_1.

Theorem 6. *Let M_1 be a minimal LFT in $\mathcal{L}_{n_1}$ with structural matrices A_1, B_1, C_1, D_1, and structural parameters l, m, n_1. Let M_2 be the LFT described in Proposition 2. The number of finite transducers $M \in \mathcal{L}_{n_2}$ which are equivalent to M_2 is $(q^{n_2} - 1)(q^{n_2} - q) \cdots (q^{n_2} - q^{r-1}) q^{(n_2 + l)(n_2 - r)}$, where $r = \mathrm{rank}(\hat{\Delta}_{M_2})$.*

Proof. The theorem follows from the next three facts, that we will prove in the remaining of this section.

1. For all matrices $\Delta_1, \Delta_2 \in \{\hat{\Delta}_M \mid M \in \mathcal{L}_{n_2} \text{ and } M \sim M_2\}$, the number of LFTs that are equivalent to M_2 and have Δ_1 as augmented diagnostic matrix is equal to the number of LFTs that are equivalent to M_2 and have Δ_2 as augmented diagnostic matrix.

2. The number of LFTs equivalent to M_2 and have $\hat{\Delta}_{M_2}$ as augmented diagnostic matrix is $q^{(n_2 + l)(n_2 - r)}$, with $r = \mathrm{rank}(\hat{\Delta}_{M_2})$.

3. The size of $\{\hat{\Delta}_M \mid M \in \mathcal{L}_{n_2} \text{ and } M \sim M_2\}$ is $(q^{n_2} - 1)(q^{n_2} - 2) \cdots (q^{n_2} - q^{r-1})$, with $r = \mathrm{rank}(\hat{\Delta}_{M_2})$.

$\quad\square$

From Corollary 5, if two LFTs M and M' are equivalent, there is an invertible matrix X such that $\Delta_{M'} = \Delta_M X$. The first of the above items is then an instance of the following result.

Theorem 7. *Let $M \in \mathcal{L}_n$. Let $S_\Delta = \{M' \in \mathcal{L}_n \mid M' \sim M \text{ and } \hat{\Delta}_{M'} = \Delta\}$. Then, for every $X \in GL_n(\mathbb{F}_q)$, $|S_{\hat{\Delta}_M}| = |S_{\hat{\Delta}_M X}|$.*

Proof. Let $f : S_{\Delta_M} \to S_{\Delta_M X}$ such that $f(M) = M'$, where M' is the transducer defined by the matrices $A' = X^{-1}AX$, $B' = X^{-1}B$, $C' = CX$ and $D' = D$. It is straightforward to see that $\hat{\Delta}_{M'} = \hat{\Delta}_M X$, and that the application f is bijective. $\quad\square$

To prove item 2, let us count the number of transducers $M \in \mathcal{L}_{n_2}$ that are equivalent to M_2 and have $\hat{\Delta}_{M_2}$ as augmented diagnostic matrix. One has to count the possible choices for the structural matrices A, B, C and D, of M, that satisfy the condition 2 of Theorem 4, and $\hat{\Delta}_{M_2} = \hat{\Delta}_M$ (which implies condition 1). The choice for D is obvious and unique from condition 2, as well as the choice for C (from condition $\hat{\Delta}_{M_2} = \hat{\Delta}_M$). How many choices does one have for A such that the condition $\hat{\Delta}_{M_2} = \hat{\Delta}_M$ is satisfied? And, how many choices for B such that $\hat{\Delta}_{M_2} = \hat{\Delta}_M$ and the second condition is satisfied, i.e., such that $\hat{\Delta}_M B_2 = \hat{\Delta}_M B$? The following result gives the number of possible choices for A, and the proof gives the form of these matrices.

Theorem 8. *Let M_1 be a minimal LFT in $\mathcal{L}_{n_1}$ with structural matrices A_1, B_1, C_1, D_1, and M_2 the LFT described in Proposition 2. There are exactly $q^{n_2(n_2-\mathrm{rank}(\Delta_{M_2}))}$ matrices $A \in \mathcal{M}_{n_2 \times n_2}(\mathbb{F}_q)$ such that $C_2 A_2^i = C_2 A^i$, for $i = 0, 1, \cdots, 2n_2 - 1$.*

Proof. Let $A \in \mathcal{M}_{n_2 \times n_2}(\mathbb{F}_q)$ be such that $C_2 A_2^i = C_2 A^i$, for $i = 0, 1, \ldots, 2n_2 - 1$. Then, $C_2 A_2^i = C_2 A_2^{i-1} A$, for $i = 0, 1, \ldots, 2n_2 - 1$.

Take $A = \begin{bmatrix} E_1 & E_2 \\ \hline E_3 & E_4 \end{bmatrix}$, with $E_1 \in \mathcal{M}_{n_1 \times n_1}(\mathbb{F}_q)$, $E_2 \in \mathcal{M}_{n_1 \times n'}(\mathbb{F}_q)$, $E_3 \in \mathcal{M}_{n' \times n_1}(\mathbb{F}_q)$, $E_4 \in \mathcal{M}_{n' \times n'}(\mathbb{F}_q)$, and $n' = n_2 - n_1$. Then, from $C_2 A_2^i = C_2 A_2^{i-1} A$, for $i \in \{1, \ldots, 2n_2 - 1\}$, one gets that $\begin{bmatrix} C_1 A_1^i & 0_{m \times n'} \end{bmatrix} = \begin{bmatrix} C_1 A_1^{i-1} E_1 & C_1 A_1^{i-1} E_2 \end{bmatrix}$, for $i \in \{1, \ldots, 2n_2 - 1\}$, i.e., $C_1 A_1^i = C_1 A_1^{i-1} E_1$, and $C_1 A_1^{i-1} E_2 = 0$, for $i \in \{1, \ldots, 2n_2 - 1\}$. This is equivalent to $\Delta_{M_1}^{(2n_2-1)} A_1 = \Delta_{M_1}^{(2n_2-1)} E_1$, and $\Delta_{M_1}^{(2n_2-1)} E_1 = 0$, or $\Delta_{M_1}^{(2n_2-1)}(A_1 - E_1) = 0$ and $\Delta_{M_1}^{(2n_2-1)} E_1 = 0$. Since M_1 is minimal, by Lemma 4 and Corollary 3, $\mathrm{rank}(\Delta_{M_1}^{(2n_2-1)}) = \mathrm{rank}(\Delta_{M_1}) = n_1 =$ number of columns of $\Delta_{M_1}^{(2n_2-1)}$. Therefore, $E_1 = A_1$ and $E_2 = 0$. Consequently, any matrix A with the same first n_1 rows as A_2 satisfies $C_2 A_2^i = C_2 A^i$, for $i = 0, 1, \ldots, 2n_2 - 2$, and those matrices A are the only ones that satisfy condition 2. Because the last $n_2 - n_1$ rows of A can be arbitrarily chosen, and A has n_2 columns, one gets that there are $q^{n_2(n_2-n_1)}$ matrices A that satisfy the required conditions. Since $n_1 = \mathrm{rank}(\Delta_{M_1}) = \mathrm{rank}(\Delta_{M_2})$ (because M_1 is minimal, and $M_1 \sim M_2$), the result follows. $\square$

Now, for each matrix A such that $\hat{\Delta}_{M_2} = \hat{\Delta}_M$, i.e., $C_2 A_2^i = C_2 A^i$, for $i = 0, 1, \ldots, 2n_2 - 1$, one wants to count the number of matrices B that satisfy $\hat{\Delta}_M B_2 = \hat{\Delta}_M B$, that is, satisfy $C_2 A^i B_2 = C_2 A^i B$, for $i = 0, 1, \ldots, 2n_2 - 1$.

Theorem 9. *Let M_1 be a minimal LFT with structural matrices A_1, B_1, C_1, D_1, and structural parameters l, m, n_1. Let M_2 be the LFT described in Proposition 2. Given a matrix A such that $\hat{\Delta}_{M_2} = \hat{\Delta}_M$, there are exactly $q^{l(n_2-\mathrm{rank}(\Delta_{M_2}))}$ matrices $B \in \mathcal{M}_{n_2 \times l}(\mathbb{F}_q)$ such that $C_2 A^i B_2 = C_2 A^i B$ for $i = 0, 1, \cdots, 2n_2 - 1$.*

Proof. Let A be a matrix such that $\hat{\Delta}_{M_2} = \hat{\Delta}_M$, and B such that $\hat{\Delta}_M B_2 = \hat{\Delta}_M B$. Then, $\hat{\Delta}_{M_2} B_2 = \hat{\Delta}_{M_2} B$. Consequently, $\Delta_{M_2} B_2 = \Delta_{M_2} B$, which is equivalent to

$\Delta_{M_2}(B_2 - B) = 0$. Since B has n_2 rows, one concludes that there are exactly $n_2 - \text{rank}(\Delta_{M_2})$ rows in B whose entries can be arbitrarily chosen to have a solution of $\Delta_{M_2}(B_2 - B) = 0$. Therefore, and since B has l columns, there are $q^{l(n_2 - \text{rank}(\Delta_{M_2}))}$ matrices B that satisfy condition 2 of Theorem 4.   □

From this one concludes that the number of transducers in $\mathcal{L}_{n_2}$ that are equivalent to M_2 and that have the same augmented diagnostic matrix is $q^{(n_2 + l)(n_2 - r)}$, where $r = \text{rank}(\hat{\Delta}_{M_2})$, which proves item 2. Item 3 is covered by the following two results together with Corollary 5.

Theorem 10. *Let $A \in \mathcal{M}_{m \times n}(\mathbb{F}_q)$ such that $\text{rank}(A) \neq n$. Then, the number of matrices $X \in GL_n(\mathbb{F}_q)$ such that $AX = A$ is $(q^n - q^{\text{rank}(A)})(q^n - q^{\text{rank}(A)+1}) \cdots (q^n - q^{n-1})$. If $\text{rank}(A) = n$, only the identity matrix satisfies this condition.*

Proof. Let $X \in GL_n(\mathbb{F}_q)$ be such that $AX = A$. Then, there are $n - \text{rank}(A)$ rows in X whose entries can be arbitrarily chosen to have a solution of $AX = A$. But, since X has to be invertible, one has $q^n - q^{\text{rank}(A)}$ possibilities for the "first" of those rows, $q^n - q^{\text{rank}(A)+1}$ for the "second", $q^n - q^{\text{rank}(A)+2}$ for the "third", and so on. Therefore, there are $(q^n - q^{\text{rank}(A)})(q^n - q^{\text{rank}(A)+1}) \cdots (q^n - q^{n-1})$ matrices X that satisfy the required condition.   □

The following result is a direct consequence of the previous Theorem and the size of $GL_n(\mathbb{F}_q)$.

Corollary 8. *Let $A \in \mathcal{M}_{m \times n}(\mathbb{F}_q)$. Then, the number of matrices of the form AX, where $X \in GL_n(\mathbb{F}_q)$ is $(q^n - 1)(q^n - q) \cdots (q^n - q^{\text{rank}(A)-1})$.*

Since augmented diagnostic matrices of *LFTs* in the same equivalence class have the same rank, Theorem 6 can be generalized to:

Corollary 9. *Let M be a LFT with structural parameters l, m, n. Then*

$$|[M]_{\sim_n}| = (q^n - 1)(q^n - q) \cdots (q^n - q^{r-1}) q^{(n+l)(n-r)}, \text{ where } r = \text{rank}(\Delta_M).$$

Given the structural matrices of a LFT, the last Corollary gives a formula to compute the number of equivalent LFTs with the same size.

6   Conclusion

We presented a way to compute the number of equivalent LFTs with the same size, by introducing a canonial form for LFTs and a method to test LFTs equivalence. This is essencial to have a LFT uniform random generator, and to get an approximate value for the number of non-equivalent injective LFTs, which is indispensable to evaluate the key space of the FAPKC systems.

In future work we plan to use the results in the last section to deduced a recurrence relation that gives the number of non-equivalent LFTs of a given size. This, together with the approximate value for the number of non-equivalent injective LFTs, will allow us to verify if random generation of LFTs is a feasible option to generate keys.

Acknowledgements. This work was partially funded by the European Regional Development Fund through the programme COMPETE and by the Portuguese Government through the FCT under projects PEst-C/MAT/UI0144/2013 and FCOMP-01-0124-FEDER-020486.

Ivone Amorim is funded by FCT grant SFRH/BD/84901/2012.

The authors gratefully acknowledge the useful suggestions and comments of the anonymous referees.

References

1. Amorim, I., Machiavelo, A., Reis, R.: On the invertibility of finite linear transducers. RAIRO - Theoretical Informatics and Applications 48, 107–125 (2014)
2. Bao, F., Igarashi, Y.: Break Finite Automata Public Key Cryptosystem. In: Fülöp, Z. (ed.) ICALP 1995. LNCS, vol. 944, pp. 147–158. Springer, Heidelberg (1995)
3. Dai, Z., Ye, D., Ou, H.: Self-Injective Rings and Linear (Weak) Inverses of Linear Finite Automata over Rings. Science in China (Series A) 42(2), 140–146 (1999)
4. Roche, E., Shabes, Y. (eds.): Finite-State Language Processing. MIT Press, Cambridge (1997)
5. Tao, R.: Invertible Linear Finite Automata. Scientia Sinica XVI(4), 565–581 (1973)
6. Tao, R.: Invertibility of Linear Finite Automata Over a Ring. In: Lepistö, T., Salomaa, A. (eds.) ICALP 1988. LNCS, vol. 317, pp. 489–501. Springer, Heidelberg (1988)
7. Tao, R.: Finite Automata and Application to Cryptography. Springer Publishing Company, Incorporated (2009)
8. Tao, R., Chen, S.: A Finite Automaton Public Key Cryptosystem and Digital Signatures. Chinese Journal of Computers 8(6), 401–409 (1985) (in Chinese)
9. Tao, R., Chen, S.: A Variant of the Public Key Cryptosystem FAPKC3. J. Netw. Comput. Appl. 20, 283–303 (1997)
10. Tao, R., Chen, S.: The Generalization of Public Key Cryptosystem FAPKC4. Chinese Science Bulletin 44(9), 784–790 (1999)
11. Tao, R., Chen, S., Chen, X.: FAPKC3: A New Finite Automaton Public Key Cryptosystem. Journal of Computer Science and Technology 12(4), 289–305 (1997)
12. Dai, Z.-D., Ye, D.F., Lam, K.-Y.: Weak Invertibility of Finite Automata and Cryptanalysis on FAPKC. In: Ohta, K., Pei, D. (eds.) ASIACRYPT 1998. LNCS, vol. 1514, pp. 227–241. Springer, Heidelberg (1998)
13. Dai, Z., Ye, D., Ou, H.: Weak Invertibility of Linear Finite Automata I, Classification and Enumeration of Transfer Functions. Science in China (Series A) 39(6), 613–623 (1996)

On the Power of One-Way Automata
with Quantum and Classical States*

Maria Paola Bianchi, Carlo Mereghetti, and Beatrice Palano

Dipartimento di Informatica, Università degli Studi di Milano
via Comelico 39, 20135 Milano, Italy
{bianchi,mereghetti,palano}@di.unimi.it

Abstract. We consider the model of one-way automata with quantum and classical states (QCFAs) introduced in [23]. We show, by a direct approach, that QCFAs with isolated cut-point accept regular languages only, thus characterizing their computational power. Moreover, we give a size lower bound for QCFAs accepting regular languages, and we explicitly build QCFAs accepting the word quotients and inverse homomorphic images of languages accepted by given QCFAs with isolated cut-point, maintaining the same cut-point, isolation, and polynomially increasing the size.

Keywords: quantum automata, regular languages, descriptional complexity.

1 Introduction

Since we can hardly expect to see a full-featured quantum computer in the near future, it is natural to investigate the simplest and most restricted model of computation where the quantum paradigm outperforms the classical one. Classically, one of the simplest model of computation is a finite automaton. Thus, *quantum finite automata* (QFAs) are introduced and investigated by several authors.

Originally, two models of QFAs are proposed: measure-once QFAs [9,16], where the probability of accepting words is evaluated by "observing" just once, at the end of input processing, and measure-many QFAs [13], having such an observation performed after each move. Several variations of these two models, motivated by different possible physical realizations, are then proposed. Thus, e.g., enhanced [19], reversible [10], Latvian [1], and measure-only QFAs [6] are introduced. Results in the literature (see, e.g., [1,3,15]) show that all these models of QFAs are strictly less powerful than deterministic finite automata (DFAs), although retaining a higher descriptional power (i.e., they can be significantly smaller than equivalent classical devices).

To enhance the low computational power of these "purely quantum" systems, *hybrid models* featuring both a quantum and a classical component are

---

* Partially supported by MIUR under the project "PRIN: Automi e Linguaggi Formali: Aspetti Matematici e Applicativi."

M. Holzer and M. Kutrib (Eds.): CIAA 2014, LNCS 8587, pp. 84–97, 2014.

studied. Examples of such hybrid systems are QFAs with open time evolution (GQFAs) [11,14], QFAs with control language (QFCs) [3,17], and QFAs with quantum and classical states (QCFAs) [23], this latter model being the one-way restriction of the model introduced in [2]. It is proved that the class of languages accepted with isolated cut-point by GQFAs and QFCs coincides with the class of regular languages, while for QCFAs it is only known that they can simulate DFAs. A relevant feature of these hybrid models is that they can naturally and directly simulate several variants of QFAs by preserving the size. This property makes each of them a good candidate as a general unifying framework within which to investigate size results for different quantum paradigms [4,5,8,18].

In this paper, we focus on the model of QCFAs. We completely characterize their computational power and study some descriptional complexity issues. It may be interesting to point out that the relevant difference between QFCs and QCFAs rely in the communication policy between the two internal components: in QCFAs a two-way information exchange between the classical and quantum parts is established, while in QFCs only the quantum component affects the dynamic of the classical one. Here, by a direct approach, we show that the two-way communication is not more powerful than one-way communication. In fact, we prove that QCFAs accept with isolated cut-point regular languages only (exactly as QFCs), thus characterizing their computational power. We obtain this result by studying properties of formal power series associated with QCFAs.

We continue the investigations on QCFAs by studying their descriptional power. Our approach for proving regularity of languages accepted with isolated cut-point by QCFAs enables us to give a lower bound for the size complexity of QCFAs, which is logarithmic in the size of equivalent DFAs, in analogy with QFCs [7]. Next, we study the size cost of implementing some language operations on QCFAs. Results for Boolean operations are provided in [23]. Here, we explicitly construct QCFAs accepting word quotients and inverse homomorphic images of languages accepted by given QCFAs with isolated cut-point, maintaining the same cut-point, isolation, and polynomially increasing the size. For other types of QFAs, these two latter operations are investigated, e.g., in [1,17].

2 Preliminaries

2.1 Linear Algebra

We quickly recall some notions of linear algebra, useful to describe the quantum world. For more details, we refer the reader to, e.g., [22]. The fields of real and complex numbers are denoted by $\mathbb{R}$ and $\mathbb{C}$, respectively. Given a complex number $z = a + ib$, we denote its *conjugate* by $z^* = a - ib$ and its *modulus* by $|z| = \sqrt{zz^*}$. We let $\mathbb{C}^{n \times m}$ and $\mathbb{C}^n$ (shorthand for $\mathbb{C}^{1 \times n}$) denote, respectively, the set of $n \times m$ matrices and n-dimensional row vectors with entries in $\mathbb{C}$. We denote by $[\mathbf{0}]_{n \times m}$ ($[\mathbf{0}]_n$) the zero matrix in $\mathbb{C}^{n \times m}$ ($\mathbb{C}^{n \times n}$). The identity matrix in $\mathbb{C}^{n \times n}$ is denoted by I_n. We let $\mathbf{0}_n$ ($\mathbf{1}_n$) be the zero vector (the vector of all ones) in $\mathbb{C}^n$. When the dimension is clear from the context, we simply write $[\mathbf{0}]$, I, $\mathbf{0}$, and $\mathbf{1}$.

We let $e_j = (0, \ldots, 0, 1, 0, \ldots, 0)$ be the characteristic vector having 1 in its jth component and 0 elsewhere. Given a vector $\varphi \in \mathbb{C}^n$, we denote by $(\varphi)_j \in \mathbb{C}$ its jth component.

Given a matrix $M \in \mathbb{C}^{n \times m}$, we let M_{ij} denote its (i, j)th entry. The *transpose* of M is the matrix $M^T \in \mathbb{C}^{m \times n}$ satisfying $M^T{}_{ij} = M_{ji}$, while we let M^* be the matrix satisfying $M^*{}_{ij} = (M_{ij})^*$. The *adjoint* of M is the matrix $M^\dagger = (M^T)^*$. For matrices $A, B \in \mathbb{C}^{n \times m}$, their *sum* is the $n \times m$ matrix $(A + B)_{ij} = A_{ij} + B_{ij}$. For matrices $C \in \mathbb{C}^{n \times m}$ and $D \in \mathbb{C}^{m \times r}$, their *product* is the $n \times r$ matrix $(CD)_{ij} = \sum_{k=1}^m C_{ik} D_{kj}$. For matrices $A \in \mathbb{C}^{n \times m}$ and $B \in \mathbb{C}^{p \times q}$, their *direct sum* and *Kronecker (or tensor or direct) product* are the $(n + p) \times (m + q)$ and $np \times mq$ matrices defined, respectively, as

$$A \oplus B = \begin{pmatrix} A & [0] \\ [0] & B \end{pmatrix}, \qquad A \otimes B = \begin{pmatrix} A_{11}B & \cdots & A_{1m}B \\ \vdots & \ddots & \vdots \\ A_{n1}B & \cdots & A_{nm}B \end{pmatrix}.$$

When operations can be performed, we have that $(A \otimes B) \cdot (C \otimes D) = AC \otimes BD$ and $(A \oplus B) \cdot (C \oplus D) = AC \oplus BD$. For vectors $\varphi \in \mathbb{C}^n$ and $\psi \in \mathbb{C}^m$, their *direct sum* is the vector $\varphi \oplus \psi = (\varphi_1, \ldots, \varphi_n, \psi_1, \ldots, \psi_m) \in \mathbb{C}^{n+m}$.

A *Hilbert space* of dimension n is the linear space $\mathbb{C}^n$ of n-dimensional complex row vectors equipped with sum and product by elements in $\mathbb{C}$, in which the *inner product* $\langle \varphi, \psi \rangle = \varphi \psi^\dagger$ is defined, for $\varphi, \psi \in \mathbb{C}^n$. The *norm* of a vector $\varphi \in \mathbb{C}^n$ is given by $\|\varphi\| = \sqrt{\langle \varphi, \varphi \rangle}$. If $\langle \varphi, \psi \rangle = 0$ (and $\|\varphi\| = 1 = \|\psi\|$), than φ and ψ are *orthogonal (orthonormal)*. Two subspaces $X, Y \subseteq \mathbb{C}^n$ are orthogonal if any vector in X is orthogonal to any vector in Y. In this case, we denote by $X \dotplus Y$ the linear space generated by $X \cup Y$. For vectors φ and ψ, $\|\varphi \otimes \psi\| = \|\varphi\| \cdot \|\psi\|$.

A matrix $M \in \mathbb{C}^{n \times n}$ is said to be *unitary* whenever $MM^\dagger = I = M^\dagger M$. Equivalently, M is unitary if and only if it preserves the norm, i.e., $\|\varphi M\| = \|\varphi\|$ for any $\varphi \in \mathbb{C}^n$. It is easy to see that, given two unitary matrices A and B, the matrices $A \oplus B$, $A \otimes B$, and AB are unitary as well.

A matrix $H \in \mathbb{C}^{n \times n}$ is said to be *Hermitian (or self-adjoint)* whenever $H = H^\dagger$. A matrix $P \in \mathbb{C}^{n \times n}$ is a *projector* if and only if P is Hermitian and idempotent, i.e., $P^2 = P$. Given the Hermitian matrix H, let $c_1, \ldots, c_s$ be its eigenvalues and $E_1, \ldots, E_s$ the corresponding eigenspaces. It is well known that each eigenvalue c_k is real, that E_i is orthogonal to E_j for $i \neq j$, and that $E_1 \dotplus \cdots \dotplus E_s = \mathbb{C}^n$. Thus, every vector $\varphi \in \mathbb{C}^n$ can be uniquely decomposed as $\varphi = \varphi_1 + \cdots + \varphi_s$ for unique $\varphi_j \in E_j$. The linear transformation $\varphi \mapsto \varphi_j$ is the projector $P(c_j)$ onto the subspace E_j. Actually, the Hermitian matrix H is biunivocally determined by its eigenvalues and projectors as $H = \sum_{i=1}^s c_i P(c_i)$. We note that $\{P(c_1), \ldots, P(c_s)\}$ is a complete set of mutually orthogonal projectors, i.e., $\sum_{i=1}^s P(c_i) = I$ and $P(c_i)P(c_j)^\dagger = [0]$ for $i \neq j$. For the Hermitian matrix $H = \sum_{i=1}^s c_i P(c_i)$, we define the *circulant matrix* built on $P(c_1), \ldots, P(c_s)$ as

$$\Xi(H) = \begin{pmatrix} P(c_1) & P(c_2) & \cdots & P(c_s) \\ P(c_2) & P(c_3) & \cdots & P(c_1) \\ \vdots & \vdots & \ddots & \vdots \\ P(c_s) & P(c_1) & \cdots & P(c_{s-1}) \end{pmatrix}.$$

The following lemma will be useful later:

Lemma 1. *Given a Hermitian matrix H, the matrix $\Xi(H)$ is unitary.*

2.2 Languages and Formal Power Series

We assume familiarity with basics in formal language theory (see, e.g., [12]). The set of all words (including the empty word ε) over a finite alphabet Σ is denoted by Σ^*. For a word $\omega \in \Sigma^*$, we let: $|\omega|$ denote its length, ω_i its ith symbol, $\omega[j] = \omega_1 \omega_2 \cdots \omega_j$ its prefix of length $0 \le j \le |\omega|$ with $\omega[0] = \varepsilon$. For any $n \ge 0$, we let $\Sigma^n = \{\omega \in \Sigma^* \mid |\omega| = n\}$.

For a language $L \subseteq \Sigma^*$ and two words $v, w \in \Sigma^*$, the *word quotient* of L with respect to v, w is the language $v^{-1}Lw^{-1} = \{x \in \Sigma^* \mid vxw \in L\}$. For two alphabets Σ, Δ, a language $L \subseteq \Delta^*$, and a homomorphism $\phi : \Sigma^* \to \Delta^*$, the *inverse homomorphic image* of L is the language $\phi^{-1}(L) = \{x \in \Sigma^* \mid \phi(x) \in L\}$. For a word $y \in \Delta^*$, we set $\phi^{-1}(y) = \{x \in \Sigma^* \mid \phi(x) = y\}$. Thus, we have $\phi^{-1}(L) = \bigcup_{y \in L} \phi^{-1}(y)$.

A *formal power series (in noncommuting variables)* with coefficients in $\mathbb{C}$ is any function $\rho : \Sigma^* \to \mathbb{C}$, usually expressed by the formal sum $\rho = \sum_{\omega \in \Sigma^*} \rho(\omega)\,\omega$. We denote by $\mathbb{C}\langle\langle \Sigma \rangle\rangle$ the set of formal power series $\rho : \Sigma^* \to \mathbb{C}$. An important subclass of $\mathbb{C}\langle\langle \Sigma \rangle\rangle$ is the class $\mathbb{C}^{\mathrm{Rat}}\langle\langle \Sigma \rangle\rangle$ of *rational series* [20].

One among possible characterizations of $\mathbb{C}^{\mathrm{Rat}}\langle\langle \Sigma \rangle\rangle$ is given by the notion of linear representation. A *linear representation of dimension m* of a formal power series $\rho \in \mathbb{C}\langle\langle \Sigma \rangle\rangle$ is a triple $(\pi, \{A(\sigma)\}_{\sigma \in \Sigma}, \eta)$, with $\pi, \eta \in \mathbb{C}^m$ and $A(\sigma) \in \mathbb{C}^{m \times m}$, such that, for any $\omega \in \Sigma^*$, we have

$$\rho(\omega) = \pi A(\omega)\eta^\dagger = \pi \left(\prod_{i=1}^{|\omega|} A(\omega_i) \right) \eta^\dagger.$$

In [21], it is shown that *a formal power series is rational if and only if it has a linear representation (of finite dimension)*.

Given a *real valued* $\rho \in \mathbb{C}\langle\langle \Sigma \rangle\rangle$ (i.e., with $\rho(\omega) \in \mathbb{R}$, for any $\omega \in \Sigma^*$) and a real cut-point λ, *the language defined by ρ with cut-point λ* is defined as the set

$$L_{\rho,\lambda} = \{\omega \in \Sigma^* \mid \rho(\omega) > \lambda\}.$$

The cut-point λ is said to be *isolated* if there exists a positive real δ such that $|\rho(\omega) - \lambda| > \delta$, for any $\omega \in \Sigma^*$.

We call *bounded series* any $\rho \in \mathbb{C}^{\mathrm{Rat}}\langle\langle \Sigma \rangle\rangle$ admitting a linear representation $(\pi, \{A(\sigma)\}_{\sigma \in \Sigma}, \eta)$ such that $\|\pi A(\omega)\| \le K$, for a fixed positive constant K and every $\omega \in \Sigma^*$. In [3], it is proved the following

Theorem 1. *Let $\rho \in \mathbb{C}^{\mathrm{Rat}}\langle\langle \Sigma \rangle\rangle$ be a real valued bounded series defining the language $L_{\rho,\lambda}$ with isolated cut-point λ. Then, $L_{\rho,\lambda}$ is a regular language.*

2.3   Finite Automata

A *deterministic finite automaton* (DFA) is a 5-tuple $\mathcal{D} = \langle S, \Sigma, \tau, s_1, F \rangle$, where S is the finite set of states, Σ the finite input alphabet, $s_1 \in S$ the initial state, $F \subseteq S$ the set of accepting states, and $\tau : S \times \Sigma \to S$ is the transition function. An input word is *accepted* by $\mathcal{D}$ if the induced computation starting from the initial state ends in some accepting state after consuming the whole input. The set $L_{\mathcal{D}}$ of all words accepted by $\mathcal{D}$ is called the accepted language. A linear representation for the DFA $\mathcal{D}$ is the 3-tuple $(\alpha, \{M(\sigma)\}_{\sigma \in \Sigma}, \beta)$, where $\alpha \in \{0,1\}^{|S|}$ is the characteristic row vector of the initial state, $M(\sigma) \in \{0,1\}^{|S| \times |S|}$ is the boolean transition matrix satisfying $(M(\sigma))_{ij} = 1$ if and only if $\tau(s_i, \sigma) = s_j$, and $\beta \in \{0,1\}^{|S| \times 1}$ is the characteristic column vector of the final states. The accepted language can now be defined as $L_{\mathcal{D}} = \{\omega \in \Sigma^* \mid \alpha M(\omega)\beta = 1\}$, where we let $M(\omega) = \prod_{i=1}^{|\omega|} M(\omega_i)$.

We introduce the model of a finite automaton with quantum and classical states [23]. In what follows, we denote by $\mathcal{U}(\mathbb{C}^n)$ ($\mathcal{O}(\mathbb{C}^n)$) the set of unitary (Hermitian) matrices on $\mathbb{C}^n$. As we will see, unitary matrices describe the evolution of the quantum component of the automaton, while Hermitian matrices represent observables to be measured.

Definition 1. *A* one-way finite automaton with quantum and classical states *(QCFA) is formally defined by the 9-tuple $\mathcal{A} = \langle Q, S, \Sigma, \Upsilon, \Theta, \tau, \pi_1, s_1, F \rangle$, where:*

- *Q is the finite set of orthonormal quantum basis states for the Hilbert space $\mathbb{C}^{|Q|}$ within which the quantum states are represented as vectors of norm 1,*
- *S is the finite set of classical states,*
- *Σ is the finite input alphabet; its extension by a right endmarker symbol $\sharp \notin \Sigma$ defines the tape alphabet $\Gamma = \Sigma \cup \{\sharp\}$,*
- *$\pi_1 \in \mathbb{C}^{|Q|}$ is the initial quantum state, satisfying $\|\pi_1\| = 1$,*
- *$s_1 \in S$ is the initial classical state,*
- *$F \subseteq S$ is the set of classical accepting states,*
- *$\Upsilon : S \times \Gamma \to \mathcal{U}(\mathbb{C}^{|Q|})$ is the mapping assigning, according to the current classical state and scanned tape symbol, a unitary transformation defining the evolution of the quantum state,*
- *$\Theta : S \times \Gamma \to \mathcal{O}(\mathbb{C}^{|Q|})$ is the mapping assigning, according to the current classical state and scanned tape symbol, a Hermitian matrix defining the observable to be measured on the quantum state,*
- *$\tau : S \times \Gamma \times \mathcal{C} \to S$ is the mapping defining the next classical state as a function of the current classical state, scanned tape symbol, and measurement outcome from a set $\mathcal{C}$.*

When addressing the *size*, we say that the QCFA $\mathcal{A}$ in Definition 1 has $|Q|$ quantum basis states and $|S|$ classical states.

Let us now explain in details how $\mathcal{A}$ works. Given an *input word* $\omega \in \Sigma^*$, we let $w = \omega\sharp$ be the associated *tape word* to be processed by $\mathcal{A}$. At any time along the computation on w, the quantum state of $\mathcal{A}$ is represented by a vector $\pi \in \mathbb{C}^{|Q|}$ with $\|\pi\| = 1$, while its classical state is an element from S. The computation starts in the quantum state π_1, in the classical state s_1, and by scanning w_1. Then, the transformations associated with symbols in w are applied in succession. Precisely, the transformation associated with a state $s \in S$ and a tape symbol $\gamma \in \Gamma$ consists of three steps:

- FIRST: the unitary transformation $\Upsilon(s, \gamma)$ is applied to the current quantum state π, yielding the new quantum state $\pi' = \pi\Upsilon(s, \gamma)$.
- SECOND: the observable $\Theta(s, \gamma) = \sum_{i=1}^{m} c_i P(s, \gamma)(c_i)$ is measured on π', leading to one among the possible measurement outcomes from the set $\mathcal{C}(s, \gamma) = \{c_1, \ldots, c_m\}$. According to quantum mechanics principles, the outcome c_i is returned with probability $p_i = \|\pi' P(s, \gamma)(c_i)\|^2$, and correspondingly the quantum state π' collapses to the quantum state $\pi' P(s, \sigma)(c_i)/\sqrt{p_i}$.
- THIRD: the current classical state s switches to $\tau(s, \gamma, c_i)$, and the tape symbol γ is consumed.

The input word ω is *accepted* by $\mathcal{A}$ if the classical state reached after processing the right endmarker $\sharp$ of the corresponding tape word w is an accepting state, i.e., it belongs to F. Otherwise, ω is rejected. Clearly, accepting ω takes place with a certain probability we are now going to explicate.

Let $\mathcal{C} = \bigcup_{s \in S, \gamma \in \Gamma} \mathcal{C}(s, \gamma)$ be the set of measurement outcomes of all observables associated with $\mathcal{A}$. Indeed, in a standard fashion, we can define τ^* as the extension to $\bigcup_{i \geq 0}(S \times \Gamma^i \times \mathcal{C}^i)$ of the classical evolution $\tau : S \times \Gamma \times \mathcal{C} \to S$. More precisely, for any $s \in S$, $w \in \Gamma^n$, $y \in \mathcal{C}^n$, we let

$$\tau^*(s, \varepsilon, \varepsilon) = s, \text{ and}$$
$$\tau^*(s, w[j], y[j]) = \tau(\tau^*(s, w[j-1], y[j-1]), w_j, y_j) \text{ for } 1 \leq j \leq n.$$

So, for a tape word $w = \omega\sharp \in \Sigma^{n-1}\sharp$, the probability that $\mathcal{A}$ accepts the corresponding input word ω can be written as

$$\mathcal{E}_{\mathcal{A}}(\omega) = \sum_{\{y \in \mathcal{C}^n \ | \ \tau^*(s_1, w, y) \in F\}} \|\pi_1 A(w, y)\|^2, \text{ with} \qquad (1)$$

$$A(w, y) = \prod_{i=1}^{n} \Upsilon(\tau^*(s_1, w[i-1], y[i-1]), w_i) P(\tau^*(s_1, w[i-1], y[i-1]), w_i)(y_i)$$

and the convention that $P(s, \gamma)(c) = [0]$ whenever $c \notin \mathcal{C}(s, \gamma)$. We maintain this convention throughout the rest of the paper. The function $\mathcal{E}_{\mathcal{A}} : \Sigma^* \to [0, 1]$ is usually known as the *stochastic event induced by* $\mathcal{A}$. We notice that, in principle, $\mathcal{A}$ may exhibit a nonzero probability of accepting non well-formed inputs, i.e., words in $\Gamma^* \setminus \Sigma^*\sharp$. However, it is easy to see that, by augmenting the classical component with two new states, we can obtain a QCFA behaving as $\mathcal{A}$ on words in $\Sigma^*\sharp$ and rejecting with certainty words in $\Gamma^* \setminus \Sigma^*\sharp$. So, without loss of generality, throughout the rest of the paper, we will always be assuming the QCFA $\mathcal{A}$ to have this latter behavior.

We let $\rho_{\mathcal{A}} \in \mathbb{C}\langle\langle \Gamma \rangle\rangle$, the *real valued formal power series associated with* $\mathcal{A}$, be defined as $\rho_{\mathcal{A}}(\omega\sharp) = \mathcal{E}_{\mathcal{A}}(\omega)$ for every $\omega \in \Sigma^*$, and yielding 0 on words in $\Gamma^* \setminus \Sigma^*\sharp$. The *language accepted by* $\mathcal{A}$ *with cut-point* λ is defined to be the set

$$L_{\mathcal{A},\lambda} = (L_{\rho_{\mathcal{A}},\lambda})\sharp^{-1} = \{\omega \in \Sigma^* \mid \mathcal{E}_{\mathcal{A}}(\omega) > \lambda\}.$$

As for formal power series, the cut-point λ is said to be *isolated* if there exists a positive real δ such that $|\mathcal{E}_{\mathcal{A}}(\omega) - \lambda| > \delta$, for any $\omega \in \Sigma^*$. Acceptance with δ-isolated $\lambda = 1/2$ is also known in the literature as *bounded error acceptance* with error probability $1/2 - \delta$. It may be verified that, by adding one quantum basis state, isolated cut-point acceptance may be turned into bounded error acceptance.

As a final observation, we note that, for the model of QCFA in Definition 1, acceptance is determined by accepting states in the classical component. Alternatively, acceptance could be settled in the quantum component through an accepting/rejecting outcome of the measurement on $\sharp$. These two models of acceptance are actually equivalent.

3   Characterizing the Power of QCFAs

The fact that any regular language can be accepted by a QCFA comes trivially, due to the presence of the classical component (see [23] for formal details). Here, we focus on the converse, and show that the language accepted by any QCFA $\mathcal{A}$ with isolated cut-point is regular. To this aim, we prove that the associated formal power series $\rho_{\mathcal{A}}$ is bounded rational, and so we can apply Theorem 1. This direct approach also enables us to state a size lower bound for QCFAs accepting regular languages with isolated cut-point.

Consider a QCFA $\mathcal{A} = \langle Q, S = \{s_1, \ldots, s_k\}, \Sigma, \Upsilon, \Theta, \tau, \pi_1, s_1, F \rangle$, with q quantum basis states, k classical states, and $\mathcal{C} = \bigcup_{s \in S, \gamma \in \Gamma} \mathcal{C}(s, \gamma)$ the set of all possible measurement outcomes. We let the linear representation of the classical component be the 3-tuple $\langle \alpha, \{T(\gamma, c)\}_{\gamma \in \Gamma, c \in \mathcal{C}}, \beta \rangle$, where $\alpha = e_1 \in \{0,1\}^k$ is the characteristic vector of the initial state s_1, $\beta \in \{0,1\}^k$ is the characteristic vector of the set F of accepting states, and $T(\gamma, c) = \sum_{i=1}^{k} e_i^T \otimes e_{\text{next}(i)}$, with $\text{next}(i) = j \Leftrightarrow s_j = \tau(s_i, \gamma, c)$, is the $k \times k$ transition matrix on $\gamma \in \Gamma$, $c \in \mathcal{C}$ induced by τ. Moreover, we let $D(s_i, \gamma, c) = e_i^T \otimes e_{\text{next}(i)}$ be the $k \times k$ matrix $T(\gamma, c)$ "restricted" to the ith row.

We let the 3-tuple $\text{Li}(\mathcal{A}) = \langle \varphi_1, \{M(\gamma)\}_{\gamma \in \Gamma}, \eta \rangle$, with $\varphi_1 \in \mathbb{C}^{q^2 k}$, $\eta \in \{0,1\}^{q^2 k}$, and $M(\gamma) \in \mathbb{C}^{q^2 k \times q^2 k}$, be defined as:

- $\varphi_1 = \alpha \otimes \pi_1 \otimes \pi_1^*$,
- $M(\gamma) = \sum_{s \in S, c \in \mathcal{C}} D(s, \gamma, c) \otimes \Upsilon(s, \gamma)P(s, \gamma)(c) \otimes \Upsilon^*(s, \gamma)P^*(s, \gamma)(c)$,
- $\eta = \sum_{j=1}^{q} \beta \otimes e_j \otimes e_j$.

We are going to prove that Li($\mathcal{A}$) *is a linear representation of the formal power series* $\rho_{\mathcal{A}}$, meaning that $\rho_{\mathcal{A}}$ *is rational*, as pointed out in Section 2.2.

We begin by the following lemma which, very roughly speaking, says that a state vector of Li($\mathcal{A}$) "embodies" the evolution of the classical part of $\mathcal{A}$ in its first components (namely, by the operator $T(w,y)$ below), while the others account for the dynamics of the quantum part (by the operator $A(w,y)$):

Lemma 2. *For any* $w \in \Gamma^n$ *and* $y \in \mathcal{C}^n$, *we let* $M(w) = \prod_{i=1}^{n} M(w_i)$ *and* $T(w,y) = \prod_{i=1}^{n} T(w_i, y_i)$. *Then, for any two vectors* $v_1, v_2 \in \mathbb{C}^q$, *we have*

$$(\alpha \otimes v_1 \otimes v_2^*)M(w) = \sum_{y \in \mathcal{C}^n} \alpha\, T(w,y) \otimes v_1\, A(w,y) \otimes (v_2\, A(w,y))^*.$$

This enables us to state

Theorem 2. *Given a* QCFA $\mathcal{A}$, *the associated formal power series* $\rho_{\mathcal{A}}$ *is rational.*

Proof. It suffices to show that Li($\mathcal{A}$) $= \langle \varphi_1, \{M(\gamma)\}_{\gamma \in \Gamma}, \eta \rangle$ is a linear representation for $\rho_{\mathcal{A}}$, i.e.:

$$\rho_{\mathcal{A}}(w) = \varphi_1 M(w)\, \eta^\dagger, \quad \text{for any } w \in \Gamma^n.$$

Indeed, by Lemma 2, we have

$$\varphi_1 M(w) \eta = \left(\sum_{y \in \mathcal{C}^n} \alpha\, T(w,y) \otimes \pi_1 A(w,y) \otimes (\pi_1 A(w,y))^* \right) \cdot \sum_{j=1}^{q} \beta^\dagger \otimes e_j{}^\dagger \otimes e_j{}^\dagger$$

$$= \sum_{y \in \mathcal{C}^n} \alpha\, T(w,y) \beta^\dagger \cdot \sum_{j=1}^{q} \left| (\pi_1 A(w,y))_j \right|^2$$

$$= \sum_{\{y \in \mathcal{C}^n \ \mid \ \tau^*(s_1, w, y) \in F\}} \| \pi_1 A(w,y) \|^2,$$

which, according to (1), is $\mathcal{E}_{\mathcal{A}}(\omega)$ if $w = \omega \sharp \in \Sigma^{n-1} \sharp$, and 0 otherwise. $\square$

To show boundedness of $\rho_{\mathcal{A}}$, we need a generalization of Lemma 1 in [3]:

Lemma 3. *For a given* $n \geq 0$, *let* $\{U(y[i-1]) \mid y \in \mathcal{C}^n, 1 \leq i \leq n\}$ *be a set of unitary matrices, and* $\{R(y[i-1])(y_i) \mid y \in \mathcal{C}^n, 1 \leq i \leq n\}$ *a set of matrices such that, for any* $0 \leq i \leq n-1$ *and any word* $\hat{y} \in \mathcal{C}^i$, *the nonzero matrices in the set* $\{R(\hat{y})(c) \mid c \in \mathcal{C}\}$ *define an observable (i.e., they form a complete set of mutually orthogonal projectors). Then, for any complex vector* π, *we get*

$$\sum_{y \in \mathcal{C}^n} \left\| \pi \prod_{i=1}^{n} U(y[i-1]) R(y[i-1])(y_i) \right\|^2 = \|\pi\|^2. \tag{2}$$

We are now ready to prove boundedness of the series associated with QCFAs:

Theorem 3. *Given a* QCFA $\mathcal{A}$, *the associated formal power series* $\rho_{\mathcal{A}}$ *is bounded.*

Proof. Consider the linear representation $\mathrm{Li}(\mathcal{A}) = \langle \varphi_1, \{M(\gamma)\}_{\gamma \in \Gamma}, \eta \rangle$ of $\rho_{\mathcal{A}}$. We show that, for any $w \in \Gamma^n$, we get $\|\varphi_1 M(w)\| \leq 1$. Indeed, we have

$$
\begin{aligned}
\|\varphi_1 M(w)\| &= \left\| \sum_{y \in \mathcal{C}^n} \alpha T(w,y) \otimes (\pi_1 A(w,y)) \otimes (\pi_1 A(w,y))^* \right\| && \text{(by Lemma 2)} \\
&\leq \sum_{y \in \mathcal{C}^n} \|\alpha T(w,y)\| \cdot \|\pi_1 A(w,y)\|^2 && \text{(by triangular inequality)} \\
&= \sum_{y \in \mathcal{C}^n} \|\pi_1 A(w,y)\|^2 = \|\pi_1\|^2 = 1 && \text{(by Lemma 3 on } A(w,y)\text{).}
\end{aligned}
$$

$\square$

In conclusion, we get our main result

Theorem 4. *The class of languages accepted by* QCFAs *with isolated cut-point coincides with the class of regular languages.*

Proof. As observed at the beginning of this section, QCFAs accept all regular languages. For the converse, Theorems 1, 2, and 3 ensures that, for any QCFA $\mathcal{A}$ and any isolated cut-point λ, the language $L_{\rho_A, \lambda}$ is regular. This, together with the fact that regular languages are closed under word quotient, clearly implies that $L_{\mathcal{A}, \lambda} = (L_{\rho_A, \lambda}) \sharp^{-1}$ is regular. $\square$

A natural question arising from Theorem 4 is the size-cost of converting a given QCFA $\mathcal{A}$ into a language-equivalent DFA. Starting from the linear representation $\mathrm{Li}(\mathcal{A})$ which has dimension $q^2 k$, we can apply the Rabin-like technique presented in [7] to get an equivalent DFA whose number of states is bounded as:

Theorem 5. *For any* QCFA $\mathcal{A}$ *with q quantum basis states, k classical states, and δ-isolated cut-point λ, there exists a m-state* DFA *accepting $L_{\mathcal{A}, \lambda}$, with*

$$
m \leq \left(1 + \frac{4\sqrt{qk}}{\delta} \right)^{q^2 k}.
$$

We quickly point out that this result can be used "the other way around", to get a size lower bound for QCFAs accepting regular languages, namely: any QCFA with q quantum states, k classical states, and δ-isolated cut point accepting a regular language whose minimal DFA has μ states, must satisfy

$$
qk \geq \left(\frac{\log(\mu)}{\log(5/\delta)} \right)^{\frac{4}{9}}.
$$

The optimality of such lower bound is an open problem. As a partial answer, we can immediately state that the optimal lower bound cannot be raised to $\omega(\log(\mu))$, since an asymptotically optimal lower bound of $\log(\mu)/(2\log(1+2/\delta))$ is obtained in [5] for measure-once quantum automata, which is easily simulated by QCFAs with the same number of quantum basis states and 3 classical states [23].

4   Size-Cost of Language Operations on QCFAs

By the characterization in the previous section, we immediately get that the class of languages accepted by QCFAs with isolated cut-point is closed under word quotients and inverse homomorphic images. Here, we are going to explicitly construct QCFAs that accept word quotients and inverse homomorphic images of regular languages defined by QCFAs. This allows us to study the cost, in terms of quantum basis states and classical states, of implementing such operations on QCFAs.

It is well known that on DFAs both word quotients and inverse homomorphisms can be easily implemented without increasing the number of states. Here, we perform such operations on QCFAs by polynomially increasing the size and preserving cut-point and isolation.

We begin by approaching the construction of QCFAs for word quotients. We construct QCFAs for accepting $\sigma^{-1}L$ and $L\sigma^{-1}$, for given $\sigma \in \Sigma$ and a language $L \subseteq \Sigma^*$ accepted by a QCFA with isolated cut-point. By iterating these constructions, one obtains a QCFA for $v^{-1}Lw^{-1}$, for given $v, w \in \Sigma^*$.

Theorem 6. *Let $L \subseteq \Sigma^*$ be a language accepted with δ-isolated cut-point λ by a QCFA A with q quantum basis states and k classical states. Then, for any given $\sigma_0 \in \Sigma$, there exists a QCFA B with at most q^2 quantum basis states and $k+1$ classical states that accepts $\sigma_0^{-1}L$ with δ-isolated cut-point λ.*

Proof. Let the QCFA $A = \langle Q, S, \Sigma, \Upsilon, \Theta, \tau, \pi_0, s_0, F \rangle$. To avoid too heavy technicalities, we assume that all observables associated with A exhibit the same set $\mathcal{C} = \{c_0, \ldots, c_{h-1}\}$ of outcomes. So, for any $s \in S$ and $\sigma \in \Sigma \cup \{\sharp\}$, we have $\Theta(s, \sigma) = \sum_{j=0}^{h-1} c_j P(s, \sigma)(c_j)$. However, our technique can be easily adapted to the general case.

We construct the QCFA $B = \langle \hat{Q}, S \cup \{\hat{s}_0\}, \Sigma, \hat{\Upsilon}, \hat{\Theta}, \hat{\tau}, \hat{\pi}_0, \hat{s}_0, F \rangle$ such that:

- $\hat{Q} = \{e_j \otimes \pi \mid \pi \in Q, e_j \in \mathbb{C}^h, 1 \leq j \leq h\}$,
- $\hat{\pi}_0 = \bigoplus_{j=0}^{h-1} \pi_0 \Upsilon(s_0, \sigma_0) P(s_0, \sigma_0)(c_j)$,
- for $s \in S$ and $\sigma \in \Sigma \cup \{\sharp\}$, we set $\hat{\Upsilon}(s, \sigma) = \bigoplus_{j=0}^{h-1} \Upsilon(s, \sigma)$, and $\hat{\Upsilon}(\hat{s}_0, \sigma) = \bigoplus_{j=0}^{h-1} \Upsilon(\tau(s_0, \sigma_0, c_j), \sigma)$,
- for $s \in S \cup \{\hat{s}_0\}$ and $\sigma \in \Sigma \cup \{\sharp\}$, we set $\hat{\Theta}(s, \sigma) = \sum_{j=0}^{h-1} \sum_{i=0}^{h-1} \hat{c}_{i,j} \hat{P}(s, \sigma)(\hat{c}_{i,j})$, with $\hat{P}(s, \sigma)(\hat{c}_{i,j}) = [0]_{(j-1)q} \oplus P(s_{l_j}, \sigma)(c_i) \oplus [0]_{(h-j)q}$ and $s_{l_j} = \tau(s_0, \sigma_0, c_j)$ if $s = \hat{s}_0$, otherwise $s_{l_j} = s$. We let $\hat{\mathcal{C}} = \{\hat{c}_{i,j} \mid 0 \leq i, j \leq h - 1\}$ be the set of the outcomes of all observables associated with B,
- for $s \in S$ and $\sigma \in \Sigma \cup \{\sharp\}$, we set $\hat{\tau}(s, \sigma, \hat{c}_{i,j}) = \tau(s, \sigma, c_i)$, and $\hat{\tau}(\hat{s}_0, \sigma, \hat{c}_{i,j}) = \tau^*(s_0, \sigma_0 \sigma, c_j c_i)$.

We describe intuitively how the QCFA B on input $w\sharp$ mimics the computation of A on input $\sigma_0 w \sharp$. The initial quantum state $\hat{\pi}_0$ consists of h blocks. Each one represents the unitary evolution of A on σ_0 from states π_0 and s_0, followed by one among the h projections associated with the observable $\Theta(s_0, \sigma_0)$. Upon reading the first input symbol, B implements in the jth block the evolution in A associated with the classical state $\tau(s_0, \sigma_0, c_j)$ and symbol ω_1, followed by a measurement yielding the result $\hat{c}_{i,j}$. Such a measurement simulates the outcome sequence $c_j c_i$ possibly obtained in A while processing the input prefix $\sigma_0 \omega_1$. From ω_2 on, the computation of A is simulated in the jth block, in which an outcome $\hat{c}_{i,j}$ corresponds to the outcome c_i in A. One may verify that the probability that B accepts ω coincides with the probability that A accepts $\sigma_0 \omega$. Clearly, B has $k+1$ classical states and $hq \leq q^2$ quantum basis states.   □

Theorem 7. *Let $L \subseteq \Sigma^*$ be a language accepted with δ-isolated cut-point λ by a QCFA A with q quantum basis states and k classical states. Then, for any given $\sigma_0 \in \Sigma$, there exists a QCFA B with at most q^2 quantum basis states and k classical states that accepts $L\sigma_0^{-1}$ with δ-isolated cut-point λ.*

Proof. Let the QCFA $A = \langle Q, S, \Sigma, \Upsilon, \Theta, \tau, \pi_0, s_0, F \rangle$. As in the previous proof, all the observables of A are assumed of the form $\Theta(s, \sigma) = \sum_{j=0}^{h-1} c_j P(s, \sigma)(c_j)$.

We construct the QCFA $B = \langle \hat{Q}, S, \Sigma, \hat{\Upsilon}, \hat{\Theta}, \hat{\tau}, \hat{\pi}_0, s_0, F \rangle$ such that:

- $\hat{Q} = \{e_j \otimes \pi \mid \pi \in Q, \, e_j \in \mathbb{C}^h, 1 \leq j \leq h\}$,
- $\hat{\pi}_0 = \pi_0 \oplus \mathbf{0}_{q(h-1)}$,
- for $s \in S$ and $\sigma \in \Sigma$, we set $\hat{\Upsilon}(s, \sigma) = \Upsilon(s, \sigma) \oplus I_{q(h-1)}$, and $\hat{\Upsilon}(s, \sharp) = \left(\bigoplus_{j=0}^{h-1} \Upsilon(s, \sigma_0) \right) \cdot \Xi(\Theta(s, \sigma_0)) \cdot \left(\bigoplus_{j=0}^{h-1} \Upsilon(\tau(s, \sigma_0, c_j), \sharp) \right)$, where $\Xi(\Theta(s, \sigma_0))$ is the unitary circulant matrix addressed in Lemma 1.
- for $s \in S$ and $\sigma \in \Sigma \cup \{\sharp\}$, we set $\hat{\Theta}(s, \sigma) = \sum_{j=0}^{h-1} \sum_{i=0}^{h-1} \hat{c}_{i,j} \hat{P}(s, \sigma)(\hat{c}_{i,j})$, with $\hat{P}(s, \sigma)(\hat{c}_{i,j}) = [\mathbf{0}]_{(j-1)q} \oplus P(s_{l_j}, \sigma)(c_i) \oplus [\mathbf{0}]_{(h-j)q}$ and $s_{l_j} = \tau(s, \sigma_0, c_j)$ if $\sigma = \sharp$, otherwise $s_{l_j} = s$. We let $\hat{C} = \{\hat{c}_{i,j} \mid 0 \leq i, j \leq h-1\}$ be the set of the outcomes of all observables associated with B,
- for $s \in S$ and $\sigma \in \Sigma$, we set $\hat{\tau}(s, \sigma, \hat{c}_{i,j}) = \tau(s, \sigma, c_i)$ and $\hat{\tau}(s, \sharp, \hat{c}_{i,j}) = \tau^*(s, \sigma_0 \sharp, c_j c_i)$.

The initial quantum state $\hat{\pi}_0$ consists of h blocks, all being zero blocks except the first being π_0. On the symbols of the tape word $\omega \sharp$ preceding the endmarker, B implements in the first block the same computation as A, leading to a state vector $\pi' \oplus \mathbf{0}_{q(h-1)}$. Upon reading $\sharp$, the application of the operator $\hat{\Upsilon}(s, \sharp)$ has the effect of storing the vector $\pi' \Upsilon(s, \sigma_0) P(s, \sigma_0)(c_j) \Upsilon(\tau(s, \sigma_0, c_j), \sharp)$ in the jth block. Moreover, the outcome $\hat{c}_{i,j}$ of the measurement on $\sharp$ in B corresponds to the outcome sequence $c_j c_i$ possibly obtained in A while processing the input suffix $\sigma_0 \sharp$. Clearly, the probability that B accepts ω coincides with the probability that A accepts $\omega \sigma_0$. The number of classical states in B remains k, while the number of quantum states is $hq \leq q^2$.   □

Let us now focus on constructing QCFAs for inverse homomorphic images. We recall that a homomorphism $\phi : \Sigma^* \to \Delta^*$ of a free monoid into another is entirely defined by the image of each symbol in Σ.

Theorem 8. *Let $L \subseteq \Sigma^*$ be a language accepted with δ-isolated cut-point λ by a QCFA A with q quantum basis states and k classical states. Then, for any given* *homomorphism $\phi : \Sigma \to \Delta^*$, with $m = \max\{|\phi(\sigma)| \mid \sigma \in \Sigma\}$, there exists a* QCFA *$B$ with at most q^{m+1} quantum basis states and $q^m k$ classical states that accepts $\phi^{-1}(L)$ with δ-isolated cut-point λ.*

Proof. For reader's ease of mind, we exhibit our construction for a homomorphism $\phi : \{a, b\} \to \{\alpha, \beta\}^*$ defined as $\phi(a) = \alpha\beta$ and $\phi(b) = \beta$, so that $m = 2$. Yet, we consider the language L to be accepted by a QCFA A with binary observables. These assumptions do not substantially affect the generality of our construction. So, let the QCFA $A = \langle Q, S, \{\alpha, \beta\}, \Upsilon, \Theta, \tau, \pi_0, s_0, F \rangle$, where all observables are assumed to have the form $\Theta(s, \sigma) = 0 \cdot P(s, \sigma)(0) + 1 \cdot P(s, \sigma)(1)$ and hence with $\mathcal{C} = \{0, 1\}$ as set of outcomes.

We construct the QCFA $B = \langle \hat{Q}, \hat{S}, \{a, b\}, \hat{\Upsilon}, \hat{\Theta}, \hat{\tau}, \hat{\pi}_0, (s_0, 0), \hat{F} \rangle$ such that:

- $\hat{Q} = \{e_j \otimes \pi \mid \pi \in Q, \ e_j \in \mathbb{C}^4, 1 \leq j \leq 4\}$,
- $\hat{S} = \{(s, j) \mid s \in S, \ 0 \leq j \leq 3\}$,
- $\hat{\pi}_0 = \pi_0 \oplus \mathbf{0}_{3q}$,

- for $(s, 0) \in \hat{S}$, we set $\hat{\Upsilon}((s, 0), a) = \begin{pmatrix} A_0 & A_1 \\ A_1 & A_0 \end{pmatrix} \cdot \begin{pmatrix} B_0 & [\mathbf{0}] \\ [\mathbf{0}] & B_1 \end{pmatrix}$, where

$$A_i = \Upsilon(s, \alpha)P(s, \alpha)(i) \oplus \Upsilon(s, \alpha)P(s, \alpha)(i),$$

$$B_i = \begin{pmatrix} \Upsilon(\tau(s, \alpha, i), \beta)P(\tau(s, \alpha, i), \beta)(0) & \Upsilon(\tau(s, \alpha, i), \beta)P(\tau(s, \alpha, i), \beta)(1) \\ \Upsilon(\tau(s, \alpha, i), \beta)P(\tau(s, \alpha, i), \beta)(1) & \Upsilon(\tau(s, \alpha, i), \beta)P(\tau(s, \alpha, i), \beta)(0) \end{pmatrix},$$

and $\hat{\Upsilon}((s, 0), b) = C \oplus C$, where

$$C = \begin{pmatrix} \Upsilon(s, \beta)P(s, \beta)(0) & \Upsilon(s, \beta)P(s, \beta)(1) \\ \Upsilon(s, \beta)P(s, \beta)(1) & \Upsilon(s, \beta)P(s, \beta)(0) \end{pmatrix};$$

for $(s, j) \subset \hat{S}$ with $j \neq 0$, we set

$$\hat{\Upsilon}((s, j), a) = \Pi^j \cdot \hat{\Upsilon}((s, 0), a), \quad \hat{\Upsilon}((s, j), b) = \Pi^j \cdot \hat{\Upsilon}((s, 0), b),$$

where $\Pi = \begin{pmatrix} 0 & 0 & 0 & 1 \\ 1 & 0 & 0 & 0 \\ 0 & 1 & 0 & 0 \\ 0 & 0 & 1 & 0 \end{pmatrix} \otimes I_q$ is the circular block permutation matrix,

- for $(s, j) \in \hat{S}$ and $\sigma \in \{a, b\}$, we set $\hat{\Theta}((s, j), \sigma) = \sum_{i=0}^{3} c_i \cdot [\mathbf{0}]_{iq} \oplus I_q \oplus [\mathbf{0}]_{(3-i)q}$,
- for $(s, j) \in \hat{S}$ and $0 \leq i \leq 3$, we set $\hat{\tau}((s, j), a, c_i) = (\tau^*(s, \alpha\beta, \text{bin}_2(i)), i)$, and $\hat{\tau}((s, j), b, c_i) = (\tau^*(s, \beta, \text{bin}_1(i)), i)$, where $\text{bin}_2(i)$ is the binary representation of i on 2 bits, while $\text{bin}_1(i) = 0^{-1}\text{bin}_2(i) \cup 1^{-1}\text{bin}_2(i)$,
- $\hat{F} = \{(s, j) \in \hat{S} \mid s \in F\}$.

The evolution matrices of the QCFA B can be regarded as block matrices with blocks of dimension $q \times q$. For $0 \leq i, j \leq 3$, the (i, j)th block of $\hat{\Upsilon}((s, i), a)$ is $\Upsilon(s, \alpha)P(s, \alpha)(j_1)\Upsilon(\tau(s, \alpha, j_1), \beta)P(\tau(s, \alpha, j_1), \beta)(j_2)$ with $j_1 j_2 = \text{bin}_2(j)$, while the (i, j)th block of $\hat{\Upsilon}((s, i), b)$ is $\Upsilon(s, \alpha)P(s, \alpha)(j)$ for $j = 0, 1$, and is $[\mathbf{0}]_q$ for $j = 2, 3$. Analogously, $\hat{\pi}_0$ consists of 4 blocks, all being zero blocks except the first being π_0. On reading a (b), the evolution matrix in B simulates the sequence of evolutions and measurements of A while processing $\alpha\beta$ (β), and stores each possible resulting quantum state in each block. Then, the observable acts on the jth block, and the outcome c_j represents the outcome sequence $\text{bin}_2(j)$ $(\text{bin}_1(j)$; notice that the possible outcomes of the measurements on b are only c_0 and c_1) in A. At any time, only one block of the quantum state of B is nonzero. This information is encoded in the classical state so that the evolution matrix in B selected by the classical state always stores in the jth block the result of the simulation of A for the outcome sequence $\text{bin}_2(j)$ $(\text{bin}_1(j))$. The function $\hat{\tau}$ mimics the transition function τ in the state first component, and stores in the second component the index of the nonzero block of the quantum state of B. One may verify that the probability that B accepts ω coincides with the probability that A accepts $\phi^{-1}(\omega)$. The number of classical states is $2^2 k = |\mathcal{C}|^m k \le q^m k$, while the number of quantum basis states is $2^2 q = |\mathcal{C}|^m q \le q^{m+1}$.   □

References

1. Ambainis, A., Beaudry, M., Golovkins, M., Kikusts, A., Mercer, M., Thérien, D.: Algebraic results on quantum automata. Theory of Comp. Sys. 39, 165–188 (2006)
2. Ambainis, A., Watrous, J.: Two-way finite automata with quantum and classical states. Theoretical Computer Science 287, 299–311 (2002)
3. Bertoni, A., Mereghetti, C., Palano, B.: Quantum computing: 1-way quantum automata. In: Ésik, Z., Fülöp, Z. (eds.) DLT 2003. LNCS, vol. 2710, pp. 1–20. Springer, Heidelberg (2003)
4. Bertoni, A., Mereghetti, C., Palano, B.: Small size quantum automata recognizing some regular languages. Theoretical Computer Science 340, 394–407 (2005)
5. Bertoni, A., Mereghetti, C., Palano, B.: Some formal tools for analyzing quantum automata. Theoretical Computer Science 356, 14–25 (2006)
6. Bertoni, A., Mereghetti, C., Palano, B.: Trace monoids with idempotent generators and measure-only quantum automata. Natural Computing 9, 383–395 (2010)
7. Bianchi, M.P., Mereghetti, C., Palano, B.: Size Lower Bounds for Quantum Automata. In: Mauri, G., Dennunzio, A., Manzoni, L., Porreca, A.E. (eds.) UCNC 2013. LNCS, vol. 7956, pp. 19–30. Springer, Heidelberg (2013)
8. Bianchi, M.P., Palano, B.: Behaviours of unary quantum automata. Fundamenta Informaticae 104, 1–15 (2010)
9. Brodsky, A., Pippenger, N.: Characterizations of 1-way quantum finite automata. SIAM J. Computing 5, 1456–1478 (2002)
10. Golovkins, M., Kravtsev, M.: Probabilistic reversible automata and quantum automata. In: Ibarra, O.H., Zhang, L. (eds.) COCOON 2002. LNCS, vol. 2387, pp. 574–583. Springer, Heidelberg (2002)
11. Hirvensalo, M.: Quantum automata with open time evolution. Int. J. Natural Computing Research 1, 70–85 (2010)
12. Hopcroft, J.E., Motwani, R., Ullman, J.D.: Introduction to Automata Theory, Languages, and Computation. Addison-Wesley, Reading (2001)
13. Kondacs, A., Watrous, J.: On the power of quantum finite state automata. In: Proc. 38th Symposium on Foundations of Computer Science (FOCS 1997), pp. 66–75 (1997)

14. Li, L., Qiu, D., Zou, X., Li, L., Wu, L., Mateus, P.: Characterizations of one-way general quantum finite automata. Theoretical Computer Science 419, 73–91 (2012)
15. Mercer, M.: Lower bounds for generalized quantum finite automata. In: Martín-Vide, C., Otto, F., Fernau, H. (eds.) LATA 2008. LNCS, vol. 5196, pp. 373–384. Springer, Heidelberg (2008)
16. Moore, C., Crutchfield, J.: Quantum automata and quantum grammars. Theoretical Computer Science 237, 275–306 (2000)
17. Mereghetti, C., Palano, B.: Quantum finite automata with control language. Theoretical Informatics and Applications 40, 315–332 (2006)
18. Mereghetti, C., Palano, B.: Quantum automata for some multiperiodic languages. Theoretical Computer Science 387, 177–186 (2007)
19. Nayak, A.: Optimal lower bounds for quantum automata and random access codes. In: Proc. 40th Symp. Found. Comp. Sci (FOCS 1999), pp. 369–376 (1999)
20. Salomaa, A., Soittola, M.: Automata theoretic aspects of formal power series. In: Texts and Monographs in Computer Science. Springer (1978)
21. Schützenberger, M.P.: On the definition of a family of automata. Information and Control 4, 245–270 (1961)
22. Shilov, G.: Linear Algebra. Prentice Hall (1971); Reprinted by Dover (1977)
23. Zheng, S., Qiu, D., Li, L., Gruska, J.: One-Way finite automata with quantum and classical states. In: Bordihn, H., Kutrib, M., Truthe, B. (eds.) Dassow Festschrift 2012. LNCS, vol. 7300, pp. 273–290. Springer, Heidelberg (2012)

On Comparing Deterministic Finite Automata and the Shuffle of Words*

Franziska Biegler[1] and Ian McQuillan[2]

[1] Machine Learning Group, TU Berlin
Germany
`franziska.biegler@tu-berlin.de`
[2] Department of Computer Science, University of Saskatchewan
Saskatoon, SK S7N 5A9, Canada
`mcquillan@cs.usask.ca`

Abstract. We continue the study of the shuffle of individual words, and the problem of decomposing a finite automaton into the shuffle on words. There is a known polynomial time algorithm to decide whether the shuffle of two words is a subset of the language accepted by a deterministic finite automaton [5]. In this paper, we consider the converse problem of determining whether or not the language accepted by a deterministic finite automaton is a subset of the shuffle of two words. We provide a polynomial time algorithm to decide whether the language accepted by a deterministic finite automaton is a subset of the shuffle of two words, for the special case when the skeletons of the two words are of fixed length. Therefore, for this special case, we can decide equality in polynomial time as well. However, we then show that this problem is coNP-Complete in general, as conjectured in [2].

1 Introduction

The shuffle operation (denoted by $\sqcup\!\sqcup$ here) on words describes the set of all words that can be obtained by interleaving the letters of the operands in all possible ways, such that the order of the letters of each operand is preserved (the operation can then be extended to languages). There have been a number of theoretical results and algorithms involving shuffle such as [10] which showed that the so-called *shuffle languages* obtained from finite languages via union, concatenation, Kleene star, shuffle and shuffle closure, are in P. In [12], it is shown that given a word w, and n other words, it is NP-Complete to decide if w is in the shuffle of the n words.

Despite the length of time since the operator was introduced [7], there remains a number of standard formal language theoretic questions involving shuffle that are unsolved. For example, there is a long-standing open problem as to whether it is decidable to decompose an arbitrary regular language into the shuffle of

---

* Research supported, in part, by the Natural Sciences and Engineering Research Council of Canada.

M. Holzer and M. Kutrib (Eds.): CIAA 2014, LNCS 8587, pp. 98–109, 2014.

two languages. Certain special cases are known to be decidable however, such as for commutative regular languages and locally testable languages, while it is undecidable for context-free languages [6].

Indeed, even the special case of the shuffle of individual words, rather than sets of words, has received considerable attention but there remains a number of yet unsolved problems. In [1], it is shown that the shuffle of individual words (with at least two letters) has a unique shuffle decomposition over words. That result was extended in [3] to show that the shuffle of two words (each with at least two letters) has a unique shuffle decomposition over arbitrary sets.

However, the complexity of taking a language as input, and determining if it has a decomposition into the shuffle of two words, remains an open question (which also depends on the method that the language uses as input). Despite this, in [5], it is shown that if a language accepted by a deterministic finite automaton (DFA) M has a decomposition into words, there is an algorithm that finds the unique decomposition into words in time linear in the lengths of the words (sublinear in the size of the automaton). However, if the input automaton is not decomposable, the algorithm cannot always determine that it is not decomposable, but will instead in those cases output two strings u and v, despite $L(M)$ not having any shuffle decomposition. As the algorithm does not have knowledge regarding whether $L(M)$ has a decomposition, one could take the output strings u and v, and test if their shuffle is equal to $L(M)$, thus testing whether $L(M)$ was itself decomposable. One way to do this would be to construct a DFA accepting $u \sqcup\!\sqcup v$, and test equality with $L(M)$, however it was shown in [4] that the size of minimal DFAs accepting the shuffle of two strings can grow exponentially in the length of the strings. Therefore, it still remains an open problem as to whether there is a polynomial time algorithm to test if the language accepted by a DFA has a decomposition into the shuffle of words.

Here, we are interested in testing inclusion between the language accepted by a DFA and the shuffle of two words. One direction of this problem, testing whether the shuffle of two strings is contained in the language accepted by a DFA has a known polynomial time algorithm [5,2]. In this paper, we investigate the complexity of the converse of this problem. It is shown that given a DFA M and words $u, v \in \Sigma^+$, the problem of deciding whether or not $L(M) \subseteq u \sqcup\!\sqcup v$ is coNP-Complete, as conjectured in [2]. However, for the special case of the problem on words u, v with fixed-length skeletons (the length of a skeleton is the number of "lettered sections"), we provide a polynomial time algorithm. This also gives a polynomial time algorithm to decide if $L(M) = u \sqcup\!\sqcup v$ for this special case. However, the exact complexity of deciding whether $L(M) = u \sqcup\!\sqcup v$ in general remains open despite the fact that we know it takes polynomial time to check $u \sqcup\!\sqcup v \subseteq L(M)$ and it is coNP-Complete to check $L(M) \subseteq u \sqcup\!\sqcup v$.

2   Preliminaries

Let $\mathbb{N}_0$ be the set of non-negative integers. An alphabet Σ is a finite, non-empty set of letters. The set of all words over Σ is denoted by Σ^*, and this set contains the empty word, λ. The set of all non-empty words over Σ is denoted by Σ^+. For $n \in \mathbb{N}_0$, let Σ^n be all words of length n over Σ.

Let Σ be an alphabet. For a word $w \in \Sigma^*$, the length of w is denoted by $|w|$. Let $w(i)$ be the i-th letter of w, let $w[i]$ be the word which is the first i characters of w, and let $w[i,j]$ be the subword between characters i and j where these are undefined if i or j are not in $\{1, \ldots, |w|\}$, or if $j < i$. The skeleton of w is λ if $w = \lambda$, and is $a_1 a_2 \cdots a_n$ where $w = a_1^{\alpha_1} a_2^{\alpha_2} \cdots a_n^{\alpha_n}, n \geq 1, \alpha_i > 0, a_i \in \Sigma, 1 \leq i \leq n, a_j \neq a_{j+1}, 1 \leq j < n$. For example, the skeleton of $aaaaabbbabbbb$ is $abab$.

Let $u, v \in \Sigma^*$. The *shuffle* of u and v is defined as $u \shuffle v = \{u_1 v_1 \cdots u_n v_n \mid u = u_1 \cdots u_n, v = v_1 \cdots v_n, u_i, v_i \in \Sigma^*, 1 \leq i \leq n\}$. For example, $aab \shuffle ba = \{aabba, aabab, ababa, abaab, baaba, baaab\}$. We say u is a *prefix* of v, written $u \leq_p v$, if $v = ux$, for some $x \in \Sigma^*$. Let $w, x \in \Sigma^*$. The left quotient of x by w, written $w^{-1}x = x_1$ if $x = wx_1$, and undefined otherwise.

We assume the reader to be familiar with deterministic finite automata (DFAs), nondeterministic finite automata (NFAs), the subset construction commonly used to convert an NFA into an equivalent DFA, and minimal DFAs. See [13,9] for an introduction and more details on finite automata.

3   Fixed-Length Skeleton Polynomial Algorithm

The purpose of this section is to give special cases on an input DFA M and $u, v \in \Sigma^+$ whereby there is a polynomial algorithm to decide whether or not $L(M) \subseteq u \shuffle v$. In particular, the main result is that when u and v have fixed-length skeletons, there is a polynomial time algorithm. We will see in the next section that in general, this problem is coNP-Complete and therefore there likely is not a polynomial time algorithm (unless $P = coNP$).

We will start by examining the complement of the problem. That is, the problem of whether there exists some $w \in L(M)$ such that $w \notin u \shuffle v$ (or whether $L(M) \not\subseteq u \shuffle v$). If we can provide a polynomial time algorithm to solve this problem for some special cases, then we can solve the problem of $L(M) \subseteq u \shuffle v$ for those cases.

Given two words $u, v \in \Sigma^+$, there is an "obvious" NFA accepting $u \shuffle v$ with $(|u| + 1) \cdot (|v| + 1)$ states, where each state is an ordered pair representing the position within both u and v. This NFA is called the *naive NFA* and is defined formally in [4].

First notice, that if we could construct a DFA accepting $u \shuffle v$ in polynomial time, then we could build a DFA accepting $(u \shuffle v)^c$ (the complement) in polynomial time, using the standard algorithm to take the complement of a DFA (Theorem 3.2, [9]). Similarly, we could build a DFA accepting $L(M) \cap (u \shuffle v)^c$ in polynomial time, using the standard algorithm for taking the intersection of two DFAs (Theorem 3.3, [9]). Moreover we could test whether this set is empty in polynomial time (Theorem 3.7, [9]).

Proposition 1. *Let M be any DFA. Let $\mathcal{F}$ be a subset of $\Sigma^+ \times \Sigma^+$ such that there exists a polynomial f from $\mathbb{N}_0 \times \mathbb{N}_0$ to $\mathbb{N}_0$ and an algorithm A converting* *any pair $(u, v) \in \mathcal{F}$ to a DFA accepting $u \sqcup v$ in time less than or equal to $f(|u|, |v|)$. Then we can test whether or not $L(M) \subseteq u \sqcup v$ in polynomial time, for any $(u, v) \in \mathcal{F}$. Similarly for testing whether $L(M) = u \sqcup v$.*

The ability to test for equality follows from the existing polynomial time algorithm to determine if $u \sqcup v \subseteq L(M)$ [5].

In particular, the algorithm A from this proposition could simply be to construct the naive NFA for $u \sqcup v$ and then to use the standard subset construction algorithm applied to the naive NFA accepting $u \sqcup v$ for $(u, v) \in \mathcal{F}$, which could produce DFAs that are polynomial in size for certain sets $\mathcal{F}$. Therefore, if there is a subset $\mathcal{F}$ of $\Sigma^+ \times \Sigma^+$ such that the subset construction applied to the naive NFAs create DFAs that are polynomial in size, then we have a polynomial time algorithm to decide if $L(M) \subseteq u \sqcup v$, for all $(u, v) \in \mathcal{F}$.

Before establishing the main result of this section, we first need the following definition and three lemmas.

Let $u, v, w \in \Sigma^+$, where $w = a_1^{\alpha_1} a_2^{\alpha_2} \cdots a_n^{\alpha_n}, a_i \neq a_{i+1}, 1 \leq i < n, \alpha_l > 0, a_i \in \Sigma, 1 \leq l \leq n$. For each l, $0 \leq l \leq n$, let

$$g(w, u, v, l) = \{(i, j) \mid a_1^{\alpha_1} a_2^{\alpha_2} \cdots a_l^{\alpha_l} \in u[i] \sqcup v[j] \text{ and either } u(i+1) = a_{l+1} \text{ or} $$
$$v(j+1) = a_{l+1} \text{ or } (i = |u| \text{ and } j = |v|)\}.$$

Note that if $i = |u|$ then $u(i+1)$ is undefined forcing $u(i+1) = a_{l+1}$ to not be true, as with the case where $j = |v|$.

For example, if $u = aabbaabb, v = aabbaaa$ and $w = aabbaabbaabbaaa$, then $g(w, u, v, 3) = \{(6, 0), (4, 2), (2, 4)\}$.

The definition of g can be rewritten recursively as follows:

Lemma 1. *For all l, $1 \leq l \leq n$,*

$$g(w, u, v, l) = \{(h, m) \mid (i, j) \in g(w, u, v, l-1), u[i+1, h] = a_l^\gamma, v[j+1, m] = a_l^\delta$$
$$\text{for some } \gamma, \delta \geq 0, \text{ either } (h, m) = (|u|, |v|) \text{ or}$$
$$u(h+1) = a_{l+1} \text{ or } v(m+1) = a_{l+1}\}.$$

Next, we will show the following three conditions are equivalent, and then use condition 1 within the decision procedure below.

Lemma 2. *The following conditions are equivalent:*

1. *$g(w, u, v, l) \neq \emptyset$ for all l, $0 \leq l \leq n$, and $(|u|, |v|) \in g(w, u, v, n)$,*
2. *$(|u|, |v|) \in g(w, u, v, n)$,*
3. *$w \in u \sqcup v$.*

This can be seen as conditions 2 and 3 are equivalent from the definition of g. Further, condition 1 implies 2 directly. And condition 3 implies 1 as $w \in u \sqcup v$ implies that for all l, $0 \leq l < n$, $a_1^{\alpha_1} \cdots a_l^{\alpha_l} a_{l+1} \in u[i] \sqcup v[j]$ for some i, j.

Next, we see that as long as the g function is of size bounded by a constant for each $w \in L(M)$, there is a polynomial algorithm.

Lemma 3. *Let k be a constant, M a DFA and $u, v \in \Sigma^+$. If, for every $w \in$
$L(M)$ where $w = a_1^{\alpha_1} \cdots a_n^{\alpha_n}, a_i \neq a_{i+1}, 1 \leq i < n, \alpha_j > 0, a_j \in \Sigma, 1 \leq j \leq n$,
and for every $l, 0 \leq l \leq n$, the inequality $|g(w, u, v, l)| \leq k$ is true, then there is
a polynomial time algorithm for deciding whether or not $L(M) \subseteq u \sqcup v$, and for
deciding whether $L(M) = u \sqcup v$.*

Proof. We will construct a logspace bounded nondeterministic Turing machine
for the algorithm, and use the fact that NLOGSPACE is a subset of P (Corollary
to Theorem 7.4 in [11]). It will decide if $L(M) \not\subseteq u \sqcup v$; that is, if there exists
w such that $w \in L(M)$ but $w \notin u \sqcup v$. We will use condition 1 of Lemma 2
combined with Lemma 1. The Turing Machine will guess a word $w \in L(M)$,
where $w = a_1^{\alpha_1} \cdots a_n^{\alpha_n}, a_i \neq a_{i+1}, 1 \leq i < n, \alpha_l > 0, a_l \in \Sigma, 1 \leq l \leq n$, and
for every l from 0 to n, writes out the list of all (i, j) such that $a_1^{\alpha_1} \cdots a_l^{\alpha_l} \in$
$u[i] \sqcup v[j]$, where either $(i, j) = (|u|, |v|)$, or $u(i+1) = a_{l+1}$, or $v(j+1) = a_{l+1}$,
which must be of size at most k. Indeed this can be done by at first writing out
the elements of $g(w, u, v, 0)$ $((0, 0)$ if either $u(1)$ or $v(1)$ is equal to a_1, and $\emptyset$
otherwise). Then, if after writing out the list $\{(i_1, j_1), \ldots, (i_q, j_q)\}$ after reading
$a_1^{\alpha_1} \cdots a_l^{\alpha_l}, l < n$, then for $a_{l+1}^{\alpha_{l+1}}$, the new list is obtained as follows: by taking
each $(i_p, j_p), 1 \leq p \leq q$, removing it from the list and adding $(i_p + \gamma, j_p + \delta)$, where
$\gamma + \delta = \alpha_{l+1}$, such that the subword of u starting at position $i_p + 1$ is a_{p+1}^γ, and the
subword of v starting at position $j_p + 1$ is a_{p+1}^δ and either $(i_p + \gamma, j_p + \delta) = (|u|, |v|)$,
or $a_{p+1}^\gamma a_{p+2}$ is a subword of u starting at position $i_p + 1$, or $a_{p+1}^\delta a_{p+2}$ is a subword
of v starting at position $v_p + 1$. This will accurately calculate each set $g(w, u, v, l)$
by Lemma 1. If we do this for all (i_p, j_p), the resulting list must be of size less
than or equal to k after each step l by the assumption and therefore we can store
the numbers in logspace. Moreover, $(|u|, |v|)$ does not appear in the final list if
and only if $w \notin u \sqcup v$ by Lemma 2.   □

These three lemmas allow to prove the main result of this section.

Theorem 1. *Let M be a DFA, and let $u, v \in \Sigma^+$ with fixed-length skeletons.
Then, there is a polynomial time algorithm to decide whether or not $L(M) \subseteq$
$u \sqcup v$, and to decide whether $L(M) = u \sqcup v$.*

Proof. Let $u = b_1^{\beta_1} b_2^{\beta_2} \cdots b_p^{\beta_p}, b_j \neq b_{j+1}, 1 \leq j < p, \beta_i > 0, b_i \in \Sigma, 1 \leq i \leq p$, and
let $v = c_1^{\gamma_1} c_2^{\gamma_2} \cdots c_q^{\gamma_q}, c_j \neq c_{j+1}, 1 \leq j < q, \gamma_i > 0, c_i \in \Sigma, 1 \leq i \leq q$, where p and
q are fixed. Let k be the maximum of p and q. If, for all $w \in L(M)$ and every l
from 1 to the length of the skeleton of w, it is true that $|g(w, u, v, l)|$ is less than
or equal to some constant, then the result follows from Lemma 3.

Let $w = a_1^{\alpha_1} a_2^{\alpha_2} \cdots a_n^{\alpha_n}, a_j \neq a_{j+1}, 1 \leq j < n, \alpha_i > 0, a_i \in \Sigma, 1 \leq i \leq$
n. Then we can build a directed acyclic graph $G = (V, E)$ such that $V =$
$\bigcup_{0 \leq l \leq n} g(w, u, v, l)$ and

$$E = \{(x, y) \mid x = (i, j) \in g(w, u, v, l) \text{ for some } l, 0 \leq l < n, y = (h, m) \in$$
$$g(w, u, v, l+1) \text{ s. t. } a_{l+1}^{\alpha_{l+1}} \in u[i+1, h] \sqcup v[j+1, m] \text{ and either}$$
$$u(h+1) = a_{l+2} \text{ or } v(m+1) = a_{l+2} \text{ or } (h = |u| \text{ and } m = |v|)\}.$$

It is clear that $w \in u \shuffle v$ if and only if there is a path from $(0,0)$ to $(|u|, |v|)$ in G from Lemma 1 and Lemma 2. Also, every $x \in V$ is reachable from $(0,0)$ via at least one path.

Let l be such that $0 \le l < n$, and let $(i,j) \in g(w, u, v, l)$. Consider $a_{l+1}^{\alpha_{l+1}}$. Assume $a_{l+1} \ne a(i+1)$ and $a_{l+1} \ne v(j+1)$. Then there is no outgoing edge from (i,j). Assume that $a_{l+1} = u(i+1)$ but $a_{l+1} \ne v(j+1)$. If $u[i+1, i+\alpha_{l+1}] = a_{l+1}^{\alpha_{l+1}}$ and either $u(i + \alpha_{l+1} + 1) = a_{l+2}$ or $v(j + 1) = a_{l+2}$ or $(i + \alpha_{l+1} = |u|$ and $j = |v|)$, then (i,j) has one outgoing edge. Otherwise (i,j) has no outgoing edges. Similarly if $a_{l+1} \ne u(i + 1)$ and $v_{l+1} = v(j + 1)$. Then the only way of having more than one outgoing edge from (i,j) is if $a_{l+1} = u(i + 1) = v(j + 1)$. But in this situation, from (i,j), there are at most two outgoing edges since all copies of a_{l+1} must be consumed from either u or v by the definition of g.

Let $(i_0, j_0) = (0,0), (i_1, j_1), \dots, (i_x, j_x)$ (with $x \le n$ and potentially $x = n$) be a sequence of vertices such that e_α connects (i_α, j_α) to $(i_{\alpha+1}, j_{\alpha+1})$ for all α, $0 \le \alpha < x$. Let $(p_0, q_0), \dots, (p_m, q_m)$ be the subsequence of $(i_0, j_0), (i_1, j_1), \dots, (i_x, j_x)$ such that there are two outgoing edges in G from $(p_\alpha, q_\alpha), 0 \le \alpha \le m$. This list is of size at most $2k$ since each one consumes one section of the skeleton of u or v. Further, it can be shown (omitted for reasons of space) that, if every such path has at most $2k$ branching points, the number of elements in $g(w, u, v, l)$ is at most $2^{2k+1} - 1$ for every l, which is a constant since k is a constant. □

4 General coNP-Completeness

The purpose of this chapter is to show that given a DFA M and words $u, v \in \Sigma^+$, the problem of determining whether or not $L(M) \subseteq u \shuffle v$ is coNP-Complete.

To show in general (for any u, v and M), this problem is not solvable in polynomial time, we need to examine pairs of words u, v whereby the DFA created from the subset construction accepting $u \shuffle v$ is not polynomial in size by Proposition 1. Indeed, we know that such automata exist, as in [4], an infinite subset of $\Sigma^+ \times \Sigma^+$ is demonstrated such that for each (u, v) in the subset, the minimal DFA accepting $u \shuffle v$ requires an exponential number of states. These word pairs are quite similar to those constructed in Theorem 2 below. The existence of minimal DFAs accepting the shuffle of two words that requires an exponential number of states is not enough information on its own though to show that testing $L(M) \subseteq u \shuffle v$ cannot be done in polynomial time, as there could in principle be an algorithm that tests this fact without first constructing the minimal DFA accepting $u \shuffle v$ (as is the case for the converse problem, which can be tested in polynomial time).

We will examine the complement of that problem; that is, the problem of whether there exists some $w \in L(M)$ such that $w \notin u \shuffle v$. We will show that this problem is NP-Complete. This implies that the problem of determining whether or not $L(M) \subseteq u \shuffle v$ is coNP-Complete, by Proposition 10.1 of [11], which states that the complement of an NP-Complete language is coNP-Complete.

Throughout the proof, we will refer to Example 1 for the purposes of intuition. It is helpful to follow the example together with the proof.

Theorem 2. *Let M be a DFA and let $u, v \in \Sigma^+$ where Σ has at least two letters. The problem of determining whether there exists $w \in L(M)$ such that $w \notin u \sqcup v$ is NP-Complete.*

Proof. It is clear that this problem is in NP since we can construct a nondeterministic Turing Machine that guesses a word in $L(M)$ and then verifies that $w \notin u \sqcup v$ (we can test membership in the naive NFA accepting $u \sqcup v$, and NFA membership testing can be done in polynomial time [8]).

Thus, we need to show that the problem is NP-hard. Let F be an instance of the satisfiability problem with $X = \{x_1, \ldots, x_p\}$ the set of Boolean variables, and $C = \{c_1, \ldots c_q\}$ the set of clauses over X where each clause in C has three literals. This problem is known as 3SAT and it is NP-Complete (Proposition 9.2 of [11]). We will also assume without loss of generality that $q \geq 2$. For a variable x, let x^+ be the literal obtained from the variable x as true (simply the variable x), and x^- be the literal obtained from the variable as false (the negation of x). If d is a truth assignment, then d is a function from X to $\{+, -\}$ (true or false). We extend d to a function on clauses, where $d(c) = +$ if c contains at least one literal that matches the sign of d applied to its variable, and $d(c) = -$ if all literals have differing signs than its variables on d.

We will first provide the construction. Although technical, we will refer throughout the proof to Example 1, located after the proof, for intuition. Next, we construct the two words u, v, and the DFA accepting the language L. The words u and v depend only on the number of variables and clauses. The language L accepts a different string for every possible truth assignment to the variables X. Each such string contains a substring for each variable consecutively, and within each such substring, another sequence of substrings for each clause consecutively.

Let $f(c_i, x_j, \alpha)$ be a function from $C \times X \times \{+, -\}$ such that

$$f(c_i, x_j, \alpha) = \begin{cases} bbbabaaa & \text{if } c_i \text{ does not contain literal } x_j^\alpha, \\ bbbbaaaa & \text{otherwise (if } c_i \text{ does contain literal } x_j^\alpha). \end{cases}$$

Then, let

$$u = (aabb)^{q-1}(aabababb(aabb)^{q-2})^p(aabb),$$
$$v = (aabb)^{q-1}(aabbaabb(aabb)^{q-2})^p(aabb),$$
$$g(x_j, \alpha) = f(c_1, x_j, \alpha)f(c_2, x_j, \alpha) \cdots f(c_q, x_j, \alpha), x_j \in X, \alpha \in \{+, -\},$$
$$y(d) = g(x_1, d(x_1)) \cdots g(x_p, d(x_p)), d \text{ a function from } X \text{ to } \{+, -\},$$
$$Y = \{y(d) \mid d \text{ a function from } X \text{ to } \{+, -\}\},$$
$$L = a(aabb)^{q-1}aaa \cdot Y \cdot bbbb(aabb)^{q-1}.$$

Below, we will show that F is satisfiable if and only if there exists a word in L that is not in $u \sqcup v$. First, notice that we can build u and v in polynomial time, and they depend only on the number of variables and clauses in F.

We will build a DFA accepting L in polynomial time in several steps. First, we can build a DFA accepting each $f(c_i, x_j, \alpha)$ which only has 9 states since each accepts only one word of length 8. We can also build all $f(c_i, x_j, \alpha)$ in $O(pq)$ time. Then, we can accept each $g(x_j, \alpha)$ in polynomial time. We can then build a DFA accepting $\{g(x_j, +), g(x_j, -)\}$ for every j. Indeed, if x_j is a variable, and l is the length of the longest common prefix of $g(x_j, +)$ and $g(x_j, -)$, then we can build a DFA that reads this common prefix and then switches to one of two states based on whether the next letter is from $g(x_j, +)$ or $g(x_j, -)$. Then, from the two different states, it reads what remains of either $g(x_j, +)$ or $g(x_j, -)$. Upon reading the last symbol of either, the DFA switches to a common final state. Thus, we can build a DFA accepting $\{g(x_j, +), g(x_j, -)\}$ in polynomial time. Next, note that $Y = \{g(x_1, +), g(x_1, -)\} \cdot \{g(x_2, +), g(x_2, -)\} \cdots \{g(x_p, +), g(x_p, -)\}$. Hence, we can also build an automaton accepting Y in polynomial time by concatenating the DFA for $\{g(x_1, +), g(x_1, -)\}$ to that of $\{g(x_2, +), g(x_2, -)\}$, and so on, for all p variables. Then we can transform the DFA accepting Y into one accepting L in polynomial time. In Example 1, we provide an instance of the 3SAT problem and show its DFA (accepting Y) created by this construction in Figure 1. Intuitively, notice that every path through the automaton corresponds to a different truth assignment. For each variable, consecutively, taking the upper path corresponds to setting that variable to be true, and the lower path corresponds to setting that variable to be false.

For a prefix w of a word in L, we let

$$h(w) = \{(i, j) \mid w \in u[i] \sqcup\!\sqcup v[j]\}.$$

We will show next that d is a satisfying truth assignment if and only if $h(a(aabb)^{q-1}aaa \cdot y(d)) = \emptyset$.

Each word of Y is composed of $y(d)$ where d is any assignment of the variables. That is, each word is of the form $g(x_1, \alpha_1) \cdots g(x_p, \alpha_p)$ where $\alpha_1, \ldots, \alpha_p$ can be any assignment of each variable to $+$ (true) or $-$ (false). Each $g(x_j, \alpha)$ is a string where we concatenate for each clause $bbbabuaa$ when x_j^α is not in the clause (either the variable is not in the clause at all, or only the negation of x_j^α is in the clause), and $bbbbaaaa$ when the literal is in the clause.

Every word in L starts with $a(aabb)^{q-1}$, and indeed, for any $q \geq 2$,

$$h(a(aabb)^{q-1}) = \{(4q - 4l, 4l - 3), (4q - 4l - 3, 4l) \mid 1 \leq l < q\},$$

(proof omitted due to space constraints). This part is essentially identical to a claim in Theorem 13 of [4]. Intuitively, in Figure 2, this can be seen visually as the set of points at the bottom diagonal of the "duplication section", where $l = 1$ occurs for the first two points, followed by $l = 2$ for the next two points, etc. Then,

$$h(a(aabb)^{q-1}aaa) = \{(4q - 4l + 2, 4l - 2) \mid 1 \leq l \leq q\}$$

(this is diagonal below the previous diagonal marked on the diagram as β_0), as the point $(4q - 4l, 4l - 3)$ gives one point two rows down and one column to the right, $(4q - 4l + 2, 4l - 2)$, while the point $(4q - 4l - 3, 4l)$ gives one point one row down and two columns to the right, $(4q - 4l - 3 + 1, 4l + 2) = (4q - 4(l + 1) + 2, 4(l + 1) - 2)$.

This paragraph will first explain the intuition regarding the rest of the proof before the formal proof (again, while referencing Example 1 and its figures). Looking at Figure 2, at the diagonal marked β_0, there is a dot for each clause, spaced evenly apart. Then, a sequence of words will be read that are each either $bbbabaaa$ or $bbbbaaaa$ between consecutive diagonal lines marked as β_k, for some k. At first, if $bbbabaaa$ is read, then the first "clause" can pass the horizontal line marked "prune x_1", as it does in the diagram, and end with a single point four rows down and four columns to the right on the next diagonal. If $bbbbaaaa$ is read, then the "clause" gets cut off instead (as the second clause does in the figure when reaching the "prune x_1" line). Then, as some word of $(bbbabaaa + bbbbaaaa)^q$ is read, each clause is being "cut off", one at a time if and only if the literal x_1^- is in the clause (by having the word $bbbbaaaa$). Any clause not at the "prune" line (either before or after the line) leads to an identical point on the next diagonal four rows down and four columns to the right when reading either $bbbabaaa$ or $bbbbaaaa$. Thus, it is only the prune line that affects whether the clause continues, and each clause reaches the "'prune x_1" line consecutively as each $f(c_i, x_1, d(x_1))$ is read, for each $i, 1 \leq i \leq q$. Moreover, since M consecutively reads an entire word, either $g(x_1, +) = f(c_1, x_1, +) \cdots f(c_q, x_1, +)$ or $g(x_1, -) = f(c_1, x_1, -) \cdots f(c_q, x_1, -)$, this enforces that x_1 is set to true or false identically for each clause. This process then continues for each variable from $x_2, \ldots, x_p$ using the consecutive prune lines. Should a "clause" continue past all prune lines, that means that all three literals in the clause were the opposite sign as the function d applied to those variables, implying that it corresponds to a non-satisfiable truth assignment. Therefore, if every word in $L(M)$ has at least one path continue past every prune line, then no possible truth assignment is satisfying. Conversely, if a clause does get cut off, that means that one of the variables in the clause has the same value as d. Therefore, if every clause has some variable set as in d (F is satisfiable), then every clause gets cut off by some prune line and $w_d \notin u \sqcup v$. This is the case in Example 1 and Figure 2.

Formally, it can be shown that (proof omitted due to space constraints)

$$h(a(aabb)^{q-1}aaa \cdot y(d)) = \{(4q - 4l + 2 + 4pq, 4l - 2 + 4pq) \mid 1 \leq l \leq q, d(c_l) = -\}.$$

This means that after reading $y(d)$, if all of the clauses are satisfied by d, then $h(a(aabb)^{q-1}aaa \cdot y(d)) = \emptyset$. Therefore, if we add any suffix to the end of $a(aabb)^{q-1}aaa \cdot y(d)$, it cannot be in $u \sqcup v$. Hence, $a(aabb)^{q-1}aaa \cdot y(d) \cdot bbbb(aabb)^{q-1} \notin u \sqcup v$.

Conversely, if after reading $y(d)$, at least one of the clauses is not satisfied by d. Then, $h(a(aabb)^{q-1}aaa \cdot y(d)) \neq \emptyset$, and there must exist some $l, 1 \leq l \leq q$ such that $d(c_l) = -$ and $(4q - 4l + 2 + 4pq, 4l - 2 + 4pq) \in h(a(aabb)^{q-1}aaa \cdot y(d))$. Notice that $|u| = |v| = 4q + 4pq$, and so for each l between 1 and q, there are $4l - 2$ letters left from u and $2 + 4(q - l)$ letters left from v. Then what remains of u is $bb(aabb)^{l-1}$ and what remains of v is $bb(aabb)^{q-l}$. But every point in this set can reach point $(|u|, |v|)$ on input $bbbb(aabb)^{q-1}$ since $(l-1) + (q-l) = q - 1$.

Hence, there exists a word in L that is not in $u \sqcup v$ if and only F is satisfiable.

$\square$

Example 1. Consider the following instance of 3SAT with clauses $c_1 = (x_1^+ \vee x_2^+ \vee x_3^+)$, $c_2 = (x_1^- \vee x_2^+ \vee x_3^-)$, $c_3 = (x_1^+ \vee x_2^- \vee x_3^-)$. From this, we can construct each $g(x_j, \alpha)$ as follows:

$$\overbrace{}^{f(c_1,x_1,+)}\overbrace{}^{f(c_2,x_1,+)}\overbrace{}^{f(c_3,x_1,+)}$$
$$g(x_1, +) = \overbrace{bbbbaaaa}\ \overbrace{bbbabaaa}\ \overbrace{bbbbaaaa}$$

$$\overbrace{}^{f(c_1,x_1,-)}\overbrace{}^{f(c_2,x_1,-)}\overbrace{}^{f(c_3,x_1,-)}$$
$$g(x_1, -) = \overbrace{bbbabaaa}\ \overbrace{bbbbaaaa}\ \overbrace{bbbabaaa}$$

$$\overbrace{}^{f(c_1,x_2,+)}\overbrace{}^{f(c_2,x_2,+)}\overbrace{}^{f(c_3,x_2,+)}$$
$$g(x_2, +) = \overbrace{bbbbaaaa}\ \overbrace{bbbbaaaa}\ \overbrace{bbbabaaa}$$

$$\overbrace{}^{f(c_1,x_2,-)}\overbrace{}^{f(c_2,x_2,-)}\overbrace{}^{f(c_3,x_2,-)}$$
$$g(x_2, -) = \overbrace{bbbabaaa}\ \overbrace{bbbabaaa}\ \overbrace{bbbbaaaa}$$

$$\overbrace{}^{f(c_1,x_3,+)}\overbrace{}^{f(c_2,x_3,+)}\overbrace{}^{f(c_3,x_3,+)}$$
$$g(x_3, +) = \overbrace{bbbbaaaa}\ \overbrace{bbbabaaa}\ \overbrace{bbbabaaa}$$

$$\overbrace{}^{f(c_1,x_3,-)}\overbrace{}^{f(c_2,x_3,-)}\overbrace{}^{f(c_3,x_3,-)}$$
$$g(x_3, -) = \overbrace{bbbabaaa}\ \overbrace{bbbbaaaa}\ \overbrace{bbbbaaaa}$$

From this, we can construct the set Y. In Figure 1, we draw the automaton accepting the set Y (and therefore, $a(aabb)^{q-1}aaa$ must be prepended and $bbbb(aabb)^{q-1}$ must be appended to transform it into a DFA accepting L).

This instance has a solution $d : x_1 \to -, x_2 \to -, x_3 \to +$. Then consider the word $w_d = a(aabb)^2aaa \cdot y(d) \cdot bbbb(aabb)^2$, where $y(d)$ is equal to

$$\overbrace{}^{f(c_1,x_1,-)}\overbrace{}^{f(c_2,x_1,-)}\overbrace{}^{f(c_3,x_1,-)}\overbrace{}^{f(c_1,x_2,-)}\overbrace{}^{f(c_2,x_2,-)}\overbrace{}^{f(c_3,x_2,-)}\overbrace{}^{f(c_1,x_3,+)}\overbrace{}^{f(c_2,x_3,+)}\overbrace{}^{f(c_3,x_3,+)}$$
$$\underbrace{\overbrace{bbbabaaa\ bbbbaaaa\ bbbabaaa}}_{g(x_1,-)}\ \underbrace{\overbrace{bbbabaaa\ bbbabaaa\ bbbbaaaa}}_{g(x_2,-)}\ \underbrace{\overbrace{bbbbaaaa\ bbbabaaa\ bbbabaaa}}_{g(x_3,+)}.$$

This word is in $L(M)$ as seen in Figure 1 by taking the lower path, then the next lower path, then the upper path. But this word is not in $u \sqcup v$ as demonstrated in Figure 2. The dots are placed at indices (i, j) where each prefix of w_d is in $u[i] \sqcup v[j]$. The additional annotation in the diagram is referred to in the proof of Theorem 2.

If instead we tried the assignment $d' : x_1 \to +, x_2 \to -, x_3 \to +$ (which is not a satisfying truth assignment), then first "clause 1" gets cut off by the x_1 prune line since $x_1^+ \in c_1$, then "clause 2" can go through the line since $x_1^+ \notin c_2$, then "clause 3" gets cut off since $x_1^+ \in c_3$. Then "clause 1" is already cut off and doesn't reach the x_2 prune line, then "clause 2" can continue as $x_2^- \notin c_2$. Then when "clause 2" reaches the x_3 prune line, it can continue since $x_3^+ \notin c_2$. So, at least one clause passes all prune lines and reading the remaining portion of w_d gives a word in $u \sqcup v$, and thus d' is not satisfiable. But if there is at least one word in L that is not in $u \sqcup v$, then this corresponds to a satisfying truth assignment. Hence, at least one word in L is not in $u \sqcup v$ if and only if there is some satisfying truth assignment.

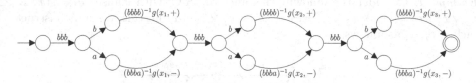

Fig. 1. The DFA accepting Y obtained from the instance of 3SAT from Example 1. We use a word on a transition as a compressed notation to represent a sequence of non-branching transitions.

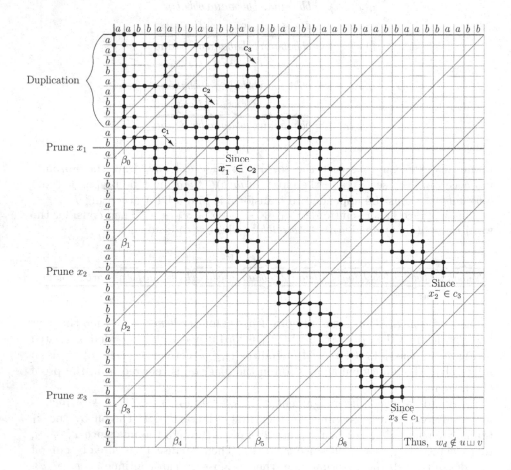

Fig. 2. The word u is labelling the vertical axis and v is labelling the horizontal axis. Considering the word w_d from Example 1, a dot is placed at (i, j) if w_d has a prefix in $u[i]$ shuffled with $v[j]$. Only a portion of the diagram is shown, and u and v continue along the axes, although there are no dots in the rest of the diagram. The lines connecting the dots demonstrates the change of states by reading individual characters.

As mentioned earlier, because testing $L(M) \not\subseteq u \, \text{ш} \, v$ is an NP-Complete problem, this implies that testing whether $L(M) \subseteq u \, \text{ш} \, v$ is a coNP-Complete problem.

Corollary 1. *Let M be a DFA and let $u, v \in \Sigma^+$, where Σ has at least two letters. The problem of determining whether $L(M) \subseteq u \, \text{ш} \, v$ is coNP-Complete.*

Despite the now known complexity of both deciding whether $L(M) \subseteq u \, \text{ш} \, v$ and $u \, \text{ш} \, v \subseteq L(M)$, the exact complexity of deciding whether or not $L(M) = u \, \text{ш} \, v$ is not immediate. In the proof of Theorem 2, had we started with u, v, M under the assumption that $u \, \text{ш} \, v \subseteq L(M)$, and shown coNP-Completeness as to whether $L(M) \subseteq u \, \text{ш} \, v$, then that would imply that testing whether $L(M) = u \, \text{ш} \, v$ would also be coNP-Complete. However, in the proof of Theorem 2, there are many words in $u \, \text{ш} \, v$ that are not in $L(M)$ and it is not clear how one could alter M to solve the problem while still creating it in polynomial time. Hence, the problem of calculating the exact complexity of testing whether $L(M) = u \, \text{ш} \, v$ remains an open problem.

References

1. Berstel, J., Boasson, L.: Shuffle factorization is unique. Theoretical Computer Science 273, 47–67 (2002)
2. Biegler, F.: Decomposition and Descriptional Complexity of Shuffle on Words and Finite Languages. Ph.D. thesis, University of Western Ontario, London, Canada (2009)
3. Biegler, F., Daley, M., Holzer, M., McQuillan, I.: On the uniqueness of shuffle on words and finite languages. Theoretical Computer Science 410, 3711–3724 (2009)
4. Biegler, F., Daley, M., McQuillan, I.: On the shuffle automaton size for words. Journal of Automata, Languages and Combinatorics 15, 53–70 (2010)
5. Biegler, F., Daley, M., McQuillan, I.: Algorithmic decomposition of shuffle on words. Theoretical Computer Science 454, 38–50 (2012)
6. Câmpeanu, C., Salomaa, K., Vágvölgyi, S.: Shuffle quotient and decompositions. In: Kuich, W., Rozenberg, G., Salomaa, A. (eds.) DLT 2001. LNCS, vol. 2295, pp. 186–196. Springer, Heidelberg (2002)
7. Ginsburg, S., Spanier, E.: Mappings of languages by two-tape devices. Journal of the ACM 12(3), 423–434 (1965)
8. Holub, J., Melichar, B.: Implementation of nondeterministic finite automata for approximate pattern matching. In: Champarnaud, J.-M., Maurel, D., Ziadi, D. (eds.) WIA 1998. LNCS, vol. 1660, pp. 92–99. Springer, Heidelberg (1999)
9. Hopcroft, J., Ullman, J.: Introduction to Automata Theory, Languages, and Computation. Addison-Wesley, Reading (1979)
10. Jędrzejowicz, J., Szepietowski, A.: Shuffle languages are in **P**. Theoretical Computer Science 250, 31–53 (2001)
11. Papadimitriou, C.M.: Computational complexity. Addison-Wesley, Reading (1994)
12. Warmuth, M., Haussler, D.: On the complexity of iterated shuffle. Journal of Computer and System Sciences 28(3), 345–358 (1984)
13. Yu, S.: Regular languages. In: Rozenberg, G., Salomaa, A. (eds.) Handbook of Formal Languages, vol. 1, pp. 41–110. Springer, Heidelberg (1997)

Minimal Partial Languages and Automata*

Francine Blanchet-Sadri[1], Kira Goldner[2], and Aidan Shackleton[3]

[1] Department of Computer Science, University of North Carolina
P.O. Box 26170, Greensboro, NC 27402–6170, USA
blanchet@uncg.edu
[2] Department of Mathematics, Oberlin College
King 205, 10 N. Professor St, Oberlin, OH 44074, USA
[3] Department of Computer Science, Swarthmore College
500 College Ave, Swarthmore, PA 19081, USA

Abstract. Partial words are sequences of characters from an alphabet in which some positions may be marked with a "hole" symbol, $\diamond$. We can create a $\diamond$-substitution mapping this symbol to a subset of the alphabet, so that applying such a substitution to a partial word results in a set of full words (ones without holes). This setup allows us to compress regular languages into smaller partial languages. Deterministic finite automata for such partial languages, referred to as $\diamond$-DFAs, employ a limited non-determinism that can allow them to have lower state complexity than the minimal DFAs for the corresponding full languages. Our paper focuses on algorithms for the construction of minimal partial languages, associated with some $\diamond$-substitution, as well as approximation algorithms for the construction of minimal $\diamond$-DFAs.

1 Introduction

Words over some finite alphabet Σ are sequences of characters from Σ and the set of all such sequences is denoted by Σ^* (we also refer to elements of Σ^* as *full words*). The *empty word* ε is the unique sequence of length zero. A *language* over Σ is a subset of Σ^*. The *regular languages* are those that can be recognized by *finite automata*. A *deterministic finite automaton*, or DFA, is a tuple $M = (Q, \Sigma, \delta, s, F)$: a set of states, an input alphabet, a transition function $\delta : Q \times \Sigma \to Q$, a start state, and a set of accept or final states. The machine M accepts w if and only if the state reached from s after reading w is in F. In a DFA, δ is defined for all state-symbol pairs, so there is exactly one computation for any word. In contrast, a *non-deterministic finite automaton*, or NFA, is a tuple $N = (Q, \Sigma, \Delta, s, F)$, where $\Delta : Q \times \Sigma \to 2^Q$ is the transition function that maps state-symbol pairs to zero or more states, and consequently may have zero or more computations on a given word. Additionally, N accepts a word w if *any* computation on w ends in an accept state. Two automata are *equivalent* if they recognize the same language, so every NFA has an equivalent DFA. In general,

---

* This material is based upon work supported by the National Science Foundation under Grant No. DMS–1060775.

NFAs allow for a more compact representation of a given language. The *state complexity* of an automaton with state set Q is $|Q|$. If a given NFA has state complexity n, the smallest equivalent DFA may require as many as 2^n states.

Partial words over Σ are sequences of characters from $\Sigma_\diamond = \Sigma \cup \{\diamond\}$, where $\diamond \notin \Sigma$ is a "hole" symbol representing an "undefined" position. A *partial language* over Σ is a subset of $\Sigma_\diamond^*$, the set of all partial words over Σ. A partial language, subset of $\Sigma_\diamond^*$, is associated with a full language, subset of Σ^*, through a $\diamond$-*substitution* $\sigma : \Sigma_\diamond^* \to 2^{\Sigma^*}$, defined such that $\sigma(a) = \{a\}$ for all $a \in \Sigma$, $\sigma(\diamond) \subseteq \Sigma$, $\sigma(uv) = \sigma(u)\sigma(v)$ for all $u, v \in \Sigma_\diamond^*$, and $\sigma(L) = \bigcup_{w \in L} \sigma(w)$ for all $L \subseteq \Sigma_\diamond^*$. A $\diamond$-substitution, then, maps a partial language to a full language and is completely defined by $\sigma(\diamond)$; e.g., if $\sigma(\diamond) = \{a, b\}$ and $L = \{\diamond a, b \diamond c\}$ then $\sigma(L) = \{aa, ba, bac, bbc\}$. By reversing this process, we can compress full languages into partial languages. We can easily extend regular languages to *regular partial languages* as the subsets of $\Sigma_\diamond^*$ that are regular when treating $\diamond$ as a character in the input alphabet. We can recognize them using *partial word DFAs*, introduced by Dassow et al. [4]. A $\diamond$-DFA $M_\sigma = (Q, \Sigma_\diamond, \delta, s, F)$, associated with some σ, is defined as a DFA that recognizes a partial language L, but that is also associated with the full language $\sigma(L)$. Balkanski et al. [1] proved that given a $\diamond$-DFA with state complexity n associated with some σ and recognizing L, the smallest DFA recognizing $\sigma(L)$ may require as many as $2^n - 1$ states.

Given classes of automata $\mathcal{A}, \mathcal{B}$ and a finite automaton A from $\mathcal{A}$, the problem $\mathcal{A} \to \mathcal{B}$-MINIMIZATION asks for an automaton B from $\mathcal{B}$ that has the lowest state complexity possible while maintaining $L(A) = L(B)$, i.e., the language that A accepts is the language that B accepts. We will abbreviate $\mathcal{A} \to \mathcal{A}$-MINIMIZATION by $\mathcal{A}$-MINIMIZATION. Now, let $\mathcal{DFA}$, $\mathcal{NFA}$, and $\diamond$-$\mathcal{DFA}$ be the class of all DFAs, NFAs, and $\diamond$-DFAs, respectively. It is known that $\mathcal{DFA}$-MINIMIZATION can be done in $O(n \log n)$ time [7], where n is the number of states in the input DFA, and that $\mathcal{DFA} \to \mathcal{NFA}$-MINIMIZATION is $\mathcal{PSPACE}$-complete [8]. Looking at $\diamond$-DFAs as DFAs over the extended alphabet $\Sigma_\diamond$ makes the minimization step easy ($\diamond$-DFAs are DFAs, so $\diamond$-$\mathcal{DFA}$-MINIMIZATION is $\mathcal{DFA}$-MINIMIZATION), and in general, $\mathcal{A} \to \diamond$-$\mathcal{DFA}$-MINIMIZATION and $\diamond$-$\mathcal{DFA} \to \mathcal{A}$-MINIMIZATION are not defined because $\diamond$-DFAs accept partial languages (not full languages). We thus define a slightly different problem for $\diamond$-DFAs: given a DFA M, MINIMAL-$\diamond$-$\mathcal{DFA}$ asks for the smallest $\diamond$-DFA (over all possible $\diamond$-substitutions) associated with $L(M)$. Using the methods from Björklund and Martens [2], it is a simple exercise to show that MINIMAL-$\diamond$-$\mathcal{DFA}$ is $\mathcal{NP}$-hard, so we discuss an approach to approximating minimal $\diamond$-DFAs. Note also that Holzer et al. [6] have recently further studied the computational complexity of partial word automata problems and have shown that many problems are $\mathcal{PSPACE}$-complete, among them is MINIMAL-$\diamond$-$\mathcal{DFA}$.

The contents of our paper are as follows: In Section 2, we set our notation and introduce the σ-minimal partial languages given a $\diamond$-substitution σ. In Section 3, we approximate minimal finite partial languages, associated with a $\diamond$-substitution σ, by describing our *Minlang* algorithm. We then prove that running *Minlang* a polynomial number of times in the size of the input with our *Redundancy*

Check algorithm outputs the unique minimal partial language corresponding to the input language given σ. We describe our *Partial Language Check* algorithm that given a $\diamond$-DFA M_σ and a finite language L, verifies that $\sigma(L(M_\sigma)) = L$ and that M_σ is a "contender" for a minimal $\diamond$-DFA for L given σ. We also discuss the algorithms' runtime. In Section 4, we adapt *Minlang* for infinite languages. Finally in Section 5, we conclude with some open problems.

2   Approximating Minimal $\diamond$-DFAs

The complexity of a minimal DFA may be exponentially larger than that of a minimal $\diamond$-DFA for the same language. Balkanski et al. [1] gave a construction whereby for any integer $n > 1$ there exists a $\diamond$-DFA with n states such that the minimal DFA for the same language has $2^n - 1$ states. Fig. 1 illustrates their construction for $n = 3$.

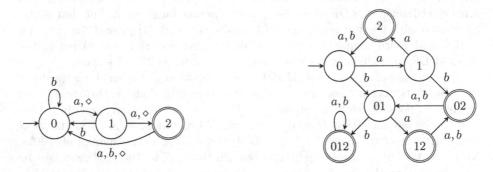

Fig. 1. Left: A 3-state $\diamond$-DFA M_σ, with $\sigma(\diamond) = \{a, b\}$. Right: The smallest DFA M satisfying $L(M) = \sigma(L(M_\sigma))$, which has $2^3 - 1 = 7$ states.

Since the problem of finding a minimal $\diamond$-DFA (over all $\diamond$-substitutions) for the full language L accepted by a given DFA is $\mathcal{NP}$-hard, we give an approximation via minimal partial languages associated with L and $\diamond$-substitutions σ, i.e., σ-*minimal partial languages*. The smallest among associated $\diamond$-DFAs thus provides an approximation for a $\diamond$-DFA with minimal state complexity for L. Before stating our definition, we recall some background material (see [3] for more information).

A partial word u is *contained* in a partial word v, denoted by $u \subset v$, if they have the same length and if a position defined in u is defined by the same letter in v (abbreviate "$u \subset v$, $u \neq v$" by "$u \sqsubset v$"). Partial word u is *compatible* with partial word v, denoted by $u \uparrow v$, if they have the same length and if a position defined in both u and v is defined by the same letter, in which case the *least upper bound* of u and v, denoted by $u \vee v$, is the partial word such that

$u \subset (u \vee v)$, $v \subset (u \vee v)$, and if $u \subseteq w$ and $v \subseteq w$ then $(u \vee v) \subseteq w$. For example, $aa \diamond b \diamond \uparrow a \diamond b \diamond \diamond$ and $(aa \diamond b \diamond \vee a \diamond b \diamond \diamond) = aabb \diamond$. Fixing a $\diamond$-substitution σ over Σ, a set $X \subseteq \Sigma_\diamond^*$ covers a partial word w if $x \uparrow w$ for all $x \in X$ and $\sigma(w) \subseteq \sigma(X)$; if X is a singleton $\{x\}$, we abbreviate "X covers w" by "x covers w".

Definition 1. *Let L be a full regular language over alphabet Σ, and let σ be a $\diamond$-substitution over Σ. The σ-minimal partial language for L is the unique partial language $L_{\min,\sigma}$ such that*

1. *$\sigma(L_{\min,\sigma}) = L$;*
2. *for all partial languages L' satisfying $\sigma(L') = L$, $|L_{\min,\sigma}| \leq |L'|$;*
3. *the partial words in $L_{\min,\sigma}$ are as weak as possible, i.e., for no partial word $w \in L_{\min,\sigma}$ does there exist x satisfying $\sigma(x) \subseteq L$ and $x \sqsubset w$.*

For each $\diamond$-substitution σ, there exists a partial language $L_{opt,\sigma}$ such that a minimal $\diamond$-DFA recognizing $L_{opt,\sigma}$ is identical to a minimal $\diamond$-DFA for $L = \sigma(L_{opt,\sigma})$. The σ-minimal partial language $L_{\min,\sigma}$ is "close" to $L_{opt,\sigma}$ and, as a result, a minimal $\diamond$-DFA recognizing $L_{\min,\sigma}$ is a "good" approximation for a minimal $\diamond$-DFA for L associated with σ. The more $\diamond$'s we have in our partial words, the more we are taking advantage of the non-determinism that the $\diamond$-DFAs embody.

For convenience of notation, when referring to a particular $\diamond$-substitution σ, we replace σ with $\sigma(\diamond)$, e.g., $\{a \diamond, \diamond b\}$ is an $\{a, b\}$-minimal partial language for $\{aa, ab, bb\}$. Note that $a \diamond$ covers both aa and ab, and $\diamond b$ covers both ab and bb.

3   Computing Minimal $\diamond$-DFAs

We describe our algorithms for approximating minimal $\diamond$-DFAs. The input and output finite languages are represented by listing their words.

3.1   Our *Minlang* Algorithm

Algorithm 1, referred to as *Minlang*, is an efficient algorithm for approximating $L_{\min,\sigma}$. Pseudocode is given below, as well as an example of its execution on the full language $L = \{aaa, aab, aac, aba, abb, aca, acb, bac, cac\}$ with $\sigma(\diamond) = \{a, b, c\}$ (see Figs. 2–3). Note that the output L_σ of *Minlang* is not necessarily σ-minimal. In our example, $L_{\min,\sigma} = \{a \diamond \diamond, a \diamond b, \diamond ac\}$ and $L_\sigma = \{aa \diamond, a \diamond a, a \diamond b, \diamond ac\}$, the partial word $aa \diamond$ being redundant. However, *Minlang* is useful as both an approximation and as a stepping stone toward the minimal partial language in the sense of Definition 1 (see Section 3.2).

For any finite language L and $\diamond$-substitution σ, the output of *Minlang* on input L and σ is a tree that can easily be converted to a $\diamond$-DFA with the start state represented by the root node, the accept states by the terminal nodes, and the transitions by the edges. Running standard DFA minimization algorithms on this $\diamond$-DFA results in an approximation of a minimal $\diamond$-DFA for L.

Proposition 1. *The language L_σ output by Minlang satisfies $\sigma(L_\sigma) = L$.*

Algorithm 1. *Minlang* Given as input a finite language L over Σ and a $\diamond$-substitution σ, computes a partial language L_σ that approximates $L_{\min,\sigma}$

---

1. put L into a prefix tree with leaf nodes marked as terminals
2. **for** each node n in a reverse level-order traversal of the tree **do**
3.   **if** parent$(n) = u$ has children on every branch of $\sigma(\diamond)$ **then**
4.   order children of u by non-decreasing height, $c_1, \ldots, c_k$
5.   initialize $\mathcal{C} = \{$all terminal paths from c_1 (including ε if c_1 is terminal node)$\}$
6.   **for** $m \in \{c_2, \ldots, c_k\}$ **do**
7.   **for all** $w \in \mathcal{C}$ **do**
8.   remove w from $\mathcal{C}$
9.   **if** there is a terminal path from m to mx such that $x \subset w$ **then**
10.   add w to $\mathcal{C}$
11.   **if** there is a terminal path from m to mx such that $w \subset x$ **then**
12.   add x to $\mathcal{C}$ for all such x
13.   **if** $\mathcal{C}$ is empty **then**
14.   break
15.   **if** $\mathcal{C}$ is non-empty **then**
16.   add a $\diamond$-transition from node u to a new node $u\diamond$ and a terminal path from node $u\diamond$ to a new node $u\diamond w$, for each $w \in \mathcal{C}$
17.   for each pair $a \in \sigma(\diamond), w \in \mathcal{C}$, start from uaw, unmark uaw as a terminal node and move upwards, deleting the path until a node is found that has more than one child or is terminal

---

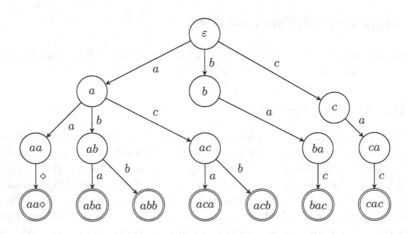

Fig. 2. Starting with the prefix tree for $L = \{aaa, aab, aac, aba, abb, aca, acb, bac, cac\}$, *Minlang* begins at the leaf nodes, compiling $\mathcal{C}$ when a node's parent has children for every letter in $\sigma(\diamond) = \{a, b, c\}$. This consolidates aa's children into a single child, $aa\diamond$. Then *Minlang* examines nodes at reverse depth 1 and their parents, finding children for every letter in $\sigma(\diamond)$ at the node a. Then $\mathcal{C} = \{a, b\}$, adding a transition from a to $a\diamond$ and from $a\diamond$ to children $a\diamond a$ and $a\diamond b$, removing the b and c branches from a, but leaving the a branch as it does not contain any words in $\mathcal{C}$.

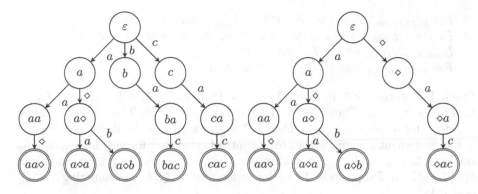

Fig. 3. Left: *Minlang* examines nodes at reverse depth 2 and their parents, finding children for every letter in $\sigma(\diamond)$ at the node ε. Then $C = \{ac\}$, adding a transition from ε to $\diamond$ and from $\diamond$ a terminal path to $\diamond ac$, removing the b and c branches from ε, but leaving the a branch as it does not contain any words in C. Right: the final tree output by *Minlang*. The words in L_σ are the labels of the terminal nodes in the tree.

Proof. First, a node u in the tree for L has a *representation* x in the tree for L_σ if $u \in \sigma(x)$. Now, the proof is by strong induction on the reverse depth, i.e., the height of the tree minus the depth of a given node. We show that for each $n \geq 1$, all nodes at reverse depth $n - 1$ or less in the tree for L_σ have their σ-image in the tree for L implying $\sigma(L_\sigma) \subseteq L$, and all nodes at reverse depth $n - 1$ or less in the tree for L have a representation in the tree for L_σ implying $L \subseteq \sigma(L_\sigma)$.

For the inductive step, consider a node u' at reverse depth n, with parent u, in the tree for L_σ. If $u' = ua$ for some $a \in \Sigma$, then by the inductive hypothesis, all nodes uav, where $v \neq \varepsilon$, in the tree for L_σ have their σ-image in the tree for L. So the σ-image of u', i.e., $\sigma(u)a$, is in the tree for L. If $u' = u\diamond$, then all nodes $u\diamond v$, where $v \neq \varepsilon$, in the tree for L_σ have their σ-image in the tree for L. So the σ-image of u', i.e., $\{\sigma(u)a \mid a \in \sigma(\diamond)\}$, is in the tree for L.

Consider a node u' at reverse depth n, with parent u, in the tree for L. First, suppose that *Minlang* finds that from u, there is a transition labeled by a in the tree for L, for each $a \in \sigma(\diamond)$. By the inductive hypothesis, each node uav, where $a \in \sigma(\diamond)$ and $v \neq \varepsilon$, in the tree for L has a representation $x\diamond y$ in the tree for L_σ. So ua, where $a \in \sigma(\diamond)$, has a representation $x\diamond$ in the tree for L_σ. Next, suppose that *Minlang* does not find such transitions from u. Set $u' = ua$ for some $a \in \Sigma$. By the inductive hypothesis, each node uav, where $v \neq \varepsilon$, in the tree for L has a representation xay in the tree for L_σ ($|x| = |u|$ and $|y| = |v|$). So, ua has a representation xa in the tree for L_σ. In either case, u' has a representation in the tree for L_σ. $\square$

The next lemma gives properties of the language output by *Minlang*.

Lemma 1. *The language L_σ output by* Minlang *satisfies the following:*

1. *For $x \in L_\sigma$, there exists some $w \in L_{\min,\sigma}$ such that $x \uparrow w$; similarly, for $w \in L_{\min,\sigma}$, there exists some $x \in L_\sigma$ such that $x \uparrow w$.*

2. For $w \in L_{\min,\sigma}$, there is no $x \in L_\sigma \setminus L_{\min,\sigma}$ such that $x \subset w$.
3. For $w \in L_{\min,\sigma}$ and $x \in L_\sigma$, if $w \subset x$, then $w = x$; consequently, if $w \in L_{\min,\sigma}$, there is no $x \in L_\sigma$ such that $w \sqsubset x$.
4. For $w \in L_{\min,\sigma}$ and $x \in L_\sigma$, if $x \uparrow w$, then $w \in L_\sigma$.

Proof. For Statement 1, by Definition 1 and Proposition 1, $\sigma(L_{\min,\sigma}) = \sigma(L_\sigma) = L$, and let $x \in L_\sigma$. Then $\sigma(x) \subseteq L$, and take $\hat{x} \in \sigma(x)$. Thus $\hat{x} \in L$, and so $\hat{x} \in \sigma(w)$ for some $w \in L_{\min,\sigma}$. Then $x \subset \hat{x}$ and $w \subset \hat{x}$, so by definition, $x \uparrow w$.

For Statement 2, suppose towards a contradiction that for $w \in L_{\min,\sigma}$, $x \in L_\sigma$ and $x \notin L_{\min,\sigma}$, we have $x \subset w$. Since $x \neq w$, we can write $x \sqsubset w$, and since $\sigma(L_\sigma) = L$ by Proposition 1, we know that $\sigma(x) \subseteq L$ contradicting Definition 1(3).

For Statement 3, we show by induction on k that for $w \in L_{\min,\sigma}$ and $x \in L_\sigma$, if $w \subset x$, then the suffix of length k of x equals the suffix of length k of w. For the inductive step, suppose towards a contradiction that $w = w'\diamond v$ and $x = x'av'$ where $a \in \Sigma_\diamond$ and $|v| = |v'| = k$. By the inductive hypothesis, $v = v'$. Since $w' \subset x'$, we have that $\{\sigma(x')b\sigma(v) \mid b \in \sigma(\diamond)\} \subseteq \sigma(w) \subseteq L$. Hence the node $\sigma(x')$ has children on every branch of $\sigma(\diamond)$, and each node $\sigma(x')bv$ is marked as terminal. Thus *Minlang* adds v to C and iterates over each child $\sigma(x')b$. It adds a $\diamond$-transition from $\sigma(x')$ to $\sigma(x')\diamond$ and a terminal path from $\sigma(x')\diamond$ to $\sigma(x')\diamond v$, and it switches the a to a $\diamond$ in the $(k+1)$th last character's index, a contradiction. Thus the suffix of length $k+1$ of x is identical to the suffix of length $k+1$ of w.

For Statement 4, we show the result by induction on the length of w. Assume that $w \in L_{\min,\sigma}$, $x \in L_\sigma$, and $x \uparrow w$. For the inductive step, suppose $w = cw'$ and $x = c'x'$, with $|c| = |c'| = 1$, such that $x' \uparrow w', c \uparrow c'$. First, suppose that $c \neq \diamond$. Consider the language $L' = \{t' \mid ct' \in L_{\min,\sigma}\}$ and the tree T' that results from applying *Minlang* to the tree for $\sigma(L')$. Then, clearly, $w' \in L'$, so by the inductive hypothesis, T' contains a terminal path to w'. Hence the tree for L_σ contains a terminal path from c to cw', thus $w = cw' \in L_\sigma$. Now, suppose that $c = \diamond$. Then $\sigma(\diamond w') \subseteq L$ implies that $d\sigma(w') \subseteq L$ for all $d \in \sigma(\diamond)$. Then if we consider each language L_{d_σ} constructed from taking the sub-tree of L_σ with d as the root node, we have that $w' \in L_{d_{\min,\sigma}}$, where $L_{d_{\min,\sigma}}$ is a minimal language such that $\sigma(L_{d_{\min,\sigma}}) = \sigma(L_{d_\sigma})$. By Statement 1, there exists some $t' \in L_{d_\sigma}$ such that $w' \uparrow t'$, so by the inductive hypothesis, L_{d_σ} contains a terminal path to w'. Then since every child $d \in \sigma(\diamond)$ contains a path to w', *Minlang* adds $\diamond w' = w$ to L_σ. □

From Lemma 1, we can easily derive the following proposition.

Proposition 2. *The language L_σ output by* Minlang *satisfies* $L_{\min,\sigma} \subseteq L_\sigma$.

Proof. Let $w \in L_{\min,\sigma}$. Then by Lemma 1(1), there exists some $x \in L_\sigma$ such that $x \uparrow w$. By Lemma 1(4), $w \in L_\sigma$. □

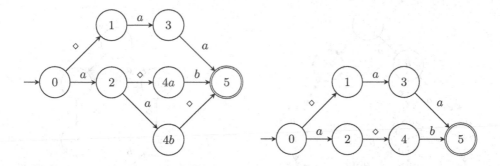

Fig. 4. Left: a minimal ◇-DFA recognizing $L_\sigma = \{\diamond aa, a\diamond b, aa\diamond\}$, where $\sigma(\diamond) = \{a, b\}$, output by *Minlang* has 8 states. Right: a minimal ◇-DFA recognizing $L_{\min,\sigma} = \{\diamond aa, a\diamond b\}$, which is also a minimal ◇-DFA for the full language $L = \sigma(L_\sigma) = \sigma(L_{\min,\sigma})$, has 7 states. Error states are omitted.

Fig. 4 illustrates a minimal ◇-DFA for L_σ, the language output by *Minlang* that has more states than a minimal ◇-DFA for the full language L as a result of redundancies in the *Minlang* approximation.

3.2 Our *Redundancy Check* Algorithm

We describe a second algorithm, referred to as *Redundancy Check*, to fine-tune the result of *Minlang*, guaranteed to output $L_{\min,\sigma}$ exactly. We prove the correctness of the algorithm and give a worst-case runtime bound. Redundancy occurs when a partial word w is already covered by some set $X \subseteq L_{\min,\sigma}$, i.e., $\sigma(w) \subseteq \sigma(X)$. In Algorithm 2, V is the set of suffixes v of partial words $u\diamond v$ in L_σ, where u is a fixed word with no holes, R_a is the set of suffixes r of partial words uar in L_σ, where $a \in \Sigma_\diamond$ and r is compatible with some element of V, and $\bar{r}$ is the part of the image $\sigma(r)$ that is left uncovered by the elements of V. Referring to Figs. 2–3, *Redundancy Check* is illustrated by Fig. 5. *Redundancy Check* maintains the relationship $L_{\min,\sigma} \subseteq L_\sigma$ while removing from L_σ any partial words not in $L_{\min,\sigma}$.

Algorithm 2. *Redundancy Check* Given as input the output L_σ of *Minlang*, computes $L_{\min,\sigma}$

---

1. **for all** $u\diamond, u \in \Sigma^*$ **do**
2. $V = \{v \mid u\diamond v \in L_\sigma\}$
3.   **for all** children ua of u for $a \in \sigma(\diamond)$ **do**
4.   compile $R_a = \{r \mid r \uparrow v$ for some $v \in V, uar \in L_\sigma\}$
5.   **for all** $r \in R_a$ **do**
6.   let $\bar{r} = \sigma(r) \setminus \{r \vee v \mid r \uparrow v$ for $v \in V\}$
7.   **if** for every $e \in \bar{r}$ there is a path $uae' \in L_\sigma$ such that $e' \subset e$ **then**
8.   delete r
9. compile L_σ from the tree (every root-to-terminal path)
10. **return** L_σ

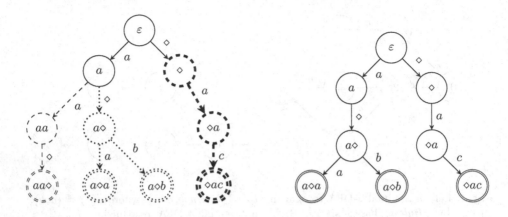

Fig. 5. Left: for $u = \varepsilon$, the thick dashed path gives the set $V = \{ac\}$ from the pseudocode of *Redundancy Check*. The thin dashed path is the only element $r = a\diamond \in R_a$. Then $\bar{r} = \{aa, ab, ac\} \setminus \{ac\} = \{aa, ab\}$. The dotted paths from a labeled $\diamond a, \diamond b$ represent $e' \sqsubset e \in \bar{r}$, hence we delete r. Right: the final language tree for the $\{a, b, c\}$-minimal partial language for L, $L_{\min,\sigma} = \{a\diamond a, a\diamond b, \diamond ac\}$.

Theorem 1. *Given as input a finite language L over Σ and a $\diamond$-substitution σ, Minlang followed by Redundancy Check returns $L_{\min,\sigma}$. The runtime is polynomial in the size of the input.*

Proof. Recall by Definition 1 that $L_{\min,\sigma}$ is a minimal partial language with its words of the weakest form such that $\sigma(L_{\min,\sigma}) = L$. We claim that *Redundancy Check* removes the elements of the output of *Minlang*, L_σ, that are redundant. It follows directly from Proposition 2 and our claim that $L_\sigma = L_{\min,\sigma}$.

To prove our claim, consider some element x that is removed by our *Redundancy Check*. Thus $x = uar$ for some $u \in \Sigma^*$, $r \in \Sigma_\diamond^*$, and $a \in \sigma(\diamond)$, and there exists $w = u\diamond y \in L_{\min,\sigma}$ such that $y \in \Sigma_\diamond^*$ and $y \uparrow r$. Then for x to be removed, $r \in R_a$, which means that $r \uparrow v$ for some $v \in V$, i.e., $u\diamond v \in L_\sigma$, which is the case since $y \in V$. Then for every $e \in \bar{r}$, there must be some path $uae' \in L_\sigma$ such that $e' \sqsubset e$. But if this is the case, then $uae' \sqsubset uae$ for all such e'. This means precisely that $\sigma(x) \subseteq \sigma(L_\sigma \setminus \{x\})$, and hence $x \notin L_{\min,\sigma}$, so we remove x.

Similarly, if $x \notin L_{\min,\sigma}$ and $x \in L_\sigma$, then there must be some minimal set $X \subseteq L_{\min,\sigma} \subseteq L_\sigma$ such that $\sigma(x) \subseteq \sigma(X)$. If we take an element of X with a hole in the leftmost position of any partial word in X, say $u\diamond v$, we have that $x = uar$ where $a \in \Sigma_\diamond$, and $r \uparrow v$ since $x \uparrow w$ for every $w \in X$. Then clearly $v \in V$ and $r \in R_a$. Any intersection between $\sigma(r)$ and $\{r \vee v\}$ is removed from $\sigma(r)$. Removing from $\sigma(r)$ all elements $r \vee z$ with $r \uparrow z$ and $z \in V$ yields the set $\bar{r}$, so every path in $\bar{r}$ contains a weaker path, and hence x is removed from L_σ. □

3.3   Our *Partial Language Check* Algorithm

We describe a third algorithm, referred to as *Partial Language Check*, that verifies if $\sigma(L(M_\sigma)) = L$ when given as input a $\diamond$-DFA $M_\sigma = (Q, \Sigma_\diamond, \delta, s, F)$, associated with a $\diamond$-substitution σ, and a finite language L over Σ. A *contender* for a minimal $\diamond$-DFA for a finite language L is a $\diamond$-DFA M_σ such that $|L(M_\sigma)| \leq |L|$.

Algorithm 3. *Partial Language Check* Given a $\diamond$-substitution σ, a $\diamond$-DFA $M_\sigma = (Q, \Sigma_\diamond, \delta, s, F)$, and a finite language L over Σ, checks whether $\sigma(L(M_\sigma)) = L$

1. run standard DFA minimization on M_σ
2. compile the list P of all paths from s to any $f \in F$ (if any grows longer than ℓ, the length of the longest word in L, terminate and **return false**)
3. compile L' from P (if $|L'|$ grows larger than $|L|$, terminate and **return false**)
4. let ℓ' be the length of the longest word in L'
5. **if** $\ell' \neq \ell$ **then return false**
6. let L'_σ be the result of running *Minlang* on L' and σ
7. **while** L'_σ continues to change with each pass (at most ℓ times) **do**
8.   run *Minlang* on L'_σ
9. run *Redundancy Check* on L'_σ
10. run *Minlang* with *Redundancy Check* on L, creating $L_{\min,\sigma}$
11. **for each** $w \in L_{\min,\sigma}$ **do**
12.   **if** $w \in L'_\sigma$ **then** delete w from L'_σ
13.   **else return false**
14. **if** L'_σ is non-empty **then return false**
15. **else return true**

Referring to the notation used in the pseudocode of Algorithm 3, for a word $w \in L_{\min,\sigma}$, the *weakest covering set* X for w is the set of those words $x \in L'_\sigma$ such that $x \uparrow w$, L'_σ contains no element z satisfying $z \sqsubset x$, and $\sigma(w) \subseteq \sigma(X)$.

Theorem 2. *Let L be a language over alphabet Σ and σ be a $\diamond$-substitution over Σ. If L' is a partial language such that $\sigma(L') = L$, the language L'_σ produced by running* Minlang *at most ℓ times on L' and then running* Redundancy Check *is equal to $L_{\min,\sigma}$, where ℓ denotes the length of the longest word in L.*

Proof. First, we claim that for any partial language L' such that $\sigma(L') = L$, if L'_σ is the language produced by running *Minlang* as many times as necessary on L', then $L_{\min,\sigma} \subseteq L'_\sigma$. To prove our claim, let $w \in L_{\min,\sigma}$. Since L'_σ is a partial language associated with L, L'_σ contains some weakest covering set X for w. We show that for all $x \in X$ and any factorizations $w = uv$ and $x = u'v'$ where $|v| = |v'|$, we have that $u' \uparrow u$ and $v' \sqsubset v$. We do this by induction on $|v|$. For the inductive step, consider the factorizations $w = uv = uay$ and $x = u'v' = u'a'y'$ where $|a| = |a'| = 1$ and $|y| = |y'|$. By the inductive hypothesis, $u'a' \uparrow ua$ and $y' \sqsubset y$, so $u' \uparrow u$. Suppose $a \neq \diamond$ or $a' = \diamond$. To have $u'a' \uparrow ua$, we must have $a' \sqsubset a$, hence $v' = a'y' \sqsubset ay = v$. Otherwise, since $X \subseteq L'_\sigma$ covers w, $u'by' \in L'_\sigma$ for all $b \in \sigma(\diamond)$, and a pass of *Minlang* clearly results in $u'\diamond y' \in L'_\sigma$. This implies $u'\diamond y' \sqsubset u'a'y'$, contradicting the fact that X is a weakest covering set for w. Thus, for every $w \in L_{\min,\sigma}$, we have $x \sqsubset w$ for all $x \in X$. By Lemma 1(3), $x = w$, so $w \in L'_\sigma$. Hence $L_{\min,\sigma} \subseteq L'_\sigma$.

Next, we claim that if ℓ is the length of the longest word in L, no more than ℓ passes of *Minlang* are required for our first claim to hold. To see this, if L' is a partial language for L with $\diamond$-substitution σ, the language tree for L' is of height ℓ. Likewise, the tree for L'_σ, the language produced by running *Minlang* on L', is of height ℓ. The only case where we require an additional pass of *Minlang* on L'_σ is when in the previous pass of *Minlang*, some partial word $u\diamond xay$, where $u, x, y \in \Sigma^*_\diamond$ and $a \in \sigma(\diamond)$, is added to L'_σ such that it is then possible to add $u\diamond x\diamond y'$ to L'_σ for some $y' \in \Sigma^*_\diamond$ with $y \subset y'$. As this newly-available addition can only occur at a strictly lower level of the tree than the previous addition, the tree is correctly minimized to depth k by the kth pass. Then, minimization is complete after at most ℓ passes.

By our two claims, $L_{\min,\sigma} \subseteq L'_\sigma$ after at most ℓ passes of *Minlang*. By Theorem 1, we have that *Redundancy Check* on L'_σ removes all redundant elements of L'_σ, resulting in simply $L_{\min,\sigma}$. $\square$

We finally prove *Partial Language Check*'s correctness and runtime.

Theorem 3. *Given as input a finite language L, a $\diamond$-substitution σ, and a $\diamond$-DFA M_σ, Partial Language Check runs in polynomial time in the size of the input. It properly verifies that $\sigma(L(M_\sigma)) = L$ and that M_σ is a contender for a minimal $\diamond$-DFA for L given σ.*

Proof. To see that *Partial Language Check* properly verifies that $\sigma(L(M_\sigma)) = L$ and that M_σ is a contender for a minimal $\diamond$-DFA for L given σ, it first minimizes M_σ to optimize runtime. It compiles all paths from the start state s to all accept states $f \in F$ and consolidates the paths into a partial language $L' = L(M_\sigma)$. However, a contender for a minimal $\diamond$-DFA for L never accepts a language larger than L, so if it finds that $|L'|$ has grown larger than $|L|$ at any given point in the compiling of L', it immediately terminates and returns false, as this $\diamond$-DFA is no longer a contender for a minimal $\diamond$-DFA for L.

If the length ℓ' of the longest word in L' is not the length ℓ of the longest word in L, then clearly L' is not a partial language for L, so it suffices to terminate and return false.

The next step is to run *Minlang* followed by *Redundancy Check* on L and *Minlang* at most ℓ times on L' followed by *Redundancy Check*. By Theorem 1 and Theorem 2, this produces the unique σ-minimal partial language for L and for $\sigma(L')$. Hence if $\sigma(L') = L$, then $L_{\min,\sigma} = L'_\sigma$. It then checks if the two languages $L_{\min,\sigma}$ and L'_σ are equal. If not, it returns false, and if so, it returns true (Lines 14–15). Hence it returns true if and only if $\sigma(L(M_\sigma)) = L$ and M_σ is a contender for a minimal $\diamond$-DFA for L given σ. $\square$

4   Adapting *Minlang* for Infinite Languages

We can extend regular expressions to partial words by adding $\diamond$ to the basic regular expressions. This leads naturally to the concept of regular partial languages as the sets of partial words that match partial regular expressions. It is possible to run a slightly modified version of *Minlang* on an infinite language L using the following process. In place of a complete list of the words in L, we use a regular expression for L along with a given $\diamond$-substitution σ.

First, convert a regular expression for our language L into a slightly modified but equivalent form: distribute out unions whenever possible and separate the expression into a list of words that are unioned together at the outer most level. Call this list L, as it is evidently equivalent. Note that the only remaining unions must be inside a Kleene star block. Denote the start of a Kleene star block with "[" and the end of it with "]".

Then, put L into a prefix tree. However, whenever we start a Kleene star block, each element that is unioned together is the child of the "[" character. The end of each element in the Kleene star block has a child to the same joint "]" node that continues on with the suffix of the block.

Next, perform the same algorithm as *Minlang* with respect to the given $\diamond$-substitution σ, except when deleting redundant paths for some word w, if $w = u\,]\,v$ and the "]" node has multiple parents, only delete the nodes relating to u according to the algorithm's requirements and break the tie from the "]" node to its parent in the path of u.

Hence *Minlang* finds all possible $\diamond$'s and removes redundancies in this tree. Then a trained traversal of the tree that matches every "[" node with its descendant balanced "]" node and unions all paths from the same "[" node to the joined "]" node yields all regular expressions with the maximum number of $\diamond$'s in place. We call the resulting list of regular expressions represented, L_σ. The modified algorithm runs in polynomial time of the input regular expression.

This modification of *Minlang* does not produce an equivalent minimal partial language L_σ for the infinite language L. First, such a definition does not make sense, as we cannot produce a language of minimal size, since any partial language for L is infinite. Thus we focus on the equivalent of Definition 1(3): for no partial word $w \in L_\sigma$ does there exist x satisfying $\sigma(x) \subseteq L$ and $x \sqsubseteq w$. We cannot guarantee that L_σ meets this criterion, as L_σ is dependent on the regular expression used for L and not on the infinite language that the regular expression represents. The problem lies in the representation of a Kleene star block. While $ab(cb)^*b \equiv a(bc)^*bb$, the regular expression $ab(cb)^*b + a(bc)^*ab + a(bc)^*cb$, that uses the former form, finds no $\diamond$'s when the modified *Minlang* is run on it. However, the regular expression $a(bc)^*bb + a(bc)^*ab + a(bc)^*cb$, that uses the latter form, finds $a(bc)^*\diamond b$.

Checking all possible configurations of a loop for every loop used in a regular expression for the language is intractable. We could use a standardized configuration of a loop, such as the unambiguous form from [5]. For a regular expression composed of regular expressions x, y, define $x(yx)^*$ to be the unambiguous form, as a DFA is easily constructed from it. This is opposed to any $u(vyu)^*v$ where $uv = x$. However, even with a standardized unambiguous form, $a(ba)^* + (ab)^*b + (ab)^*c$ finds no $\diamond$'s, while $(ab)^*a + (ab)^*b + (ab)^*c$ finds $(ab)^*\diamond$.

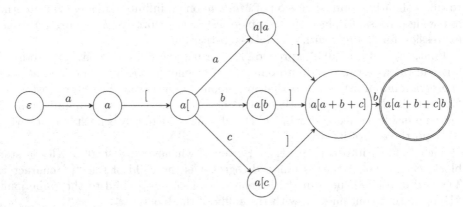

Fig. 6. A language tree representing the regular expression $a(a+b+c)^*b$. Running the modified *Minlang* produces the desired regular expression $a\diamond^*b$.

5   Conclusion and Open Problems

The choice of a $\diamond$-substitution σ can vastly change the state complexity of a minimal $\diamond$-DFA, associated with σ, for a given DFA. Fig. 7 illustrates different $\diamond$-substitutions resulting in different state complexities for minimal $\diamond$-DFAs, associated with them. An open problem is to develop computational techniques for selecting an *optimal* $\diamond$-substitution σ for a given DFA M, that is, optimality occurs when a minimal $\diamond$-DFA for $L(M)$, associated with σ, has the same state complexity as a minimal $\diamond$-DFA for $L(M)$ over all possible $\diamond$-substitutions. Because a solution to the σ-CHOICE problem is defined in terms of a solution to the

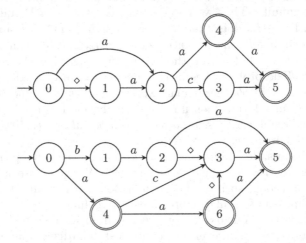

Fig. 7. Top: A minimal $\diamond$-DFA, associated with $\sigma(\diamond) = \{a, b\}$, having 7 states including the error state, a sink non-accept state. Bottom: A minimal $\diamond$-DFA for the same full language, associated with $\sigma(\diamond) = \{a, c\}$, having 8 states including the error state.

MINIMAL-$\diamond$-$\mathcal{DFA}$ problem, which is $\mathcal{NP}$-hard, it does not make sense to define or attempt to solve the σ-CHOICE problem separately from the MINIMAL-$\diamond$-$\mathcal{DFA}$ problem.

Another open problem is the one of extending $\diamond$-DFAs. In light of the understanding that $\diamond$-DFAs are weakly non-deterministic, it makes sense to ask whether meaningful extensions of the class $\diamond$-$\mathcal{DFA}$ exist, and what properties those extensions might have. In particular, what would happen if we created additional $\diamond$-like symbols, say $\diamond_1, \ldots, \diamond_k$?

A World Wide Web server interface has been established at

$$\texttt{www.uncg.edu/cmp/research/planguages2}$$

for automated use of a program that given a $\diamond$-substitution σ and a full language L, computes the σ-minimal partial language for L. This is our own implementation, we do not use any known automata library.

References

1. Balkanski, E., Blanchet-Sadri, F., Kilgore, M., Wyatt, B.J.: Partial word DFAs. In: Konstantinidis, S. (ed.) CIAA 2013. LNCS, vol. 7982, pp. 36–47. Springer, Heidelberg (2013)
2. Björklund, H., Martens, W.: The tractability frontier for NFA minimization. Journal of Computer and System Sciences 78, 198–210 (2012)
3. Blanchet-Sadri, F.: Algorithmic Combinatorics on Partial Words. Chapman & Hall/CRC Press, Boca Raton (2008)
4. Dassow, J., Manea, F., Mercaş, R.: Regular languages of partial words. Information Sciences 268, 290–304 (2014)
5. Groz, B., Maneth, S., Staworko, S.: Deterministic regular expressions in linear time. In: 31th ACM Symposium on Principles of Database Systems, PODS 2012, pp. 49–60 (2012)
6. Holzer, M., Jakobi, S., Wendlandt, M.: On the computational complexity of partial word automata problems. IFIG Research Report 1404, Institut für Informatik, Justus-Liebig-Universität Gießen, Arndtstr. 2, D-35392 Gießen, Germany (May 2014)
7. Hopcroft, J.E.: An n log n algorithm for minimizing states in a finite automaton. Tech. rep., DTIC Document (1971)
8. Jiang, T., Ravikumar, B.: Minimal NFA problems are hard. SIAM Journal on Computing 22, 1117–1141 (1993)

Large Aperiodic Semigroups[*]

Janusz Brzozowski[1] and Marek Szykuła[2]

[1] David R. Cheriton School of Computer Science, University of Waterloo
Waterloo, ON, N2L 3G1, Canada
brzozo@uwaterloo.ca
[2] Institute of Computer Science, University of Wrocław
Joliot-Curie 15, 50-383 Wrocław, Poland
msz@cs.uni.wroc.pl

Abstract. We search for the largest syntactic semigroup of a star-free language having n left quotients; equivalently, we look for the largest transition semigroup of an aperiodic finite automaton with n states.

We first introduce *unitary* semigroups generated by transformations that change only one state. In particular, we study *complete* unitary semigroups which have a special structure, and we show that each maximal unitary semigroup is complete. For $n \geqslant 4$ there exists a complete unitary semigroup that is larger than any aperiodic semigroup known to date.

We then present even larger aperiodic semigroups, generated by transformations that map a non-empty subset of states to a single state; we call such transformations and semigroups *semiconstant*. In particular, we examine semiconstant *tree* semigroups which have a structure based on full binary trees. The semiconstant tree semigroups are at present the best candidates for largest aperiodic semigroups.

Keywords: aperiodic, monotonic, nearly monotonic, partially monotonic, semiconstant, transition semigroup, star-free language, syntactic complexity, unitary.

1 Introduction

The *state complexity* of a regular language is the number of states in a complete minimal deterministic finite automaton (DFA) accepting the language [14]. An equivalent notion is that of *quotient complexity*, which is the number of left quotients of the language [1]; we prefer quotient complexity since it is a language-theoretic notion. The usual measure of complexity of an operation on regular languages [1,14] is the quotient complexity of the result of the operation as a function of the quotient complexities of the operands. This measure has some serious disadvantages, however. For example, as shown in [5], in the class of star-free languages all common operations have the same quotient complexity as they do in the class of arbitrary regular languages with two small exceptions.

[*] This work was supported by the Natural Sciences and Engineering Research Council of Canada grant No. OGP000087 and by Polish NCN grant DEC-2013/09/N/ST6/01194.

Thus quotient complexity fails to differentiate between the very special class of star-free languages and the class of all regular languages.

It has been suggested that other measures of complexity—in particular, the *syntactic complexity* of a regular language, which is the cardinality of its syntactic semigroup [12]—may also be useful [2]. Syntactic complexity is the same as the cardinality of the *transition semigroup* of a minimal DFA accepting the language, and it is this latter representation that we use here. The *syntactic complexity of a class* of languages is the size of the largest syntactic semigroups of languages in that class as a function of the quotient complexities of the languages. Since the syntactic complexity of star-free languages is considerably smaller than that of regular languages, this measure succeeds in distinguishing the two classes.

The class of *star-free* languages is the smallest class obtained from finite languages using only boolean operations and concatenation, but no star. By Schützenberger's theorem [13] we know that a language is star-free if and only if the transition semigroup of its minimal DFA is *aperiodic*, meaning that it contains no non-trivial subgroups. Star-free languages and the DFAs that accept them were studied by McNaughton and Papert in 1971 [11].

Two aperiodic semigroups, monotonic and partially monotonic, were studied by Gomes and Howie [8]. Their results were adapted to finite automata in [4], where nearly monotonic semigroups were also introduced; they are larger than the partially monotonic ones and were the largest aperiodic semigroups known to date for $n \leqslant 7$. For $n \geqslant 8$ the largest aperiodic semigroups known to date were those generated by DFAs accepting $\mathcal{R}$-trivial languages [3]. The syntactic complexity of $\mathcal{R}$-trivial languages is $n!$. As to aperiodic semigroups, tight upper bounds on their size were known only for $n \leqslant 3$.

The following are the main contributions of this paper:

1. Using the method of [10], we have enumerated all aperiodic semigroups for $n = 4$, and we have shown that the maximal aperiodic semigroup has size 47, while the maximal nearly monotonic semigroup has size 41. Although this may seem like an insignificant result, it provided us with strong motivation to search for larger semigroups.
2. We studied semigroups generated by transformations that change only one state; we call such transformations and semigroups *unitary*. We characterized unitary semigroups and computed their maximal sizes up to $n = 1,000$. For $n \geqslant 4$ the maximal unitary semigroups are larger than any previously known aperiodic semigroup.
3. For each n we found a set of DFAs whose inputs induce *semiconstant tree* transformations that send a non-empty subset of states to a single state, and have a structure based on full binary trees. For $n \geqslant 4$, there is a semiconstant tree semigroup larger than the largest complete unitary semigroup. We computed the maximal size of these transition semigroups up to $n = 500$.
4. We derived formulas for the sizes of complete unitary and semiconstant tree semigroups. We also provided recursive formulas characterizing the maximal complete unitary and semiconstant tree semigroups; these formulas lead to efficient algorithms for computing the forms and sizes of such semigroups.

Our results about aperiodic semigroups are summarized in Tables 1 and 2 for small values of n. Transformation 1 is the identity; it can be added to unitary and semiconstant transformations without affecting aperiodicity. The classes are listed in the order of increasing size when n is large.

There are two more classes of syntactic semigroups that have the same complexity as the semigroups of finite languages [4]: those of cofinite and reverse definite languages. The lower bound for definite languages ([4]) is the same as the tight upper bound for $\mathcal{J}$-trivial languages ([3]), but it is not known whether this is also an upper bound for definite languages.

Omitted proofs can be found in [6].

Table 1. Large aperiodic semigroups

$n:$	1	2	3	4	5	6	7	8
Monotonic $\binom{2n-1}{n}$	1	3	10	35	126	462	1,716	6,435
Part. mon. $e(n)$	–	2	8	38	192	1,002	5,336	28,814
Near. mon. $e(n)+n-1$	–	3	10	41	196	1,007	5,342	28,821
Finite $(n-1)!$	1	1	2	6	24	120	720	5,040
$\mathcal{J}$-trivial $\lfloor e(n-1)! \rfloor$	1	2	5	16	65	326	1,957	13,700
$\mathcal{R}$-trivial $n!$	1	2	6	24	120	720	5,040	40,320
Complete unitary with 1	–	3	10	45	270	1,737	13,280	121,500
Semiconstant tree with 1	–	3	10	47	273	1,849	14,270	126,123
Aperiodic	1	3	10	47	?	?	?	?

Table 2. Large aperiodic semigroups continued

$n:$	9	10	11	12	13
Monotonic	24,310	92,378	352,716	1,352,078	5,200,300
Part. mon.	157,184	864,146	4,780,008	26,572,086	148,321,344
Near. mon.	157,192	864,155	4,780,018	26,572,097	148,321,352
Finite	40,320	362,880	3,628,800	39,916,800	479,001,600
$\mathcal{J}$-trivial	109,601	986,410	9,864,101	108,505,112	1,302,061,345
$\mathcal{R}$-trivial	362,880	3,628,800	39,916,800	479,001,600	6,227,020,800
Comp. unit., 1	1,231,200	12,994,020	151,817,274	2,041,564,499	29,351,808,000
Sc. tree, 1	1,269,116	14,001,630	169,410,933	2,224,759,334	31,405,982,420
Aperiodic	?	?	?	?	?

2   Terminology and Notation

Let Σ be a finite alphabet. The elements of Σ are *letters* and the elements of Σ^* are *words*, where Σ^* is the free monoid generated by Σ. The empty word is denoted by ε, and the set of all non-empty words is Σ^+, the free semigroup generated by Σ. A *language* is any subset of Σ^*.

Suppose $n \geqslant 1$. Without loss of generality we assume that our basic set under consideration is $Q = \{0, 1, \ldots, n-1\}$. A *deterministic finite automaton (DFA)* is a quintuple $\mathcal{D} = (Q, \Sigma, \delta, 0, F)$, where Σ is a finite non-empty *alphabet*, $\delta \colon Q \times \Sigma \to Q$ is the *transition function*, $0 \in Q$ is the *initial state*, and $F \subseteq Q$ is the set of *final states*. We extend δ to $Q \times \Sigma^*$ and to $2^Q \times \Sigma^*$ in the usual way. A DFA $\mathcal{D}$ *accepts* a word $w \in \Sigma^*$ if $\delta(0, w) \in F$. The *language accepted* by $\mathcal{D}$ is $L(\mathcal{D}) = \{w \in \Sigma^* \mid \delta(0, w) \in F\}$.

By the *language of a state* q of $\mathcal{D}$ we mean the language $L_q(\mathcal{D})$ accepted by the DFA $(Q, \Sigma, \delta, q, F)$. A state is *empty* (also called *dead* or a *sink*) if its language is empty. Two states p and q of $\mathcal{D}$ are *equivalent* if $L_p(\mathcal{D}) = L_q(\mathcal{D})$. Otherwise, states p and q are *distinguishable*. A state q is *reachable* if there exists a word $w \in \Sigma^*$ such that $\delta(0, w) = q$. A DFA is *minimal* if all its states are reachable and pairwise distinguishable.

A *transformation* of Q is a mapping of Q into itself. Let t be a transformation of Q; then qt is the *image* of $q \in Q$ under t. If P is a subset of Q, then $Pt = \{qt \mid q \in P\}$. An arbitrary transformation has the form $t = \begin{pmatrix} 0 & 1 & \cdots & n-2 & n-1 \\ p_0 & p_1 & \cdots & p_{n-2} & p_{n-1} \end{pmatrix}$, where $p_q = qt$ for $q \in Q$. We also use $t = [p_0, \ldots, p_{n-1}]$ as a simplified notation. The *composition* of two transformations t_1 and t_2 of Q is a transformation $t_1 \circ t_2$ such that $q(t_1 \circ t_2) = (qt_1)t_2$ for all $q \in Q$. We usually write $t_1 t_2$ for $t_1 \circ t_2$.

Let $\mathcal{T}_Q$ be the set of all n^n transformations of Q; then $\mathcal{T}_Q$ is a monoid under composition. The *identity* transformation $\mathbf{1}$ maps each element to itself, that is, $q\mathbf{1} = q$ for all $q \in Q$. A *permutation* of Q is a mapping of Q *onto* itself. For $k \geqslant 2$, a transformation (permutation) t of a set $P = \{q_0, q_1, \ldots, q_{k-1}\} \subseteq Q$ is a *k-cycle* if $q_0 t = q_1, q_1 t = q_2, \ldots, q_{k-2} t = q_{k-1}, q_{k-1} t = q_0$. A k-cycle is denoted by $(q_0, q_1, \ldots, q_{k-1})$. If a transformation t of Q acts like a k-cycle on some $P \subseteq Q$, we say that t has a *k-cycle*. A transformation has a *cycle* if it has a k-cycle for some $k \geqslant 2$. For $p \neq q$, a *transposition* is the 2-cycle (p, q). A transformation is *aperiodic* if it contains no cycles. A transformation semigroup is aperiodic if it contains only aperiodic transformations.

In any DFA $\mathcal{D}$, each word $w \in \Sigma^*$ induces a transformation t_w of Q defined by $q t_w = \delta(q, w)$ for all $q \in Q$. The set of all transformations of Q induced in $\mathcal{D}$ by non-empty words is the *transition semigroup* of $\mathcal{D}$, a subsemigroup of $\mathcal{T}_Q$. A DFA is *aperiodic* if its transition semigroup is aperiodic. If $\mathcal{D}$ is minimal, its transition semigroup is isomorphic to the *syntactic semigroup* of the language $L(\mathcal{D})$ [11,12]. A language is regular if and only if its syntactic semigroup is finite. The size of the syntactic semigroup of a language is called its *syntactic complexity*. We deal only with transition semigroups and view syntactic complexity as the size of the transition semigroup.

If T is a set of transformations, then $\langle T \rangle$ is the semigroup generated by T. If $\mathcal{D} = (Q, \Sigma, \delta, 0, F)$ is a DFA, the transformations induced by letters of Σ are called *generators of the transition semigroup* of $\mathcal{D}$, or simply *generators of $\mathcal{D}$*.

3 Unitary and Semiconstant DFAs

We now define a new class of aperiodic DFAs among which are found the largest transition semigroups known to date. We also study several of its subclasses.

A *unitary* transformation t, denoted by $(p \to q)$, has $p \neq q$, $pt = q$ and $rt = r$ for all $r \neq p$. A DFA is *unitary* if each of its generators is unitary. A semigroup is *unitary* if it has a set of unitary generators.

A *constant* transformation t, denoted by $(Q \to q)$, has $pt = q$ for all $p \in Q$. A transformation t is *semiconstant* if it maps a non-empty subset P of Q to a single element q and leaves the remaining elements of Q unchanged. It is denoted by $(P \to q)$. A constant transformation is semiconstant with $P = Q$, and a unitary transformation $(p \to q)$ is semiconstant with $P = \{p\}$. A DFA is *semiconstant* if each of its generators is semiconstant. A semigroup is *semiconstant* if it has a set of semiconstant generators.

For each $n \geqslant 1$ we shall define several DFAs. Let $m, n_1, n_2, \ldots, n_m$ be positive natural numbers. Also, let $n = n_1 + \cdots + n_m$, and for each i, $1 \leqslant i \leqslant m$, define r_i by $r_i = \sum_{j=1}^{i-1} n_j$. For $i = 1, \ldots, m$, let $Q_i = \{r_i, r_i + 1, \ldots, r_{i+1} - 1\}$; thus the cardinality of Q_i is n_i. Let $Q = Q_1 \cup \cdots \cup Q_m = \{0, \ldots, n - 1\}$; the cardinality of Q is n. The sequence $(n_1, n_2, \ldots, n_m)$ is called the *distribution* of Q.

A binary tree is *full* if every vertex has either two children or no children. Let Δ_Q be a full binary tree with m leaves labeled $Q_1, \ldots, Q_m$ from left to right. To each node $v \in \Delta_Q$, we assign the union $Q(v)$ of all the sets Q_i labeling the leaves in the subtree rooted at v.

With each full binary tree we can associate different distributions. A full binary tree Δ_Q with a distribution attached is denoted by $\Delta_Q(n_1, n_2, \ldots, n_m)$ and is called the *structure* of Q. This structure will uniquely determine the transition function δ of the DFAs defined below.

We can denote the structure of Q as a binary expression. For example, the expression $((3, 2), (4, 1))$ denotes the full binary tree in which the leaves are labeled Q_1, Q_2, Q_3, and Q_4, where $|Q_1| = 3, |Q_2| = 2, |Q_3| = 4, |Q_4| = 1$, and the interior nodes are labeled by $Q_1 \cup Q_2$, $Q_3 \cup Q_4$ and $Q_1 \cup Q_2 \cup Q_3 \cup Q_4$. On the other hand, the expression $(((3, 2), 4), 1)$ has interior nodes labeled $Q_1 \cup Q_2$, $Q_1 \cup Q_2 \cup Q_3$ and $Q_1 \cup Q_2 \cup Q_3 \cup Q_4$.

Definition 1 (Transformations)

Type 1: *Suppose $n > 1$ and $(n_1, n_2, \ldots, n_m)$ is a distribution of Q. For all $i = 1, \ldots, m$ and $q, q + 1 \in Q_i$ Type 1 transformations are the unitary transformations $(q \to q + 1)$ and $(q + 1 \to q)$.*

Type 2: *Suppose $n > 1$ and $(n_1, n_2, \ldots, n_m)$ is a distribution of Q. If $1 \leqslant i \leqslant m - 1$ and $i < j \leqslant m$, for each $q \in Q_i$ and $p \in Q_j$, $(q \to p)$ is a Type 2 transformation.*

Type 3: *Suppose $n > 1$ and $\Delta_Q(n_1, n_2, \ldots, n_m)$ is a structure of Q. For each internal node w the semiconstant transformation $(Q(w) \to \min(Q(w)))$ is of Type 3.*

Type 4: *The identity transformation $\mathbf{1}$ on Q is of Type 4.*

In the following DFAs the transition function is defined by a set of transformations and the alphabet consists of letters inducing these transformation.

Definition 2 (DFAs). *Suppose* $n > 1$.

1. *If there is no* $i \in \{1, \ldots, m-1\}$ *such that* $|Q_i| = |Q_{i+1}| = 1$, *then any DFA of the form* $\mathcal{D}_u(n_1, \ldots, n_m) = (Q, \Sigma_u, \delta_u, 0, \{n-1\})$, *where* δ_u *has all the transformations of Types 1 and 2, is a* complete unitary *DFA.*
2. $\mathcal{D}_{ui}(n_1, \ldots, n_m) = (Q, \Sigma_{ui}, \delta_{ui}, 0, \{n-1\})$ *is* $\mathcal{D}_u(n_1, \ldots, n_m)$ *with* **1** *added.*
3. *Any DFA* $\mathcal{D}_{sct}(\Delta_Q(n_1, \ldots, n_m)) = (Q, \Sigma_{sct}, \delta_{sct}, 0, \{n-1\})$, *where* δ_{sct} *has all the transformations of Types 1, 2 and 3, is a* semiconstant tree *DFA.*
4. $\mathcal{D}_{scti}(\Delta_Q(n_1, \ldots, n_m)) = (Q, \Sigma_{scti}, \delta_{scti}, 0, \{n-1\})$ *is* $\mathcal{D}_{sct}(\Delta_Q(n_1, \ldots, n_m))$ *with* **1** *added.*

Following [7], we define a *bipath (bidirectional path)* to be a graph (V, E), where $V = \{v_0, \ldots, v_{k-1}\}$ for some $k \geqslant 1$, and for each $v_q, v_{q+1} \in V$ there are two edges (v_q, v_{q+1}) and (v_{q+1}, v_q). If $k = 1$, $(\{v_0\}, \emptyset)$ is a trivial bipath. If we ignore self-loops, each edge in the graph uniquely determines a unitary transformation, and the states in each Q_i in $\mathcal{D}_u(n_1, \ldots, n_m)$ constitute a *bipath*. Also, the graph of $\mathcal{D}_u(n_1, \ldots, n_m)$ is a sequence $(Q_1, \ldots, Q_m)$ of bipaths, where there are transitions from every q in Q_i to every p in Q_j, if $i < j$.

Example 1. Figure 1 shows three examples of unitary DFAs. In Fig. 1 (a) we have DFA $\mathcal{D}_u(3)$, where the letter a_{pq} induces the unitary transformation $(p \to q)$. In Fig. 1 (b) we present $\mathcal{D}_u(3)$, where only the transitions between *different* states are included to simplify the figure. Also, the letter labels are deleted because they are easily deduced. Next, in Figs. 1 (c) and (d), we have the DFAs $\mathcal{D}_u(3, 1)$ and $\mathcal{D}_u(2, 2, 2)$, respectively. We shall return to these examples later.

Remark 1. All four DFAs of Definition 2 are minimal as is easily verified.

4   Unitary Semigroups

We study unitary semigroups because their generators are the simplest. We begin with three previously studied special semigroups.

Monotonic Semigroups [4,8,9]

A transformation t of Q is *monotonic* if there exists a total order $\leqslant$ on Q such that, for all $p, q \in Q$, $p \leqslant q$ implies $pt \leqslant qt$. A DFA is *monotonic* if each of its generators is monotonic. A semigroup is *monotonic* if it has a set of monotonic generators. We assume that $\leqslant$ is the usual order on integers.

The following result of [8] is somewhat modified for our purposes:

Proposition 1 (Gomes and Howie). *The set* M *of all* $\binom{2n-1}{n} - 1$ *monotonic transformations other than* **1** *is an aperiodic semigroup generated by* $G_M = \{(q \to q+1) \mid 0 \leqslant q \leqslant n-2\} \cup \{(q \to q-1) \mid 1 \leqslant q \leqslant n-1\}$, *and no smaller set of unitary transformations generates* M.

Corollary 1. *The transition semigroup of* $\mathcal{D}_{ui}(n)$ *is the semigroup* $M \cup \{1\}$ *of all monotonic transformations.*

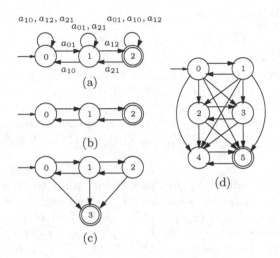

Fig. 1. Unitary DFAs: (a) $\mathcal{D}_u(3)$; (b) $\mathcal{D}_u(3)$ simplified; (c) $\mathcal{D}_u(3,1)$; (d) $\mathcal{D}_u(2,2,2)$

Figure 1 (b) shows $\mathcal{D}_u(3)$ and $\mathcal{D}_{ui}(3)$, if **1** is added. The transition semigroup of $\mathcal{D}_{ui}(3)$ has ten elements and is the largest aperiodic semigroup for $n = 3$ [4].

Partially Monotonic Semigroups [4,8]

A *partial transformation* t of Q is a partial mapping of Q into itself. If t is defined for $q \in Q$, then qt is the image of q under t; otherwise, we write $qt = \square$. By convention, $\square t = \square$. The *domain* of t is the set $dom(t) = \{q \in Q \mid qt \neq \square\}$. A partial transformation is *monotonic* if there exists an order $\leqslant$ on Q such that for all $p, q \in dom(t)$, $p \leqslant q$ implies $pt \leqslant qt$.

We start with all partial transformations of $Q\setminus\{n-1\}$ and add state $(n-1)$ for the undefined value $\square$. The resulting transformations are *partially monotonic*. The next result follows from [8]:

Proposition 2. *For $n \geqslant 2$, the DFA $\mathcal{D}_{ui}(n-1,1) = (Q, \Sigma_{ui}, \delta_{ui}, 0, \{n-1\})$ has the following properties:*

1. Each of the $3n-4$ transformations of $\mathcal{D}_{ui}(n-1,1)$ is partially monotonic. Thus $\mathcal{D}_{ui}(n-1,1)$ is partially monotonic, and hence aperiodic.

2. The transition semigroup PM_Q of $\mathcal{D}_{ui}(n-1,1)$ consists of all the $e(n)$ partially monotonic transformations of Q, where $e(n) = \sum_{k=0}^{n-1} \binom{n-1}{k}\binom{n+k-2}{k}$.

*3. Each generator is idempotent, and $3n-4$ is the smallest number of idempotent generators of PM_Q. Moreover, each generator except **1** is unitary, and $3n-5$ is the smallest number of unitary generators of $PM_Q \setminus \{\mathbf{1}\}$.*

For $n \geqslant 4$ the semigroup of all partially monotonic transformations is larger than the semigroup of all monotonic transformations.

General Unitary Semigroups

A set $\{t_0, \ldots, t_{k-1}\}$ of unitary transformations is *k-cyclic* if it has the form $t_0 = (q_0 \to q_1)$, $t_1 = (q_1 \to q_2), \ldots, t_{k-2} = (q_{k-2} \to q_{k-1})$, $t_{k-1} = (q_{k-1} \to q_0)$, where the q_i are distinct.

Lemma 1. *Let T be a set of unitary transformations.*

1. If T has a k-cyclic subset $\{t_0, \ldots, t_{k-1}\}$ with $k \geqslant 3$, then $\langle T \rangle$ is not aperiodic.

2. If T contains a subset $T_6 = \{t_{01}, t_{10}, t_{12}, t_{13}, t_{21}, t_{31}\}$ where $t_{i,j} = (q_i \to q_j)$ and $q_0, q_1, q_2, q_3 \in Q$, then $\langle T \rangle$ is not aperiodic.

Theorem 1. *If $\mathcal{D} = (Q, \Sigma, \delta, 0, F)$ is unitary, the following are equivalent:*

1. $\mathcal{D}$ is aperiodic.

2. The set of generators of $\mathcal{D}$ does not contain any k-cyclic subsets with $k \geqslant 3$, and does not contain any sets of type T_6.

3. Every strongly connected component of $\mathcal{D}$ is a bipath.

Proof. $1 \Rightarrow 2$: This follows from Lemma 1.

$2 \Rightarrow 3$: Consider a strongly connected component C. If $|C| = 1$, the claim holds. Otherwise, suppose $p \in C$ and $(p \to q)$ is a transition. Then there must also be a directed path from q to p. If the last transition in that path is $(r \to p)$, where $r \neq q$, then the set of generators must contain a k-cyclic subset with $k \geqslant 3$, which is a contradiction. Hence the transition $(q \to p)$ must be present.

Next, suppose that there are transitions $(p \to q)$, $(p \to r)$, and $(p \to s)$. By the argument above there must also be transitions $(q \to p)$, $(r \to p)$, and $(r \to s)$. But then the set of generators contains a subset of type T_6, which is again a contradiction.

It follows that every strongly connected component is a bipath, and the graph of the transitions of $\mathcal{D}$ is a loop-free connection of such bipaths.

$3 \Rightarrow 1$: Since a bipath is monotonic, it is aperiodic by Proposition 1. By Schützenberger's theorem [13], the language of all words taking any state of the bipath to any other state of that bipath is star-free. Since the graph of $\mathcal{D}$ is a loop-free connection of bipaths, the language of all words taking any state of $\mathcal{D}$ to any other state of $\mathcal{D}$ is star-free. Hence $\mathcal{D}$ is aperiodic. $\square$

A unitary DFA is *complete* if the addition of any unitary transition results in a DFA that is not aperiodic.

Theorem 2. *A maximal aperiodic unitary semigroup is isomorphic to the transition semigroup of a complete unitary DFA $\mathcal{D}_u(n_1, \ldots, n_m)$, where $(n_1, \ldots, n_m)$ is some distribution of Q.*

Proof. We know that an aperiodic unitary DFA $\mathcal{D}$ is a loop-free connection of bipaths. Let $Q_1, \ldots, Q_m$ be the bipaths of $\mathcal{D}$. There exists a linear ordering $<$ of them, such that there is no transformation $(p \to q)$ for $q \in Q_i, p \in Q_j, i < j$. If all possible transformations $(q \to p)$ for $q \in Q_i, p \in Q_j, i < j$ are present, then $\mathcal{D}$ is isomorphic to $\mathcal{D}_u(n_1, \ldots, n_m)$. Otherwise we can add more unitary transformations of Type 2 and obtain a larger semigroup.   □

For each distribution $(n_1, \ldots, n_m)$, we calculate the size of the transition semigroup of $\mathcal{D}_{ui}(n_1, \ldots, n_m)$.

Theorem 3. *The cardinality of the transition semigroup of $\mathcal{D}_{ui}(n_1, \ldots, n_m)$ is*

$$\prod_{i=1}^{m} \left(\binom{2n_i - 1}{n_i} + \sum_{h=0}^{n_i - 1} \left(\sum_{j=i+1}^{m} n_j \right)^{n_i - h} \binom{n_i}{h} \binom{n_i + h - 1}{h} \right). \tag{1}$$

Note that each factor of the product in Theorem 3 depends only on n_i and on the sum $k = n_{i+1} + \cdots + n_m$. Hence if $\mathcal{D}_{ui}(n_1, \ldots, n_m)$ is maximal, then $\mathcal{D}_{ui}(n_2, \ldots, n_m)$ is also maximal and so on. Consequently, we have

Corollary 2. *Let $m_{ui}(n)$ be the cardinality of the largest transition semigroup of DFA $\mathcal{D}_{ui}(n_1, \ldots, n_m)$ with n states. If we define $m_{ui}(0) = 1$, then for $n > 0$*

$$m_{ui}(n) = \max_{j=1,\ldots,n} \left(m_{ui}(n-j) \left(\binom{2j-1}{j} + \sum_{h=0}^{j-1} (n-j)^{j-h} \binom{j}{h} \binom{j+h-1}{h} \right) \right). \tag{2}$$

This leads directly to a dynamic algorithm taking $O(n^3)$ time for computing $m_{ui}(n)$ and the distributions $(n_1, \ldots, n_m)$ yielding the maximal unitary semigroups. This holds assuming constant time for computing the internal terms in the summation and summing them, where, however, the numbers can be very large. The precise complexity depends on the algorithms used for multiplication, exponentiation and calculation of binomial coefficients.

We were able to compute the maximal $\mathcal{D}_{ui}$ up to $n = 1,000$. Here is an example of the maximal one for $n = 100$: $\mathcal{D}_{ui}(12, 11, 10, 10, 9, 8, 8, 7, 6, 5, 5, 4, 3, 2)$; its syntactic semigroup size exceeds 2.1×10^{160}. Compare this to the previously known largest semigroup of an $\mathcal{R}$-trivial language; its size is $100!$ which is approximately 9.3×10^{157}. On the other hand, the maximal possible syntactic semigroup of any regular language for $n = 100$ is 10^{200}.

Asymptotic Lower Bound

We were not able to compute the tight asymptotic bound on the maximal size of unitary semigroups. However, we computed a lower bound which is larger than $n!$, the previously known lower bound for the size of aperiodic semigroups.

Theorem 4. *For even n the size of the maximal unitary semigroup is at least*

$$\frac{n!(n+1)!}{2^n ((n/2)!)^2}.$$

For $n = 100$ the bound exceeds 7.5×10^{158}. Larger lower bounds can also be found using increasing values of j in $\mathcal{D}_{ui}(j, j, \ldots, j)$, but the complexity of the calculations increases, and such bounds are not tight.

5   Semiconstant Semigroups

Nearly Monotonic Semigroups [4]

Let K_Q be the set of all constant transformations of Q, and $NM_Q = PM_Q \cup K_Q$. We call the transformations in NM_Q *nearly monotonic* with respect to the usual order on integers. For $n \geqslant 4$ the semigroup of all nearly monotonic transformations is larger than that of all partially monotonic ones.

Semiconstant Tree Semigroups

An example of a maximal semiconstant tree DFA for $n = 6$ is $\mathcal{D}_{scti}((2,2),2)$; its transition semigroup has 1,849 elements. For $n \geqslant 4$, the maximal semiconstant tree semigroup is the largest aperiodic semigroup known.

Definition 3. *Let $\mathcal{A} = (Q_A, \Sigma_A, \delta_A, q_A, F_A)$ and $\mathcal{B} = (Q_B, \Sigma_B, \delta_B, q_B, F_B)$ be DFAs, where $Q_A \cap Q_B = \emptyset$, and $\Sigma_A \cap \Sigma_B = \emptyset$. The semiconstant sum of $\mathcal{A}$ and $\mathcal{B}$ is denoted by $\mathcal{C} = (\mathcal{A}, \mathcal{B})$ and is the DFA $(Q_C, \Sigma_C, \delta_C, q_A, F_B)$, where $Q_C = Q_A \cup Q_B$. For each transition t in δ_A, we have a transition t' in δ_C such that $qt' = qt$ for $q \in Q_A$ and $qt' = q$ otherwise. Dually, we have transitions defined by t in δ_B. Moreover, we have a unitary transformation $(p \to q)$ for each $p \in Q_A, q \in Q_B$, and a constant transformation $(Q_C \to q_A)$.*

Lemma 2. *The semiconstant sum $\mathcal{C} = (\mathcal{A}, \mathcal{B})$ is minimal if and only if every state of $\mathcal{A}$ is reachable from q_A, the states of $\mathcal{B}$ are pairwise distinguishable, and F_B is non-empty.*

For $m > 1$, each $\mathcal{D}_{scti}(\Delta_Q(n_1, \ldots, n_m))$ is a semiconstant sum of two smaller semiconstant tree DFAs $\mathcal{D}_{scti}(\Delta_{Q_{left}}(n_1, \ldots, n_r))$, defined by the left subtree of the root of $\Delta_Q(n_1, \ldots, n_m)$, and $\mathcal{D}_{scti}(\Delta_{Q_{right}}(n_{r+1}, \ldots, n_m))$, defined by the right subtree.

Lemma 3. *If $\mathcal{A}$ and $\mathcal{B}$ are aperiodic, so is their semiconstant sum.*

Proof. Suppose that $\langle (\mathcal{A}, \mathcal{B}) \rangle$ contains a cycle t. This cycle cannot include both a state from $\mathcal{A}$ and a state from $\mathcal{B}$, since the only way to map a state from $\mathcal{B}$ to a state from $\mathcal{A}$ in $(\mathcal{A}, \mathcal{B})$ is by a constant transformation, and a constant transformation cannot be used as a generator of a cycle. Hence all the cyclic states must be either in Q_A or Q_B, which contradicts the assumption that $\mathcal{A}$ and $\mathcal{B}$ are aperiodic. $\square$

An DFA is *transition-complete* if it is aperiodic and adding any transition to it destroys aperiodicity.

Lemma 4. *If $\mathcal{A}$ and $\mathcal{B}$ are transition-complete, so is their semiconstant sum.*

Corollary 3. *All semiconstant tree DFAs of the form $\mathcal{D}_{scti}(\Delta_Q(n_1, \ldots, n_m))$ are transition-complete.*

In order to count the size of the semigroup of a semiconstant sum, we extend the concept of partial transformations to k-partial transformations.

Definition 4. *A k-partial transformation of Q is a transformation of Q into $Q \cup \{\square_1, \square_2, \ldots, \square_k\}$, where $\square_1, \square_2, \ldots, \square_k$ are pairwise distinct, and distinct from all $q \in Q$.*

Let $\mathcal{A} = (Q, \Sigma, \delta, s, F)$ be a DFA, and let t be a k-partial transformation of Q. We say that t is *consistent* for $\mathcal{A}$ if there exists t' in δ such that if $qt \in Q$, then $qt = qt'$ for all $q \in Q$.

The set of consistent k-partial transformations of a semigroup describes its potential for forming a large number of transformations, when used in a semiconstant sum. For a fixed $n \geqslant 6$, there exist semigroups with smaller cardinalities than the maximal ones, but with larger numbers of consistent k-partial transformations for some k. Thus k-partial transformations are useful for finding such non-maximal semigroups, as they can result in larger semigroups when used in compositions.

The transition semigroup of $\mathcal{A}$ can be characterized by a function $f_{\mathcal{A}} \colon \mathbb{N} \to \mathbb{N}$ counting all consistent k-partial transformations for a given k. For example, for $k = 1$, $f_{\mathcal{A}}$ is the number of all consistent partial transformations for $\mathcal{A}$. For a DFA $\mathcal{A} = \mathcal{D}_{ui}(n_1, \ldots, n_m)$, $f_{\mathcal{A}}(1)$ is the size of the semigroup of $\mathcal{D}_{ui}(n_1, \ldots, n_m, 1)$.

From Theorem 3 we know that the number of consistent k-partial transformations for a bipath of size n having an identity transformation is $m_{bi}(n, k) = \binom{2n-1}{n} + \sum_{h=0}^{n-1} k^{n-h} \binom{n}{h} \binom{n+h-1}{h}$.

Theorem 5. *Let $\mathcal{A}$ and $\mathcal{B}$ be strongly connected DFAs with n and m states, respectively. Let $f_{\mathcal{A}}(k)$ and $f_{\mathcal{B}}(k)$ be the functions counting their consistent k-partial transformations. Then the function f_C counting the consistent k-partial transformations of the semiconstant sum $C = (\mathcal{A}, \mathcal{B})$ is $f_C(k) = f_{\mathcal{A}}(m+k) f_{\mathcal{B}}(k) + n(k+1)^n ((k+1)^m - k^m)$.*

Corollary 4. *The number of k-partial transformations of $\mathcal{D}_{scti}(\Delta_Q(n_1, \ldots, n_m))$ of size n is:*

$$f_{\mathcal{D}}(k) = \begin{cases} m_{bi}(n, k), & \text{if } m = 1; \\ f_{\mathcal{D}_{left}}(r + k) f_{\mathcal{D}_{right}}(k) + \ell(k+1)^\ell ((k+1)^r - k^r), & \text{if } m > 1, \end{cases}$$

where $\mathcal{D}_{left}$ is the DFA defined by $\Delta_{Q_{left}}(n_1, \ldots, n_i)$, the left subtree of the tree $\Delta_Q(n_1, \ldots, n_m)$, $\mathcal{D}_{right}$ is defined by $\Delta_{Q_{right}}(n_{i+1}, \ldots, n_m)$, the right subtree of $\Delta_Q(n_1, \ldots, n_m)$, and ℓ, r are the numbers of states in $\mathcal{D}_{left}$ and $\mathcal{D}_{right}$, respectively.

Proof. This follows from Theorems 3 and 5. $\square$

The size of the semigroup of DFA $\mathcal{D}_{scti}(n_1, \ldots, n_m)$ is $f_{\mathcal{D}}(0)$.

Corollary 5. *Let $m_{scti}(n, k)$ be the maximal number of k-partial transformations of a semiconstant DFA $\mathcal{D}_{scti}(n_1, \ldots, n_m)$ with n states. Then*

$$m_{scti}(n, k) = \max \begin{cases} m_{bi}(n, k) \\ \max\limits_{s=1,\ldots,n-1} \left\{ \begin{array}{l} m_{scti}(n - s, s + k) m_{scti}(s, k) \\ + (n - s)(k+1)^{n-s}((k+1)^s - k^s) \end{array} \right\} \end{cases}. \quad (3)$$

The maximal size of semigroups of the DFAs $\mathcal{D}_{scti}$ with n states is $m_{scti}(n, 0)$. Instead of a bipath and the value $m_{bi}(n, k)$ we could use any strongly connected automaton with an aperiodic semigroup. If such a semigroup would have a larger number of k-partial transformations than our semiconstant tree DFAs for some k, then we could obtain even larger aperiodic semigroups.

The corollary results directly in a dynamic algorithm working in $O(n^3)$ time (assuming constant time for arithmetic operations and computing binomials) for computing $m_{scti}(n, 0)$, and the distribution with the full binary tree yielding the maximal semiconstant tree semigroup.

We computed the maximal semiconstant tree semigroups up to $n = 500$. For $n = 100$, for example, one of the maximal DFAs is

$$\mathcal{D}_{scti} \;\; (((((((2,2),(2,2)),((2,2),(2,2))),(((2,2),(2,2)),((2,2),3))),$$
$$((((2,2),3),(3,3)),((3,3),(3,3)))),((((3,2),(3,2)),((3,2),(2,2))),$$
$$((2,2),(2,2)))),(((3,3),(3,2)),((2,2),2))),$$

and its syntactic semigroup size exceeds 3.3×10^{160}.

References

1. Brzozowski, J.: Quotient complexity of regular languages. J. Autom. Lang. Comb. 15(1/2), 71–89 (2010)
2. Brzozowski, J.: In search of the most complex regular languages. Internat. J. Found. Comput. Sci. 24(6), 691–708 (2013)
3. Brzozowski, J., Li, B.: Syntactic complexity of $\mathcal{R}$- and $\mathcal{J}$-trivial languages. In: Jurgensen, H., Reis, R. (eds.) DCFS 2013. LNCS, vol. 8031, pp. 160–171. Springer, Heidelberg (2013)
4. Brzozowski, J., Li, B., Liu, D.: Syntactic complexities of six classes of star-free languages. J. Autom. Lang. Comb. 17(2-4), 83–105 (2012)
5. Brzozowski, J., Liu, B.: Quotient complexity of star-free languages. Internat. J. Found. Comput. Sci. 23(6), 1261–1276 (2012)
6. Brzozowski, J., Szykuła, M.: Large aperiodic semigroups (2014), http://arxiv.org/abs/1401.0157
7. Diestel, R.: Graph Theory, Graduate Texts in Mathematics, 4th edn., vol. 173. Springer, Heidelberg (2010), http://diestel-graph-theory.com
8. Gomes, G., Howie, J.: On the ranks of certain semigroups of order-preserving transformations. Semigroup Forum 45, 272–282 (1992)
9. Howie, J.M.: Products of idempotents in certain semigroups of transformations. Proc. Edinburgh Math. Soc. 17(2), 223–236 (1971)
10. Kisielewicz, A., Szykuła, M.: Generating small automata and the Černý conjecture. In: Konstantinidis, S. (ed.) CIAA 2013. LNCS, vol. 7982, pp. 340–348. Springer, Heidelberg (2013)
11. McNaughton, R., Papert, S.A.: Counter-Free Automata, MIT Research Monographs, vol. 65. MIT Press (1971)
12. Pin, J.E.: Syntactic semigroups. In: Handbook of Formal Languages. Word, Language, Grammar, vol. 1, pp. 679–746. Springer, New York (1997)
13. Schützenberger, M.: On finite monoids having only trivial subgroups. Inform. and Control 8, 190–194 (1965)
14. Yu, S.: State complexity of regular languages. J. Autom. Lang. Comb. 6, 221–234 (2001)

On the Square of Regular Languages[*]

Kristína Čevorová[1], Galina Jirásková[2], and Ivana Krajňáková[3]

[1] Mathematical Institute, Slovak Academy of Sciences
Štefánikova 49, 814 73 Bratislava, Slovakia
cevorova@mat.savba.sk
[2] Mathematical Institute, Slovak Academy of Sciences
Grešákova 6, 040 01 Košice, Slovakia
jiraskov@saske.sk
[3] Institute of Computer Science, Faculty of Science, P.J. Šafárik University
Jesenná 5, 040 01 Košice, Slovakia
ivana.krajnakova.vt@gmail.com

Abstract. We show that the upper bound $(n - k) \cdot 2^n + k \cdot 2^{n-1}$ on the state complexity of the square of a regular language recognized by an n-state deterministic finite automaton with k final states is tight in the ternary case for every k with $1 \le k \le n-2$. Using this result, we are able to define a language that is hard for the square operation on languages accepted by alternating finite automata. In the unary case, the known upper bound for square is $2n - 1$, and we prove that each value in the range from 1 to $2n - 1$ may be attained by the state complexity of the square of a unary language with state complexity n whenever $n \ge 5$.

1 Introduction

Square is an operation on formal languages which is defined as $L^2 = L \cdot L = \{uv \mid u \in L \text{ and } v \in L\}$. It is known that if a regular language L is recognized by an n-state deterministic finite automaton (DFA), then the language L^2 is recognized by a DFA of at most $n \cdot 2^n - 2^{n-1}$ states [11]. This upper bound follows from the upper bound $m \cdot 2^n - 2^{n-1}$ on the state complexity of the concatenation $K \cdot L = \{uv \mid u \in K \text{ and } v \in L\}$ of languages K and L recognized by m-state and n-state DFAs, respectively [9, 14]; here, the state complexity of a regular language is the smallest number of states in any DFA recognizing this language.

Yu *et al.* [14] proved that the upper bound for concatenation is tight in the ternary case by describing languages over a three-letter alphabet that meet this upper bound for their concatenation. The binary witnesses have been presented already in [9], however no proof has been given here. The tightness of this upper bound in the binary case is proved in [5].

In [14] it is shown that the upper bound $m \cdot 2^n - 2^{n-1}$ for concatenation cannot be met if the first language is accepted by an m-state DFA that has more than one final state. In such a case, the upper bound is $(m - k) \cdot 2^n + k \cdot 2^{n-1}$, where k is the number of final states in the DFA for the first language [14].

---

[*] Research supported by grant APVV-0035-10.

M. Holzer and M. Kutrib (Eds.): CIAA 2014, LNCS 8587, pp. 136–147, 2014.
© Springer International Publishing Switzerland 2014

The tightness of these bounds has been studied in [4], where binary witnesses are described for every k with $1 \leq k \leq n-1$. Later these results have been useful for defining languages that are hard for concatenation of languages accepted by alternating finite automata (AFAs). The known upper bound for alternating finite automata is $2^m + n + 1$ [3], and the authors of [3] wrote: "...we show that $2^m + n + 1$ states suffice for an AFA to accept the concatenation of two languages accepted by AFA with m and n states, respectively. We conjecture that this number is actually necessary in the worst case, but have no proof."

This open problem is almost solved in [6] by taking binary languages K^R and L^R accepted by 2^m-state and 2^n-state DFAs, respectively, both with half of states final, that meet the upper bound for concatenation in [4]. Then, as shown in [6], the languages K and L are accepted by m-state and n-state AFAs, respectively, and every AFA for the language $K \cdot L$ requires at least $2^m + n$ states.

Motivated by the same problem for the square operation on alternating finite automata, we study this operation in more detail in this paper. The upper bound $n \cdot 2^n - 2^{n-1}$ on the state complexity of the square of a language recognized by an n-state DFA is known to be tight in the binary case. Rampersad [11] described a language over a binary alphabet recognized by an n-state DFA with one final state whose square meets this upper bound.

As in the case of concatenation, this upper bound cannot be met by a language accepted by an n-state DFA that has more than one final state. Here, the upper bound for concatenation gives the upper bound $(n - k) \cdot 2^n + k \cdot 2^{n-1}$ on the state complexity of the square of a language recognized by an n-state DFA with k final states. In the first part of our paper, we show that these upper bounds are tight in the ternary case for every k with $1 \leq k \leq n - 2$. We are not able to prove the tightness in the case of $k = n - 1$, and we conjecture that in this case, the upper bound cannot be met. The binary case remains open as well.

Using these results, we are able to describe a language L accepted by an n-state AFA such that every AFA for the language L^2 needs at least $2^n + n$ states. This is smaller just but one than the upper bound $2^n + n + 1$ which follows from the known upper bound $2^m + n + 1$ for concatenation of AFA languages [3].

In the second part of the paper, we study the square operation on unary regular languages. In the unary case, the known upper bound on the state complexity of the square of a language recognized by an n-state unary DFA is $2n - 1$ [11]. We are interested in the question which values in the range from 1 to $2n - 1$ may be attained by the state complexity of the square of a unary language with state complexity n. We prove that for every n with $n \geq 5$, the hierarchy of possible complexities is contiguous with no gaps in it. For every n and α with $n \geq 5$ and $1 \leq \alpha \leq 2n - 1$, we are able to define a unary language L with state complexity n such that the state complexity of the language L^2 is α. This is in contrast to the results for the star of unary languages [2], where there are at least two gaps of length n of values in the range from 1 to $(n - 1)^2 + 1$ that cannot be attained by the star of any unary language with state complexity n.

We first recall some basic definitions; for further details, the reader may refer refer to [12, 13].

A *nondeterministic finite automaton* (NFA) is a quintuple $A = (Q, \Sigma, \delta, I, F)$, where Q is a finite set of states, Σ is a finite alphabet, $\delta \colon Q \times \Sigma \to 2^Q$ is the transition function which is extended to the domain $2^Q \times \Sigma^*$ in the natural way, $I \subseteq Q$ is the set of initial states, and $F \subseteq Q$ is the set of final states. The language accepted by A is the set $L(A) = \{w \in \Sigma^* \mid \delta(I, w) \cap F \neq \emptyset\}$. An NFA A is *deterministic* (and complete) if $|I| = 1$ and $|\delta(q, a)| = 1$ for each q in Q and each a in Σ. In such a case, we write $q \cdot a = q'$ instead of $\delta(q, a) = \{q'\}$.

The state complexity of a regular language L, $\mathrm{sc}(L)$, is the number of states in the minimal DFA for L. It is well known that a DFA is minimal if all its states are reachable from its initial state, and no two of its states are equivalent.

The *concatenation* of two languages K and L is the language $K \cdot L = \{uv \mid u \in K \text{ and } v \in L\}$. The square of a language L is the language $L^2 = L \cdot L$.

The *reverse* of a string w is defined by $\varepsilon^R = \varepsilon$ and $(wa)^R = aw^R$ for a string w and a symbol a. The reverse of a language L is the language $L^R = \{w^R \mid w \in L\}$.

A language is called *unary* (*binary*, *ternary*) if it is defined over an alphabet containing one (two, three, respectively) symbols.

2   Square for Automata with k Final States

In this section we consider languages over an alphabet of at least two symbols. The state complexity of concatenation of regular languages accepted by an m-state and an n-state DFAs is known to be $m \cdot 2^n - 2^{n-1}$ [9, 14]. However, if the first automaton has k final states, then the upper bound for concatenation is $(m - k) \cdot 2^n + k \cdot 2^{n-1}$ [14], and it is known to be tight in the binary case for every k with $1 \leq k \leq n - 1$ [4].

It follows that the upper bound on the complexity of square is $n \cdot 2^n - 2^{n-1}$. A binary witness language meeting this bound is presented in [11]. If a language is accepted by an n-state DFA with k final states, then the upper bound is $(n - k) \cdot 2^n + k \cdot 2^{n-1}$. For the sake of completeness, we give a simple alternative proof here.

Lemma 1. *Let $n \geq 2$ and $1 \leq k \leq n - 1$. If a language L is accepted by an n-state DFA with k final states, then $\mathrm{sc}(L^2) \leq (n - k) \cdot 2^n + k \cdot 2^{n-1}$.*

Proof. Let L be a language accepted by a DFA $A = (Q, \Sigma, \cdot, 0, F)$, where $Q = \{0, 1, \ldots, n - 1\}$ and $|F| = k$. Construct an NFA N for the language L^2 from the DFA A as follows. Take two copies of the DFA A; the states in the first copy are labeled by $q_0, q_1, \ldots, q_{n-1}$, and the states of the second copy are labeled by $0, 1, \ldots, n - 1$. For each state q_i and each symbol a, add the transition on a from q_i to the initial state 0 of the second copy whenever $i \cdot a \in F$. The initial state of the NFA N is q_0 if $0 \notin F$, otherwise N has two initial states q_0 and 0. The final states of N are final states in the second copy, thus states in F.

Consider the subset automaton of the NFA N. Each reachable subset of the subset automaton is of the form $\{q_i\} \cup S$, where $S \subseteq \{0, 1, \ldots, n - 1\}$. Moreover, if $i \in F$, then S must contain the state 0. It follows that the number of reachable states in the subset automaton is at most $(n - k) \cdot 2^n + k \cdot 2^{n-1}$.   □

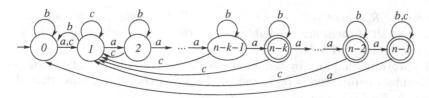

Fig. 1. A DFA A of a language meeting the bound $(n - k) \cdot 2^n + k \cdot 2^{n-1}$ for square

Our next aim is to show that the bounds $(n - k) \cdot 2^n + k \cdot 2^{n-1}$ can be met by languages over a three-letter alphabet assuming that $1 \leq k \leq n - 2$. We are not able to prove the tightness in the case of $k = n - 1$, and we conjecture that in this case, the bound $2^n + (n - 1) \cdot 2^{n-1}$ cannot be met.

Lemma 2. *Let $n \geq 3$ and $1 \leq k \leq n - 2$. There exists a ternary regular language L accepted by an n-state DFA with k final states and such that $\mathrm{sc}(L^2) = (n - k) \cdot 2^n + k \cdot 2^{n-1}$.*

Proof. Let L be the language accepted by the DFA $A = (Q, \{a, b, c\}, \cdot, 0, F)$ shown in Fig. 1, in which $Q = \{0, 1, \ldots, n - 1\}$, $F = \{i \mid n - k \leq i \leq n - 1\}$, and
  $q \cdot a = (q + 1) \bmod n$;
  $q \cdot b = q$ if $q \neq 1$ and $1 \cdot b = 0$;
  $q \cdot c = 1$ if $q \neq n - 1$ and $(n - 1) \cdot c = n - 1$;
notice that the automaton A restricted to the alphabet $\{a, b\}$ and with $k = 1$ is the Rampersad's witness automaton meeting the upper bound $n \cdot 2^n - 2^{n-1}$ on the state complexity of the square of regular languages [11].

Construct an NFA N for the language L^2 as described in the proof of Lemma 1. The NFA N is shown in Fig. 2; to keep the figure transparent, we omitted the transitions on c going to states q_1 and 1.

Our goal is to show that the subset automaton corresponding to the NFA N has $(n - k) \cdot 2^n + k \cdot 2^{n-1}$ reachable and pairwise distinguishable states.

To this aim consider the following family of subsets of the states of N:

$$\mathcal{R} = \{\{q_i\} \cup S \mid 0 \leq i \leq n - k - 1, \ S \subseteq \{0, 1, \ldots, n - 1\}\}$$
$$\cup \ \{\{q_i\} \cup T \mid n - k \leq i \leq n - 1, \ T \subseteq \{0, 1, \ldots, n - 1\} \text{ and } 0 \in T\}.$$

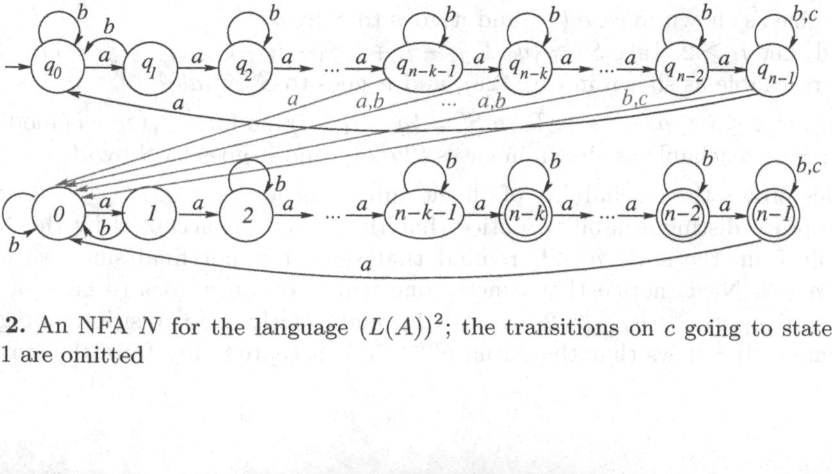

Fig. 2. An NFA N for the language $(L(A))^2$; the transitions on c going to states q_1 and 1 are omitted

The family $\mathcal{R}$ consists of $(n - k) \cdot 2^n + k \cdot 2^{n-1}$ subsets, and we are going to show that all of them are reachable and pairwise distinguishable in the subset automaton of the NFA N.

We prove reachability by induction on the size of subsets. The initial state of the subset automaton is $\{q_0\}$, and the following transitions show that all the subsets in $\mathcal{R}$ of size at most two are reachable:

$$\{q_0\} \xrightarrow{a} \{q_1\} \xrightarrow{a} \cdots \xrightarrow{a} \{q_{n-k-1}\} \xrightarrow{a} \{q_{n-k},0\} \xrightarrow{ab} \{q_{n-k+1},0\} \xrightarrow{ab} \cdots \xrightarrow{ab} \{q_{n-1},0\},$$

$$\{q_{n-1},0\} \xrightarrow{a} \{q_0,1\} \xrightarrow{b} \{q_0,0\},$$

$$\{q_0,1\} \xrightarrow{(ab)^{j-1}} \{q_0,j\} \text{ where } 2 \le j \le n-1, \text{ and}$$

$$\{q_0,(j-i) \bmod n\} \xrightarrow{a^i} \{q_i,j\} \text{ where } 1 \le i \le n-k-1,\ 0 \le j \le n-1.$$

Now let $2 \le t \le n$, and assume that each subset in $\mathcal{R}$ of size t is reachable in the subset automaton. Let us show that then also each subset in $\mathcal{R}$ of size $t+1$ is reachable.

To this aim let $S = \{q_i, j_1, j_2, \ldots, j_t\}$, where $0 \le j_1 < j_2 < \cdots < j_t \le n-1$, be a subset in $\mathcal{R}$ of size $t+1$. Consider several cases:

(1) Let $n - k \le i \le n - 1$, so $j_1 = 0$. We show that the set S is reachable by induction on i.

(1a) If $i = n - k$, then S is reached from $\{q_{n-k-1}, j_2 - 1, j_3 - 1, \ldots, j_t - 1\}$ by a, and the latter set is reachable by induction on t.

(1b) Suppose $i > n - k$. If $j_2 \ge 2$, then the set S is reached from the set $\{q_{i-1}, 0, j_2 - 1, \ldots, j_t - 1\}$ by ab. If $j_2 = 1$, then the set S is reached from the set $\{q_{i-1}, n - 1, 0, j_3 - 1, \ldots, j_t - 1\}$ by a. Both sets containing q_{i-1} are reachable by induction on i.

(2) Let $i = 0$. There are four subcases:

(2a) Let $j_1 = 0$ and $j_2 = 1$. Take $S' = \{q_{n-1}, n - 1, 0, j_3 - 1, \ldots, j_t - 1\}$. Then S' is reachable as shown in case (1), and it goes to S by a.

(2b) Let $j_1 = 0$ and $j_2 \ge 2$. Take $S' = \{q_{n-1}, 0, j_2 - 1, j_3 - 1, \ldots, j_t - 1\}$. Then S' is reachable as shown in case (1), and it goes to S by ab.

(2c) Let $j_1 = 1$. Take $S' = \{q_{n-1}, 0, j_2 - 1, j_3 - 1, \ldots, j_t - 1\}$. Then S' is reachable as shown in case (1), and it goes to S by a.

(2d) Let $j_1 \ge 2$. Take $S' = \{q_0, 1, j_2 - j_1 + 1, j_3 - j_1 + 1, \ldots, j_t - j_1 + 1\}$. Then S' is reachable as shown in case (2c), and it goes to S by $(ab)^{j_1 - 1}$.

(3) Let $1 \le i \le n - k - 1$. Take $S' = \{q_0, (j_1 - i) \bmod n, \ldots, (j_t - i) \bmod n\}$. Then S' is reachable as shown in cases (2a-2d), and it goes to S by a^i.

This proves the reachability of all the subsets in $\mathcal{R}$.

To prove distinguishability, notice that the string c is accepted by the NFA N only from the state $n - 1$; remind that state 1 is not final since we have $k \le n - 2$. Next, notice that exactly one transition on a goes to each of the states in $\{q_1, q_2, \ldots, q_{n-1}, 1, 2, \ldots, n - 1\}$, and exactly one transition on c goes to state 0. It follows that the string $a^{n-1-i}c$ is accepted only from the state i, where $0 \leq i \leq n - 2$, the string $ca^{n-1}c$ is accepted only from the state q_{n-1}, and finally the string $a^{n-1-i}ca^{n-1}c$ is accepted only from the state q_i, where $0 \leq i \leq n-2$. Fig. 3 illustrates this for $n = 5$ and $k = 3$. This means that all the states in the subset automaton of the NFA N are pairwise distinguishable since two distinct subsets must differ in a state q of N, and the string that is accepted only from q distinguishes the two subsets. This proves distinguishability, and concludes the proof.   □

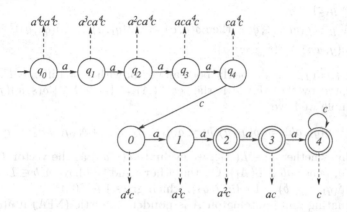

Fig. 3. The strings accepted only from the corresponding states; $n = 5$ and $k = 3$. Notice that exactly one transition on a goes to each of the states in $\{q_1, q_2, \ldots, q_{n-1}, 1, 2, \ldots, n - 1\}$, and exactly one transition on c goes to state 0. The unique acceptance of appropriate strings follows from these facts.

As a corollary of the two lemmata above, we get the following result.

Theorem 1 (Square: k Final States). *Let $n \geq 3$ and $1 \leq k \leq n - 2$. Let L be a language over an alphabet Σ accepted by an n-state DFA with k final states. Then $\mathrm{sc}(L^2) \leq (n - k) \cdot 2^n + k \cdot 2^{n-1}$, and the bound is tight if $|\Sigma| \geq 3$.*   □

2.1 An Application

In this subsection we show how the witness languages described in Lemma 2 can be used to define languages that almost meet the upper bound on the square operation on alternating finite automata.

First, let us give some basic definitions and notations. For details and all unexplained notions, the reader may refer to [1, 3, 6–8, 12, 13].

An *alternating finite automaton* (AFA) is a quintuple $A = (Q, \Sigma, \delta, s, F)$, where Q is a finite non-empty set of states, $Q = \{q_1, \ldots, q_n\}$, Σ is an input alphabet, δ is the transition function that maps $Q \times \Sigma$ into the set $\mathcal{B}_n$ of boolean functions, $s \in Q$ is the initial state, and $F \subseteq Q$ is the set of final states. For example, let $A_1 = (\{q_1, q_2\}, \{a, b\}, \delta, q_1, \{q_2\})$, where transition function δ is given in Table 1.

Table 1. The transition function of the alternating finite automaton A_1

δ	a	b
q_1	$q_1 \wedge q_2$	1
q_2	q_2	$q_1 \vee \overline{q_2}$

The transition function δ is extended to the domain $\mathcal{B}_n \times \Sigma^*$ as follows: For all g in $\mathcal{B}_n$, a in Σ, and w in Σ^*,

$$\delta(g, \varepsilon) = g;$$
$$\text{if } g = g(q_1, \ldots, q_n), \text{ then } \delta(g, a) = g(\delta(q_1, a), \ldots, \delta(q_n, a));$$
$$\delta(g, wa) = \delta(\delta(g, w), a).$$

Next, let $f = (f_1, \ldots, f_n)$ be the boolean vector with $f_i = 1$ iff $q_i \in F$. The language accepted by the AFA A is the set $L(A) = \{w \in \Sigma^* \mid \delta(s, w)(f) = 1\}$. In our example we have

$$\delta(s, ab) = \delta(q_1, ab) = \delta(\delta(q_1, a), b) = \delta(q_1 \wedge q_2, b) = 1 \wedge (q_1 \vee \overline{q_2}) = q_1 \vee \overline{q_2}.$$

To determine whether $ab \in L(A_1)$, we evaluate $\delta(s, ab)$ at the vector $f = (0, 1)$. We obtain 0, hence $ab \notin L(A_1)$. On the other hand, we have $abb \in L(A_1)$ since $\delta(s, abb) = \delta(q_1 \vee \overline{q_2}, b) = 1 \vee (\overline{q_1} \wedge q_2)$, which gives 1 at $(0, 1)$.

An alternating finite automaton A is nondeterministic (NFA) if $\delta(q_k, a)$ are of the form $\bigvee_{i \in I} q_i$. If $\delta(q_k, a)$ are of the form q_i, then the automaton A is deterministic (DFA).

Recall that the state complexity of a regular language L, $\mathrm{sc}(L)$, is the smallest number of states in any DFA accepting L. Similarly, the alternating state complexity of a language L, in short $\mathrm{asc}(L)$, is defined as the smallest number of states in any AFA for L. The following results are well known.

Lemma 3 ([1, 3, 6, 7]). *If L is accepted by an AFA of n-states, then L^R is accepted by a DFA of 2^n states. If $\mathrm{sc}(L^R) = 2^n$ and the minimal DFA for L^R has 2^{n-1} final states, then $\mathrm{asc}(L) = n$.* $\square$

It follows that $\mathrm{asc}(L) \geq \log(\mathrm{sc}(L^R))$. Using the results given by Lemma 2 and Lemma 3, we get a language that almost meets the upper bound on the complexity of the square operation on AFAs.

Theorem 2 (Square on AFAs). *Let L be a language over an alphabet Σ with $\mathrm{asc}(L) = n$. Then $\mathrm{asc}(L^2) \leq 2^n + n + 1$. The bound $2^n + n$ is met if $|\Sigma| \geq 3$.*

Proof. The upper bound follows from the upper bound $2^m + n + 1$ on the concatenation of AFA languages [3]. Now let L^R be the ternary witness for square from Lemma 2 with 2^n states and 2^{n-1} final states. Then, by Lemma 3, we have $\mathrm{asc}(L) = n$. By Lemma 2, we get

$$\mathrm{sc}((L^2)^R) = \mathrm{sc}((L^R)^2) = 2^{n-1} \cdot 2^{2^n} + 2^{n-1} \cdot 2^{2^n - 1} \geq 2^{n-1} \cdot 2^{2^n}(1 + 1/2).$$

By Lemma 3, we have $\mathrm{asc}(L^2) \geq \lceil \log(2^{n-1} \cdot 2^{2^n}(1 + 1/2)) \rceil = 2^n + n$, which proves the theorem. $\square$

The above result for square complements the results on the complexity of basic operations on AFA languages obtained in [6]. The following table summarizes these results, and compares them to the known results for DFAs [9, 11, 14].

	union	intersection	concatenation	reversal	star	square
AFAs	$m + n + 1$	$m + n + 1$	$\geq 2^m + n$	$\geq 2^n$	$\geq 2^n$	$\geq 2^n + n$
			$\leq 2^m + n + 1$	$\leq 2^n + 1$	$\leq 2^n + 1$	$\leq 2^n + n + 1$
DFAs	mn	mn	$m \cdot 2^n - 2^{n-1}$	2^n	$3/4 \cdot 2^n$	$n \cdot 2^n - 2^{n-1}$

3   Square of Languages over Unary Alphabet

A unary alphabet is fundamentally different from the general case. It has a close relation to the number theory – the length of strings is their only property that really matters in complexity questions. From this point of view, unary languages are nothing else than subsets of natural numbers. Instead of writing $a^n \in L$, we will write $n \in L$. The square operation is then, in fact, the sum of two numbers in the language. Let us start with some basic definitions and notations.

For integers i and j with $i \leq j$, let $[i, j] = \{i, i + 1, \ldots, j\}$.

A DFA $A = (Q, \{a\}, \delta, q_0, F)$ for a unary language is uniquely given by less information than an arbitrary DFA. Identify states with numbers from the interval $[0, n - 1]$ via $q \sim \min\{i \mid \delta(q_0, a^i) = q\}$. Then A is unambiguously given by the number of states n, the set of final numbers F, and the "loop" number $\ell = \delta(q_0, a^n)$. This allows us to freely interchange states and their ordinal numbers and justifies the notation convention used by Nicaud [10], where (n, ℓ, F) denotes a unary automaton with n states, the loop number ℓ, and the set of final states F. Nicaud also provided the following characterization of minimal unary automata.

Theorem 3 ([10, Lemma 1]). *A unary automaton (n, ℓ, F) is minimal if and only if both conditions below are true:*
(1) its loop is minimal, and
(2) states $n - 1$ and $\ell - 1$ do not have the same finality (that is, exactly one of them is final).   □

Finite and cofinite languages are always regular, and if they are unary, then it is easy to determine their state complexity.

Proposition 1. *Let L be a unary language. If L is cofinite, then we have $\mathrm{sc}(L) = \max\{m \mid m \notin L\} + 2$. If L is finite, then $\mathrm{sc}(L) = \max\{m \mid m \in L\} + 2$.*   □

Proposition 2. *If a language is (co)finite, then also its square is (co)finite.*   □

If $\varepsilon \in L$, then every string w in L can be written as εw. This leads to the following observation.

Proposition 3. *If $\varepsilon \in L$, then $L \subseteq L^2$.*   □

3.1   Finite Unary Languages

Interestingly, if we consider only finite languages with state complexity n, then we cannot get any other complexity for square but $2n - 2$.

Lemma 4. *Let L be a finite unary regular language with $\mathrm{sc}(L) = n$. Then $\mathrm{sc}(L^2) = 2n - 2$.*

Proof. By Proposition 1, the greatest number in L is $n - 2$. It follows that the greatest member of L^2 is the number $2n - 4$. Hence L^2 is also finite and $\mathrm{sc}(L^2) = 2n - 4 + 2 = 2n - 2$.   □

3.2   General Unary Languages

If we take a unary language with state complexity n, the state complexity of its square will be between 1 and $2n - 1$ [11]. But could it be *anywhere* between these two bounds? The next result shows that the answer is yes if $n \geq 5$.

Theorem 4. *Let $n \geq 5$ and $1 \leq \alpha \leq 2n - 1$. There exists a unary language L such that $\mathrm{sc}(L) = n$ and $\mathrm{sc}(L^2) = \alpha$.*

Proof. We will provide a witness for every liable combination of n and α. The proof is structured to four main cases depending on α:

1. $\alpha = 2$ (the proof works for $n \geq 6$),
2. $\alpha = 2n - 2$ (the proof works for $n \geq 2$),
3. $1 \leq \alpha \leq n - 1$ and $\alpha \neq 2$ (the proof works for $n \geq 8$),
4. $n \leq \alpha \leq 2n - 1$ and $\alpha \neq 2n - 2$ (the proof works for $n \geq 2$).

All witnesses uncovered by these general proofs are part of Table 2 which is an overview of the situation for $n < 5$ and $\alpha \leq n$. If the combination of n and α is liable, one witness is listed, non-existence is indicated by the symbol −.

Table 2. Witnesses for liable combinations of small values of n and α; $2 \leq n \leq 7$ and $1 \leq \alpha \leq n$

α	$n = 2$	$n = 3$	$n = 4$	$n = 5$	$n = 6$	$n = 7$
1	−	$(3, 0, \{0, 1\})$	$(4, 3, \{0, 1, 3\})$	$(5, 4, \{0, 1, 2, 4\})$	$(6, 5, \{0, 1, 2, 3, 5\})$	$(7, 6, \{0, 1, 2, 3, 4, 6\})$
2	$(2, 0, \{0\})$	−	−	$(5, 1, \{0, 2\})$	$(6, 0, \{0, 2\})$	$(7, 5, \{0, 2, 6\})$
3		$(3, 0, \{0\})$	$(4, 2, \{1, 2\})$	$(5, 0, \{0, 2, 3\})$	$(6, 5, \{0, 2, 3, 5\})$	$(7, 6, \{0, 2, 3, 4, 6\})$
4			$(4, 0, \{0\})$	$(5, 0, \{0, 3, 4\})$	$(6, 2, \{0, 3, 4, 5\})$	$(7, 0, \{0, 3, 4, 5\})$
5				$(5, 0, \{0\})$	$(6, 2, \{0, 2, 5\})$	$(7, 6, \{0, 1, 4, 6\})$
6					$(6, 0, \{0\})$	$(7, 1, \{0, 1, 3\})$
7						$(7, 0, \{0\})$

1. Let $\alpha = 2$ and $n \geq 6$. The construction of witnesses depends on the parity of n. If n is even, then we take the language recognized by the witness DFA $A = (n, 0, F)$, where $F = \{i \in [0, n-1] \mid i \text{ is even and } i \neq n-2\}$. If n is odd, then the witness DFA is $A = (n, n-2, \{i \in [0, n-1] \mid i \text{ is even and } i \neq n-3\})$.

We claim that in both cases $L(A)^2$ is the language of even numbers with the corresponding minimal DFA $(2, 0, \{0\})$. We give the proof for n even; the proof for n odd has only slight technical differences.

We first show that A is minimal. The second condition of Theorem 3 is fulfilled vacantly. The first condition – the minimality of the loop – is satisfied as well: Any equivalent loop must be of even length as not to accept strings of different parity. Since there is exactly one even non-accepting state, it cannot be equivalent with any other state, and the loop is unfactorizable.

Now we show that $L(A)^2$ is the language of all even numbers. Since $L(A)$ contains only even numbers and the sum of two even numbers is even, $L(A)^2$ contains only even numbers. Let us show that $L(A)^2$ contains all even numbers.

By Proposition 3, we have $L(A) \subseteq L(A)^2$. All even numbers missing in $L(A)$ are in the form $kn + (n-2)$. But these numbers are in $L(A)^2$, since $kn + (n-2) = (2 + kn) + (n-4)$, which is a sum of two numbers in $L(A)$; recall, that $2 \neq n-2$, since $n \geq 6$.

2. Let $\alpha = 2n - 2$ and $n \geq 2$. By Lemma 4, every finite language of complexity n is a witness in this case; for example, we can take the language $\{n - 2\}$.

From now on, our strategy is based on Proposition 1. All our languages will be cofinite, so their complexity is easily determined by answering the question – how long is the longest string not contained in this language?

3. Let $1 \leq \alpha \leq n - 1$, $\alpha \neq 2$, and $n \geq 8$. Technically, this case is further divided to subcases $\alpha = 1$, $\alpha = 3$, $\alpha = 4$, $5 \leq \alpha \leq n$ where $\alpha \neq n-1$, and $\alpha = n - 1$. However, the main idea of the construction is always the same, so we provide only the witness automata in Fig. 4, and one exemplary proof in the case of $5 \leq \alpha \leq n$ where $\alpha \neq n-1$. For this case, consider the language L accepted by the unary automaton $A = (n, n-1, F)$, where $F = [\alpha-1, n-3] \cup \{0, 1, n-1\}$.

First, let us show that the numbers greater than $\alpha - 2$ are in L^2. Since $\varepsilon \in L$, the language L is a subset of L^2 by Proposition 3. The only number greater than $\alpha - 2$ that is not in L is $n - 2$. However, we have $n - 2 = (n-3) + 1$, which is the sum of two numbers in L. Therefore, the number $n - 2$ is in L^2. It follows that all numbers greater than $\alpha - 2$ are in L^2.

Now let us show that $\alpha - 2$ is not in L^2. The only numbers in L that are smaller than $\alpha - 1$ are 0 and 1. The sum of any two of them is at most 2, which is less than $\alpha - 2$. Thus, by Proposition 1, the state complexity of L^2 is α.

4. $n \leq \alpha \leq 2n - 1$ and $\alpha \neq 2n - 2$, $n \geq 2$. Consider the unary language $L = \{i \mid i \geq n - 1\}$. Then $\mathrm{sc}(L) = n$ and $\mathrm{sc}(L^2) = 2n - 1$ since the greatest number that is not in L^2 is $2n - 3$. By adding an arbitrary number different than $n - 2$ to the language L, we get a language with the same state complexity as L. But the state complexity of the square of the resulting language will be

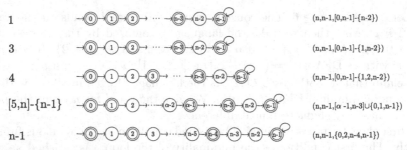

Fig. 4. The construction of witnesses for n and α with $1 \le \alpha \le n$

different. Let $m = \alpha - n$. Then $0 \le m \le n - 1$ and $m \ne n - 2$. Let us see what happens, if we add the number m to L.

Let $L_m = L \cup \{m\}$. Then we have $L_m^2 = \{2m\} \cup \{m + i \mid i \ge (n - 1)\}$. Since $m \ne n - 2$, we have $2m \ne m + n - 2$, and therefore the greatest number that is not in L_m^2 is $m + n - 2$. It follows that $\mathrm{sc}(L_m^2) = m + n - 2 + 2 = \alpha$.   □

4   Conclusions

We considered the square operation on regular languages. In the unary case, the state complexity of square is $2n - 1$ [11]. We proved that each value in the range from 1 to $2n - 1$ may be attained by the state complexity of the square of a unary language with state complexity n whenever $n \ge 5$.

Next, we studied the square operation on languages over an alphabet of at least two symbols. The known upper bound in this case is $n \cdot 2^n - 2^{n-1}$, and it is known to be tight in the binary case [11]. We investigated the square for languages accepted by automata with more final states. The upper bound on the state complexity of the square of a language accepted by an n-state DFA with k final states is $(n - k) \cdot 2^n + k \cdot 2^{n-1}$. We showed that these upper bounds are tight in the ternary case assuming that $1 \le k \le n - 2$.

The case of $k = n - 1$ remains open, and we conjecture that the upper bound $2^n + (n - 1) \cdot 2^{n-1}$ cannot be met in this case. The binary case is open as well.

As an application, we were able to define a ternary language L accepted by an n-state alternating finite automaton such that every alternating finite automaton for the language L^2 requires at least $2^n + n$ states. This is smaller just by one than the known upper bound $2^n + n + 1$ [3]. Our result on the square operation complements the results on the complexity of union, intersection, concatenation, star, and reversal on AFA languages obtained in [6].

References

1. Brzozowski, J., Leiss, E.: On equations for regular languages, finite automata, and sequential networks. Theoret. Comput. Sci. 10, 19–35 (1980)
2. Čevorová, K.: Kleene star on unary regular languages. In: Jurgensen, H., Reis, R. (eds.) DCFS 2013. LNCS, vol. 8031, pp. 277–288. Springer, Heidelberg (2013)

3. Fellah, A., Jürgensen, H., Yu, S.: Constructions for alternating finite automata. Int. J. Comput. Math. 35, 117–132 (1990)
4. Jirásek, J., Jirásková, G., Szabari, A.: State complexity of concatenation and complementation. Internat. J. Found. Comput. Sci. 16, 511–529 (2005)
5. Jirásková, G.: State complexity of some operations on binary regular languages. Theoret. Comput. Sci. 330, 287–298 (2005)
6. Jirásková, G.: Descriptional complexity of operations on alternating and boolean automata. In: Hirsch, E.A., Karhumäki, J., Lepistö, A., Prilutskii, M. (eds.) CSR 2012. LNCS, vol. 7353, pp. 196–204. Springer, Heidelberg (2012)
7. Leiss, E.: Succinct representation of regular languages by boolean automata. Theoret. Comput. Sci. 13, 323–330 (1981)
8. Leiss, E.: On generalized language equations. Theoret. Comput. Sci. 14, 63–77 (1981)
9. Maslov, A.N.: Estimates of the number of states of finite automata. Soviet Math. Doklady 11, 1373–1375 (1970)
10. Nicaud, C.: Average state complexity of operations on unary automata. In: Kutyłowski, M., Wierzbicki, T., Pacholski, L. (eds.) MFCS 1999. LNCS, vol. 1672, pp. 231–240. Springer, Heidelberg (1999)
11. Rampersad, N.: The state complexity of L^2 and L^k. Inform. Process. Lett. 98, 231–234 (2006)
12. Sipser, M.: Introduction to the theory of computation. PWS Publishing Company, Boston (1997)
13. Yu, S.: Regular languages. In: Rozenberg, G., Salomaa, A. (eds.) Handbook of Formal Languages, vol. I, pp. 41–110. Springer, Heidelberg (1997)
14. Yu, S., Zhuang, Q., Salomaa, K.: The state complexity of some basic operations on regular languages. Theoret. Comput. Sci. 125, 315–328 (1994)

Unary Languages Recognized by Two-Way One-Counter Automata[*]

Marzio De Biasi[1] and Abuzer Yakaryılmaz[2,3,**]

[1] marziodebiasi@gmail.com
[2] University of Latvia, Faculty of Computing
Raina bulv. 19, Rīga, 1586, Latvia
[3] National Laboratory for Scientific Computing
Petrópolis, RJ, 25651-075, Brazil
abuzer@lncc.br

Abstract. A two-way deterministic finite state automaton with one counter (2D1CA) is a fundamental computational model that has been examined in many different aspects since sixties, but we know little about its power in the case of unary languages. Up to our knowledge, the only known unary nonregular languages recognized by 2D1CAs are those formed by strings having exponential length, where the exponents form some trivial unary regular language. In this paper, we present some nontrivial subsets of these languages. By using the input head as a second counter, we present simulations of two-way deterministic finite automata with linearly bounded counters and linear–space Turing machines. We also show how a fixed-size quantum register can help to simplify some of these languages. Finally, we compare unary 2D1CAs with two–counter machines and provide some insights about the limits of their computational power.

1 Introduction

A finite automaton with one counter is a fundamental model in automata theory. It has been examined in many different aspects since sixties [8]. One recent significant result, for example, is that the equivalence problem of deterministic one-way counter automata is NL-complete [2]. After introducing quantum automata [21,15] at the end of the nineties, quantum counter automata have also been examined (see a very recent research work in [28]).

A *counter* is a very simple working memory which can store an arbitrary long integer that can be incremented or decremented; but only a single bit of information can be retrieved from it: whether its value is zero or not. It is a well-known fact that a two-way deterministic finite automaton with two counters is universal [18,19,22]. Any language recognized by a two-way deterministic finite automaton with one counter (2D1CA), on the other hand, is in deterministic

[*] The full paper is at http://arxiv.org/abs/1311.0849
[**] Abuzer Yakaryılmaz was partially supported by CAPES, ERC Advanced Grant MQC, and FP7 FET project QALGO.

M. Holzer and M. Kutrib (Eds.): CIAA 2014, LNCS 8587, pp. 148–161, 2014.
© Springer International Publishing Switzerland 2014

logarithmic space (L) [24].[1] Replacing the counter of a 2D1CA with a stack, we get a two-way deterministic pushdown automaton (2DPDA), that can recognize more languages [7]. Similarly, nondeterminism also increases the class of the languages recognized by 2D1CAs [4].

Unary or tally languages, defined over a single letter alphabet, have deserved special attention. When the input head is not allowed to move to the left (*one-way head*), it is a well-known fact that unary nondeterministic pushdown automata can recognize only regular languages [9]. The same result was shown for bounded-error probabilistic pushdown automata, too [13]. Currently, we do not know whether "quantumness" can add any power. Their alternating versions were shown to be quite powerful: they can recognize any unary language in deterministic exponential time with linear exponent [3]. But, if we replace the stack with a counter, only a single family of unary nonregular languages [6] is known: $\text{UPOWER}(k) = \left\{ a^{k^n} \mid n \geq 1 \right\}$, for a given integer $k \geq 2$. In the case of two-way head, we know that the unary encoding of every language in deterministic polynomial time (P) can be accepted by 2DPDAs [20]; however we do not know whether 2DPDAs are more powerful than 2D1CAs (see also [11]) on unary languages like in the case of binary languages. Any separation between L and P can of course answer this question positively, but, it is still one of the big open problems in complexity theory. On the other hand, researchers also proposed some simple candidate languages not seemingly accepted by any 2D1CA [11,23], e.g. $\text{USQUARE} = \left\{ a^{n^2} \mid n \geq 1 \right\}$. Although it was shown that two–counter machines (2CAs) cannot recognize USQUARE if the input counter is initialized with n^2 (i.e. no Gödelization is allowed) [12,26], up to our knowledge, there is not any known nondeterministic, alternating, or probabilistic one-counter automaton for it. We only know that USQUARE can be recognized by exponential expected time 2D1CAs augmented with a fixed-size quantum register [28] or realtime private alternating one-counter automata [5]. Apart from this open problem, we do not know much about which nonregular unary languages can be recognized by 2D1CAs. In this paper, we provide some answers to this question. In his seminal paper [18], Minsky showed that the emptiness problem for 2D1CAs on unary languages is undecidable. In his proof, he presented a simulation of two-way deterministic finite automaton with two counters on the empty string by a 2D1CA using its input head as a second counter. We use a similar idea but as a new programming technique for 2D1CAs on unary languages that allows to simulate multi-counter automata and space bounded Turing machines operating on unary or general alphabets. A 2D1CA can take the input and the working memory of the simulated machine as the exponent of some integers encoded on unary inputs. Thus, once the automaton becomes sure about the correctness of the encoding, it can start a two-counter simulation of the given machine, in which the second counter is implemented by the head on the unary input. Based on this idea, we will present several new nonregular unary languages recognized by 2D1CAs. Our technique

---

[1] Since a 2D1CA using super-linear space on its counter should finally enter an infinite loop, any useful algorithm can use at most linear space on a counter, which can be simulated by a logarithmic binary working tape.

can be applicable to nondeterministic, alternating, and probabilistic cases in a straightforward way. We also show that using a constant-size quantum memory can help to replace the encoding on binary alphabets with unary alphabets. Finally we compare unary 2D1CAs with 2CAs and provide some insights about the limits of their computational power.

2   Background

Throughout the paper, Σ denotes the input alphabet and the extra symbols ¢ and \$ are the *end-markers* (the tape alphabet is $\tilde{\Sigma} = \Sigma \cup \{¢, \$\}$). For a given string w, w^r is the reverse of w, $|w|$ is the length of w, and w_i is the ith symbol of w, where $1 \leq i \leq |w|$. The string ¢w\$ is represented by $\tilde{w}$. Each counter model defined in the paper has a two-way finite read-only input tape whose squares are indexed by integers. Any given input string, say $w \in \Sigma^*$, is placed on the tape as $\tilde{w}$ between the squares indexed by 0 and $|w| + 1$. The tape has a single head, and it can stay in the same position ($\downarrow$) or move to one square to the left ($\leftarrow$) or to the right ($\rightarrow$) in one step. It must always be guaranteed that the input head never leaves $\tilde{w}$. A counter can store an integer and has two observable states: *zero* (0) or *nonzero* ($\pm$), and can be updated by a value from $\{-1, 0, +1\}$ in one step. Let $\Theta = \{0, \pm\}$.

A *two-way deterministic one-counter automaton* (2D1CA) is a two-way deterministic finite automaton with a counter. Formally, a 2D1CA $\mathcal{D}$ is a 6-tuple

$$\mathcal{D} = (S, \Sigma, \delta, s_1, s_a, s_r),$$

where S is the set of states, $s_1 \in S$ is the initial state, $s_a, s_r \in S$ ($s_a \neq s_r$) are the accepting and rejecting states, respectively, and δ is the transition function governing the behaviour of $\mathcal{D}$ in each step, i.e.

$$\delta : S \setminus \{s_a, s_r\} \times \tilde{\Sigma} \times \Theta \rightarrow S \times \{\leftarrow, \downarrow, \rightarrow\} \times \{-1, 0, +1\}.$$

Specifically, $\delta(s, \sigma, \theta) \rightarrow (s', d_i, c)$ means that when $\mathcal{D}$ is in state $s \in S \setminus \{s_a, s_r\}$, reads symbol $\sigma \in \tilde{\Sigma}$, and the state of its counter is $\theta \in \Theta$, then it updates its state to $s' \in S$ and the position of the input head with respect to $d_i \in \{\leftarrow, \downarrow, \rightarrow\}$, and adds $c \in \{-1, 0, +1\}$ to the value of the counter. In order to stay on the boundaries of $\tilde{w}$, if $\sigma = ¢$ then $d_i \in \{\downarrow, \rightarrow\}$ and if $\sigma = \$$ then $d_i \in \{\downarrow, \leftarrow\}$.

At the beginning of the computation, $\mathcal{D}$ is in state s_1, the input head is placed on symbol ¢, and the value of the counter is set to zero. A configuration of $\mathcal{D}$ on a given input string is represented by a triple (s, i, v), where s is the state, i is the position of the input head, and v is the value of the counter. The computation ends and the input is accepted (resp. rejected) by $\mathcal{D}$ when it enters s_a (resp. s_r).

For any $k > 1$, a *two-way deterministic k-counter automaton* (2DkCA) is a generalization of a 2D1CA and is equipped with k counters; in each transition, it checks the states of all counters and then updates their values. Moreover, we call a counter *linearly bounded* if its value never exceeds $O(|w|)$, where w is the given input. But restricting this bound to $|w|$ does not change the computational power of any kind of automaton having linearly bounded counters, i.e. the value of any counter can be compressed by any rational number by using extra control states. A *two-counter automaton* (2CA) is a 2D2CA over a unary alphabet and without the input tape: the length of the unary input is placed in one of the counters at the beginning of the computation. We underline that the length of the unary input a^n is placed in the counter as it is: indeed if we allow a suitable encoding of the input (by Gödelization, e.g. setting its initial value to 2^n) a 2CA can simulate any Turing machine [18,26].

We replace "D" that stands for *deterministic* in the abbreviations of deterministic machines with "N", "A", and "P" for representing the abbreviations of their *nondeterministic, alternating,* and *probabilistic* counterparts.

We finish the section with some useful technical lemmas.

Lemma 1. *2D1CAs can check whether a given string is a member of language* UPOWER(k) $= \{a^{k^n} \mid n \geq 1\}$, *with* $k \geq 2$.

Lemma 2. *For any given* $p \in \mathbb{Z}^+$, *there exists a 2D1CA* $\mathcal{D}$ *that can set the value of its counter to* M *if its initial value is* $M \cdot p^n$ *provided that the length of the input is at least* $M \cdot p^{n-1}$, *where* $M \in \mathbb{Z}^+$, $p \nmid M$, *and* $n > 0$.

Lemma 3. *The language* $L = \{a^{2^j 3^k} \mid j, k \geq 0\}$ *can be recognized by a 2D1CA.*

Lemma 4. *For any given* $p > 1$, *a 2D2CA* $\mathcal{D}$ *with values* $M > 0$ *and* 0 *in its counters can test whether* p *divides* M *without moving the input head and, after testing, it can recover the values of the counters.*

3   Main Results

We start with the simulation of linearly bounded multi–counter automata on unary languages and establish a direct connection with logarithmic-space unary languages. Secondly, we present the simulation of linear–space Turing machines on binary languages. Then we generalize this simulation for Turing machines that use more space and for Turing machines without any resource bound. Thirdly, we present our quantum result. We finish the section comparing unary 2D1CAs and 2CAs.

3.1   Simulation of Multi-counter Automata on Unary Alphabet

We assume that all linearly bounded counters do not exceed the length of the input. Let $L \subseteq \{a\}^*$ be a unary language recognized by a 2D2CA $\mathcal{M}$ with linearly bounded counters and $w = a^n$ be the given input that is placed on the input tape (between the two end-markers as $\mathcal{c}a^n\$$ and indexed from 0 to $|w|+1$). We can represent the configurations of $\mathcal{M}$ on w with a state, an integer, and a Boolean variable as follows:

$$(s, 2^i 3^{n-i} 5^{c_1} 7^{c_2}, OnDollar), \tag{1}$$

where

- s is the current state,
- $OnDollar = true$ means that the input head is on \$,
- $OnDollar = false$ means that the input head is on the ith square, and,
- c_1 (resp. c_2) represents the value of the first (resp. second) counter.

By using $OnDollar$ variable, we do not need to set i to $(n+1)$ and this will simplify the languages that we will define soon. Note that we are using two exponents, i.e. $2^i 3^{n-i}$, to store the position of the input head. In this way, we can implicitly store the length of the given input (n).

Lemma 5. *A 2D2CA, say $\mathcal{M}'$, can simulate $\mathcal{M}$ on w without using its input head, if its first counter is set to $2^0 3^n 5^0 7^0$.*

Proof. For any $p \in \{2,3,5,7\}$, $\mathcal{M}'$, by help of the second counter, can easily increase the exponent of p by 1, test whether the exponent of p is zero or not, and decrease the exponent of p by 1 if it is not zero. Moreover, $\mathcal{M}'$ can keep the value of $OnDollar$, which is $false$ at the beginning, by using its control states. Note that when the exponent of 3 is zero and the input head of $\mathcal{M}$ is moved to the right, the value of $OnDollar$ is set to $true$; and, whenever the input head of $\mathcal{M}$ leaves the right end-marker, the value of $OnDollar$ is set to $false$ again. During both operations, the exponents of 2 and 3 remain the same. Thus, $\mathcal{M}'$ can simulate $\mathcal{M}$ on w and it never needs to use its input head. $\square$

Note that, during the simulation given above, $2^i 3^{n-i}$ is always less than 3^n for any $i \in \{0, \dots, n\}$, and so, the values of both counters never exceed $3^n 5^n 7^n$.

Now, we build a 2D1CA, say $\mathcal{M}''$, simulating the computation of $\mathcal{M}'$ on some specific unary inputs. Let $u \subseteq \{a\}^*$ be the given input.

1. $\mathcal{M}''$ checks whether the input is of the form $3^n 5^n 7^n = 105^n$ for a nonnegative integer n (Lemma 1). If not, it rejects the input.
2. $\mathcal{M}''$ sets its counter to $2^0 3^n 5^0 7^0$ (Lemma 2). Then, by using its input head as the second counter, it simulates $\mathcal{M}'$ which actually simulates M on a^n (Lemma 5). $\mathcal{M}''$ accepts (resp. rejects) the input if $\mathcal{M}$ ends with the decision of "acceptance" (resp. "rejection").

Thus, we can obtain that if $L \subseteq \{a\}^*$ can be recognized by a 2D2CA with linearly bounded counters, then $\{a^{105^n} \mid a^n \in L\}$ is recognized by a 2D1CA. Actually, we can replace 105 with 42 by changing the representation given in Equation 1 as:

$$(s, 5^i 7^{n-i} 2^{c_1} 3^{c_2}, OnDollar),$$

where $5^i 7^{n-i}$ is always less than 7^n for any $i \in \{0, \dots, n\}$.

Theorem 1. *If $L \subseteq \{a\}^*$ can be recognized by a 2D2CA with linearly bounded counters, then $\{a^{42^n} \mid a^n \in L\}$ is recognized by a 2D1CA.*

Based on this theorem, we can easily show some languages recognized by 2D1CAs, e.g.

$$\left\{ a^{42^{n^2}} \mid n \geq 0 \right\} \text{ and } \left\{ a^{42^p} \mid p \text{ is a prime} \right\}.$$

We can generalize our result for 2DkCAs with linearly bounded counters in a straightforward way.

Theorem 2. *Let $k > 2$ and $p_1, \ldots, p_{k+1}$ be some different prime numbers such that one of them is greater than the $(k+1)^{th}$ prime number. If $L \subseteq \{a\}^*$ can be recognized by a 2DkCA with linearly bounded counters, then*

$$\left\{ a^{(p_1 \cdot p_2 \cdots p_{k+1})^n} \mid a^n \in L \right\}$$

is recognized by a 2D1CA.

Proof. Let $P = \{p_1, \ldots, p_{k+1}\}$. Since one prime number in P, say p_{k+1}, is greater than the $(k+1)^{th}$ prime number, there should be a prime number not in P, say p'_{k+1}, that is not greater than the $(k+1)^{th}$ prime number. We can use the representation given in Equation 1 for a configuration of the 2DkCA:

$$\left(s, p_{k+1}^i (p'_{k+1})^{n-i} p_1^{c_1} p_2^{c_2} \cdots p_k^{c_k}, OnDollar \right).$$

$p_{k+1}^i (p'_{k+1})^{n-i}$ is always less than p_{k+1}^n, and so, the integer part of the configuration is always less than $(p_1 \cdot p_2 \cdots p_{k+1})^n$. As described before, a 2D1CA can check whether the length of the input is a power of $(p_1 \cdot p_2 \cdots p_{k+1})$, and, if so, it can simulate the computation of the 2DkCA on the input. The 2D1CA needs to simulate k counters instead of 2 counters but the technique is essentially the same.   □

The simulation given above can be easily generalized for nondeterministic, alternation, and probabilistic models. The input check and the initialization of the simulation are done deterministically. Therefore, the computation trees of the simulated and simulating machines have the same structure for the well-formed inputs, i.e. the inputs not rejected by the initial input check.

Theorem 3. *Let $k \geq 2$ and $p_1, \ldots, p_{k+1}$ be some different prime numbers such that one of them is greater than the $(k+1)^{th}$ prime number. If $L \subseteq \{a\}^*$ can be recognized by a 2NkCA (resp. 2AkCA, bounded-error 2PkCA, or unbounded-error 2PkCA) with linearly bounded counters, then*

$$\left\{ a^{(p_1 \cdot p_2 \cdots p_{k+1})^n} \mid a^n \in L \right\}$$

is recognized by a 2N1CA (resp. 2A1CA, bounded-error 2P1CA, or unbounded-error 2P1CA).

Now, we establish the connection with logarithmic-space unary languages. The following two easy lemmas are a direct consequence of the fact that, over unary alphabet, a linear bounded counter can be simulated by the head position and vice versa.

Lemma 6. *Any two-way automaton with k-heads on unary inputs can be simulated by a two-way automaton with k-linearly bounded counters, where $k > 1$.*

The reverse simulation holds even on generic alphabets.

Lemma 7. *Any two-way automaton with k-linearly bounded counters can be simulated by a two-way automaton with $(k+1)$-heads, where $k > 1$.*

Both simulations work for deterministic, nondeterministic, alternating, and bounded- and unbounded-error probabilistic models.

Fact 1. *[10,14,16] The class of languages recognized by two-way multi-head deterministic, nondeterministic, alternating, bounded-error probabilistic, and unbounded-error probabilistic finite automata are*

$$\mathsf{L, NL, AL(= P), BPL,} \ and \ \mathsf{PL,}$$

(deterministic, nondeterministic, alternating, bounded-error probabilistic, and unbounded-error probabilistic logarithmic space) respectively.

Based on this fact, the last two lemmas, and the other results in this section, we can obtain the following theorem.

Theorem 4. *Let L be any unary language in L (resp., NL, P, BPL, and PL). Then there is an integer p, product of some primes, such that*

$$\{a^{p^n} \mid a^n \in L\}$$

can be recognized by a 2D1CA (resp., 2N1CA, 2A1CA, bounded-error 2P1CA, and unbounded-error 2P1CA).

3.2 Simulation of Turing Machines on Binary and General Alphabets

Let $\mathcal{N}$ be a single-tape single-head DTM (deterministic Turing machine) working on a binary alphabet $\Sigma = \{a, b\}$. Note that its tape alphabet also contains the blank symbol #. We assume that the input is written between two blank symbols for DTMs. We define some restrictions on $\mathcal{N}$:

- There can be at most one block of non-blank symbols.
- The tape head is placed on the right end-marker at the beginning of the computation which makes easier to explain our encoding used by the 2D3CA given below. Note that this does not change the computational power of the DTMs.

A configuration of $\mathcal{N}$ on a given input, say $w \in \{a, b\}^*$, can be represented as usv, where $uv \in \#\{a, b\}^*\#$ represents the current tape content and s is the current state. Moreover, the tape head is on the last symbol of $\#u$. The initial configuration is $\#w\#s_1$, where s_1 is the initial state. Here v is the empty string. Note that u can never be the empty string.

By replacing a with 0, and b and # with 1s, we obtain a binary number representation of u and v – we will denote these binary numbers by u and v, respectively. Now, we give a simulation of $\mathcal{N}$ by a 2D3CA, say $\mathcal{N}'$, on w.[2] $\mathcal{N}'$ does not have a tape but can simulate it by using three counters. During the simulation, $\mathcal{N}'$ keeps u and v^r on its two counters. If $\mathcal{N}'$ knows the state and the symbol under head, it can update the simulating tape. $\mathcal{N}'$ can keep the state of $\mathcal{N}$ by its internal states and can easily check whether:

---

[2] We refer the reader to [17] for a general theory of simulations.

- u equals to 1 or is bigger than 1;
- v^r equals to 0, equals to 1, or bigger than 1;
- the last digit of u is zero or one; and
- the last digit of v^r is zero or one.

Based on these checks, $\mathcal{N}'$ can simulate the corresponding change on the tape (in a single step of $\mathcal{N}$) with the values of the counters.

By a suitable encoding, two counters can simulate $k > 2$ counters. Let $p_1, p_2, \ldots, p_k$ be co-prime numbers. The values of all k counters, say $c_1, c_2, \ldots, c_k$, can be represented as $p_1^{c_1} p_2^{c_2} \cdots p_k^{c_k}$. A counter can hold this value, and, by using the second counter, $\mathcal{N}'$ can check if c_i is equal to zero (Lemma 4) and it can simulate an appropriate increment/decrement operation on c_i, where $1 \leq i \leq k$. Therefore, we can conclude that a 2D2CA, say $\mathcal{N}''$, can simulate $\mathcal{N}$ on w by using prime numbers $\{2, 3, 5\}$ for encoding, if its first counter is set to 3^{1w1}. Here $\mathcal{N}''$ can use the exponents of 3 and 5 for keeping the content of the tape and the exponent of 2 to simulate the third counter.

Let's assume that $\mathcal{N}$ uses exactly $|w| + 2$ space, i.e. the tape head never leaves the tape squares initially containing $\#w\#$. That is, the binary value of the tape is always less than twice of $1w1$, which is $1w10$, during the whole computation. Then the values of the counters can never exceed $5^{1w10} 2^{1w10}$ or $25^{1w1} 4^{1w1}$, where the exponents are the numbers in binary. Note that the whole tape is kept by the exponent of 3 and 5, and so, their product is always less than 5^{1w10}.

Theorem 5. *If $L \subseteq \{a, b\}^*$ can be recognized by a DTM, say $\mathcal{N}$, in space $|w| + 2$ with binary work alphabet, then*

$$\left\{ a^{100^{1w1}} \mid w \in L \right\}$$

can be recognized by a 2D1CA, say $\mathcal{N}'''$.

Proof. $\mathcal{N}'''$ rejects the input if it is not of the form $\{a^{100^n}\}$ (Lemma 1), where $n > 0$. Then, it sets its first counter to 4^n (Lemma 2). $\mathcal{N}'''$ rejects the input, if n is not of the form $1w1$ for some $w \in \{a, b\}$. We know that a 2D2CA can easily do this check if one of its counter is set to n, i.e. it needs to check n is odd and $n \notin \{0, 1, 2\}$. So, $\mathcal{N}'''$ can implement this test by using its input head as the second counter.

As described above, if its first counter is set to 3^{1w1}, the 2D2CA $\mathcal{N}''$ can simulate $\mathcal{N}$ on a given input w. Due to the space restriction on $\mathcal{N}$, we also know that the counter values (of the 2D2CA's) never exceed 100^{1w1}. So, $\mathcal{N}'''$ needs only to set its counter value to 3^{1w1}. $\mathcal{N}'''$ firstly sets its counter to $4^{1w1} 3^0$, and then transfers $1w1$ to the exponent of 3. $\qquad\square$

Remark that the language recognized by $\mathcal{N}'''$ can also be represented as

$$\left\{ a^{10^{1w10}} \mid w \in L \right\}.$$

This representation is more convenient when considering DTMs working on bigger alphabets.

Corollary 1. *Let $k > 2$ and $L \subseteq \{a_1, \ldots, a_k\}^*$ be a language recognized by a DTM in space $|w| + 2$ with a work alphabet having $k' \geq k$ elements. Then*

$$\left\{ a^{10^{1w10}} \mid w \in L \right\}$$

can be recognized by a 2D1CA, where $w \in \{a_1, \ldots, a_k\}^*$ and $1w10$ is a number in base-k'.

Proof. The proof is almost the same by changing base-2 with base-k'. Additionally, the 2D1CA needs to check whether each digit of w is less than k. □

We know that 2D1CAs can recognize $\texttt{POWER} = \left\{ a^n b a^{2^n} \mid n > 0 \right\}$ [23]. Therefore, by using a binary encoding, we can give a simulation of exponential space DTMs where the exponent is linear. Here, the input is supposed to be encoded into the exponent of the first block of a's and the working memory in the second block of a's.

Theorem 6. *Let $k > 2$ and $L \subseteq \{a_1, \ldots, a_k\}^*$ be a language recognized by a DTM in space $2^{|w|}$ with a work alphabet having $k' \geq k$ elements. Then*

$$\left\{ a^{10^x} b a^{2^{(10^x)}} \mid x = 1w10 \text{ and } w \in L \right\}$$

can be recognized by a 2D1CA, where $w \in \{a_1, \ldots, a_k\}^*$ and $x = 1w10$ is a number in base-k'.

We can generalize this result for any arbitrary space-bounded DTMs. It is not hard to show that, for any $z > 1$, 2D1CAs can recognize

$$\texttt{POWER(z)} = \left\{ a^n b a^{exp(n)} b a^{exp^2(n)} b \cdots b a^{exp^z(n)} \mid n > 0 \right\}.$$

Corollary 2. *Let $z > 1$ and $k > 2$ and $L \subseteq \{a_1, \ldots, a_k\}^*$ be a language recognized by a DTM in space $exp^z(|w|)$ with a work alphabet with $k' \geq k$ elements. Then*

$$\left\{ a^{10^x} b a^{exp(10^x)} b a^{exp^2(10^x)} b \cdots b a^{exp^z(10^x)} \mid x = 1w10 \text{ and } w \in L \right\}$$

can be recognized by a 2D1CA, where $w \in \{a_1, \ldots, a_k\}^*$ and $x = 1w10$ is a number in base-k'.

Note that, similar to the previous section, all of the above results are valid if we replace deterministic machines with nondeterministic, alternating, or probabilistic ones.

Now, we present a more general result.

Theorem 7. *Let L be a recursive enumerable language and $\mathcal{T}$ be a DTM recognizing it (note that $\mathcal{T}$ may not halt on some non-members). The language*

$$L_{\mathcal{T}} = \left\{ a^{2^{1w} 3^{S(w)}} \mid w \in L \right\},$$

where $S(w)$ is a sufficiently big number that depends on w, can be recognized by a two way deterministic one counter automaton $\mathcal{D}$.

Proof. We use a slight variation of the 2DCA simulation of a DTM given above. Informally the $a^{3^{S(w)}}$ part of the input gives $\mathcal{D}$ enough space to complete its simulation, i.e. decide the membership of $w \in L$ using its head position as a second counter, being sure that its value never exceeds the size of the input.

First we show that if $S(w)$ is large enough then a 2D1CA $\mathcal{D}$ can recognize the language $L_{\geq \mathcal{T}}$:

$$L_{\geq \mathcal{T}} = \left\{ a^{2^{1_w} 3^k} \mid w \in L \text{ and } k \geq S(w) \right\}.$$

$\mathcal{D}$ checks that the input is in the correct format $a^{2^{1_w} 3^k}$ (Lemma 3), then it simulates $\mathcal{T}$ on w like showed in the proof of Theorem 5. During its computation, if $\mathcal{D}$ reaches the right end-marker, then it stops and rejects.

Suppose that on input w the Turing machine $\mathcal{T}$ does not halt: it visits an infinite number of empty cells of its tape or it enters an infinite loop. In both cases, the value of $S(w)$ is irrelevant, and $\mathcal{D}$ will never accept the input $a^{2^{1_w} 3^{S(w)}}$: in the first case for all values of $S(w)$ $\mathcal{D}$ will hit the right end-marker and will reject; in the second case, if $S(w)$ is too low and $\mathcal{D}$ has not enough space to simulate $\mathcal{T}$ in the loop area of the tape it will hit the right end-marker and reject, if $S(w)$ is large enough, $\mathcal{D}$ will also enter the endless loop.

Now suppose that the Turing machine $\mathcal{T}$ accepts (resp. rejects) w then there are two possibilities: *a)* during its computation the 2D1CA (that uses the head position like a second counter) never reaches the right end-marker; in this case it can correctly accept (resp. reject) the input; or *b)* during its computation the 2D1CA reaches the right end-marker (informally it has not enough space) and cannot correctly decide the membership of $w \in L$; but in this case we are sure that there exists a larger value $S(w) = s' > s$ that assures enough space to end the computation. Also for every $k \geq S(w)$, $\mathcal{T}$ will correctly accept each string in

$$\left\{ a^{2^{1_w} 3^k} \mid w \in L \right\}.$$

We can slightly modify $\mathcal{D}$ and narrow the language it recognizes to exactly $L_{\mathcal{T}}$, i.e. making it accept each string in:

$$\left\{ a^{2^{1_w} 3^{S(w)}} \mid w \in L \right\},$$

but reject each string in:

$$\{ a^{2^{1_w} 3^i} \mid w \in L \text{ and } i \neq S(w) \}.$$

We can divide the natural numbers as follows:

$$[0, 3N) \quad [3N, 9N) \quad [9N, 27N) \quad \ldots \quad [3^{k-1}N, 3^k N) \quad [3^k N, 3^{k+1} N) \quad \ldots$$

Let M be the maximum value of the second counter of $\mathcal{D}$ during the simulation of the DTM on w (for each member of L, such value exists). M must be in one of the above intervals, let's say in $[3^{k-1}N, 3^k N)$. It is obvious that $3M$ must be in the next interval $[3^k N, 3^{k+1} N)$.

The second counter can use the set $\{+3, 0, -3\}$ instead of $\{+1, 0, -1\}$ for update operations, i.e. the head moves three steps left or right instead of a single step, and using the internal states we can allow it to exceed the input length up to three times its value: when the head reaches the right end-marker it can keep track that it has made one *"fold"* and continues moving towards the left; thereafter, if it reaches the left end-marker, it records that it has made two folds and continues move rightward, and so on. When, after a fold, it hits the last end-marker again it can decrease the number of folds and continue. Let $\mathsf{FOLD} \in \{0, 1, 2, 3\}$ store the number of folds. When FOLD becomes 3, then the 2D1CA rejects the input immediately, i.e. the counter value reaches the value of three times of the input length.

On input $a^{2^{1^w} 3^k}$, the second counter, that changes its value by $\{+3, 0, -3\}$, will exceed $3^k N$ but will never try to exceed $3^{k+1} N$ ($3^k N \le 3M < 3^{k+1} N$). So, FOLD must be 1 at least once and never becomes 3. Therefore, the 2D1CA accepts the input if the simulation ends with the decision of "acceptance" and FOLD takes a non-zero value at least once but never takes the value of 3.

If the input is $a^{2^{1^w} 3^{k-i}}$, for some positive integer i, then the second counter must need to exceed $3^k N$, so, the FOLD value takes 3 before simulation terminates and the 2D1CA rejects the input.

If the input is $a^{2^{1^w} 3^{k+i}}$, for some positive integer i, then the second counter can be at most $3^{k+1} - 1$, so the FOLD value never takes the value of 1 during the simulation and the 2D1CA rejects the input. Thus, we can be sure that such k has a minimum value and it corresponds to $S(w)$ in the language $L_{\mathcal{T}}$.   □

Note that if the language L recognized by $\mathcal{T}$ is recursive, then the same 2D1CA $\mathcal{D}$ described in Theorem 7 is a decider for $L_{\mathcal{T}}$.

3.3   A Quantum Simplification

Ambainis and Watrous [1] showed that augmenting a two-way deterministic finite automata (2DFAs) with a fixed-size quantum register[3] makes them more powerful than 2DFAs augmented with a random number generator. Based on a new programming technique given for fixed-size quantum registers [29], it was shown that 2D1CAs having a fixed-size quantum register can recognize $\{a^{n3^n} \mid n \ge 1\}$, $\{a^{2^n 3^{2^n}} \mid n \ge 1\}$, or any similar language by replacing bases 2 or 3 with some other integers for any error bound [27,25]. Therefore, we can replace

---

[3] It is a constant-size quantum memory whose dimension does not depend on the input length. The machine can apply to the register some quantum operators (unitary operators, measurements, or superoperators) determined by the classical configuration of the machine. If the operator is a measurement or a superoperator, then there can be more than one outcome and the next classical transition is also determined by this outcome, which makes the computation probabilistic. However, as opposed to using a random number generator, the machine can store some information on the quantum register and some pre-defined branches can disappear during the computation due to the interference of the quantum states, which can give some extra computational power to the machine.

binary encoding with a unary one for Theorem 6 by enhancing a 2D1CA with a fixed-size quantum register.

Theorem 8. *Let $k > 2$ and $L \subseteq \{a_1, \ldots, a_k\}^*$ be a language recognized by a DTM in space $3^{|w|}$ with a work alphabet having $k' \geq k$ elements. Then*

$$\left\{ a^{10^x 3^{(10^x)}} \mid x = 1w10 \text{ and } w \in L \right\}$$

can be recognized by a 2D1CA augmented with a fixed-size quantum register for any error bound, where $x = 1w10$ is a number in base-k' and $w \in \{a_1, \ldots, a_k\}^*$.

Proof. Here the input check can be done by the help of the quantum register by using the corresponding quantum algorithms given in [27]. Then, our standard deterministic simulation is implemented.   □

3.4   Unary 2D1CAs versus Two-Counter Machines

Minsky [18] showed that, for any given recursive language L defined over $\mathbb{N}$,

$$\mathtt{UMINSKY}(L) = \{a^{2^x} \mid x \in L\}$$

can be recognized by a 2CA[4]. It is clear that $\mathtt{UMINSKY}(L)$ is recursive enumerable if and only if L is recursive enumerable. Moreover, any language L not recognizable by any $s(n)$-space DTM, $\mathtt{UMINSKY}(L)$ cannot be recognized by any $\log(s(n))$-space DTM, for any $s(n) \in \Omega(n)$. On the other hand, any language recognized by a 2D1CA is in L (see Footnote 1). Therefore, there are many recursive and non-recursive languages recognized by 2CAs but not by 2D1CAs.

Neverthless we believe that 2CAs and unary 2D1CAs are incomparable, i.e. there also exist languages recognizable by a 2D1CA but not by any 2CA. Let $k > 1$, $\Sigma = \{a_0, \ldots, a_{k-1}\}$ be the alphabet, and $r_k : \mathbb{N} \to \Sigma^*$ be a function mapping $n = (d_l \cdots d_1 d_0)_k$, k-ary representation of n, to

$$r_k(n) = \begin{cases} a_{d_0} a_{d_1} \cdots a_{d_l}, & \text{if } n > 0 \\ \varepsilon, & \text{if } n = 0 \end{cases}.$$

Both 2D1CAs and 2CAs can calculate $r_k(n)$ symbol by symbol on the input a^n, and the following is immediate:

Lemma 8. *If R is a regular languages over an alphabet of k symbols, then a 2D1CA can decide the language $L = \{a^n \mid r_k(n) \in R\}$.*

Proof. Suppose that $\mathcal{F}$ is a DFA that decides R; after transferring the input to the counter, a 2D1CA can calculate incrementally the digits $d_0, d_1, \ldots, d_l$ up to the final fixed digit: it repeatedly divides the counter by k, and d_i is the remainder of the division; so it can simulate the transition of $\mathcal{F}$ on symbol a_{d_i}, and accept or reject accordingly when it reaches the last digit.   □

---

[4] The definition used by Minsky is a little different than ours but they are equivalent.

Hence both 2D1CAs and 2CAs can recognize the whole class of unary languages:

$$\mathcal{C} = \{L \mid L = \{a^n \mid r_k(n) \in R\} \text{ and } R \text{ is a regular language}$$
$$\text{over an alphabet of size } k\}$$

As an example the family of unary non regular languages $\{a^{2^n}\}$ is contained in $\mathcal{C}$. But, we conjecture that the following language cannot be recognized by 2CAs:

$$L_\oplus = \{a^n \mid |r_2(n)| + |r_3(n)| \equiv 0 \mod 2\},$$

i.e. the binary representation and the ternary representation of n have both even or odd length. A 2D1CA can easily decide $L_\oplus$: after calculating if the length of the binary representation of n is odd or even, it can recover the input using the tape endmarkers, and then check if the length of the ternary representation of n is the same. But there is no way for a 2CA to recover the input, so it should calculate the binary and ternary representations of n in parallel, which seems impossible.

Acknowledgements. We thank Alexander Okhotin, Holger Petersen, and Klaus Reinhardt for their answers to our questions on the subject matter of this paper. We also thank anonymous referees for their very helpful comments.

References

1. Ambainis, A., Watrous, J.: Two–way finite automata with quantum and classical states. Theoretical Computer Science 287(1), 299–311 (2002)
2. Böhm, S., Göller, S., Jancar, P.: Equivalence of deterministic one-counter automata is NL-complete. In: STOC, pp. 131–140. ACM (2013)
3. Chandra, A.K., Kozen, D.C., Stockmeyer, L.J.: Alternation. Journal of the ACM 28(1), 114–133 (1981)
4. Chrobak, M.: Nondeterminism is essential for two-way counter machines. In: Chytil, M.P., Koubek, V. (eds.) MFCS 1984. LNCS, vol. 176, pp. 240–244. Springer, Heidelberg (1984)
5. Demirci, H.G., Hirvensalo, M., Reinhardt, K., Say, A.C.C., Yakaryılmaz, A.: Classical and quantum realtime alternating automata. In: NCMA (to appear, 2014)
6. Ďuriš, P.: Private communication (October 2013)
7. Ďuriš, P., Galil, Z.: Fooling a two way automaton or one pushdown store is better than one counter for two way machines. Theoretical Computer Science 21, 39–53 (1982)
8. Fischer, P.C., Meyer, A.R., Rosenberg, A.L.: Counter machines and counter languages. Mathematical Systems Theory 2(3), 265–283 (1968)
9. Ginsburg, S., Rice, H.G.: Two families of languages related to ALGOL. Journal of the ACM 9(3), 350–371 (1962)
10. Hartmanis, J.: On non-determinancy in simple computing devices. Acta Informatica 1, 336–344 (1972)
11. Ibarra, O.H., Jiang, T., Trân, N.Q., Wang, H.: On the equivalence of two-way pushdown automata and counter machines over bounded languages. International Journal of Foundations of Computer Science 4(2), 135–146 (1993)

12. Ibarra, O.H., Trân, N.Q.: A note on simple programs with two variables. Theoretical Computer Science 112(2), 391–397 (1993)
13. Kaneps, J., Geidmanis, D., Freivalds, R.: Tally languages accepted by Monte Carlo pushdown automata. In: Rolim, J.D.P. (ed.) RANDOM 1997. LNCS, vol. 1269, pp. 187–195. Springer, Heidelberg (1997)
14. King, K.N.: Alternating multihead finite automata. Theoretical Computer Science 61(2-3), 149–174 (1988)
15. Kondacs, A., Watrous, J.: On the power of quantum finite state automata. In: FOCS 1997: Proceedings of the 38th Annual Symposium on Foundations of Computer Science, pp. 66–75 (1997)
16. Macarie, I.I.: Multihead two-way probabilistic finite automata. Theory of Computing Systems 30(1), 91–109 (1997)
17. van Emde Boas, P.: Machine models and simulations. In: Handbook of Theoretical Computer Science, vol. A, pp. 1–66 (1990)
18. Minsky, M.: Recursive unsolvability of Post's problem of "tag" and other topics in theory of Turing machines. Annals of Mathematics 74(3), 437–455 (1961)
19. Minsky, M.: Computation: Finite and Infinite Machines. Prentice-Hall (1967)
20. Monien, B.: Deterministic two-way one-head pushdown automata are very powerful. Information Processing Letters 18(5), 239–242 (1984)
21. Moore, C., Crutchfield, J.P.: Quantum automata and quantum grammars. Theoretical Computer Science 237(1-2), 275–306 (2000)
22. Morita, K.: Universality of a reversible two-counter machine. Theoretical Computer Science 168(2), 303–320 (1996)
23. Petersen, H.: Two-way one-counter automata accepting bounded languages. SIGACT News 25(3), 102–105 (1994)
24. Petersen, H.: Private communication (June 2012)
25. Reinhardt, K., Yakaryılmaz, A.: The minimum amount of useful space: New results and new directions. In: Developments in Language Theory (to appear, 2014), arXiv:1405.2892
26. Schroeppel, R.: A two counter machine cannot calculate 2^n. Technical Report AIM-257. MIT (1972)
27. Yakaryılmaz, A.: Log-space counter is useful for unary languages by help of a constant-size quantum register. Technical Report arXiv:1309.4767 (2013)
28. Yakaryılmaz, A.: One-counter verifiers for decidable languages. In: Bulatov, A.A., Shur, A.M. (eds.) CSR 2013. LNCS, vol. 7913, pp. 366–377. Springer, Heidelberg (2013)
29. Yakaryılmaz, A., Say, A.C.C.: Succinctness of two-way probabilistic and quantum finite automata. Discrete Mathematics and Theoretical Computer Science 12(2), 19–40 (2010)

A Type System for Weighted Automata and Rational Expressions

Akim Demaille[1], Alexandre Duret-Lutz[1], Sylvain Lombardy[2],
Luca Saiu[3,1], and Jacques Sakarovitch[3]

[1] LRDE, EPITA, France
{akim,adl}@lrde.epita.fr
[2] LaBRI, Institut Polytechnique de Bordeaux, France
Sylvain.Lombardy@labri.fr
[3] LTCI, CNRS / Télécom-ParisTech, France
{saiu,sakarovitch}@telecom-paristech.fr

Abstract. We present a type system for automata and rational expressions, expressive enough to encompass weighted automata and transducers in a single coherent formalism. The system allows to express useful properties about the applicability of operations including binary heterogeneous functions over automata.

We apply the type system to the design of the VAUCANSON 2 platform, a library dedicated to the computation with finite weighted automata, in which genericity and high efficiency are obtained at the lowest level through the use of template metaprogramming, by letting the C++ template system play the role of a static type system for automata. Between such a low-level layer and the interactive high-level interface, the type system plays the crucial role of a mediator and allows for a cleanly-structured use of dynamic compilation.

1 Introduction

VAUCANSON[1] is a free software[2] *platform* dedicated to the computation of and with *finite automata*. It is designed with several use cases in mind. First and foremost it must support experiments by automata theory researchers. As a consequence, *genericity* and *flexibility* have been goals since day one: automata and transducers must support any kind of semiring of weights, and labels must not be restricted to just letters. In order to demonstrate the computational qualities of algorithms, *performance* must also be a main concern. To enforce this we aim, eventually, at applying VAUCANSON to linguistics, whose problems are known for their size; on this standpoint we share goals with systems such as OpenFST [2]. Finally our platform should be *easy to use* by teachers and students in language theory courses (a common goal with FAdo [3]), which also justifies our focus on rational expressions.

---

[1] Work supported by ANR Project 10-INTB-0203 VAUCANSON 2.

[2] http://vaucanson.lrde.epita.fr

M. Holzer and M. Kutrib (Eds.): CIAA 2014, LNCS 8587, pp. 162–175, 2014.

Among our goals flexibility and efficiency are potentially in conflict. The main objective of this work is demonstrating how to reconcile them, and how to use a type system to manage such complexity.

Aiming at both efficiency and flexibility essentially dictates the architecture: the software needs to be rigidly divided into *layers*, varying in comfort and speed.

The bottom layer (named *static*) is a C++ library. For the sake of efficiency the classical object-oriented run-time method dispatch (associated to the C++ `virtual` keyword) is systematically avoided, instead achieving compile-time code generation by using `template` metaprogramming [1]. This results in a *closed world*: new types of automata require the compilation of dedicated code.

At the opposite end of the spectrum, the topmost layer is based on IPython [6]. It is visual (automata are displayed on-screen) and, most importantly, interactive: the user no longer needs to write a C++ or even a Python program, and instead just interacts with the system using Python as a command language. In such a high-level environment the closed-world restriction would be unacceptable, resulting as it would in error messages such as "this type of automaton is not supported; please recompile and rerun". To address this issue VAUCANSON uses on-the-fly generation and compilation of code, relying on our type system in a fundamental way.

This paper builds on top of ideas introduced last year [4][3]. However, in that work contexts were partitioned and entities of different types could not be mixed together. In particular algorithms such as the union of automata were "homogenous": operands had all the same type, which was that of the result. The contribution of this paper is to introduce support for heteregeneous types: the definition of a type calculus, its implementation and, to gain full benefit from it, dynamic code generation.

This paper is structured as follows. In Sec. 2 we describe the types of weighted automata, rational expressions and their components. Then, in Sec. 3, we study how types relate to one another and how to type operations over automata. We introduce the implementation counterpart of types in Sec. 4, which also explains how run-time compilation reconciles performances and flexibility. Sec. 5 discusses the pros and cons of the current implementation.

2 Typing Automata and Rational Expressions

Computing with weighted automata or rational expressions entails reasoning about *types*. We should have a system strong enough to detect some unmet preconditions (for instance applying subset construction on an automaton weighted in $\mathbb{Z}$), and at the same time expressive enough to encompass many different kinds of automata, including transducers.

---

[3] Names and notations have slightly changed. We now name "Value/ValueSet" the core design principle in Vaucanson, rather than "Element/ElementSet". For consistency with POSIX regular expression syntax, curly braces now denote power: 'a{2}' means aa instead of $a \cdot 2$, which is now written 'a<2>'. Similarly, 'a(*min,max)' is now written 'a{min,max}'.

2.1   Weighted Automata

Usually a weighted automaton $\mathcal{A}$ is defined as a sextuple $(A, \mathbb{K}, Q, I, F, E)$, A being an alphabet (a finite set of symbols), $\mathbb{K}$ a semiring, Q a finite set of states, I/F initial/final (partial) functions $Q \to \mathbb{K}$, and E a (partial) function in $Q \times A \times Q \to \mathbb{K}$. With such a definition, the generalization to transducers involves turning the sextuple into a septuple by adding a second *output* alphabet, changing the transition function domain to also take output labels into account, among the rest. Independently from transducers, definitions also need variants for many alternative cases, such as admitting the empty word as an input or output label. In VAUCANSON this variability is captured by *contexts*, each composed of one *LabelSet* and one *WeightSet*.

Different *LabelSets* model multiple variations on *labels*, members of a monoid:

letterset. Fully defined by an alphabet A, its labels being just letters. It is simply denoted by A. It corresponds to the usual definition of an NFA.

nullableset. Denoted by $A^?$, also defined by an alphabet A, its labels being either letters or the empty word. This corresponds to what is often called ε-NFAs.

wordset. Denoted by A^*, also defined by an alphabet A, its labels being (possibly empty) words on this alphabet.

oneset Denoted by $\{1\}$, containing a single label: 1, the empty word.

tupleset. Cartesian product of LabelSets, $L_1 \times \cdots \times L_n$. This type implements the concept of transducers with an arbitrary number of "tapes".

In the implementation LabelSets define the underlying monoid operations, and a few operators such as comparison.

A *WeightSet* is a semiring whose operations determine how to combine weights when evaluating words. Examples of WeightSets include $\langle \mathbb{B}, \vee, \wedge \rangle$, the family $\langle \mathbb{N}, +, \times \rangle$, $\langle \mathbb{Z}, +, \times \rangle$, $\langle \mathbb{Q}, +, \times \rangle$, $\langle \mathbb{R}, +, \times \rangle$ and tropical semirings such as $\langle \mathbb{Z} \cup \{\infty\}, \min, + \rangle$; moreover tuplesets also allow to combine WeightSets, making weight tuples into weights.

In the implementation a WeightSet defines the semiring operations and comparison operators, plus some feature tests such as "star-ability" [5].

We may finally introduce contexts, and the definition of automata used in VAUCANSON — a triple corresponding to its type (context), its set of states and its set of transitions.

Definition 1 (Context). *A context C is a pair (L, W), denoted by $L \to W$, where:*

– L is a LabelSet, a subset of a monoid;

– W is a WeightSet, a semiring.

Definition 2 ((Typed, Weighted) Automaton). *An automaton $\mathcal{A}$ is a triple (C, Q, E) where:*

– $C = L \to W$ is a context;

– Q is a finite set of states;

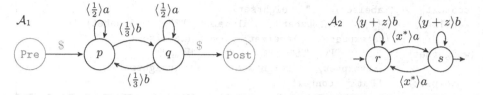

Fig. 1. Two (typed) automata: $\mathcal{A}_1$, whose context is $C_1 = \{a, b, c\} \to \mathbb{Q}$, and $\mathcal{A}_2$, whose context is $C_2 = \{a, b, d\} \to \mathsf{RatE}[\{x, y, z\} \to \mathbb{B}]$, i.e., with rational expressions as weights. In $\mathcal{A}_1$ we reveal the Pre and Post hidden states.

 – E *is a (partial) function whose domain represents the set of transitions, in:*
$$(Q \times L \times Q) \cup (\{\mathtt{Pre}\} \times \{\$\} \times Q) \cup (Q \times \{\$\} \times \{\mathtt{Post}\}) \to (W \setminus \{0\}).$$

Notice that the initial and final functions are embedded in the definition of E through two special states —the *pre-initial* and *post-final* states Pre and Post— and a special label not part on L and only occurring on *pre-transitions* (transitions from Pre) and *post-transitions* (transitions from Post). This somewhat contrived definition actually results in much simpler data structures and algorithms: with a unique Pre and a unique Post there is no need to deal with initial and final weights in any special way. On Fig. 1, automaton $\mathcal{A}_1$ is drawn with explicit Pre and Post states, while $\mathcal{A}_2$ is drawn without them.

2.2 Rational Expressions

Definition 3 ((Typed, Weighted) Rational Expression). *A rational expression $\mathcal{E}$ is a pair (C, E) where:*

 – $C = L \to W$, *is a context,*
 – E *is a term built from the following abstract grammar*

$$\mathsf{E} := 0 \mid 1 \mid \ell \mid \mathsf{E} + \mathsf{E} \mid \mathsf{E} \cdot \mathsf{E} \mid \mathsf{E}^* \mid \langle w \rangle \mathsf{E} \mid \mathsf{E} \langle w \rangle$$

where $\ell \in L$ is any label, and $w \in W$ is any weight.

The set of rational expressions of type $L \to W$ is denoted by $\mathsf{RatE}[L \to W]$, and called a *ratexpset*. With a bit of caution rational expressions can be used as weights, as exemplified by automaton $\mathcal{A}_2$ in Fig. 1: equipped with the sum of rational expressions as sum, their concatenation as product, 0 as zero, and 1 as unit, it is very close to being a semiring[4].

Rational expressions may also serve as labels, yielding what is sometimes named *Extended Finite Automata* [3], a convenient internal representation to perform, for example, state elimination, a technique useful to extract a rational expression from an automaton. So, just like tuplesets, ratexpsets can be used as either a WeightSet or a LabelSet.

---

[4] Ratexpset do *not* constitute a semiring for lack of, for instance, equality between two rational expressions; however rational expressions provide an acceptable approximation of *rational series* [7, Chap. III], the genuine corresponding semiring.

```
⟨context⟩   ::= ⟨labelset⟩ "→" ⟨weightset⟩
⟨labelset⟩  ::= "{1}" | ⟨alphabet⟩ | ⟨alphabet⟩ "?" | ⟨alphabet⟩ "*"
             | ⟨ratexpset⟩ | ⟨labelset⟩ ×···× ⟨labelset⟩
⟨weightset⟩ ::= "𝔹" | "ℕ" | "ℤ" | "ℚ" | "ℝ" | "ℤ_min"
             | ⟨ratexpset⟩ | ⟨weightset⟩ ×···× ⟨weightset⟩
⟨ratexpset⟩ ::= "RatE" ⟨context⟩
```

Fig. 2. A Grammar of Types

Fig. 2 shows the precise relation among the different entities introduced up to this point: LabelSets, WeightSets, contexts, ratexpsets.

3 The Type System

3.1 Operations on Automata

Several binary operations on automata exist: union, concatenation, product, shuffle and infiltration products, to name a few. To demonstrate our purpose we consider the simplest one, i.e., the union of two automata, whose behavior is the sum of the behavior of each operand.

Definition 4 ((Homogeneous) Union of Automata). *Let* $\mathcal{A}_1 = (C, Q_1, E_1)$ *and* $\mathcal{A}_2 = (C, Q_2, E_2)$ *be two automata of the same type* C. $\mathcal{A}_1 \cup \mathcal{A}_2$ *is the automaton* $(C, Q_1 \cup Q_2, E_1 \cup E_2)$.

Def. 4 is simple, but has the defect of requiring the two argument automata to have exactly the same type. Overcoming this restriction and making operations such as automata union more widely applicable is a particularly stringent requirement in an interactive system (Sec. 4.3).

Automata union can serve as a good example to convey the intuition of heterogeneous operation typing: if its two operands have LabelSets with different alphabets, the result LabelSet should have their *union* as alphabet; if one operand is an NFA and the other a ε-NFA, their union should also be a ε-NFA. It is also reasonable to define the union between an automaton with spontaneous transitions only (oneset) and an NFA (letterset) as a ε-NFA (nullableset) — a type different from *both* operands', and intuitively "more general" than either.

Much in the same way, some WeightSets are straightforward to embed into others: $\mathbb{Z}$ into $\mathbb{Q}$, and even $\mathbb{Q}$ into $\mathsf{RatE}[L \to \mathbb{Q}]$. Then, let two automata have weights in $\mathbb{Q}$ and $\mathsf{RatE}[L \to \mathbb{Z}]$; their union should have weights in *the least WeightSet that contains both* $\mathbb{Q}$ *and* $\mathsf{RatE}[L \to \mathbb{Z}]$, which is to say $\mathsf{RatE}[L \to \mathbb{Q}]$. Once more the resulting type is new: it does not match the type of either operand.

3.2 The Hierarchy of Types

The observations above can be captured by introducing a *subtype* relation as a partial order on LabelSets, WeightSets and contexts, henceforth collectively denoted as *ValueSets*. We write $V_1 <: V_2$ to mean that V_1 is a subtype of V_2; in this case each element of V_1 may be used wherever an element of V_2 would be expected, and we have in particular that $V_1 \subseteq V_2$. Notice that this makes our relation reflexive, so for every ValueSet V we have that $V <: V$.

For simplicity we will focus on free monoids only. Let A, B be any alphabets such that $A \subseteq B$. Then we define:

$$\{1\} <: A^? \qquad\qquad A <: A^? \qquad\qquad A^? <: A^*$$
$$A <: B \qquad\qquad A^? <: B^? \qquad\qquad A^* <: B^*$$

For WeightSets, if the WeightSet W_1 is a sub-semiring of W_2, it trivially holds that $W_1 <: W_2$; therefore $\mathbb{N} <: \mathbb{Z} <: \mathbb{Q} <: \mathbb{R}$. The WeightSet $\mathbb{B}$, as the WeightSet of language recognizers, is worthy of special treatment; in particular it is convenient to allow heterogeneous operations between automata over $\mathbb{B}$ and automata over other WeightSets, which yields:

$$\mathbb{B} <: \mathbb{N} <: \mathbb{Z} <: \mathbb{Q} <: \mathbb{R} \qquad \mathbb{B} <: \mathbb{Z}_{\min} \tag{1}$$

This allows for instance to restrict the domain of a series realized by a weighted automaton to the rational language described by a Boolean automaton. For this reason it is desirable to have $\mathbb{B}$ at the bottom of the WeightSet hierarchy, so that it can be promoted to any other WeightSet simply by mapping false to the WeightSet zero, and true to its unit. However such conversion requires care and should not be used blindly; in particular converting an ambiguous Boolean automaton to another WeightSet leads in general to an automaton which does *not* realize the characteristic series of the language recognized by the original.

A context C_1 is a subtype of a context C_2 if C_1 has a LabelSet and a WeightSet which are respectively subtypes of the LabelSet and WeightSet of C_2.

$$(L_1 \rightarrow W_1) <: (L_2 \rightarrow W_2) \quad \text{iff} \quad L_1 <: L_2 \text{ and } W_1 <: W_2 \tag{2}$$

As of today tuples of ValueSets do not mix with other values:

$$(V_1 \times \cdots \times V_n) <: (V_1' \times \cdots \times V_n') \quad \text{iff} \quad (V_i <: V_i') \text{ for all } 1 \leq i \leq n \tag{3}$$

Interestingly, rational expressions can play the role of both labels and weights:

$$\mathsf{RatE}[C_1] <: \mathsf{RatE}[C_2] \qquad\quad \text{iff} \quad C_1 <: C_2$$
$$L_1 <: \mathsf{RatE}[L_2 \rightarrow W_2] \quad \text{iff} \quad L_1 <: L_2 \tag{4}$$
$$W_1 <: \mathsf{RatE}[L_2 \rightarrow W_2] \quad \text{iff} \quad W_1 <: W_2$$

The subtype relations between LabelSets are summarized in Fig. 3. If two LabelSets L_1 and L_2 admit a least upper bound (*resp.* a greatest lower bound), we call it the join (*resp.* the meet) of these two LabelSets and we denote it by $L_1 \vee L_2$ (*resp.* the $L_1 \wedge L_2$). The cases where no join or meet exists correspond in practice to compilation errors about undefined cases. The join and meet operations extend naturally to other ValueSets such as WeightSets, tuples,

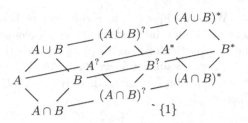

Fig. 3. The Hasse diagram of the LabelSets generated by the two alphabets A and B showing, for instance, that $A^? \vee B = (A \cup B)^?$

contexts and rational expressions, as per Equations (1) to (4)). For instance, for any LabelSet L_1, L_2 and any WeightSet W_1, W_2:

$$\mathsf{RatE}[L_1 \to W_1] \vee L_2 := \mathsf{RatE}[(L_1 \vee L_2) \to W_1]$$
$$\mathsf{RatE}[L_1 \to W_1] \vee W_2 := \mathsf{RatE}[L_1 \to (W_1 \vee W_2)]$$
$$\mathsf{RatE}[L_1 \to W_1] \vee \mathsf{RatE}[L_2 \to W_2] := \mathsf{RatE}[(L_1 \to W_1) \vee (L_2 \to W_2)]$$

At this point we are ready to describe typing for binary operations on heterogeneous automata more formally. An operation on two automata with contexts $L_1 \to W_1$ and $L_2 \to W_2$ will yield a result with context $(L_1 \vee L_2) \to (W_1 \vee W_2)$. As an example we can extend Def. 4 into:

Definition 5 (Heterogeneous Union of Automata). *Let $\mathcal{A}_1 = (C_1, Q_1, E_1)$ and $\mathcal{A}_2 = (C_2, Q_2, E_2)$ be two automata. $\mathcal{A}_1 \cup \mathcal{A}_2 := (C_1 \vee C_2, Q_1 \cup Q_2, E_1 \cup E_2)$.*

3.3   Type Restriction

The specific semantics of some binary operations let us characterize the result type more precisely. For instance spontaneous-transition-removal applied to an automaton with LabelSet $A^?$ returns a *proper* automaton, i.e., an automaton with LabelSet A. Another interesting example is the product of automata labeled by letters[5], whose behavior is the Hadamard product of series of the behavior of each operand, if the WeightSet is commutative.

Definition 6 (Product of Automata). *Let $\mathcal{A}_1 = ((L_1 \to W_1), Q_1, E_1)$ and $\mathcal{A}_2 = ((L_2 \to W_2), Q_2, E_2)$ be two automata, where L_1 and L_2 are lettersets. $\mathcal{A}_1 \& \mathcal{A}_2$ is the accessible part of the automaton $(C_\&, Q_\&, E_\&)$ where $C_\& = (L_1 \wedge L_2) \to (W_1 \vee W_2)$, $Q_\& = Q_1 \times Q_2$, and*

$$((q_1, q_2), \ell, (q_1', q_2')) \in \mathsf{Dom}(E_\&) \quad \textit{iff} \begin{cases} (q_1, \ell, q_1') \in \mathsf{Dom}(E_1), \\ (q_2, \ell, q_2') \in \mathsf{Dom}(E_2); \end{cases}$$
$$E_\&((q_1, q_2), \ell, (q_1', q_2')) = E_1(q_1, \ell, q_1') \cdot E_2(q_2, \ell, q_2').$$

---

[5] The product operation can actually be extended to nullablesets, using a more complex algorithm related to weighted transducer composition.

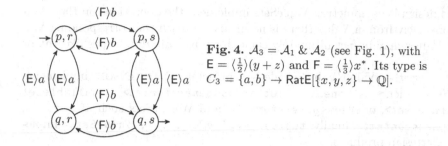

Fig. 4. $\mathcal{A}_3 = \mathcal{A}_1 \,\&\, \mathcal{A}_2$ (see Fig. 1), with $\mathsf{E} = \langle \frac{1}{2} \rangle (y + z)$ and $\mathsf{F} = \langle \frac{1}{3} \rangle x^*$. Its type is $C_3 = \{a, b\} \to \mathsf{RatE}[\{x, y, z\} \to \mathbb{Q}]$.

Like for other binary operations it would be correct to describe the type of the result of a product as the join of its operand types; however in this case the specific operation semantics permits us to be more precise: a product result transition is created if and only if labels match in the two argument automata, and therefore the result LabelSet happens to lie in the *meet* of the argument LabelSets. By contrast, each weight is computed as the product of argument weights, in general belonging to two different WeightSets: the WeightSet of the product hence lies in the *join* of the argument WeightSets.

Fig. 4 shows the heterogeneous product of $\mathcal{A}_1$ and $\mathcal{A}_2$ from Fig. 1.

4   Implementation Facet

4.1   The Value/ValueSet Design Principle

The implementation of VAUCANSON closely follows its algebraic design illustrated in Sec. 2 in terms of labels, weights, automata and rational expressions. Other entities not shown here also exist, such as polynomials.

In a typical object-oriented implementation each of these concepts would be implemented as a class, possibly templated. For instance a Boolean weight would be an instance of some class `boolean_weight` having a `bool` attribute. However some of these concepts require run-time meta-data; for instance a letterset needs a set of letters, so a `letter_label` would aggregate not only a `char` for the label, but also *the whole alphabet*, as a `char` vector. As a context aggregates a LabelSet and a WeightSet it requires run-time meta-data as well, and since rational expressions can also be used as weights, they, too, depend on run-time meta-data. Therefore weights and LabelSets both need to be associated to meta-data at run time.

However it would result in an unacceptable penalty to have every instance carry even a mere pointer to meta-data such as an alphabet (a simple `char` label, because of alignment, would then require at least eight bytes, a 8× space penalty on a 32-bit architecture!). To cut this Gordian knot, as a design principle, we split traditional values into *Value/ValueSet* pairs. The value part is but the implementation of a datum; the *ValueSet*, on the other hand, stores *only one copy* of the meta-data related to the type (such as the alphabet) and performs the operations on values (such as + for $\mathbb{Z}$ and `min` for $\mathbb{Z}_{\min}$) without relying on dynamic dispatch.

This design is asymmetric: ValueSets implement the operations on their Values; conversely from a Value there is no means to reach the corresponding ValueSet. Values may in fact ultimately come down to plain data types like int or char.

Following the Value/ValueSet design principle, VAUCANSON implements LabelSets such as oneset, letterset<*generatorset*>[6], nullableset <*generatorset*>, wordset<*generatorset*>, and WeightSets such as b, z, ..., ratexpset<*context*>; finally, tupleset<*ValueSet*$_1$, ..., *ValueSet*$_n$> implements Cartesian products.

4.2 Computations on Types

Two different sets of routines are needed to support heterogeneous operations such as the product and sum of automata or rational expressions: first a computation on types based on join and meet, then a conversion of values to these types.

The computation of joins and meets on basic types is straightforward.

```
r join(const r&, const b&) { return r(); }
r join(const r&, const z&) { return r(); }
r join(const r&, const q&) { return r(); }
```

The code snippet above states that $\mathbb{R} \vee W := \mathbb{R}$ for $W \in \{\mathbb{B}, \mathbb{Z}, \mathbb{Q}\}$. Composite types such as rational expressions, tuples or even contexts follow the same pattern, but are computed recursively.

Some features new to C++11 let us express the product context computation (as per Def. 6) quite cleanly, as follows:

```
template <typename LhsLabelSet, typename LhsWeightSet,
          typename RhsLabelSet, typename RhsWeightSet>
auto product_ctx(const context<LhsLabelSet, LhsWeightSet>& lhs,
                 const context<RhsLabelSet, RhsWeightSet>& rhs)
  -> context<decltype(meet(lhs.LabelSet(), rhs.LabelSet())),
             decltype(join(lhs.WeightSet(), rhs.WeightSet()))>
{
  auto ls = meet(lhs.LabelSet(), rhs.LabelSet());
  auto ws = join(lhs.WeightSet(), rhs.WeightSet());
  return {ls, ws};
}
```

Two WeightSets are involved in the process of value conversions: the source one, which is used below as a key to select the proper conv routine, and the destination one (r in the following example). Type conversion may require runtime computation such as the floating-point division below, or even something more substantial like the construction of a rational expression in other cases.

---

[6] *generatorset* provides type and value information on the monoid generators; in practice this corresponds to the type of characters and the alphabet, as a vector of characters of the appropriate type.

```
class r {
  using value_t = float;
  ...
  value_t conv(b, b::value_t v) { return v; }
  value_t conv(z, z::value_t v) { return v; }
  value_t conv(q, q::value_t v) { return value_t(v.num)/value_t(v.den); }
  ...
};
```

The process generalizes in a natural way to the case of composite types.

The join, meet and conv functions are used in the implementation of binary operations such as the product, shown below as an example[7]; the idea is to first compute the result type ctx, and then use it to create the result automaton res.

```
template <typename Ctx1, typename Ctx2>
auto product(const automaton<Ctx1>& lhs, const automaton<Ctx2>& rhs)
  -> ...
{
  auto ctx = product_ctx(lhs.context(), rhs.context());
  auto res = make_automaton(ctx);
  auto ws = res.WeightSet();   // a shorthand to the resulting WeightSet.
  ...
  return res;
}
```

The core of the algorithm consists in an iteration over each reachable left-right pair of states (lhs_src, rhs_src); for each pair of transitions with the same label from lhs_src and rhs_src, it adds a transition from the source state pair to the destination state pair, with the same label and the product of weights as weight.

```
for (auto lhs_trans : lhs.out(lhs_src))
  for (auto rhs_trans : rhs.out(rhs_src, lhs_trans.label))
  {
    auto weight = ws.mul(ws.conv(lhs.WeightSet(), lhs_trans.weight),
                         ws.conv(rhs.WeightSet(), rhs_trans.weight));
    res.add_transition({lhs_src, rhs_src}, {lhs_trans.dst, rhs_trans.dst},
                       lhs_trans.label, weight);
  }
```

Three WeightSets play a role in the computation of the resulting weight: first ws.conv(lhs.WeightSet(), lhs_trans.weight) promotes the left-hand side weight from its original WeightSet lhs.WeightSet() to the resulting one ws, and likewise for the second weight; finally the resulting WeightSet multiplies the weights (ws.mul(...)). For instance in Fig. 4 there is a transition from state (p, r) to state (p, s) with label a, and whose weight is the product of $\frac{1}{2}$ and $(y + z)$. The conversion of the first weight corresponds to '$C_3.W$.conv($C_1.W$, $\frac{1}{2}$)', which results in $\langle \frac{1}{2} \rangle 1$; likewise for the second weight: '$C_3.W$.conv($C_2.W$,

---

[7] In the following code excerpts some details have been omitted for clarity.

$\langle \mathbf{1} \rangle (y+z)) = \langle \frac{1}{1} \rangle (y+z)$'. The resulting WeightSet, C_3 then multiplies them: '$C_3.W.\texttt{mul}(\langle \frac{1}{2} \rangle 1, \ \langle \frac{1}{1} \rangle (y+z)))$', i.e., $\langle \frac{1}{2} \rangle (y+z)$.

4.3   On-the-Fly Compilation

Code snippets shown so far are all part of the *static* layer, the statically-typed, lowest-level Application Program Interface (API) of VAUCANSON, which strictly follows the Value/ValueSet principle. As long as this API is used the compiler will take care of generating the appropriate versions of the routine for the types at hand, with no run-time overhead. Programming at this level however offers little flexibility: the program is written and then compiled, period. Moreover, types have to be explicitly spelled out in the program.

On top of this static layer, the *dyn* API takes care of the template parameter book-keeping, memory allocation and deallocation, and even re-unites split objects: for example a `dyn::ratexp` aggregates both a (static-level) rational expression and its (static-level) ratexpset. By design *dyn* only includes a handful of types such as `dyn::context`, `dyn::automaton`, `dyn::weight` and `dyn::label`: all the wide variety of static-level entities is collapsed into a few categories of objects carrying their own run-time type information (exposed to the user as `dyn::context` objects), so that operations can automatically perform their own conversions without exposing the user to the type system.

The *static/dyn bridge* works with registries, one per algorithm. They play a role similar to *virtual tables* in C++: to select the precise implementation of an algorithm that corresponds to the effective type of the operands. These registries are just dictionaries, mapping each given list of argument types to the corresponding specific (static) implementation. This mechanism and other details on the static/dyn bridge have been described in a previous work [4, Sec. 4.2]; its complete treatment is beyond the scope of this paper.

Several commonly-used basic contexts are precompiled — in other words registries are initially loaded for some specific types. However, not only the number of contexts is too large to permit a "complete" precompilation (24: 4 basic LabelSets times 6 WeightSets), but tupleset and ratexpset also let the user define an *unbounded* number of composite ones. Moreover, as demonstrated in Fig. 4, some operation results belong to contexts that were not even in the operands. For these reasons only some select contexts can be precompiled, which will certainly frustrate some users.

On top of dyn VAUCANSON offers IPython support (see Fig. 5). IPython is an enhanced interactive Python environment [6]. Thanks to specific hooks, entities such as rational expressions feature nice LaTeX-based rendering, and automata are rendered as pictures. This binding of dyn features the familiar Python object-oriented flavor as in "$automaton.\texttt{minimize()}$", and operator overloading as in "$automaton \ \& \ automaton$". In such an interactive environment (similar to what formal mathematics environments offer), working with just a finite, predefined set of types would be unacceptable.

To address these limitations VAUCANSON's *dyn* layer features run-time code generation, compilation, and dynamic object loading. The code generation is

```
In [4]: ctx = vcsn.context("lal_char(ab)_ratexpset<lal_char(xyz)_b>"); ctx
```

Out[4]: $\{a, b\} \rightarrow \mathsf{RatE}[\{x, y, z\} \rightarrow \mathbb{B}]$

```
In [5]: r2 = ctx.ratexp('(<x*>b)*<y+z>a(<x*>b+<y+z>a(<x*>b)*<y+z>a)*'); r2
```

Out[5]: $\left(\langle x^* \rangle b\right)^* \langle y + z \rangle a \left(\langle x^* \rangle b + \langle y + z \rangle a \left(\langle x^* \rangle b\right)^* \langle y + z \rangle a\right)^*$

```
In [6]: a2 = r2.derived_term().minimize()
        a1|a2
```

Out[6]:

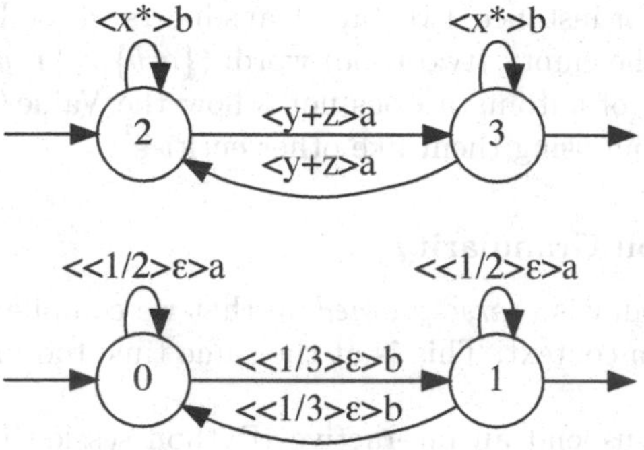

Fig. 5. The computation of $\mathcal{A}_1 \cup \mathcal{A}_2$ in the *IPython notebook* interface of VAUCANSON. The symbol ε denotes the empty word. Weights in $\mathbb{Q}$ such as $\frac{1}{2}$ have been lifted into the WeightSet of C_3: $\langle \frac{1}{2} \rangle \varepsilon \in \mathsf{RatE}[\{x, y, z\} \rightarrow \mathbb{Q}]$.

a simple translation from the context object into C++ code instantiating the existing algorithms for a given context and then entering into the appropriate registries. Once the context-plugin is successfully compiled and linked, it is loaded into the program via `dlopen`; however, differently from the usual practice, we do not need `dlsym` calls to locate plugin functions one by one; rather, when a plugin is loaded, its initialization code simply adds its functions to the registries. In other words a plugin compiled on the fly and loaded at run-time is treated exactly like precompiled contexts.

Because a call in IPython eventually resolves into a call in the static library, one benefits from both flexibility *and* efficiency — when C++ algorithms take most of the time; of course Python-heavy computations would be a different matter.

5   Future Work and Improvements

The subtype relation among semirings we introduced is natural; however a closer look at these definitions reveals that several mechanisms are involved, which may deserve more justification.

5.1   Syntactic and Semantic Improvements of Contexts

Contexts proved to be a powerful concept; however some early design decisions resulted in limitations, to be lifted in the near future.

First, the concrete syntax the user must use to define the context is cumbersome. For instance $C_3 = \{a, b\} \rightarrow \mathsf{RatE}[\{x, y, z\} \rightarrow \mathbb{Q}]$ has to be written `lal_char(ab)_ratexpset<lal_char(xyz)_q>` (see Fig. 5); a syntax closer to the mathematical notation would be an improvement.

Second, the *quantifiers* '?' and '*' should probably apply to an entire LabelSet, and not just to an alphabet like in Fig. 2:

```
⟨labelset⟩ ::= "{1}" | ⟨alphabet⟩ | ⟨labelset⟩ "?" | ⟨labelset⟩ "*"
             | ⟨ratexpset⟩ | ⟨labelset⟩ × ··· × ⟨labelset⟩
```

which would allow to define, for instance, two-tape transducers whose labels are either a couple of letters, or the empty (two-tape) word: $(\{a, b\} \times \{x, y\})^?$.

Third, our implementation of automata does not follow the Value/ValueSet pattern, which prevents us from using them like other entities.

5.2 Dynamic Compilation Granularity

The compilation of plugins today is *coarse-grained*, in that we compile "all" the existing algorithms for a given context. This is at the same time too much, and not enough.

It is too much as it may suspend an interactive IPython session for half a minute even on a fast laptop, to compile and load the given context library; caching compiled code however makes this cost a one-time penalty.

It is not enough because algorithms such as union have an open set of possible signatures. The resulting type of the union of two automata might not be precompiled, in which case, for lack of support for the resulting context, the computation would fail. An unpleasant but effective workaround consists in warning the system, at runtime, that a given context will be needed.

To address both shortcomings we plan to support fine-grained plugins able to generate, compile and load code for *one* function with *one* signature.

6 Conclusion

We presented a type system for weighted automata and rational expressions —a novel feature to the best of our knowledge— currently implemented in our VAUCANSON 2 system, but not coupled to any particular platform.

Types lie at the very foundation of our design. At the lowest level, where performance concerns are strong, we follow the *Value/ValueSet* principle and types parameterize C++ template structures and functions; there a calculus on types based on a subtype relation allows to define operations on automata of different types and handles value conversions. At a higher level types make up the bridge between the static low-level API and a dynamic one built on top of it. Finally, run-time translation of types into C++ code allows to compile, generate, and load plugins during interactive sessions, for instance under IPython.

VAUCANSON 2 is free software. Its source code is available at http://vaucanson. lrde.epita.fr, along with virtual machine images to let users experiment and play with the system without need for an installation.

Acknowledgements. We wish to thank the anonymous reviewers for their helpful and constructive suggestions.

References

1. Alexandrescu, A.: Modern C++ Design: Generic Programming and Design Patterns Applied. Addison-Wesley (2001)
2. Allauzen, C., Riley, M.D., Schalkwyk, J., Skut, W., Mohri, M.: OpenFst: A general and efficient weighted finite-state transducer library. In: Holub, J., Žďárek, J. (eds.) CIAA 2007. LNCS, vol. 4783, pp. 11–23. Springer, Heidelberg (2007), http://www.openfst.org
3. Almeida, A., Almeida, M., Alves, J., Moreira, N., Reis, R.: FAdo and GUItar: Tools for automata manipulation and visualization. In: Maneth, S. (ed.) CIAA 2009. LNCS, vol. 5642, pp. 65–74. Springer, Heidelberg (2009)
4. Demaille, A., Duret-Lutz, A., Lombardy, S., Sakarovitch, J.: Implementation concepts in Vaucanson 2. In: Konstantinidis, S. (ed.) CIAA 2013. LNCS, vol. 7982, pp. 122–133. Springer, Heidelberg (2013)
5. Lombardy, S., Sakarovitch, J.: The validity of weighted automata. Int. J. of Algebra and Computation 23(4), 863–914 (2013)
6. Pérez, F., Granger, B.E.: IPython: a system for interactive scientific computing. Computing in Science and Engineering 9(3), 21–29 (2007), http://ipython.org
7. Sakarovitch, J.: Elements of Automata Theory. Cambridge University Press (2009); Corrected English translation of Éléments de théorie des automates, Vuibert (2003)

Bounded Prefix-Suffix Duplication*

Marius Dumitran[1], Javier Gil[2], Florin Manea[3], and Victor Mitrana[2]

[1] Faculty of Mathematics and Computer Science, University of Bucharest
Str. Academiei 14, 010014 Bucharest, Romania
dmarius1@yahoo.com
[2] Department of Organization and Structure of Information
Polytechnic University of Madrid, Crta. de Valencia km. 7 – 28031 Madrid, Spain
jgil@eui.upm.es, victor.mitrana@upm.es
[3] Department of Computer Science, Christian-Albrechts University of Kiel
Christian-Albrechts-Platz 4, 24118 Kiel, Germany
flm@informatik.uni-kiel.de

Abstract. We consider a restricted variant of the prefix-suffix duplication operation, called bounded prefix-suffix duplication. It consists in
the iterative duplication of a prefix or suffix, whose length is bounded
by a constant, of a given word. We give a sufficient condition for the
closure under bounded prefix-suffix duplication of a class of languages.
Consequently, the class of regular languages is closed under bounded
prefix-suffix duplication; furthermore, we propose an algorithm deciding whether a regular language is a finite k-prefix-suffix duplication
language. An efficient algorithm solving the membership problem for
the k-prefix-suffix duplication of a language is also presented. Finally, we
define the k-prefix-suffix duplication distance between two words, extend
it to languages and show how it can be computed for regular languages.

1 Introduction

Treating sets of chromosomes and genomes as languages raises the possibility
that the structural information contained in biological sequences can be generalized and investigated by formal language theory methods [13]. Thus, the
interpretation of duplication as a formal operation on words has inspired a series of works in the area of formal languages opened by [3,14] and continued by
several other papers, e.g., [10] and the references therein. In [6] one considers
duplications that appear at the both ends of the words only, called prefix-suffix
duplications, inspired by the case of telomeric DNA. In this context, one inves-
tigates the class of languages that can be defined by the iterative application
of the prefix-suffix duplication to a word and tries to compare it to other well
studied classes of languages. It is shown that the languages of this class have a
rather complicated structure even if the initial word is rather simple.

Several problems remained unsolved in the aforementioned paper. This is the
mathematical motivation for the work presented here. By considering a weaker

---

* Florin Manea's work is supported by the *DFG* grant 596676. Victor Mitrana's work
is partially supported by the Alexander von Humboldt Foundation.

M. Holzer and M. Kutrib (Eds.): CIAA 2014, LNCS 8587, pp. 176–187, 2014.
© Springer International Publishing Switzerland 2014

variant of the prefix-suffix duplication, called bounded prefix-suffix duplication, we are able to solve, in this new setting, some of the problems that remained unsolved in [6]. Another motivation is related to the biochemical reality that inspired the definition of this operation. It seems more practical and closer to the biological reality to consider that the factor added by the prefix-suffix duplication cannot be arbitrarily long. One should note that the investigation we pursue here is not aimed to tackle real biological facts and provide solutions for them. In fact, its aim is to provide a better understanding of the structural properties of strings obtained by prefix-suffix duplication as well as specific tools for the manipulation of such strings.

We give a brief description of the contents of this work. We first define a restricted variant of the prefix-suffix duplication called bounded prefix-suffix duplication. It consists in the duplication of a prefix or suffix whose length is bounded by a constant of a given word. We give sufficient conditions for a family of languages to be closed under bounded prefix-suffix duplication. Consequently, we show that every language generated by applying iteratively the bounded prefix-suffix duplication to a word is regular. We also propose an algorithm deciding whether there exists a finite set of words generating a given regular language w.r.t. bounded-prefix-suffix duplication.

We show that the membership problem for the language obtained by applying iteratively k-prefix-suffix duplications from a language recognizable in $\mathcal{O}(f(n))$ time can be solved in $\mathcal{O}(nk \log k + n^2 f(n))$ time. In particular, when considering the k-prefix-suffix duplication language generated by a word x, this problem can be solved in $\mathcal{O}(n \log k)$ time, if $|x| \geq k$, and $\mathcal{O}(nk \log k)$ time in the general case.

We then define the k-prefix-suffix duplication distance between two given words as the minimal number of k-prefix-suffix duplications applied to one of them in order to get the other one and show how it can be efficiently computed. This distance is extended to languages and we propose an algorithm for efficiently computing the k-prefix-suffix duplication distance between two regular languages.

2   Preliminaries

We assume the reader to be familiar with fundamental concepts of formal language theory and complexity theory which can be found in many textbooks, e.g., [12] and [11], respectively.

We start by summarizing the notions used throughout this work. An *alphabet* is a finite and nonempty set of symbols. The cardinality of a finite set A is written $|A|$. Any finite sequence of symbols from an alphabet V is called a *word* over V. The set of all words over V is denoted by V^* and the empty word is denoted by ε; also V^+ is the set of non-empty words over V, V^k is the set of all words over V of length k, while $V^{\leq k}$ is the set of all words over V of length at most k. Given a word w over an alphabet V, we denote by $|w|$ its length, If $w = xyz$ for some $x, y, z \in V^*$, then x, y, z are called prefix, subword, suffix, respectively, of w. For a word w, $w[i..j]$ denotes the subword of w starting at position i and ending at position j, $1 \leq i \leq j \leq |w|$; by convention, $w[i..j] = \varepsilon$ if $i > j$. If $i = j$, then $w[i..j]$ is the i-th letter of w which is simply denoted by $w[i]$. A *period* of a word w over V is a positive integer p such that $w[i] = w[j]$ for all i and j with $i \equiv j$ (mod p). By $per(w)$ (called *the period of* w) we denote the smallest period of w. If $per(w) < |w|$ and $per(w)$ divides $|w|$, then w is a repetition; otherwise, w is called primitive. A primitively rooted square is a word w that has the form xx for some primitive word x.

We say that the pair $_w(i,p)$ is a *duplication* (*repetition*) in w starting at position i in w if $w[i..i + p - 1] = w[i + p..i + 2p - 1]$. Analogously, the pair $(i, p)_w$ is a duplication in w ending at position i in w if $w[i - 2p + 1..i - p] = w[i-p+1..i]$. In both cases, p is called the length of the duplication. Furthermore, the pair $_w(i, p)_w$ is a duplication in w having the middle at position i in w if $w[i - p + 1..i] = w[i + 1...i + p]$.

Despite that the prefix-suffix operation introduced in [6] is a purely mathematical one and the biological reality is just a source of inspiration, it seems rather unrealistic to impose no restriction on the length of the prefix or suffix which is duplicated. The restriction considered in this paper concerns the length of all prefixes and suffixes that are duplicated to the current word. They cannot be longer than a given constant. This restricted variant of prefix-suffix duplication is called *bounded prefix-suffix duplication*. Formally, given a word $x \in V^*$ and a positive integer k, we define:

– k-*prefix duplication*, namely $PD_k(x) = \{ux \mid x = uy$ for some $u \in V^+, |u| \leq k\}$. The k-*suffix duplication* is defined analogously, that is $SD_k(x) = \{xu \mid x = yu$ for some $u \in V^+, |u| \leq k\}$.

– k-*prefix-suffix duplication*, namely $PSD_k(x) = PD_k(x) \cup SD_k(x)$.

These operations are naturally extended to languages L by

$$PD_k(L) = \bigcup_{x \in L} PD_k(x), \quad SD_k(L) = \bigcup_{x \in L} SD_k(x), \quad PSD_k(L) = \bigcup_{x \in L} PSD_k(x).$$

We further define, for each $\Theta \in \{PD, SD, PSD\}$:

$$\Theta_k^0(x) = \{x\}, \Theta_k^{n+1}(x) = \Theta_k^n(x) \cup \Theta_k(\Theta_k^n(x)), \text{ for } n \geq 0, \ \Theta_k^*(x) = \bigcup_{n \geq 0} \Theta_k^n(x).$$

Furthermore, $PSD_k^*(L) = \bigcup_{x \in L} PSD_k^*(x)$. A language $L \subseteq V^*$ is called a bounded prefix-suffix duplication language if $L = PSD_k^*(x)$ for some $x \in V^*$ and $k > 0$. A prefix-suffix duplication language is defined analogously, see [6]. When duplications of arbitrary factors within the word are permitted, we obtain an (arbitrary) duplication language, see, e.g., [3].

In this paper, we show a series of results of algorithmic nature. All the time complexity bounds we obtain in this context hold for the RAM with logarithmic memory-word size. In the algorithmic problems we approach, we are usually given as input one or more words. These words are assumed to be over an integer alphabet; that is, if w is the input word, and has length n, then we assume that its letters are integers from the set $\{1, \ldots, n\}$. See a discussion about this assumption in [9]. If the input to our problems is a language, then we assume that this language is specified by a procedure deciding it (e.g., if the language is regular, then we assume that we are given a DFA accepting it).

We recall basic facts about the data structures we use. For a word u, with $|u| = n$, over $V \subseteq \{1, \ldots, n\}$ we can build in linear time a suffix array structure as well as data structures allowing us to retrieve in constant time the length of the longest common prefix of any two suffixes $u[i..n]$ and $u[j..n]$ of u, denoted $LCP(i, j)$. These structures are called LCP data structures in the following. For details, see, e.g., [8,9]. Similarly, one can construct in linear time data structures allowing us to retrieve in constant time the length of the longest common suffix of any two prefixes $u[1..i]$ and $u[1..j]$ of u, denoted $LCS(i, j)$.

We also use a linear data structure, called deque (double-ended queue, see [15]). This is a doubly linked list for which elements can be added to or removed from either the front or back. Finally, tries are complete trees whose edges are labeled with letters of an alphabet V, and ordered according to an (existing) order of the letters of this alphabet; each path of a trie corresponds to a word over V.

3   Bounded Prefix-Suffix Duplication as a Formal Operation on Languages

We start with some language theoretical properties of the class of duplication languages. By combining the results from [1] and [4] (rediscovered in [3] and [14] for arbitrary duplication languages), and [6] we recall the following result.

Theorem 1.
1. An arbitrary duplication language is regular if and only if it is a language over an alphabet with at most two symbols.
2. A prefix-suffix duplication language is context-free if and only if it is a language over the unary alphabet.

Whether or not every arbitrary duplication language is recognizable in polynomial time is open while every prefix-suffix duplication language is in **NL**.

We say that a class $\mathcal{L}$ of languages is closed under bounded prefix-suffix duplication if $PSD_k^*(L) \in \mathcal{L}$ for any $L \in \mathcal{L}$ and $k \geq 1$.

Theorem 2. *Every nonempty class of languages closed under union with regular languages, intersection with regular languages, and substitution with regular languages, is closed under bounded prefix-suffix duplication.*

Proof. Let $\mathcal{L}$ be a family of languages having all the required closure properties. By [7], $\mathcal{L}$ is closed under inverse morphism. Let $L \subseteq V^*$, with $|V| = m$, be a language from $\mathcal{L}$, and k be a positive integer. We define the alphabet

$$U = V \cup \{p_1, p_2, \ldots, p_{m^k}\} \cup \{s_1, s_2, \ldots, s_{m^k}\},$$

and the morphism $h : U^* \longrightarrow V^*$ defined by $h(a) = a$ for any $a \in V$ and $h(p_i) = h(s_i) =$ the ith word of length k over V in the lexicographic order, for all $1 \leq i \leq m^k$. Further, let F be the finite language defined by $F = \{x \in L \mid |x| \leq 2k - 1\}$ and

$$E = (L \cup PSD_k^{2k}(F)) \cap \{x \in V^+ \mid |x| \geq 2k\}.$$

As $PSD_k^{2k}(F)$ is a finite language and $\mathcal{L}$ is closed under union with regular languages and intersection with regular languages, it follows that E is still in $\mathcal{L}$. The following relation is immediate:

$$PSD_k^*(L) = PSD_k^*(E) \cup PSD_k^{2k}(F).$$

It is rather easy to prove that

$$PSD_k^*(E) = \sigma(h^{-1}(E) \cap \{p_1, p_2, \ldots, p_{m^k}\} V^* \{s_1, s_2, \ldots, s_{m^k}\}),$$

where σ is a substitution defined by $\sigma(p_i) = PD_k^*(x_i)$ and $\sigma(s_i) = SD_k^*(x_i)$, where x_i is the ith word of length k over V in the lexicographic order.

Each language $PD_k^*(x_i)$ can be generated by a prefix grammar [5], hence it is regular. Analogously, each language $SD_k^*(x_i)$ is regular. Consequently, σ is a substitution with regular languages. By the closure properties of $\mathcal{L}$, $PSD_k^*(E)$ belongs to $\mathcal{L}$, hence $PSD_k^*(L)$ is also in $\mathcal{L}$.   □

Much differently from the statements of Theorem 1 we have:

Corollary 1. *Every bounded prefix-suffix duplication language is regular.*

A language L is said to be a multiple k-prefix-suffix duplication language if there exists a language E such that $L = PSD_k^*(E)$. If E is finite, then L is said to be a finite k-prefix-suffix duplication language. Note that given a regular language L and a positive integer k, a necessary condition such that $L = PSD_k^*(E)$ holds, for some set E, is $L = PSD_k^*(L)$. By Theorem 2 a finite automaton accepting $PSD_k^*(L)$ can effectively be constructed and so the above equality can be algorithmically checked. However, if the equality holds, we cannot infer anything about the finiteness of E. The problem is completely solved by the next theorem.

Theorem 3. *Let L be a regular language which is a multiple k-prefix-suffix duplication language for some positive integer k. There exists a unique minimal (with respect to inclusion) regular language E, which can be algorithmically computed, such that $L = PSD_k^*(E)$. In particular, one can algorithmically decide whether L is a finite k-prefix-suffix duplication language.*

Proof. Let $L \subseteq V^*$ be a multiple k-prefix-suffix duplication language accepted by the deterministic finite automaton $A = (Q, V, f, q, F)$. We define the language

$$M_k(L) = \{x \in L \mid \text{ there is no } y \in L \text{ such that } x \in PSD_k(y)\}.$$

As $L = PSD_k^*(L)$, it follows that

$$M_k(L) = \{x \in L \mid \text{ there is no } y \in L, y \neq x \text{ such that } x \in PSD_k^*(y)\}.$$

Claim. *If $PSD_k^*(E) = L$ for some $E \subseteq L$, then the following statements hold:*
(i) $M_k(L) \subseteq E$, and
(ii) $PSD_k^*(M_k(L)) = L$.

Proof of the claim. (i) Let $x \in M_k(L) \subseteq L$; there exists $y \in E$ such that $x \in PSD_k^*(y)$. By the definition of $M_k(L)$, it follows that $x = y$.

(ii) Clearly, $PSD_k^*(M_k(L)) \subseteq L$. Let $y \in L$; there exists $x \in L$ such that $y \in PSD_k^*(x)$. We may choose x such that $x \in PSD_k(z)$ for no $z \in L$. Thus, $x \in M_k(L)$, and $y \in PSD_k^*(M_k(L))$, which concludes the proof of the claim.

Clearly, $M_k(L) = L \setminus PSD_k(L)$; hence $M_k(L)$ is regular and can effectively be constructed.

In order to check whether L is a finite k-prefix-suffix duplication language we first compute $M_k(L)$. Then we check whether $M_k(L)$ is finite. Finally, by Theorem 2, the language $PSD_k^*(M_k(L))$ is regular and can be effectively computed, therefore the equality $PSD_k^*(M_k(L)) = L$ can be algorithmically checked.   □

3.1   Membership Problem

In the sequel, we will make use of the following classical result from [2]. It is known that the number of primitively rooted square factors of length at most $2k$ that occur in a word w at a position is $\mathcal{O}(\log k)$. Moreover, one can construct the list of primitively rooted squares of length at most $2k$ occurring in w in $\mathcal{O}(n \log k)$ time. Each square is represented in the list by the starting position and the length of their root, and the list is ordered increasingly by the starting position of the squares; when more squares share the same starting position they are ordered by the length of the root. Moreover, one can store an array of n pointers, where the ith such pointer gives the memory location of the list of the primitively rooted squares occurring at position i. A similar list, where the squares are ordered by their ending position, can be computed in the same time. Further, we develop our main algorithmic tools.

Lemma 1. *Given $w \in V^*$, of length n, and an integer $k \leq n$, we can identify all prefixes $w[1..i]$ of w such that $w \in SD_k^*(w[1..i])$ in $\mathcal{O}(n \log k)$ time.*

Proof. We propose an algorithm that computes an array $S[\cdot]$, defined by $S[i] = 1$ if $w \in SD_k^*(w[1..i])$, and $S[i] = 0$, otherwise. This algorithm has a preprocessing phase, in which all the primitively rooted squares with root of length at most k occurring in w are computed. This preprocessing takes $\mathcal{O}(n \log k)$ time.

Now, we describe the computation of the array S. Initially, all the positions of this array are initialized to 0, except $S[n]$, which is set to 1. Clearly, this is correct, as $w \in SD_k^*(w[1..n]) = SD_k^*(w)$. Further, we update the values in the array S using a dynamic programming approach. That is, for i from n to 1, if $S[i] = 1$, then we go through all the primitively rooted squares $(w[j + 1..i])^2$, $|i - j| \leq k$, that end at position i in w. For each such factor $w[j + 1..i]$ we set $S[j] = 1$. Indeed, $w[1..i]$ can be obtained from $w[1..j]$ by appending $w[j + 1..i]$ (which is known to be a suffix of $w[1..j]$); as we already know that w can be obtained by suffix duplication from $w[1..i]$, it follows that w can be obtained by suffix duplication from $w[1..j]$. The processing for each i takes $\mathcal{O}(\log k)$ time.

It is not hard to see that our algorithm works correctly. Assume that $w \in SD_k^*(w[1..j])$ for some $j < n$. Let us consider the longest sequence of suffix duplication steps (or, for short, derivation) that produces w starting from $w[1..j]$. Say that this derivation has $s \geq 2$ steps, so it can be described by a sequence of indices $j_1 = j < j_2 < \ldots < j_s = n$ such that $w[1..j_{i+1}] \in SD_k(w[1..j_i])$ for $1 \leq i \leq s - 1$. We can show that $w[j_i + 1..j_{i+1}]$ is primitive for all i. Otherwise, $w[j_i + 1..j_{i+1}] = t^\ell$ for some word t and $\ell \geq 2$, so we can replace in the original derivation the duplication that produces $w[1..j_{i+1}]$ from $w[1..j_i]$ by other ℓ duplication steps in which t factors are added to $w[1..j_i]$. This leads to a sequence with more than s duplications steps producing w from $w[1..j]$, a contradiction. Now, it is immediate that, in our algorithm, $S[j_s]$ is set to 1 in the first step. Assuming that for some i we already have $S[j_{i+1}] = 1$, when considering the value j_{i+1} in the main loop of our algorithm, as $w[j_i + 1..j_{i+1}]^2$ is a primitively rooted square ending on position j_{i+1}, we will set $S[j_i] = 1$. In the end, we will also have $S[j] = S[j_1] = 1$, so our algorithm works properly. $\square$

Lemma 2. *Given $w \in V^*$, of length n, we can identify all suffixes $w[j..n]$ of w such that $w \in PD_k^*(w[j..n])$ in $\mathcal{O}(n \log k)$ time.*

The proof is similar to the one of Lemma 1, and it is left to the reader. The output of the algorithm will be an array $P[\cdot]$, defined by $P[j] = 1$ if $w \in PD_k^*(w[j..n])$, and $P[j] = 0$, otherwise.

The next lemma shows a way to compute the factors of length at least k, from which w can be obtained by iterated prefix or suffix duplication.

Lemma 3. *Given $w \in V^*$ of length n and a list F of factors of w of length greater than or equal to k, given by their starting and ending position, ordered by their starting position, and in case of equality by their ending position, we can check whether there exists $x \in F$ such that $w \in PSD_k^*(x)$ in time $\mathcal{O}(n \log k + |F|)$.*

Proof. The main remark of this lemma is that, if $w[i..j]$ is longer than k, then $w \in PSD_k^*(w[i..j])$ if and only if $w[1..j] \in PD_k^*(w[i..j])$ and $w = w[1..n] \in SD_k^*(w[1..j])$. Equivalently, we have $w \in PSD_k^*(w[i..j])$ if and only if $w[1..n] \in PD_k^*(w[i..n])$ and $w[1..n] \in SD_k^*(w[1..j])$.

This remark suggests the following approach: we first identify all the suffixes $w[j..n]$ of w such that $w \in PD_k^*(w[j..n])$ and all the prefixes $w[1..i]$ of w such that $w \in SD_k^*(w[1..i])$; this takes $\mathcal{O}(n \log k)$, by Lemmas 1 and 2. Now, for every factor $w[i..j]$ in list F, we just check whether $S[i] = P[j] = 1$ (that is, $w \in PD_k^*(w[i..n]) \cap SD_k^*(w[1..j])$); if so, we decide that $w \in PSD_k^*(w[i..j])$. $\square$

Building on the previous lemmas, we can now solve the membership problem for $PSD_k^*(L)$ languages, provided that we know how to solve the membership problem for L on the RAM with logarithmic word size model.

Theorem 4. *If the membership problem for the language L can be decided in $\mathcal{O}(f(n))$ time, then the membership problem for $PSD_k^*(L)$ can be decided in $\mathcal{O}(nk \log k + n^2 f(n))$.*

Proof. Assume that we are given a word w, of length n; we want to test whether $w \in PSD_k^*(L)$ or not. For simplicity, we assume that L is constant (i.e., its description, given as a procedure deciding L in $\mathcal{O}(f(n))$ time, is not part of the input). If L was given as part of the input, then we can use exactly the same algorithm, but one should add to the final time complexity the time needed to read the description of L and effectively construct a procedure deciding L in $\mathcal{O}(f(n))$ time.

First, let us note that we can identify trivially in $\mathcal{O}(n^2 f(n))$ the factors of w that are in L. More precisely, we can produce a list F of factors of w that are contained in L, specified by their starting and ending position, ordered by their starting position, and, in case of equality by their ending position. The list F can be easily split, in $\mathcal{O}(|F|)$ time, into two lists: F_1, containing the factors of length at least k, and F_2, the list of factors of length less than k. It is worth noting that $|F| \in \mathcal{O}(n^2)$. By Lemma 3 it follows that we can decide in time $\mathcal{O}(n \log k + |F_1|)$ whether $w \in PSD_k^*(x)$ for some $x \in F_1$.

It remains to test whether $w \in PSD_k^*(x)$ for some $x \in F_2$. The main remark we make in this case is that there exists $x \in F_2$ such that $w \in PSD_k^*(x)$ if and only if there exists $y \in PSD_k^*(x)$ such that $k \leq |y| \leq 2k$ and $w \in PSD_k^*(y)$. Therefore, we will produce the list F_3 of words $z \in \cup_{x \in F_2} PSD_k^*(x)$ such that z is a factor of w and $k \leq |z| \leq 2k$.

In order to compute F_3 we can use the $\mathcal{O}(|u|^2 \log |u|)$ algorithm proposed in [6] to decide whether a word u is contained in $PSD^*(v)$. In that algorithm, one first marks the factors of u that are equal to v. Further, for each possible length ℓ of the factors of u, from 1 to $|u|$, and for each $i \leq n$ where a factor of length ℓ of u may start, one checks whether $u[i..i + \ell - 1]$ can be obtained by prefix (respectively, suffix duplication) from a shorter suffix (respectively, prefix), that was already known (i.e., marked) to be in $PSD^*(v)$, such that in the last step of duplication a primitive root x of a primitively rooted square prefix x^2 of $u[i..i + \ell - 1]$ was appended to the shorter suffix (respectively, a primitive root x of a primitively rooted square suffix x^2 of $u[i..i + \ell - 1]$ was appended to the shorter prefix). Each time we found a factor of w that can be obtained in this way from one of its marked prefixes or suffixes, we marked it as part of $PSD_k^*(v)$ and continued the search with the next factor of w.

In our case, we can pursue the same strategy: taking w in the role of u, and having already marked the words of F_2 (which are factors of w) just like we did with the occurrences of v, we run the algorithm described above, but only for $\ell \leq 2k$. Note that the primitive roots of primitively rooted square suffixes or prefixes of factors $w[i..i + \ell - 1]$ with $\ell \leq 2k$ have length at most k; hence, each duplication that is made towards obtaining such a factor is, in fact, a k-prefix-suffix duplication. In this manner we obtain the factors of w of length at most $2k$ that are from $PSD_k^*(F_2)$. The time needed to obtain these factors is $\mathcal{O}(nk \log k)$. We store this set of factors in F_3 just like before: the factors are specified by their starting and ending position, ordered by their starting position, and, in case of equality by their ending position. The set F_3 may have up to $\mathcal{O}(nk)$ factors, as each of them has length at most $2k$.

By Lemma 3, we can decide in time $\mathcal{O}(n \log k + |F_3|) = \mathcal{O}(nk)$ whether $w \in PSD_k^*(F_3)$. Accordingly, adding the time needed to compute F_3 from F_2, it follows that we can decide in time $\mathcal{O}(nk \log k)$ whether $w \in PSD_k^*(F_2)$. Hence, we can decide whether $w \in PSD_k^*(L)$ in $\mathcal{O}(nk \log k + n^2 f(n))$ time. $\qquad\square$

In fact, there are classes of languages for which a better bound than the one in Theorem 4 can be obtained. If L is context-free (respectively, regular) the time needed to decide whether $w \in PSD_k^*(L)$ is $\mathcal{O}(n^3)$ (respectively, $\mathcal{O}(nk \log k + n^2)$), where $|w| = n$. Indeed, F has always at most n^2 elements, and in the case of context-free (or regular) languages it can be obtained in $\mathcal{O}(n^3)$ time (respectively, $\mathcal{O}(n^2)$) by the Cocke-Younger-Kasami algorithm (respectively, by running a DFA accepting L on all suffixes of w, and storing the factors accepted by the DFA). When L is a singleton, the procedure is even more efficient.

Corollary 2. *Given two words w and x, with $|w| \geq |x|$, we can decide whether $w \in PSD_k^*(x)$ in time $\mathcal{O}(|w|k \log k)$. If $|x| \geq k$, then we can decide whether $w \in PSD_k^*(x)$ in time $\mathcal{O}(|w| \log k)$.*

Proof. Assume that $|w| = n$ and $|x| = m$. First, note that the list F of all occurrences of x in w can be obtained in linear time $\mathcal{O}(n + m)$, using, e.g., the Knuth-Morris-Pratt algorithm [16], and $|F| \in \mathcal{O}(n)$.

For the first part, we follow the same general approach as in Theorem 4. If $|x| < k$, we produce the list of all the factors longer than k, but of length at most $2k$, that can be derived from x. This list is produced in $\mathcal{O}(nk \log k)$ time. Therefore, the total complexity of the algorithm is $\mathcal{O}(nk \log k)$, in this case.

The second result follows now immediately from Lemma 3, as F contains only words of length at least k.   □

4   Bounded Prefix-Suffix Duplication Distances

Given two words x, w and $k \geq 1$, the k-prefix-suffix duplication distance between x and w is defined by
$$\delta_k(x, w) = \inf\{\ell \mid x \in PSD_k^\ell(w) \text{ or } w \in PSD_k^\ell(x)\}.$$
By definition, the k-prefix-suffix duplication distance between two words is equal to ∞ if the longer word cannot be derived from the shorter. In a similar fashion, we can define k-suffix duplication distance or k-prefix duplication distance between x and w as the minimum number of k-suffix duplication, respectively, k-prefix duplication steps, needed to transform x into w or w into x.

Theorem 5. *Given $k \geq 1$, let x and w be two words of respective length m and n, $n > m$. If $m \geq k$, then $\delta_k(x, w)$ can be computed in $\mathcal{O}(n \log k)$. If $m < k$, then $\delta_k(x, w)$ can be computed in $\mathcal{O}(nk \log k)$.*

The k-prefix-suffix duplication distance between two words can be extended to the k-prefix-suffix duplication distance between a word x and a language L by $\delta_k(x, L) = \min\{\delta_k(x, y) \mid y \in L\}$. Moreover, one can canonically define the distance between languages: for two languages L_1, L_2 and a positive integer k, we set $\delta_k(L_1, L_2) = \min\{\delta_k(x, y) \mid x \in L_1, y \in L_2\}$.

Theorem 6. *Given two regular languages L_1 and L_2 over an alphabet V, recognised by deterministic finite automata with sets of states Q and S, respectively, and a positive integer $k \geq 1$, one can algorithmically compute $\delta_k(L_1, L_2)$ in $\mathcal{O}((k + N)M^2|V|^{2k})$, where $M = \max\{|Q|, |S|\}$ and $N = \min\{|Q|, |S|\}$.*

Proof. Let us assume that both L_1 and L_2 are given by the minimal deterministic finite automata accepting them, namely A_1 and, respectively, A_2. Let $A_1 = (Q, V, \delta', q_0, Q_f)$ and $A_2 = (S, V, \delta'', s_0, S_f)$. As a rule, we denote the states of Q and S by q and s, respectively, with or without indices.

Before starting the main proof, let us briefly explain a series of implementation details. We work with 5-tuples (q, s_1, s_2, w_1, w_2) and 4-tuples (s_1, s_2, w_1, w_2), where $w_1, w_2 \in V^*, |w_1| = |w_2| \leq k$; moreover, whenever $|w_1| < k$ then $w_1 = w_2$.

A set T of 5-tuples as above is implemented as a 3-dimensional array M_T, where $M_T[q][s_1][s_2]$ contains a representation of the set $\{(w_1, w_2) \in V^* \times V^* \mid (q, s_1, s_2, w_1, w_2) \in T\}$ which is implemented using a trie data structure essentially storing all possible words of length k, augmented with suffix links. Using this representation we can check in constant time whether or not a certain pair of words (given as pair of nodes of the trie we construct) is in the set. The same strategy may be used for implementing a set R of 4-tuples.

For a word $w \in V^*$, we denote by $pref_k(w)$ the longest prefix of length at most k of w; similarly, let $suf_k(w)$ be the longest suffix of length at most k of w.

The algorithm that computes $\delta_k(L_1, L_2)$ has two similar main parts. In the first one, we compute the minimum value d_1 such that there exists a word $x \in L_2$ with $x \in PSD_k^{d_1}(L_1)$. In the second part, we compute, using exactly the same procedure, the minimum value d_2 such that there exists a word $y \in L_1$ with $y \in PSD_k^{d_2}(L_2)$. Then, we conclude that $\delta_k(L_1, L_2) = \min\{d_1, d_2\}$. Hence, it suffices to describe how the minimum value d_1 such that there exists a word $x \in L_2$ with $x \in PSD_k^{d_1}(L_1)$ is computed.

As a preprocessing phase of our algorithm, we compute in $\mathcal{O}(k|Q|^2|V|^k)$ time (in a naive manner), for each $q_1 \in Q$ and $w \in V^{\leq k}$ all states q_2 such that $\delta'(q_2, w) = q_1$ and the state $q_3 = \delta(q_1, w)$. Provided that we use the same idea of storing words as labels of nodes from the trie (the label of w being denoted $\#(w)$), we can store this information in space $\mathcal{O}(|Q|^2|V|^k)$, so that we can obtain in constant time, for q_1 and $\#(w)$, the states q_2 and q_3 defined above. We then process the automaton A_2 in a similar manner, in time $\mathcal{O}(|S|^2|V|^k)$.

We present now the main part of our algorithm. First, we compute the set
$$R_0 = \{(s_1, s_2, w_1, w_2) \mid \text{there exists } w \in L_1 \text{ such that } \delta''(s_1, w) = s_2,$$
$$pref_k(w) = w_1, suf_k(w) = w_2\}.$$
This computation is done as follows. We compute iteratively the sets T_s^i, $i \geq 1$, each one containing the tuples (q, s, s_1, w_1, w_2) for which there exists a word w of length i, with $pref_k(w) = w_1$, $suf_k(w) = w_2$, $\delta'(q_0, w) = q$ and $\delta''(s, w) = s_1$, but there exists no word w' shorter than w with the same properties. Clearly, in such a 5-tuple, $|w_1| = |w_2|$ and if $|w_1| < k$ then $w_1 = w_2$. We can implement the union (over all values of i) of the sets T_s^i by marking in a trie storing all the words of length k over V the nodes corresponding to the words of this set. The sets T_s^i are computed as long as they are non-empty; clearly, if T_s^i is empty, then the sets T_s^j are empty, for all $j \geq i$. On the other hand, as the number of all the tuples (q, s, s_1, w_1, w_2) as above is upper bounded by $2|Q||S||V|^{2k}$, there exists i_0 such that $T_s^i = \emptyset$ when $i \geq i_0$ and $T_s^{i_0-1} \neq \emptyset$. It is not hard to see that T_s^{i+1} can be computed in time $\mathcal{O}(k|T_s^i|)$, given the elements of T_s^i.

Indeed, for each 5-tuple $(q, s, s_1, w_1, w_2) \in T_s^i$ and letter $a \in V$, we compute the 5-tuple $(\delta'(q, a), s, \delta''(s_1, a), pref_k(w_1 a), suf_k(w_2 a))$; note that the nodes of the trie corresponding to the words $pref_k(w_1 a)$ and $suf_k(w_2 a)$ can be obtained in $\mathcal{O}(1)$ time, by knowing the nodes corresponding to w_1 and w_2 and using their suffix links. Then, if the new tuple does not belong to $\bigcup_{i=1}^{i} T_s^i$, we add it to T_s^{i+1}; by maintaining another trie-structure for $\bigcup_{i=1}^{i} T_s^i$, we obtain that checking whether an element is in this set or adding an element to it is done in $\mathcal{O}(1)$ time. To efficiently go through the elements of T_s^i, we store them in a linked list.

We now set $\hat{T}_s = \bigcup_{i=1}^{i_0} T_s^i$. It follows that $\hat{T}_s$ is computed in $\mathcal{O}(|Q||S||V|^{2k})$ time. Therefore, $R_0 = \{(s_1, s_2, w_1, w_2) \mid (q, s_1, s_2, w_1, w_2) \in \bigcup_{s \in S} \hat{T}_s, q \in Q_f\}$. Clearly, it takes $\mathcal{O}(|Q||S|^2|V|^{2k})$ time to compute R_0. We now set $\hat{R}_j = \bigcup_{i=0}^{j} R_i$ and iteratively compute the sets $R_j, j = 1, 2, \ldots$ as follows:

- $R_{j+1} = (R_{j+1}^1 \cup R_{j+1}^2) \setminus \hat{R}_j$,
- $R_{j+1}^1 = \{(s_1, s', w_1', w_2') \mid$ there exist $(s_1, s_2, w_1, w_2) \in R_j$, and $w' \in V^*$ a suffix of w_2, such that $\delta''(s_2, w') = s', pref_k(w_1 w') = w_1', suf_k(w_2 w') = w_2'\}$,
- $R_{j+1}^2 = \{(s', s_2, w_1', w_2') \mid$ there exist $(s_1, s_2, w_1, w_2) \in R_j$, and $w' \in V^*$ a prefix of w_1, such that $\delta''(s', w') = s_1, pref_k(w' w_1) = w_1', suf_k(w' w_2) = w_2'\}$.

Actually, $(s_1, s_2, w_1, w_2) \in R_j$ if and only if there exists a word w which can be obtained by applying j times the k-prefix-suffix duplication to a word from L_1 such that $pref_k(w) = w_1$, $suf_k(w) = w_2$, and $\delta''(s_1, w) = s_2$; furthermore, there is no word w' that fulfils the same conditions and can be obtained by applying less than j times the k-prefix-suffix duplication to the words of L_1. Clearly, all the elements of these sets fulfil the conditions allowing us to use again a trie implementation for the union of the sets. Using this implementation, and additionally storing each R_j as a list, the time needed to compute the set R_{j+1} is upper bounded by $\mathcal{O}(k|R_j|)$. Indeed, first we construct R_{j+1}^2: for each tuple $(s_1, s_2, w_1, w_2) \in R_j$ and prefix x of w_1, we use the precomputed data structures to obtain the state s such that $\delta'(s, x) = s_1$ and decide that $(s, s_2, pref_k(x w_1), suf_k(x w_2))$ should be added to R_{j+1} (but only if it is not already in other $R_{j'}$ with $j' < j + 1$). To implement this efficiently, we consider the prefixes of x in increasing order with respect to the length, and so we will get the node corresponding to xa in the trie in $\mathcal{O}(1)$ time when we know the node corresponding to x. Then we construct R_{j+1}^1: for each tuple (s_1, s_2, w_1, w_2) and for each suffix x of w_2, we use the precomputed data structures to obtain $s = \delta'(s_2, x)$ and decide that $(s_1, s, pref_k(w_1 x), suf_k(w_2 x))$ should be added to R_{j+1} (again, only if it is not in other $R_{j'}$ with $j' < j + 1$). This time we consider the suffixes x of w_2 in decreasing order with respect to their length; in this way, we get the node corresponding to x from the node corresponding to ax in $\mathcal{O}(1)$ time using the suffix links. The sets R_j are computed until either one meets a value j_0 such that $(s_0, s, w_1, w_2) \in R_{j_0}$ for some $s \in S_f$ and $w_1, w_2 \in V^{\leq k}$, or $R_j = \emptyset$. As the number of all 4-tuples that may appear in all the sets R_j is bounded by $\mathcal{O}(|S|^2|V|^{2k})$, the computation of the sets R_j ends after at most $\mathcal{O}(k|S|^2|V|^{2k})$ steps. It is clear that if the process of computing the sets R_j

ends by reaching the value j_0 mentioned above, then we conclude that $d_1 = j_0$. Otherwise, $d_1 = \infty$ holds. The correctness of the computation of d_1 follows immediately from the discussions above.

Consequently, the total time needed to compute d_1 is $\mathcal{O}(|V|^k + k|Q|^2|V|^k + 2k|S|^2|V|^{2k} + |Q||S|^2|V|^{2k}) = \mathcal{O}(k|Q|^2|V|^k + |Q||S|^2|V|^{2k})$. We can use the same procedure to compute d_2, just by changing the roles of L_1 and L_2. Then, we return as $\delta_k(L_1, L_2) = \min\{d_1, d_2\}$. The time needed to compute this distance is $\mathcal{O}((k+N)M^2|V|^{2k})$, where $M = \max\{|Q|, |S|\}$ and $N = \min\{|Q|, |S|\}$. □

Note that if V is a constant size alphabet, then the previous result provides a cubic algorithm computing the distance between two regular languages. The following corollary follows from Theorem 6, for $L_1 = \{x\}$ and $L_2 = L$.

Corollary 3. *Given a word x, a regular language L accepted by a DFA with q states, and a positive integer $k \geq 1$, one can algorithmically compute $\delta_k(x, L)$ in $\mathcal{O}((k + |N|)|M|^2|V|^{2k})$ time, where $M = \max\{q, |x|\}$ and $N = \min\{q, |x|\}$.*

References

1. Bovet, D.P., Varricchio, S.: On the regularity of languages on a binary alphabet generated by copying systems. Inform. Proc. Letters 44(3), 119–123 (1992)
2. Crochemore, M.: An optimal algorithm for computing the repetitions in a word. Inf. Process. Lett. 12(5), 244–250 (1981)
3. Dassow, J., Mitrana, V., Păun, G.: On the regularity of duplication closure. Bull. European Assoc. Theor. Comput. Sci. 68, 133–136 (1999)
4. Ehrenfeucht, A., Rozenberg, G.: On the separating power of EOL systems. RAIRO Inform. Theor. 17(1), 13–22 (1983)
5. Frazier, M., David Page Jr., C.: Prefix grammars: an alternative characterization of the regular languages. Inf. Process. Lett. 2, 67–71 (1994)
6. Garcia Lopez, J., Manca, F., Mitrana, V.: Prefix-suffix duplication. J. Comput. Syst. Sci. (in press) doi:10.1016/j.jcss.2014.02.011
7. Ginsburg, S.: Algebraic and automata-theoretic properties of formal languages. North-Holland Pub. Co. (1975)
8. Gusfield, D.: Algorithms on strings, trees, and sequences: computer science and computational biology. Cambridge University Press, New York (1997)
9. Kärkkäinen, J., Sanders, P., Burkhardt, S.: Linear work suffix array construction. J. ACM 53(6), 918–936 (2006)
10. Leupold, P.: Reducing repetitions. In: Prague Stringology Conf., pp. 225–236 (2009)
11. Papadimitriou, C.: Computational Complexity. Addison-Wesley (1994)
12. Rozenberg, G., Salomaa, A. (eds.): Handbook of Formal Languages, vol. I–III. Springer, Berlin (1997)
13. Searls, D.B.: The computational linguistics of biological sequences. In: Artificial Intelligence and Molecular Biology, pp. 47–120. MIT Press, Cambridge (1993)
14. Wang, M.-W.: On the irregularity of the duplication closure. Bull. European Assoc. Theor. Comput. Sci. 70, 162–163 (2000)
15. Knuth, D.: The Art of Computer Programming, 3rd edn. Fundamental Algorithms, vol. 1, pp. 238–243. Addison-Wesley (1997), Section 2.2.1: Stacks, Queues, and Deques, ISBN 0-201-89683-4
16. Knuth, D., Morris, J.H., Pratt, V.: Fast pattern matching in strings. SIAM J. Comput. 6(2), 323–350 (1977)

Recognition of Labeled Multidigraphs by Spanning Tree Automata

Akio Fujiyoshi

Department of Computer and Information Sciences, Ibaraki University
4-12-1 Nakanarusawa, Hitachi, Ibaraki, 316-8511, Japan
fujiyosi@mx.ibaraki.ac.jp

Abstract. In this paper, we study tree automata recognizing labeled multidigraphs. We define that a labeled multidigraph is accepted by a tree automaton if and only if the graph has a spanning tree accepted by the tree automaton. We call this automaton a spanning tree automaton. The membership problem of labeled multidigraphs for a spanning tree automaton is NP-complete because the Hamiltonian path problem can be easily reduced to it. However, it will be shown that the membership problem is solvable in linear time for graphs of bounded tree-width.

1 Introduction

A spanning tree of a graph is a tree that consists of all of the vertices and some or all of the edges of the graph. Checking if a graph has a particular spanning tree has many practical applications. Since a set of particular trees can be defined by a tree automaton, we define that a graph is accepted by a tree automaton if and only if the graph has a spanning tree accepted by the tree automaton. We call this automaton a spanning tree automaton. In this paper, the membership problem of labeled multidigraphs for a spanning tree automaton is studied, that is, the problem to determine, for any fixed tree automaton, whether a given graph is accepted by the tree automaton.

The motivation of this study is to establish a robust and efficient recognition method for mathematical OCR [2,6,10,9]. As shown in Fig. 1, a mathematical OCR system constructs a labeled multidigraph representing the adjacency relation of mathematical symbols from a scanned image. The vertex labels represent mathematical symbols, while the edge labels represent types of the adjacency relation of mathematical symbols. From the labeled multidigraph, we want to obtain the spanning tree representing proper connections of mathematical symbols, which should be syntactically reasonable. In order to define the syntax of mathematical formulae and verify candidates of the spanning tree, we make use of spanning tree automata.

In [7,8], spanning tree automata for directed acyclic graphs (DAGs) were studied. It was shown that the membership problem of DAGs for a spanning tree automaton is NP-complete by the reduction from the Boolean satisfiability problem (SAT). A linear-time algorithm solving the membership problem of

M. Holzer and M. Kutrib (Eds.): CIAA 2014, LNCS 8587, pp. 188–199, 2014.

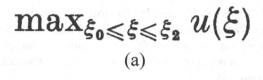

(a)

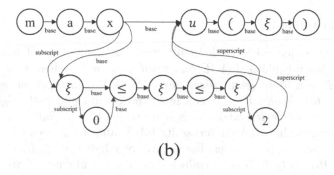

(b)

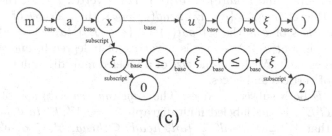

(c)

Fig. 1. (a) A scanned image, (b) a labeled multidigraph representing the adjacency relation of mathematical symbols, and (c) the spanning tree representing proper connections of mathematical symbols

DAGs of tree-width at most 2 was presented. This paper extends the results of DAGs to labeled multidigraphs.

The membership problem of labeled multidigraphs for a spanning tree automaton is NP-complete because the membership problem of DAGs is NP-complete [7,8] and DAGs are a special case of labeled multidigraphs. However, the proof of NP-completeness of the membership problem of DAGs is rather complicated. So this paper presents a simpler proof. It will be shown that the membership problem of non-labeled simple graphs is NP-complete by reducing the Hamiltonian path problem to the problem.

For labeled multidigraphs of bounded tree-width, we present a positive result. It will be shown that there exists a linear-time algorithm that solves the membership problem of labeled multidigraphs of bounded tree-width by using one theorem from Courcelle [4,1,5]. Courcelle's theorem states that all graph properties definable in monadic second-order logic (MSOL) are linear-time decidable for any graphs with bounded tree-width. A sentence in MSOL describing the property "a labeled multidigraph has a spanning tree accepted by a tree automaton" will be presented.

This paper is organized as follows: In Section 2, some definition are given; in Section 3, spanning tree automata are introduced and some basic properties of them are shown; in Section 4, the membership problem of labeled multidigraphs for spanning tree automata is studied; and in Section 5, the conclusion is drawn and future work discussed.

2   Preliminaries

In this paper, we deal with labeled multidigraphs, defined as follows: A *labeled multidigraph* is a 6-tuple $G = (V, E, tail, head, \Sigma, \Delta, \sigma, \delta)$, where V is a finite set of *vertices*, E is a finite set of *edges*, $tail : E \to V$ is a function assigning to each edge its *tail*, $head : E \to V$ is a function assigning to each edge its *head*, Σ is a finite set of *vertex labels*, Δ is a finite set of *edge labels*, $\sigma : V \to \Sigma$ is a function assigning to each vertex its label, and $\delta : E \to \Delta$ is a function assigning to each edge its label. For a pair of edges $e, e' \in E$, e and e' are *multiple* if $tail(e) = tail(e')$ and $head(e) = head(e')$, and e and e' are *symmetric* if $tail(e) = head(e')$ and $head(e) = tail(e')$. For a vertex $v \in V$, the *incoming edges* of v is the set $in(v) = \{e \in E \mid head(e) = v\}$, the *indegree* of v is $|in(v)|$, the *outgoing edges* of v is the set $out(v) = \{e \in E \mid tail(e) = v\}$, and the *outdegree* of v is $|out(v)|$. A *source* is a vertex of indegree 0, and a *sink* is a vertex of outdegree 0. We define the *size* of a labeled multidigraph G as $|V \cup E|$, the number of vertices and edges.

Let $E' \subseteq E$ be a subset of edges. The *edge-induced subgraph* of G by E', denoted by $G[E']$, is the labeled multidigraph $G' = (V', E', tail', head', \Sigma, \Delta, \sigma', \delta')$ such that $V' \subseteq V$, $tail' \subseteq tail$, $head' \subseteq head$, $\sigma' \subseteq \sigma$, $\delta' \subseteq \delta$ and every vertex in V' has at least one incoming or outgoing edge in E'. The *spanning subgraph* of G by E', denoted by $G\langle E'\rangle$, is the labeled multidigraph $(V, E', tail', head', \Sigma, \Delta, \sigma, \delta')$ such that $tail' \subseteq tail$, $head' \subseteq head$ and $\delta' \subseteq \delta$.

Let $G = (V, E, tail, head, \Sigma, \Delta, \sigma, \delta)$ be a labeled multidigraph. G is *acyclic* if there is not a subset of edges $E' \subseteq E$ such that $E' \neq \emptyset$ and every vertex of $G[E']$ has indegree 1 and outdegree 1. For a pair of distinct vertices $u, v \in V$, a *simple directed path* of G from u to v is an edge-induced subgraph $G[E']$ for some $E' \subseteq E$ such that $G[E']$ is acyclic and every vertex of $G[E']$ has indegree 1 and outdegree 1 except that u has indegree 0 and outdegree 1 and v has indegree 1 and outdegree 0.

A *labeled rooted tree* is a labeled multidigraph $T = (V, E, tail, head, \Sigma, \Delta, \sigma, \delta)$ such that T is acyclic, T has exactly one source, and there is a unique simple directed path from the source to every other vertex. The source of a tree is also called the *root*, while the sinks are also called *leaves*. An ordered tree can be seen as a special labeled rooted tree such that $\Delta = \{1, 2, \ldots, max_d\}$, max_d is the maximum outdegree of vertices, and the outgoing edges of each vertex are uniquely labeled as $1, 2, 3, \ldots$.

For a labeled multidigraph $G = (V, E, tail, head, \Sigma, \Delta, \sigma, \delta)$, a *spanning tree* of G is a spanning subgraph $G\langle E'\rangle$ for some $E' \subseteq E$ such that $G\langle E'\rangle$ is a labeled rooted tree.

Let $\mathcal{X} = \{x_1, x_2, \ldots\}$ be a fixed countable set of *variables*.

Example 1. The following is an example of a labeled multidigraph: $G = (V, E,$ *tail, head,* $\Sigma, \Delta, \sigma, \delta)$, *where*

$V = \{v_1, v_2, v_3, v_4\},$

$E = \{e_1, e_2, e_3, e_4, e_5, e_6, e_7, e_8, e_9\},$

$tail = \{(e_1, v_1), (e_2, v_2), (e_3, v_1), (e_4, v_1), (e_5, v_3), (e_6, v_2), (e_7, v_4), (e_8, v_4), (e_9, v_3)\},$

$head = \{(e_1, v_1), (e_2, v_1), (e_3, v_2), (e_4, v_3), (e_5, v_1), (e_6, v_4), (e_7, v_2), (e_8, v_3), (e_9, v_4)\},$

$\Sigma = \{acc, rej\},$

$\Delta = \{\varepsilon, 0, 1\},$

$\sigma = \{(v_1, acc), (v_2, rej), (v_3, rej), (v_4, rej)\},$ *and*

$\delta = \{(e_1, \varepsilon), (e_2, 1), (e_3, 1), (e_4, 0), (e_5, 0), (e_6, 0), (e_7, 0), (e_8, 1), (e_9, 1)\}.$

One of spanning trees of G *is* $T = (V, E', tail', head', \Sigma, \Delta, \sigma, \delta')$, *where* $E' = \{e_3, e_4, e_6\}, tail' = \{(e_3, v_1), (e_4, v_1), (e_6, v_2)\}, head' = \{(e_3, v_2), (e_4, v_3), (e_6, v_4)\},$ *and* $\delta' = \{(e_3, 1), (e_4, 0), (e_6, 0)\}$. *The root of* T *is* v_1.

G *and* T *are illustrated as (a) and (b) in Fig. 2.*

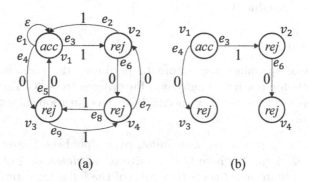

| (a) | (b) |

Fig. 2. (a) a labeled multidigraph, and (b) one of spanning trees of it

Example 2. The following is an example of an ordered tree: $T = (V, E, tail, head,$ $\Sigma, \Delta, \sigma, \delta)$, *where*

$V = \{v_1, v_2, v_3, v_4, v_5, v_6, v_7, v_8, v_9\},$

$E = \{e_1, e_2, e_3, e_4, e_5, e_6, e_7, e_8\},$

$tail = \{(e_1, v_1), (e_2, v_1), (e_3, v_2), (e_4, v_4), (e_5, v_4), (e_6, v_3), (e_7, v_3), (e_8, v_8)\},$

$head = \{(e_1, v_2), (e_2, v_3), (e_3, v_4), (e_4, v_5), (e_5, v_6), (e_6, v_7), (e_7, v_8), (e_8, v_9)\},$

$\Sigma = \{\wedge, \vee, \neg, T, F\},$

$\Delta = \{1, 2\},$

$\sigma = \{(v_1, \wedge), (v_2, \neg), (v_3, \vee), (v_4, \vee), (v_5, F), (v_6, F), (v_7, T), (v_8, \neg), (v_9, T)\},$ *and*

$\delta = \{(e_1, 1), (e_2, 2), (e_3, 1), (e_4, 1), (e_5, 2), (e_6, 1), (e_7, 2), (e_8, 1)\}.$

The root of T is v_1. T is illustrated in Fig. 3.

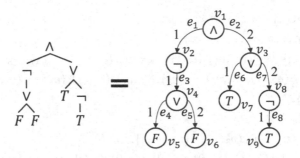

Fig. 3. Representation of an ordered tree by a labeled multidigraph

3   Spanning Tree Automaton

In this section, the definition of spanning tree automata and some basic properties of them are introduced.

3.1   Definition

The definition of spanning tree automata is almost the same as well-known nondeterministic top-down tree automata for ordered trees [3]. The difference is that edge labels can be arbitrarily specified for the transition rules of a spanning tree automaton.

Definition 1. A *spanning tree automaton* over alphabets Σ and Δ is a five-tuple $\mathcal{A} = (Q, \Sigma, \Delta, q_0, R)$ where Q is a finite set of *states*, $q_0 \in Q$ is the *initial state*, and R is a finite set of *transition rules* of the following form:

$$q(f(c_1(x_1), \ldots, c_n(x_n))) \to f(c_1(q_1(x_1)), \ldots, c_n(q_n(x_n))),$$

where $n \geq 0$, $q, q_1, \ldots, q_n \in Q$, $f \in \Sigma$, $c_1, \ldots, c_n \in \Delta$, and $x_1, \ldots, x_n \in \mathcal{X}$. The number n is called the *width* of a transition rule. When $n = 0$, we write $q(f) \to f$ instead of $q(f()) \to f()$.

Let $T = (V, E, tail, head, \Sigma, \Delta, \sigma, \delta)$ be a labeled rooted tree, and let $r \in V$ be the root of T. Let $\mathcal{A} = (Q, \Sigma, \Delta, q_0, R)$ be a spanning tree automaton. For $\mathcal{A}$, a *state mapping* on T is a function $\mu : V \to Q$. A state mapping μ on T is *acceptable* if $\mu(r) = q_0$ and, for each $v \in V$, a transition rule $q(f(c_1(x_1), \ldots, c_n(x_n))) \to f(c_1(q_1(x_1)), \ldots, c_n(q_n(x_n)))$ is in R for some $n \geq 0$, $\mu(v) = q$, $\sigma(v) = f$, and v has exactly n outgoing edges $e_1, \ldots, e_n$ such that $\delta(e_i) = c_i$ and $\mu(head(e_i)) = q_i$ for each $1 \leq i \leq n$. $\mathcal{A}$ accepts T if there exists an acceptable state mapping on T.

Let $G = (V, E, tail, head, \Sigma, \Delta, \sigma, \delta)$ be a labeled multidigraph, and let $\mathcal{A}$ be a spanning tree automaton. $\mathcal{A}$ *accepts* G if G has a spanning tree T and $\mathcal{A}$ accepts T. A set $\mathcal{S}$ of labeled multidigraphs is *recognizable* if there exists a spanning tree automaton $\mathcal{A}$ such that $\mathcal{S} = \{G \mid G \text{ is accepted by } \mathcal{A}\}$.

Example 3. The following is an example of a spanning tree automaton: $\mathcal{A} = (Q, \Sigma, \Delta, q_T, R)$, *where* $Q = \{q_T, q_F\}$, $\Sigma = \{\wedge, \vee, \neg, T, F\}$, $\Delta = \{1, 2\}$, *and* R *consists of transition rules*

$$q_F(\wedge(1(x_1), 2(x_2))) \rightarrow \wedge(q_F(1(x_1)), q_F(2(x_2))),$$
$$q_F(\wedge(1(x_1), 2(x_2))) \rightarrow \wedge(q_T(1(x_1)), q_F(2(x_2))),$$
$$q_F(\wedge(1(x_1), 2(x_2))) \rightarrow \wedge(q_F(1(x_1)), q_T(2(x_2))),$$
$$q_T(\wedge(1(x_1), 2(x_2))) \rightarrow \wedge(q_T(1(x_1)), q_T(2(x_2))),$$
$$q_F(\vee(1(x_1), 2(x_2))) \rightarrow \vee(q_F(1(x_1)), q_F(2(x_2))),$$
$$q_T(\vee(1(x_1), 2(x_2))) \rightarrow \vee(q_T(1(x_1)), q_F(2(x_2))),$$
$$q_T(\vee(1(x_1), 2(x_2))) \rightarrow \vee(q_F(1(x_1)), q_T(2(x_2))),$$
$$q_T(\vee(1(x_1), 2(x_2))) \rightarrow \vee(q_T(1(x_1)), q_T(2(x_2))),$$
$$q_F(\neg(1(x_1))) \rightarrow \neg(q_T(1(x_1))), \quad q_T(\neg(1(x_1))) \rightarrow \neg(q_F(1(x_1))),$$
$$q_T(T) \rightarrow T, \quad and \quad q_F(F) \rightarrow F.$$

The following is a state mapping on T *in Example 2 for* $\mathcal{A}$: $\mu = \{(v_1, q_T), (v_2, q_T), (v_3, q_T), (v_4, q_F), (v_5, q_F), (v_6, q_F), (v_7, q_T), (v_8, q_F), (v_9, q_T)\}$. μ *is illustrated in Fig. 4. Because* μ *is an acceptable state mapping, the labeled rooted tree* T *is accepted by* $\mathcal{A}$.

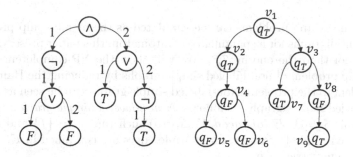

Fig. 4. A labeled rooted tree and a state mapping on it for $\mathcal{A}$

3.2   Basic Properties

Concerning closedness of recognizable sets labeled multidigraphs, the properties of spanning tree automata are the same as the case of DAGs [7,8]. Since the same discussions as in [7,8] hold, we omit proofs.

Proposition 1. *The class of recognizable sets of labeled multidigraphs is closed under union.*

Proposition 2. *The class of recognizable sets of labeled multidigraphs is not closed under intersection.*

Proposition 3. *The class of recognizable sets of labeled multidigraphs is not closed under complementation.*

The emptiness problem is also the same as the case of DAGs [7,8].

Proposition 4. *For a spanning tree automaton $\mathcal{A}$, we can determine whether there exists a labeled multidigraph accepted by $\mathcal{A}$ in linear time in the size of $\mathcal{A}$.*

4   Membership Problem

The membership problem of labeled multidigraphs for a spanning tree automaton is the problem to determine, for any fixed spanning tree automaton, whether a given labeled multidigraph is accepted by the spanning tree automaton.

4.1   NP-Completeness

The membership problem is NP-complete because the membership problem of directed acyclic graphs (DAGs) for a spanning tree automaton has been shown to be NP-complete [7,8] and, obviously, DAGs are a special case of labeled multidigraphs.

Since many graph problems can be simulated as the membership problem of labeled multidigraphs for a spanning tree automaton, this paper presents a much simpler proof than the one in [7,8]. We will show the NP-completeness of the membership problem of non-labeled simple graphs by reducing the Hamiltonian path problem to the problem. Non-labeled simple graphs can be seen as a special case of labeled multidigraphs as follows: A *non-labeled simple graph* is a labeled multidigraph $S = (V, E, tail, head, \Sigma, \Delta, \sigma, \delta)$ such that $\Sigma = \{f\}$ and $\Delta = \{c\}$ for some fixed symbols f and c, no multiple edges are contained in E, and each edge has a symmetric edge.

Theorem 1. *The membership problem of non-labeled simple graphs for a spanning tree automaton is NP-complete.*

Proof. Consider the spanning tree automaton $\mathcal{A} = (Q, \Sigma, q_0, R)$, where $Q = \{q_0\}$, $\Sigma = \{f\}$, and $R = \{q_0(f(c(x_1))) \to f(q_0(c(x_1))), \ q_0(f) \to f\}$.

It is clear that $\mathcal{A}$ accepts labeled rooted trees whose vertices have outdegree at most 1. Therefore, a non-labeled simple graph is accepted by $\mathcal{A}$ if and only if there exists a Hamiltonian path in the graph. Since the Hamiltonian path problem is NP-hard, this problem is also NP-hard.

On the other hand, given a non-labeled simple graph S, we can nondeterministically obtain a spanning tree T of S and check if T is accepted by $\mathcal{A}$ in polynomial time. Thus the problem is in the class NP. Therefore, the problem is NP-complete.   □

Corollary 1. *The membership problem of labeled multidigraphs for a spanning tree automaton is NP-complete.*

Proof. NP-hardness is clear from Theorem 1. For any labeled multidigraph G, we can nondeterministically obtain a spanning tree T of G and check if T is accepted by $\mathcal{A}$ in polynomial time. Thus the problem is in the class NP. Therefore, the problem is NP-complete.   □

4.2   Linear-Time Solvability of the Membership Problem for Graphs of Bounded Tree-Width

We will show the linear-time solvability of the membership problem for graphs of bounded tree-width by using one theorem from Courcelle [4,1,5].

Courcelle's Theorem . *Every graph property definable in monadic second-order logic (MSOL) can be decided in linear time on graphs of bounded treewidth.*

Let $G = (V, E, tail, head, \Sigma, \Delta, \sigma, \delta)$ be a labeled multidigraph, and let $\mathcal{A} = (Q, \Sigma, \Delta, q_0, R)$ be a spanning tree automaton. Without loss of generality, we may assume that $Q = \{1, \ldots, m\}$, $q_0 = 1$, and $Q \cap \Sigma = \emptyset$. Let $r_{\max}$ be the maximum width of transition rules of $\mathcal{A}$.

We will describe the property "a labeled multidigraph G has a spanning tree T, and T is accepted by the tree automaton $\mathcal{A}$" in monadic second-order logic. For that purpose, the property has to be described with propositional logic connectives ($\wedge$, $\vee$, $\neg$, $\Rightarrow$, and $\Leftrightarrow$), individual variables ($e, e_1, e_2, r, v, v_1, v_2, \ldots$), set variables ($E', C, P, P_1, P_2, V, V_1, V_2, \ldots$), the membership symbol ($\in$), the equals sign ($=$), existential ($\exists$) and universal ($\forall$) quantifiers over both individual variables and set variables, and relations representing the graph G.

For Courcelle's theorem, the following two types of representation of a directed graph are allowed:

MS$_1$: a domain V with a binary relation $E \subseteq V \times V$.
MS$_2$: a domain $V \cup E$ with unary relations (i.e., sets) V and E, and binary relations $tail \subseteq E \times V$ and $head \subseteq E \times V$.

MS$_2$ is more expressive than MS$_1$. Since multidigraphs can be only described by MS$_2$, we will take MS$_2$.

For each $f \in \Sigma$, let $V_f = \{v \in V \mid \sigma(v) = f\}$, the set of vertices whose label is f. For each $c \in \Delta$, let $E_c = \{e \in E \mid \delta(e) = c\}$, the set of edges whose label is c. The labeling functions σ and δ can be represented by these sets for all labels in Σ and Δ. Thus, the labeled multidigraph G will be represented by the sets V and E, the binary relations $tail$ and $head$, and the sets V_f and E_c for all $f \in \Sigma$ and $c \in \Delta$.

The property is described by the following MS$_2$-sentence ϕ:

$$\phi := \exists E'(\text{subset}(E', E) \wedge \text{spanning-tree}(E') \wedge \text{accept}(E'))$$

(There exists a subset of edges $E' \subseteq E$ such that $G\langle E'\rangle$ is a spanning tree of G and $G\langle E'\rangle$ is accepted by the spanning tree automaton $\mathcal{A}$.)

$$\text{subset}(E', E) := \forall e(e \in E' \Rightarrow e \in E)$$

(E' is a subset of E.)

$$\text{spanning-tree}(E') := \text{one-source}(E') \wedge \text{acyclic}(E') \wedge \text{unique-path}(E')$$

($G\langle E'\rangle$ has exactly one source, $G\langle E'\rangle$ is acyclic, and there is a unique simple directed path from the source to every other vertex in $G\langle E'\rangle$.)

$$\text{one-source}(E') := \exists r(\text{in}_0(r, E')$$
$$\wedge \forall v((v \in V \wedge v \neq r) \Rightarrow \exists e(e \in E' \wedge head(e) = v)))$$

(There exists a vertex of indegree 0, and the indegree of every other vertex is more than or equal to 1.)

$$\text{acyclic}(E') := \neg \exists C(\exists e(e \in C) \wedge \text{subset}(C, E')$$
$$\wedge \forall v(\text{member}(v, C) \Rightarrow (\text{in}_1(v, C) \wedge \text{out}_1(v, C))))$$

(There does not exist a non-empty subset of edges $C \subseteq E'$ such that every vertex of $G[C]$ has indegree 1 and outdegree 1.)

$$\text{member}(v, C) := \exists e(e \in C \wedge (head(e) = v \vee tail(e) = v))$$

(v is a vertex of $G[C]$.)

$$\text{unique-path}(E') := \exists r(\text{in}_0(r, E') \wedge \forall v((v \neq r \wedge v \in V)$$
$$\Rightarrow \exists P(\text{subset}(P, E') \wedge \text{path}(r, v, P)) \wedge \neg \exists P_1, P_2(P_1 \neq P_2$$
$$\wedge \text{subset}(P_1, E') \wedge \text{path}(r, v, P_1) \wedge \text{subset}(P_2, E') \wedge \text{path}(r, v, P_2))))$$

($G\langle E'\rangle$ has a source r and, for every vertex $v \in V$, if $v \neq r$, then there is a unique simple directed path from r to v.)

$$\text{path}(r, v, P) := \text{acyclic}(P) \wedge \text{in}_0(r, P) \wedge \text{out}_1(r, P) \wedge \text{in}_1(v, P) \wedge \text{out}_0(v, P)$$
$$\wedge \forall u((\text{member}(u, P) \wedge u \neq r \wedge u \neq v) \Rightarrow (\text{in}_1(u, P) \wedge \text{out}_1(u, P)))$$

($G[P]$ is a simple directed path from r to v.)

$$\text{in}_0(v, E') := v \in V \wedge \neg \exists e(e \in E' \wedge head(e) = v)$$

(The indegree of a vertex $v \in V$ is 0 in $G\langle E'\rangle$.)

$$\text{in}_1(v, E') := v \in V \wedge \exists e(e \in E' \wedge head(e) = v)\wedge$$
$$\neg \exists e_1, e_2(e_1 \in E' \wedge e_2 \in E' \wedge e_1 \neq e_2 \wedge head(e_1) = v \wedge head(e_2) = v)$$

(The indegree of a vertex $v \in V$ is 1 in $G\langle E'\rangle$.)

$$\text{out}_0(v, E') := v \in V \wedge \neg\exists e(e \in E' \wedge tail(e) = v)$$

(The outdegree of a vertex $v \in V$ is 0 in $G\langle E'\rangle$.)

$$\text{out}_1(v, E') := v \in V \wedge \exists e(e \in E' \wedge tail(e) = v)$$
$$\wedge \neg\exists e_1, e_2(e_1 \in E' \wedge e_2 \in E' \wedge e_1 \neq e_2 \wedge tail(e_1) = v \wedge tail(e_2) = v)$$

(The outdegree of a vertex $v \in V$ is 1 in $G\langle E'\rangle$.)

$$\text{accept}(E') := \exists V_1, \ldots, V_m(\text{vertex}(V_1, \ldots, V_m) \wedge \text{one-in}(V_1, \ldots, V_m)$$
$$\wedge \forall r(\text{in}_0(r, E') \Rightarrow r \in V_1) \wedge \text{state-mapping}(V_1, \ldots, V_m, E'))$$

(There exist subsets of vertices $V_1, \ldots, V_m \subseteq V$ such that each vertex is in exactly one of them, the root is in V_1, and there is an acceptable state-mapping μ such that $V_i = \{v \in V \mid \mu(v) = i\}$ for each $1 \leq i \leq m$.)

$$\text{vertex}(V_1, \ldots, V_m) := \forall v \left(\bigwedge_{1 \leq i \leq m} (v \in V_i \Rightarrow v \in V) \right)$$

($V_1, \ldots, V_m$ are subsets of V.)

$$\text{one-in}(V_1, \ldots, V_m) :=$$
$$\forall v \left(v \in V \Rightarrow \left(\bigvee_{1 \leq i \leq m} v \in V_i \wedge \bigwedge_{1 \leq i < j \leq m} (v \notin V_i \vee v \notin V_j) \right) \right)$$

(Each vertex $v \in V$ is in in one of the sets $V_1, \ldots, V_m$ but not in two of them.)

$$\text{state-mapping}(V_1, \ldots, V_m, E') :=$$
$$\forall v \left(\bigvee_{0 \leq n \leq r_{\max}} \exists e_1, \ldots, e_n \left(\text{outgoing}_n(v, e_1, \ldots, e_n, E') \right. \right.$$
$$\wedge \bigvee_{q(f(c_1(x_1)), \ldots, c_n(x_n))) \to f(c_1(q_1(x_1)), \ldots, c_n(q_n(x_n))) \in R} \left(v \in V_f \wedge v \in V_q \right.$$
$$\left. \left. \left. \wedge \bigwedge_{0 \leq i \leq n} (e_i \in E_{c_i} \wedge head(e_i) \in V_{q_i}) \right) \right) \right)$$

(For each $v \in V$, v has exactly n outgoing edges $e_1, \ldots, e_n \in E'$ for some $0 \leq n \leq r_{\max}$, a transition rule $q(f(c_1(x_1)), \ldots, c_n(x_n))) \to f(q_1(c_1(x_1)), \ldots,$

$q_n(c_n(x_n)))$ is in R, the rule can be applied to v, and appropriate state can be assigned to each $head(e_i)$.)

$$outgoing_n(v, e_1, \ldots, e_n, E') :=$$
$$\bigwedge_{1 \le i \le n} (e_i \in E' \wedge tail(e_i) = v) \wedge \bigwedge_{1 \le i < j \le n} e_i \ne e_j$$
$$\wedge \neg \exists e \left(e \in E' \wedge tail(e) = v \wedge \bigwedge_{1 \le i \le n} e \ne e_i \right)$$

(v has exactly n outgoing edges $\{e_1, \ldots, e_n\} \subseteq E'$.)

Clearly, the sentence ϕ means that there exists a set of edges $E' \subseteq E$ such that $G(E')$ is a spanning tree of G and $G(E')$ is accepted by the spanning tree automaton $\mathcal{A}$. By Courcelle's theorem, the following theorem holds.

Theorem 2. *If inputs are restricted to labeled multidigraphs of bounded tree-width, then the membership problem for a spanning tree automaton is solvable in linear time on the size of an input graph.*

5 Conclusion and Future Work

We have studied the membership problem of labeled multidigraphs for spanning tree automata. The NP-completeness of the membership problem was shown by reducing the Hamiltonian path problem to the problem. The linear-time solvability of the membership problem for graphs of bounded tree-width was shown by using one theorem from Courcelle [4,1,5].

As a future work, we want to construct a polynomial-time algorithm solving the membership problem for graphs of small tree-width since a linear-time algorithm obtained by Courcelle's theorem is unusable in practice because of a big hidden constant.

References

1. Arnborg, S., Lagergren, J., Seese, D.: Easy problems for tree-decomposable graphs. J. Algorithms 12(2), 308–340 (1991)
2. Chan, K.F., Yeung, D.Y.: Mathematical expression recognition: a survey. Int. J. Document Analysis and Recoginition 3(1), 3–15 (2000)
3. Comon, H., Dauchet, M., Gilleron, R., Jacquemard, F., Lugiez, D., Löding, C., Tison, S., Tommasi, M.: Tree automata techniques and applications (2007), http://www.grappa.univ-lille3.fr/tata (release October 12, 2007)
4. Courcelle, B.: The monadic second-order logic of graphs i. recognizable sets of finite graphs. Information and Computation, 12–75 (1990)
5. Courcelle, B., Engelfriet, J.: Graph Structure and Monadic Second-Order Logic - A Language-Theoretic Approach, Encyclopedia of mathematics and its applications, vol. 138. Cambridge University Press (2012)

6. Eto, Y., Suzuki, M.: Mathematical formula recognition using virtual link network. In: Proceedings of the 6th International Conference on Document Analysis and Recognition (ICDAR 2001), pp. 430–437 (2001)
7. Fujiyoshi, A.: Recognition of a spanning tree of directed acyclic graphs by tree automata. In: Maneth, S. (ed.) CIAA 2009. LNCS, vol. 5642, pp. 105–114. Springer, Heidelberg (2009)
8. Fujiyoshi, A.: Recognition of directed acyclic graphs by spanning tree automata. Theor. Comput. Sci. 411(38-39), 3493–3506 (2010)
9. Fujiyoshi, A., Suzuki, M.: Minimum spanning tree problem with label selection. IEICE Trans. Inf. & Syst. 94(2), 233–239 (2011)
10. Fujiyoshi, A., Suzuki, M., Uchida, S.: Verification of mathematical formulae based on a combination of context-free grammar and tree grammar. In: Autexier, S., Campbell, J., Rubio, J., Sorge, V., Suzuki, M., Wiedijk, F. (eds.) AISC/Calculemus/MKM 2008. LNCS (LNAI), vol. 5144, pp. 415–429. Springer, Heidelberg (2008)

Reset Thresholds of Automata
with Two Cycle Lengths

Vladimir V. Gusev and Elena V. Pribavkina*

Institute of Mathematics and Computer Science
Ural Federal University, Ekaterinburg, Russia
{vl.gusev,elena.pribavkina}@gmail.com

Abstract. We present several series of synchronizing automata with multiple parameters, generalizing previously known results. Let p and q be two arbitrary co-prime positive integers, $q > p$. We describe reset thresholds of the colorings of primitive digraphs with exactly one cycle of length p and one cycle of length q. Also, we study reset thresholds of the colorings of primitive digraphs with exactly one cycle of length q and two cycles of length p.

1 Introduction

A *complete deterministic finite automaton* $\mathscr{A}$, or simply *automaton*, is a triple $\langle Q, \Sigma, \delta \rangle$, where Q is a finite *set of states*, Σ is a finite *input alphabet*, and $\delta : Q \times \Sigma \mapsto Q$ is a totally defined *transition function*. Following standard notation, by Σ^* we mean the set of all finite words over the alphabet Σ, including the empty word ε. The function δ naturally extends to the free monoid Σ^*; this extension is still denoted by δ. Thus, via δ, every word $w \in \Sigma^*$ acts on the set Q. For each $v \in \Sigma^*$ and each $q \in Q$ we write $q \cdot v$ instead of $\delta(q, v)$ and let $Q \cdot v = \{q \cdot v \mid q \in Q\}$.

An automaton $\mathscr{A}$ is called *synchronizing*, if there is a word $w \in \Sigma^*$ which brings all states of the automaton $\mathscr{A}$ to a particular one, i.e. $|Q \cdot w| = 1$. Any such word w is said to be a *reset* (or *synchronizing*) *word* for the automaton $\mathscr{A}$. The minimum length of reset words for $\mathscr{A}$ is called the *reset threshold* of $\mathscr{A}$.

Synchronizing automata serve as transparent and natural models of error-resistant systems in many applied areas (robotics, coding theory). At the same time, synchronizing automata surprisingly arise in some parts of pure mathematics (algebra, symbolic dynamics, combinatorics on words). See recent surveys by Sandberg [10] and Volkov [13] for more details on the theory and applications of synchronizing automata.

One of the most important and natural questions related to synchronizing automata is the following: given n, how big can the reset threshold of an automaton with n states be? In 1964 Černý exhibited a series of automata with n states

* Authors are supported by the Presidential Program for Young Researchers, grant MK-3160.2014.1 and by the Russian Foundation for Basic Research, grant 13-01-00852.

M. Holzer and M. Kutrib (Eds.): CIAA 2014, LNCS 8587, pp. 200–210, 2014.

whose reset threshold equals $(n-1)^2$ [4]. Soon after he conjectured, that this series represents the worst possible case, i.e. the reset threshold of every n-state synchronizing automaton is at most $(n-1)^2$. This hypothesis has become known as the Černý conjecture. In spite of its simple formulation and many researchers' efforts, the Černý conjecture remains unresolved for about fifty years. Moreover, no upper bound of magnitude $O(n^2)$ for the reset threshold of a synchronizing n-state automaton is known so far. The best known upper bound on the reset threshold of a synchronizing n-state automaton is the bound $\frac{n^3-n}{6}$ found by Pin [8] in 1983.

In an attempt to understand why the Černý conjecture is so difficult to resolve, researchers started to look for *slowly synchronizing automata*, i.e. automata with n states and reset threshold close to $(n-1)^2$. First series of such automata were presented in [2]. The number of known series of slowly synchronizing automata was significantly increased in [1]. In the latter paper the constructions are based on the observed connection between slowly synchronizing automata and primitive digraphs with large exponent.

A digraph D is said to be *primitive*, if there is a positive integer t such that for every pair of vertices u and v there is a path form u to v of length t. The smallest t with this property is called the *exponent* of the digraph D. Equivalently, if M is the adjacency matrix of D, then t is the smallest number such that M^t is positive. For additional results on the well-established field of primitive digraphs we refer a reader to [3].

The *underlying digraph* $\mathcal{D}(\mathscr{A})$ of an automaton $\mathscr{A}$ has Q as the set of vertices, and (u, v) is an edge if $u \cdot x = v$ for some letter $x \in \Sigma$. A *coloring* of a digraph D is an automaton $\mathscr{A}$ such that $\mathcal{D}(\mathscr{A})$ is isomorphic to D. Proposition 2 [1] states, that the reset threshold of an arbitrary n-state strongly connected synchronizing automaton is greater than the exponent of the underlying digraph minus n. At the same time, the Road Coloring theorem [12] states that any primitive digraph has at least one synchronizing coloring. Thus, n-state slowly synchronizing automata can be constructed from the well-known examples [5] of primitive digraphs on n vertices with exponents close $(n-1)^2$. This idea was presented and explored in [1]. In the present paper we generalize several series of slowly synchronizing automata presented in [1]. Namely, $\mathscr{W}_n$, $\mathscr{D}'_n$ and $\mathscr{D}''_n$.

Another motivation for the present paper comes from the following facts. Computational experiments of Trahtman [11] revealed that not every positive integer in $\{1, \ldots, (n-1)^2\}$ may serve as the reset threshold of some automaton with n states over a binary alphabet. For example, there is no automaton with nine states over a binary alphabet with the reset threshold in the range from 59 to 63. Similar gaps were found for automata with the number of states ranging from 6 to 10. These results were confirmed in [1]. Moreover, a second gap was presented, i.e. there are no 9-state automata over a binary alphabet with the reset threshold from 53 to 55. For 11-state automata a third gap, along with the first two, was found in the course of computational experiments of Kisielewicz and Szykuła [6]. This brings up the following natural question: given n, which positive integers are reset thresholds of n-state automata? Surprisingly, the set

E_n of all possible exponents of primitive digraphs on a fixed number n of vertices has similar gaps [5] as the set R_n of all possible reset thresholds of n-state automata. Furthermore, for every n the set E_n is fully described [3, p. 83]. We hope that study of this similarity could shed light on properties of R_n. The following statement [7] plays the key role in the description of E_n: if the exponent of a primitive digraph D is at least $\frac{(n-1)^2+1}{2} + 2$, then D has cycles of exactly two different lengths. This motivates our choice in the present paper to focus on automata whose underlying digraphs have exactly two different cycle lengths.

Let p and q be two arbitrary co-prime positive integers, $q > p$. In section 2 we describe reset thresholds of the colorings of primitive digraphs with exactly one cycle of length p and one cycle of length q. In section 3 we study reset thresholds of the colorings of primitive digraphs with exactly one cycle of length q and two cycles of length p.

2   Wielandt-Type Automata

We start with recalling the following elementary and well-known number-theoretic result.

Theorem 1 ([9, Theorem 2.1.1]). *Given two positive co-prime integers p and q, the largest integer that is not expressible as a non-negative integer combination of p and q, is $(p-1)(q-1) - 1$.*

Let us fix two positive co-prime integers p and q. Without loss of generality, we assume $p < q$. Let n be a positive integer, $n < p+q$. We define a Wielandt-type automaton $\mathscr{W}(n,q,p)$ as follows (see Fig. 2). The state set $Q = \{0, 1, \ldots, n-1\}$, $\Sigma = \{a, b\}$, and the transitions are defined in the following way:

$0 . a = q$ if $n > q$, and $0 . a = q - p + 1$ if $n = q$; $0 . b = 1$;
$i . x = i + 1$ for $1 \leq i < n - 1$ and $i \neq q - 1$ for each $x \in \Sigma$;
$(q - 1) . x = 0$ for each $x \in \Sigma$;
if $n > q$, then $(n - 1) . x = n - p + 1$ for each $x \in \Sigma$.

In case $q = n$, $p = n - 1$ we obtain Wielandt automaton $\mathscr{W}_n$ considered in [1]. It is not hard to observe, that every strongly connected n-state automaton whose underlying digraph has exactly one cycle of length p and exactly one cycle of length q is isomorphic to $\mathscr{W}(n,q,p)$.

First let us consider the case $n = q$ (see Fig. 1).

Lemma 1. *Let $\mathscr{A}$ be a strongly connected synchronizing automaton, whose cycles have lengths p and q. If $\gcd(p,q) = 1$, then $\mathrm{rt}(\mathscr{A}) \geq (p-1)(q-1)$. Moreover, if there are states s, t, and a positive integer ℓ such that:*
(i) there is a shortest synchronizing word w which resets the automaton $\mathscr{A}$ to s,
(ii) $t . u = s$ for each word u of length ℓ,
then $\mathrm{rt}(\mathscr{A}) \geq (p-1)(q-1) + \ell$.

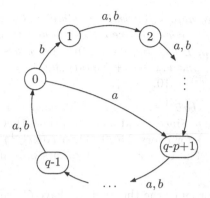

Fig. 1. The Wielandt-type automaton $\mathscr{W}(q,q,p)$

Proof. Let $\mathscr{A} = \langle Q, \Sigma, \delta \rangle$. We prove the first part of the lemma. Consider a synchronizing word w having shortest possible length. Let $s = Q \cdot w$ be the state to which the automaton is synchronized. Note, that the word uw is synchronizing for every $u \in \Sigma^*$, and $Q \cdot uw = s$. In particular, we have $s \cdot w = s \cdot uw = s$. Thus the word w, as well as the word uw, for every word u, labels a path in the automaton $\mathscr{A}$ from the state s to itself. Every such path can be decomposed into cycles of lengths p and q. Hence the number $|w|$, as well as $|w| + k$, for each positive integer k, can be represented as a non-negative combination of the numbers p and q. Thus, by theorem 1, we have $\mathrm{rt}(\mathscr{A}) \geq (p-1)(q-1)$.

Assume now that in addition there exist a state t and a positive integer ℓ such that $t \cdot u = s$ for each word u of length ℓ. Suppose, contrary to our claim, that $|w| < (p-1)(q-1) + \ell$. Let $u \in \Sigma^*$ be an arbitrary word such that $|uw| = (p-1)(q-1) + \ell - 1$. As before, the word uw synchronizes the automaton $\mathscr{A}$ to the state s. But after applying its prefix of length ℓ to the state t we end up in the state s. Hence there is a path of length $(p-1)(q-1) - 1$ from s to itself. But this number can not be represented as a non-negative combination of p and q by theorem 1. A contradiction.

Theorem 2. *The reset threshold of the Wielandt-type automaton* $\mathscr{W}(q,q,p)$ *equals* $(p-1)(q-1) + q - p$.

Proof. Any shortest reset word w for this automaton resets it to the state $q-p+1$, since it is the only state which is a common end of two different edges with the same label. Note, that any word of length $q - p$ brings the state 1 to the state $q - p + 1$. Lemma 1 implies that the reset threshold of $\mathscr{W}(q,q,p)$ is at least $(p-1)(q-1) + q - p$.

Let us check that the word $w = a^{q-p}(ba^{q-1})^{p-2}ba^{q-p}$ synchronizes $\mathscr{W}(q,q,p)$. After applying the prefix a^{q-p} we end up in the cycle C of length p:

$$Q \cdot a^{q-p} = \{0, q-p+1, q-p+2, \ldots, q-1\}.$$

Next, we show that that the word $(ba^{q-1})^{p-2}$ brings C to a two-element set. We state this fact as a separate lemma:

Lemma 2. *Let $\mathscr{A}$ be an automaton with the state set Q over the alphabet $\Sigma = \{a, b\}$. Let $q > p$ be two co-prime positive integers, and let r denote the remainder of the division of q by p. Let $C = \{0, 1, \ldots, p-1\}$ be a subset of Q such that $0 . a = 1$, $0 . ba^{q-1} = 0$, and $i . x \equiv i + 1 \bmod p$ for $1 \leq i \leq p - 1$ and for all $x \in \Sigma$. Then $C . (ba^{q-1})^{p-2} = \{0, p - r\}$.*

Proof. First note, that $i . ba^{q-1} \equiv i + r \bmod p$ for each state $i \neq 0$. Consider the equation $i + rx \equiv 0 \bmod p$. Since r and p are co-prime, this equation has unique solution in $\{1, \ldots, p-1\}$ for every $i \neq 0$. Then $i . (ba^{q-1})^x = 0$. If $x \neq p - 1$, then $i . (ba^{q-1})^{p-2} = 0$. The case $x = p - 1$ occurs only if $i = r$. In this case $r . (ba^{q-1})^{p-2} = p - r$.

Returning back to the proof of the theorem, we have $C . (ba^{q-1})^{p-2} = \{0, q - r\}$ (instead of $\{0, p - r\}$, since here the numeration of the states in the cycle C is different from that in lemma 2). The word ba^{q-p} brings the latter set to the singleton $q - p + 1$.

Let us consider now the general case of the Wielandt-type automaton $\mathscr{W}(n, q, p)$ (see Fig. 2). It is rather easy to see, that given a synchronizing automaton $\mathscr{B}$ and

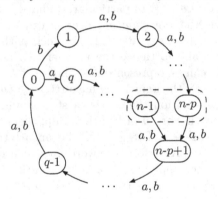

Fig. 2. The Wielandt-type automaton $\mathscr{W}(n, q, p)$

a congruence ρ, the factor automaton $\mathscr{B}/\rho$ is also synchronizing, and $\mathrm{rt}(\mathscr{B}/\rho) \leq \mathrm{rt}(\mathscr{B})$. In particular, consider the following congruence σ on $\mathscr{B}$: for two states s and t we have $s\sigma t$ if and only if $s . x = t . x$ for each $x \in \Sigma$.

Lemma 3. *If $\mathscr{B}$ is synchronizing, then $\mathscr{B}/\sigma$ is also synchronizing, and*

$$\mathrm{rt}(\mathscr{B}/\sigma) \leq \mathrm{rt}(\mathscr{B}) \leq \mathrm{rt}(\mathscr{B}/\sigma) + 1.$$

Proof. The inequality $\mathrm{rt}(\mathscr{B}/\rho) \leq \mathrm{rt}(\mathscr{B})$ is trivial. The states of $\mathscr{B}/\sigma$ are congruence classes $[s]^\sigma$ of the states s of the automaton $\mathscr{B}$. Let us consider a synchronizing word w for the automaton $\mathscr{B}/\sigma$. For every pair of states s and s' of the original automaton $\mathscr{B}$ we have $s . w \sigma s' . w$. But this means that $s . wx = s' . wx$ for any letter $x \in \Sigma$, thus, the word wx resets the automaton $\mathscr{B}$. Thus we have $\mathrm{rt}(\mathscr{B}) \leq \mathrm{rt}(\mathscr{B}/\sigma) + 1$.

Lemma 4. *If $n > q$, then $\mathscr{W}(n, q, p)/\sigma$ is equal to $\mathscr{W}(n-1, q, p)$, and*

$$\mathrm{rt}(\mathscr{W}(n, q, p)) = \mathrm{rt}(\mathscr{W}(n-1, q, p)) + 1.$$

Proof. Let w be a word of minimal length, synchronizing the automaton $\mathscr{W}(n, q, p)$. As in the proof of theorem 2, the word w resets $\mathscr{W}(n, q, p)$ to the state $n - p + 1$. On the last step w brings the states $\{n-1, n-p\}$ to the state $n-p+1$. Hence $w = w'x$, where $x \in \Sigma$, and w' brings the automaton $\mathscr{W}(n, q, p)$ to the set $\{n-1, n-p\}$. But these two states form the unique non-trivial σ-class (see Fig. 2). Thus the factor automaton $\mathscr{W}(n, q, p)/\sigma$ is equal to the Wielandt-type automaton $\mathscr{W}(n-1, q, p)$. Moreover, it is synchronized by w'. Thus, $\mathrm{rt}(\mathscr{W}(n-1, q, p)) \leq \mathrm{rt}(\mathscr{W}(n, q, p)) - 1$. On the other hand, by lemma 3 we have $\mathrm{rt}(\mathscr{W}(n-1, q, p)) \geq \mathrm{rt}(\mathscr{W}(n, q, p)) - 1$. Therefore, we get the required equality.

Theorem 3. *The reset threshold of the Wielandt-type automaton $\mathscr{W}(n, q, p)$ is equal to $(p-1)(q-1) + n - p$.*

Proof. Since there are $n - q$ states on the path from the state 0 to $n - p + 1$, lemma 4 can be applied $n - q$ times to obtain the Wielandt-type automaton $\mathscr{W}(q, q, p)$. By theorem 2, its reset threshold equals $(p-1)(q-1) + q - p$. Each time lemma 4 is applied, the reset threshold is decreased strictly by 1. Thus the reset threshold of the automaton $\mathscr{W}(n, q, p)$ is equal to $(p-1)(q-1) + n - p$.

3 Dulmage-Mendelsohn-Type Automata

As in the previous section, let q and p be two co-prime positive integers, and $q > p$. Let k be a positive integer such that $k < \min\{p, q-p+1\}$. Here we consider Dulmage-Mendelsohn-type automata, which are the colorings of the following primitive digraph $D(q, p, k)$ (see Fig. 3). Its vertex set is $\{0, \ldots, q-1\}$, the set of edges is $\{(i, (i+1) \bmod q) \mid 0 \leq i < q\} \cup \{(0, q-p+1), (k, (q-p+k+1) \bmod q)\}$. Note, that $D(q, p, k)$ has exactly one cycle of length q and two cycles of length p. The digraph $D(q, p, k)$ has only two non-isomorphic colorings $\mathscr{D}^{aa}(q, p, k)$ and $\mathscr{D}^{ab}(q, p, k)$ (see Fig. 4).

Lemma 5. *(i) Any shortest synchronizing word of the automaton $\mathscr{D}^{ab}(q, p, k)$ synchronizes it to the state $q - p + 1$.*
(ii) Any shortest synchronizing word of the automaton $\mathscr{D}^{aa}(q, p, k)$ synchronizes it to the state $q - p + 1$ when $k < q - p$.

Proof. Part (i). Let $t = q - p + k + 1$. Note, that $t = k \cdot b = (q - p + k) \cdot a = (q-p+k) \cdot b$. Any shortest synchronizing word w can synchronize the automaton $\mathscr{D}^{ab}(q, p, k)$ either to $q - p + 1$ or t. Suppose, that w synchronizes $\mathscr{D}^{ab}(q, p, k)$ to the state t. By lemma 1 we have $|w| \geq (p-1)(q-1)$. Moreover, $(p-1)(q-1) > k$. Consider the suffix v of w of length k. It is easy to see, that the full preimage $t \cdot v^{-1}$ of the state t under the action of the word v is equal to $\{1, q-p+1\}$. If $k = q - p$, then the two incoming edges to the state $q - p + 1$ are labeled by the letter a, while the only incoming edge to the state 1 is labeled by the letter b.

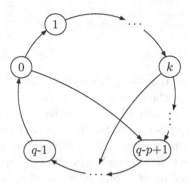

Fig. 3. Digraph $D(q, p, k)$

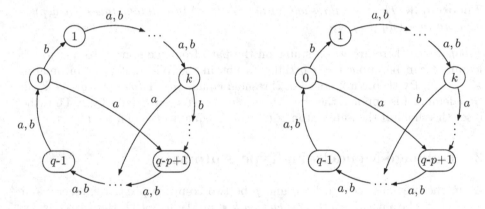

Fig. 4. Two Dulmage-Mendelsohn-type automata $\mathscr{D}^{aa}(q, p, k)$ and $\mathscr{D}^{ab}(q, p, k)$

A contradiction. If $k \neq q - p$, then the set $\{1, q - p + 1\}$ was necessarily obtained from the set $\{0, q - p\}$ by applying the letter b. But $\{0, q - p\} . a = q - p + 1$. Therefore, we can replace the suffix of w of length $k + 1$ by the letter a, in order to obtain a shorter synchronizing word. A contradiction. Hence the word w synchronizes the automaton $\mathscr{D}^{ab}(q, p, k)$ to the state $q - p + 1$.

The proof of the part (ii) of the lemma is analogous to the part (i) with only minor changes.

Theorem 4. *The reset threshold of the Dulmage-Mendelsohn-type automaton $\mathscr{D}^{ab}(q, p, k)$ is equal to $(p - 1)(q - 1) + q - p - k$.*

Proof. Let w be a reset word for the automaton $\mathscr{D}^{ab}(q, p, k)$ having minimal length. By lemma 5 the word w synchronizes the automaton to the state $q - p + 1$. Note, that any word of length $q - p - k$ brings the state $k + 1$ to the state $q - p + 1$. Lemma 1 implies $|w| \geq (p - 1)(q - 1) + q - p - k$.

First let us assume that $k = q - p$. In this case it remains to prove that the word $w_1 = (ba^{q-1})^{p-2} ba^{q-p}$ is synchronizing. Let C be the cycle $\{0, q - p + 1, q - p +$

$2, \ldots, q-1\}$. Note, that the word ba^{q-1} maps all the states, that do not belong to C, to the set $C.ba^{q-1}$. Namely, $k \cdot ba^{q-1} = (t-1) \cdot ba^{q-1}$, where $t = k \cdot b$; $(k-1) \cdot ba^{q-1} = (q-1) \cdot ba^{q-1}, (k-2) \cdot ba^{q-1} = (q-2) \cdot ba^{q-1}, \ldots, 1 \cdot ba^{q-1} = (q-k+1 = p+1) \cdot ba^{q-1}$. Thus it is enough to consider the action of the word w_1 on the cycle C. By lemma 2 we have $C.(ba^{q-1})^{p-2} = \{0, q-r\}$, where r is the remainder of the division of q by p. But then it is easy to see, that $0 \cdot ba^{q-p} = (q-r) \cdot ba^{q-p} = q-p+1$.

Now assume that $k < q-p$. Let us show that the word

$$w_2 = ba^{q-p-k-1}(ba^{q-1})^{p-2}ba^{q-p}$$

is synchronizing. All the states in the range from k to $q-p$ are mapped into the cycle C under the action of the prefix $ba^{q-p-k-1}$. This prefix maps the remaining states lying outside the cycle C, i.e. $1, 2, \ldots, k-1$, to the states ranging from $q-p-k+1$ to $q-p-1$. Namely, $(k-i) \cdot ba^{q-p-k-1} = q-p-i$ for $1 \le i \le k-1$. The action of the word ba^{q-1} on the states in $\{q-p-k+1, \ldots, q-p-1\}$ coincides with the action of this word on some states in the cycle C. More precisely, we have $(q-p-i) \cdot ba^{q-1} = (q-i) \cdot ba^{q-1}$ for $1 \le i \le k-1$, provided that for no such i we have $q-p-i = k$. If $q-p-i = k$ for some i, then we have $k \cdot ba^{q-1} = (t-1) \cdot ba^{q-1}$, where $t = k \cdot b$. In both cases the condition $k < p$ implies that all the resulting states $t-1, q-1, \ldots, q-k+1$ lie on the cycle C. Hence the word w_2 brings the automaton $\mathscr{D}^{ab}(q, p, k)$ into the subset of $C.(ba^{q-1})^{p-2}ba^{q-p}$. As we have already seen, the latter set is the singleton $q-p+1$.

Theorem 5. *The reset threshold of the Dulmage-Mendelsohn-type automaton $\mathscr{D}^{aa}(q, p, k)$ equals $(p-1)(q-1)+q-p-k$ if $k < q-p$, and $(p-1)(q-1)+2(q-p)$ if $k = q-p$.*

Proof. First let us assume that $k < q-p$. Let w be reset word for the automaton $\mathscr{D}^{aa}(q, p, k)$ having minimal possible length. Lemma 5 implies that the word w brings the automaton to the state $q-p+1$. Note, that any word of length $q-p-k$ brings the state $k+1$ to the state $q-p+1$. Thus by lemma 1 we have $|w| \ge (p-1)(q-1)+q-p-k$.

Let us prove that the word $w_1 = a^{q-p-k}(ba^{k-1}ba^{q-k-1})^{p-2}ba^{k-1}ba^{q-p-k}$ is synchronizing. Consider the cycle $C = \{0, q-p+1, q-p+2, \ldots, q-1\}$. Note, that the prefix a^{q-p-k} maps the states, ranging from $k+1$ to $q-p$, to the states in C. Consider now the action of the prefix a^{q-p-k} on the states from 1 to k. If $q-p-k+1 > k$, then all these states are mapped to some states in C. If $q-p-k+1 \le k$, then these states are mapped into $C \cup \{q-p-k+1, \ldots, k\}$. Next, for each state t from $q-p-k+1$ to k we present a state t' from C such that $t \cdot ba^{k-1} = t' \cdot ba^{k-1}$. If $t \ne k$, then it is easy to check that $t' = q-p+t$. Since $q-p-k+1 > 1$, we have $t' > q-p+1$. Hence the state $t' \in C$. If $t = k$, then $t' = k+p$ (recall, that $k+p < q$). The state $k+p$ belongs to C. Indeed, from $q-p-k+1 \le k$ and $k < p$ we obtain $k+p > 2k \ge q-p+1$. Hence the word w_1 brings the automaton $\mathscr{D}^{aa}(q, p, k)$ into the subset of $C.(ba^{k-1}ba^{q-k-1})^{p-2}ba^{k-1}ba^{q-p-k}$. Thus it remains to show, that the latter set is a singleton. The argument is similar to the proof of lemma 2. Instead of the word ba^{q-1} we use the word $v = ba^{k-1}ba^{q-k-1}$. First we note, that the word v fixes the state 0. The word v moves all the other states in C except $q-k$ along the cycle in the same way as the word ba^{q-1} does in lemma 2. The state $q-k$ leaves the cycle after applying the prefix $ba^{k-1}b$, but it can be easily seen that $(q-k).ba^{k-1}ba^{q-k-1} = (q-k).ba^{k-1}aa^{q-k-1}$. Thus we may treat the state $q-k$ as if it never left the cycle C. Following the argument in lemma 2, we conclude, that $C.v^{p-2} = \{0, q-r\}$, where r is the remainder of the division of q by p. Finally, we observe that $0.ba^{k-1}ba^{q-p-k} = (q-r).ba^{k-1}ba^{q-p-k} = q-p+1$.

Consider now the case $k = q - p$. Let w be a synchronizing word for the automaton $\mathscr{D}^{aa}(q,p,k)$ having minimal possible length. Since the incoming edges to the state $q - p + 1$ have different labels, the word w necessarily resets the automaton to the state $q - p + 1 + k$. For convenience, let t denote the state $q - p + 1 + k$. Every word of length k brings the state $q - p + 1$ to the state t. Therefore, by lemma 1 we have $|w| \geq (p - 1)(q - 1) + k$. Suppose $|w| = (p - 1)(q - 1) + k + i$ for some $0 \leq i \leq k - 1$. Consider the states $q - i$ (the state 0, if $i = 0$) and $q - p - i$. The prefix of w of length $k + 1 + i$ will bring one of these states to the state t depending on the $(i + 1)$st letter. The remaining $(p - 1)(q - 1) - 1$ letters of w will move the state t to itself. But this path is a combination of cycles of lengths p and q, which is impossible by theorem 1. Consequently, $|w| \geq (p - 1)(q - 1) + 2k = (p - 1)(q - 1) + 2(q - p)$.

Let us prove that the word $w_2 = a^{q-p}(ba^{k-1}ba^{q-k-1})^{p-2}ba^{k-1}ba^{q-p}$ is synchronizing. The prefix a^{q-p} brings all the states lying outside the cycle $C = \{0, q-p+1, q-p+2, \ldots, q-1\}$ into C. Arguing as in the previous case we conclude, that $C.(ba^{k-1}ba^{q-k-1})^{p-2} = \{0, q-r\}$. It easy to see, that $0.ba^{k-1}ba^{q-p} = (q - r).ba^{k-1}ba^{q-p} = t$.

We can partially generalize this result as we did in theorem 3 for the case of more than q states. We consider a primitive digraph $D_\lambda(q, p, k)$ presented on Fig. 5, where $1 \leq \lambda < p$. For convenience, we set $D_0(q, p, k) = D(q, p, k)$. Its colorings are denoted by $\mathscr{D}_\lambda^{aa}(q, p, k)$ and $\mathscr{D}_\lambda^{ab}(q, p, k)$.

Lemma 6. If $1 \leq \lambda < p$ and $z \in \{a, b\}$, then $\mathscr{D}_\lambda^{az}(q, p, k)/\sigma$ is equal to $\mathscr{D}_{\lambda-1}^{az}(q, p, k)$, and

$$\mathrm{rt}(\mathscr{D}_\lambda^{az}(q, p, k)) = \mathrm{rt}(\mathscr{D}_{\lambda-1}^{az}(q, p, k)) + 1.$$

Proof. Let w be a word synchronizing the automaton $\mathscr{D}_\lambda^{az}(q, p, k)$ having minimal length. Then w resets the automaton either to the state s, or to the state t. Let x be the last letter of w, so that $w = w'x$. The word w' brings the automaton $\mathscr{D}_\lambda^{az}(q, p, k)$ either to the set $\{q + \lambda - 1, s - 1\}$, or $\{q + 2\lambda - 1, t - 1\}$. These two pairs of states form the two non-trivial σ-classes. Hence the factor automaton $\mathscr{D}_\lambda^{az}(q, p, k)/\sigma$ is equal to $\mathscr{D}_{\lambda-1}^{az}(q, p, k)$, and it is synchronized by w'. Thus $\mathrm{rt}(\mathscr{D}_\lambda^{az}(q, p, k)/\sigma) \leq \mathrm{rt}(\mathscr{D}_\lambda^{az}(q, p, k)) - 1$. On the other hand, by lemma 3 we have $\mathrm{rt}(\mathscr{D}_\lambda^{az}(q, p, k)/\sigma) \geq \mathrm{rt}(\mathscr{D}_\lambda^{az}(q, p, k)) - 1$, and we get the required equality.

Theorem 6. If $1 \leq \lambda < p$, then
(i) $\mathrm{rt}(\mathscr{D}_\lambda^{ab}(q, p, k)) = (p - 1)(q - 1) + q - p - k + \lambda;$

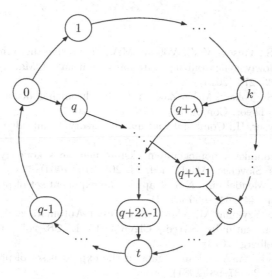

Fig. 5. The digraph $D_\lambda(q, p, k)$

(ii) $\mathrm{rt}(\mathscr{D}_\lambda^{aa}(q, p, k)) = (p-1)(q-1) + q - p - k + \lambda,$ *if* $k < q - p;$
(iii) $\mathrm{rt}(\mathscr{D}_\lambda^{aa}(q, p, k)) = (p-1)(q-1) + 2(q-p) + \lambda,$ *if* $k = q - p.$

Proof. Since there are λ states both on the path from the state 0 to s, and from k to t, and $k \le k - p$, lemma 6 can be applied λ times. Each time lemma 6 is applied, the reset threshold is decreased strictly by one. In the end, from the automaton $\mathscr{D}_\lambda^{ab}(q, p, k)$ we obtain the automaton $\mathscr{D}_0^{aa}(q, p, k)$, whose reset threshold is known by theorem 4. Therefore, we have $rt(\mathscr{D}_\lambda^{ab}(q, p, k)) = (p-1)(q-1) + q - p - k + \lambda$. In an analogous way from the automaton $\mathscr{D}_0^{aa}(q, p, k)$ we obtain the automaton $\mathscr{D}_\lambda^{aa}(q, p, k)$. Applying theorem 5, we obtain $rt(\mathscr{D}_\lambda^{aa}(q, p, k)) = (p-1)(q-1) + q - p - k + \lambda$ in case $k < q - p$, and $rt(\mathscr{D}_\lambda^{aa}(q, p, k)) = (p-1)(q-1) + 2(q-p) + \lambda$ if $k = q - p$.

The case of non-equal number of states on the paths from 0 to s and from k to t can be treated in a similar way. Suppose there are λ states on one of these paths, and $\lambda + \mu < p$ states on the other. We can apply lemma 6 λ times. Each time the reset threshold decreases strictly by one. In the end we obtain an automaton in which either the path from 0 to s is an edge, and there are μ states on the path from k to t, or, vice versa, the path from k to t is an edge, and there are μ states on the path from 0 to s. We can continue taking quotients of this automaton by σ, and on each step analyse, whether the reset threshold strictly decreases by one or remains unchanged. In the end we obtain a Dulmage-Mendelsohn-type automaton, whose reset threshold is known. But this analysis is quite technical with many cases to consider. Thus, we don't present here explicit proofs and formulas.

References

1. Ananichev, D.S., Gusev, V.V., Volkov, M.V.: Primitive digraphs with large exponents and slowly synchronizing automata. Journal of Mathematical Sciences (US) 192(3), 263–278 (2013)
2. Ananichev, D.S., Volkov, M.V., Zaks, Y.I.: Synchronizing automata with a letter of deficiency 2. Theor. Comput. Sci. 376, 30–41 (2007)
3. Brualdi, R., Ryser, H.: Combinatorial matrix theory. Cambridge University Press (1991)
4. Černý, J.: Poznámka k homogénnym eksperimentom s konečnými automatami. Mat. -Fyz. Cas. Slovensk. Akad. Vied. 14, 208–216 (1964)
5. Dulmage, A.L., Mendelsohn, N.S.: Gaps in the exponent set of primitive matrices. Illinois J. Math. 8(4), 642–656 (1964)
6. Kisielewicz, A., Szykuła, M.: Generating Small Automata and the Černý Conjecture. In: Konstantinidis, S. (ed.) CIAA 2013. LNCS, vol. 7982, pp. 340–348. Springer, Heidelberg (2013)
7. Lewin, M., Vitek, Y.: A system of gaps in the exponent set of primitive matrices. Illinois J. Math. 25, 87–98 (1981)
8. Pin, J.-E.: On two combinatorial problems arising from automata theory. Ann. Discrete Math. 17, 535–548 (1983)
9. Ramírez Alfonsín, J.L.: The diophantine Frobenius problem. Oxford University Press (2005)
10. Sandberg, S.: Homing and synchronizing sequences. In: Broy, M., Jonsson, B., Katoen, J.-P., Leucker, M., Pretschner, A. (eds.) Model-Based Testing of Reactive Systems. LNCS, vol. 3472, pp. 5–33. Springer, Heidelberg (2005)
11. Trahtman, A.N.: Notable trends concerning the synchronization of graphs and automata. Electr. Notes Discrete Math. 25, 173–175 (2006)
12. Trahtman, A.N.: The Road Coloring Problem. Israel J. Math. 172, 51–60 (2009)
13. Volkov, M.V.: Synchronizing automata and the Černý conjecture. In: Martín-Vide, C., Otto, F., Fernau, H. (eds.) LATA 2008. LNCS, vol. 5196, pp. 11–27. Springer, Heidelberg (2008)

On the Ambiguity, Finite-Valuedness, and Lossiness Problems in Acceptors and Transducers *

Oscar H. Ibarra

Department of Computer Science
University of California, Santa Barbara, CA 93106, USA
ibarra@cs.ucsb.edu

Abstract. We prove new decidability and undecidability results concerning the finite-ambiguity problem in acceptors, and the finite-valuedness and lossiness problems in transducers. The acceptors and transducers we study have infinite memory.

Keywords: acceptors, transducers, ambiguous, finite-valued, lossy.

1 Introduction

It is well-known that it is undecidable, given an NPDA whose stack makes only one reversal (or, equivalently, a linear CFG), whether it is unambiguous. It is also known that its unbounded ambiguity problem is undecidable [17]. Here, we show that it is undecidable, given a nondeterministic counter automaton (NCA), whether it is unambiguous (resp., has unbounded ambiguity). We also show that in the special case when the counter of the NCA makes at most r-reversals (i.e., alternations between increasing and decreasing modes) for a given $r \geq 1$, determining whether it is unambiguous (resp., k-ambiguous for a given k) is decidable. However, deciding if an r-reversal NCA has unbounded ambiguity is open.

We then turn our attention to transducers. We study the questions of "finite-valuedness" and "finite-lossiness" of a transducer and their connections to the ambiguity of the underlying acceptor (i.e., the acceptor obtained by deleting the outputs).

A transducer T of a given type is k-valued ($k \geq 1$) if every accepted input u is mapped into at most k distinct outputs. T is finite-valued if it is k-valued for some k. Similarly, a transducer is k-lossy ($k \geq 1$) if for every output v, there are at most k distinct accepted inputs that are mapped into v. T is finitely-lossy if it is k-lossy for some k.

We prove decidable and undecidable results concerning the finite-valuedness and finite-lossiness problems for transducers with infinite memory. Similar problems have been investigated before, e.g., for nondeterministic finite transducers (NFTs) [1,14,16]) and visibly pushdown transducers [4]. In [16], e.g., it was

---

* Supported in part by NSF Grant CCF-1117708.

M. Holzer and M. Kutrib (Eds.): CIAA 2014, LNCS 8587, pp. 211–225, 2014.

shown that finite-valuedness of NFTs is decidable. In [4], it was shown that it is decidable if a visibly pushdown transducer is k-valued for a given k. We show a similar result for 1-ambiguous pushdown transducers. (We note that as language acceptors, 1-ambiguous pushdown automata are more powerful than visibly pushdown automata.) The question of whether a transducer with infinite memory is finite-valued (i.e., k-valued for some k) has not been addressed before, as far we know. Here, we exhibit such a class with decidable finite-valuedness problem. The result concerns linear context-free grammars (LCFGs) with outputs, where we show that the finite-valuedness problem for LCFGs with outputs (thus, a finite set of output strings is associated with the application of each rule) is decidable if the LCFG is finitely-ambiguous. If the LCFG is ambiguous, the problem becomes undecidable. The decidability of finite-valuedness generalizes to finitely-ambiguous nonterminal-bounded CFGs with outputs.

Finally, we give a strong undecidability result concerning the equivalence problem for nondeterministic 2-tape finite automata. Note that a binary relation is accepted by a 2-tape finite automaton if and only if the relation is defined by an NFT.

We will use the following notation throughout the paper: NPDA for nondeterministic pushdown automaton; DPDA for deterministic pushdown automaton; NCA for an NPDA that uses only one stack symbol in addition to the bottom of the stack, which is never altered (thus, the stack is a counter); DCA for deterministic NCA; NFA for nondeterministic finite automaton; DFA for deterministic finite automaton; 2NFA for two-way NFA (with end markers). Formal definitions can be found in the book [9].

A counter is an integer variable that can be incremented by 1, decremented by 1, left unchanged, and tested for zero. It starts at zero and cannot store negative values. Thus, a counter is a pushdown stack on unary alphabet, in addition to the bottom of the stack symbol which is never altered.

An automaton (NFA, NPDA, NCA, etc.) can be augmented with multiple counters, where the "move" of the machine also now depends on the status (zero or non-zero) of the counters, and the move can update the counters. See [10] for formal definitions. It is well known that a DFA augmented with two counters is equivalent to a Turing machine (TM) [13].

In this paper, we will restrict the augmented counter(s) to only "reverse" once, i.e., once it decrements, it can no longer increment. Thus, e.g., each counter in an NPDA with 1-reversal counters makes only one reversal. Note that a counter that makes r reversals can be simulated by $\lceil \frac{r+1}{2} \rceil$ 1-reversal counters.

2   Ambiguity in Acceptors

An acceptor M of any type (e.g., NFA, NPDA, etc.) is k-ambiguous ($k \geq 1$) if every accepted string can be accepted in at most k distinct accepting computations. Note that 1-ambiguous is the same as unambiguous. M is finitely-ambiguous if it is k-ambiguous for some k; otherwise, it has unbounded ambiguity.

In this section, we look at the unambiguity and unbounded ambiguity problems concerning 1-reversal NPDAs (i.e., once the stack pops, it can no longer push) and NCAs. It is known and easy to prove (using the undecidability of the Post Correspondence Problem) that it is undecidable, given a 1-reversal NPDA M, whether M is k-ambiguous for any k. In [17] it was shown that unbounded ambiguity for linear context-free grammars (which are equivalent to 1-reversal NPDAs) is undecidable. Thus:

Theorem 1. *It is undecidable, given a 1-reversal NPDA M and an integer $k \geq 1$, whether M is k-ambiguous (resp., has unbounded ambiguity).*

Here, we show that k-ambiguity and unbounded ambiguity for NCAs are also undecidable.

Theorem 2. *It is undecidable, given an NCA M and an integer $k \geq 1$, whether M is k-ambiguous.*

Proof. We first consider the case $k = 1$. The proof uses the undecidability of the halting problem for 2-counter machines. A close look at the proof in [13] of the undecidability of the halting problem for 2-counter machines, where initially one counter has value d_1 and the other counter is zero, reveals that the counters behave in a regular pattern. The 2-counter machine operates in phases in the following way. Let c_1 and c_2 be its counters. The machine's operation can be divided into phases, where each phase starts with one of the counters equal to some positive integer d_i and the other counter equal to 0. During the phase, the positive counter decreases, while the other counter increases. The phase ends with the first counter having value 0 and the other counter having value d_{i+1}. Then in the next phase the modes of the counters are interchanged. Thus, a sequence of configurations corresponding to the phases will be of the form:

$$(q_1, d_1, 0), (q_2, 0, d_2), (q_3, d_3, 0), (q_4, 0, d_4), (q_5, d_5, 0), (q_6, 0, d_6), \ldots$$

where the q_i's are states, with q_1 the initial state, and $d_1, d_2, d_3, \ldots$ are positive integers. Note that the second component of the configuration refers to the value of c_1, while the third component refers to the value of c_2. We assume, w.l.o.g., that $d_1 = 1$.

Let Z be a 2-counter machine. We assume that if Z halts, it does so in a unique state q_h. Let Z's state set be Q, and 1 be a new symbol.

In what follows, α is any sequence of the form $I_1 I_2 \cdots I_{2m}$ (thus we assume that the length is even), where $I_i = 1^k q$ for some $k \geq 1$ and $q \in Q$, represents a possible configuration of Z at the beginning of phase i, where q is the state and k is the value of counter c_1 (resp., c_2) if i is odd (resp., even).

Define L_{odd} to be the set of all strings α such that

1. $\alpha = I_1 I_2 \cdots I_{2m}$;
2. $m \geq 1$;
3. $I_1 = 1^{d_1} q_1$, where $d_1 = 1$ and q_1 is the initial state;
4. $I_{2m} = 1^v q_h$ for some positive integer v;
5. for odd j, $1 \leq j \leq 2m - 1$, $I_j \Rightarrow I_{j+1}$, i.e., if Z begins in configuration I_j, then after one phase, Z is in configuration I_{j+1};

Similarly, define L_{even} analogously except that the condition "$I_j \Rightarrow I_{j+1}$" now applies to *even* values of j, $2 \leq j \leq 2m - 2$.

Now, let $L = L_{odd} \cup L_{even}$. Let Σ be the alphabet over which L is defined.

We can construct an NCA M accepting $L \subseteq \Sigma^*$ as follows. Given input x, M nondeterministically executes (1) or (2) below:

(1) M checks that x is in L_{odd} deterministically by simulating the 2-counter machine Z as follows: M reads $x = I_1 I_2 \cdots I_{2m} = 1^{d_1} q_1 1^{d_2} q_2 \cdots 1^{d_{2m}} q_{2m}$ and verifies that $d_1 = 1$, q_1 is the initial state, q_{2m} is the halting state q_h, and for odd j, $1 \leq j \leq 2m - 1$, $I_j \Rightarrow I_{j+1}$. To check that $I_j \Rightarrow I_{j+1}$, M reads the segment $1^{d_j} q_j$ and stores 1^{d_j} in its counter (call it c) and remembers the state q_j in its finite control. This represents the configuration of Z when one of its two counters, say c_1, has value d_j, the other counter, say c_2, has value 0, and its state is q_j. Then, starting in state q_j, M simulates the computation of Z by decrementing c (which is simulating counter c_1 of Z) and reading the input segment $1^{d_{j+1}}$ until c becomes zero and at which time, the input head of M should be on q_{j+1}. Thus, the process has just verified that counter c_2 of Z has value 1^{d_J+1}, counter c_1 has value 0, and the state is q_{j+1}.

(2) M checks that x is in L_{even}, i.e., for even j, $2 \leq j \leq 2m - 2$, $I_j \Rightarrow I_{j+1}$. M operates in a similar way as described in (1).

Clearly, M is 2-ambiguous, and it is 1-ambiguous (i.e., unmbiguous) if and only if the 2-counter machine Z does not halt, which is undecidable. Note that if Z halts, there is exactly one string in $L(M)$ that is accepted in two distinct ways. Now for any $k \geq 1$, we can construct an NCA M_k, which on any input x, has k computation paths: One path simulates M and the other $k - 1$ paths simply accept x in $k - 1$ different accepting states. Then M_k is $(k + 1)$-ambiguous, and it is k-ambiguous if and only if Z does not halt, which is undecidable. $\square$

Theorem 3. *It is undecidable, given an NCA M, whether it has unbounded ambiguity.*

Proof. Let Σ, L, and M be as in the proof of Theorem 2. Let $\#$ be a new symbol not in Σ, and $L' = \{x_1 \# \cdots \# x_n \# \mid n \geq 1, x_1, \ldots, x_n \text{ in } \Sigma^*, \text{ there is a } 1 \leq p \leq n \text{ such that } x_p \text{ is in } L\}$.

We construct an NCM M' which accepts L' as follows. M' on input z checks that z has the valid format, i.e, $z = x_1 \# \cdots \# x_n \#$ for some $n \geq 1, x_1, \ldots, x_n$ in Σ^*, and nondeterministically selects a p and simulates M on x_p and when M accepts x_p, M' accepts z. Clearly, M' is unambiguous if M is unambiguous (i.e., the 2-counter machine does not halt). If M is ambiguous, then M' will accept the string $(x_p\#)^n$ for any n in linearly distinct ways. It follows that M' has unbounded ambiguity if and only if M is ambiguous (i.e., not unambiguous), which is undecidable by Theorem 2. $\square$

In the above proof, M is either unambiguous or linearly ambiguous, but determining which one is undecidable. A simple modification in the construction yields:

Corollary 1. *It is undecidable, given an NCA M which can only be either unambiguous or exponentially ambiguous, whether it is the former or the latter.*

Proof. In the proof of Theorem 3, define $L' = \{x_1\# \cdots \#x_n\# \mid n \geq 1, x_1, \ldots, x_n$ in Σ^*, for each $1 \leq i \leq n, x_i$ is in $L\}$.   □

One interesting case is when the the counter of the NCA makes only one reversal. More generally, consider a 2NFA augmented with 1-reversal counters. Assume that it is finite-crossing in the sense that there is a given integer c such that the input head crosses the boundary between any two adjacent input symbols at most c times. Then, we have:

Theorem 4. *It is decidable, given a finite-crossing 2NFA M augmented with 1-reversal counters and an integer $k \geq 1$, whether it is k-ambiguous.*

Proof. We first prove the case when $k = 1$. Assume M has n 1-reversal counters. Given M, we construct a new finite-crossing 2NFA M' with two sets of n counters and one special 1-reversal counter, C. M', when given a input x (with left and right end markers), operates as follows. M' simulates M while at the same time uses counter C to count the number of moves M makes (by incrementing C). At some point during the simulation, M' stops incrementing C, which has now some value $t \geq 0$, and records in its finite control the "rule", r, M uses in step $t+1$. M' continues the simulation of M. If M accepts x, M' then moves its input head to the left end marker and carries out a second simulation of M using the second set of 1-reversal counters while decrementing C to count the number of steps M makes in this second run. When C becomes zero, M' checks that the rule r', M uses in the next step is different from r. M' continues the simulation and when M accepts, M' accepts. It follows that M is 1-ambiguous if and only if $L(M') = \varnothing$, which is decidable, since the emptiness problem for finite-crossing 2NFAs augmented with 1-reversal counters is decidable [6].

The proof above generalizes for $k \geq 1$. Now M' has to make $k + 1$ runs (simulations) and uses $k(k + 1)$ 1-reversal counters to verify that there are at least $k + 1$ distinct accepting computations. We omit the details.   □

We do not know if unbounded ambiguity is decidable for finite-crossing 2NFAs augmented with 1-reversal counters. In fact, even for a special case, the question is open:

Open: Is it decidable, given an NFA augmented with a single 1-reversal counter, whether it has unbounded ambiguity?

If we consider the unambiguity and unbounded ambiguity problems of languages (instead of machines), the following was shown by Alan Finkel [5]: Let $\Sigma = \{a, b, c\}$ and d be a symbol not in Σ. Let $L_M = L(M)$ be a language over Σ^* which is accepted by an NCA M. Let $V = \{a^n b^m c^p \mid n, m, p \geq 1, n = m$ or $m = p\}$ and $W_M = L_M d\Sigma^* \cup \Sigma^* dV'$, where $V' = V^+$. Using the fact that the universality problem for languages accepted by NCAs is undecidable [9], W_M, which is clearly accepted by a an NCA, is either unambiguous or exponentially ambiguous, but determining which is the case is undecidable.

Now we can change V' above to $V' = \{x_1 \cdots x_n \mid n \geq 1$, each x_i in $a^+b^+c^+$, there is a $1 \leq p \leq n$ such that x_p is in $V\}$. Then W_M (with this V') can be accepted by a 1-reversal NCA, and the above result still holds with "exponentially-ambiguous" replaced with "linearly-ambiguous". This is in contrast to our Theorem 4 which states that k-ambiguity (for any k) for finite-crossing 2NFAs augmented with 1-reversal counters is decidable.

3   Finite-Valuedness in Transducers

A transducer T is an acceptor with outputs. So, for example, an NPDT is a nondeterministic pushdown automaton with outputs. So the transitions are rules of the form:

$$(q, a, Z) \rightarrow (p, x, y)$$

where q, p are states, a is an input symbol or ε, Z is the top of the stack symbol, x is ε or a string of stack symbols, and y is an output string (possibly ε). In this transition, T in state q, reads a, 'pops' Z and writes x on the stack (if $x \neq \varepsilon$, the rightmost symbol of x becomes the top of the stack), outputs string y, and enters state p.

We say that (u, v) is a transduction accepted by T if, when started in the initial state q_0, with input u, and the top of the stack is the initial stack symbol Z_0, T enters an accepting state after reading u and producing v. The set of transductions accepted by T is denoted by $L(T)$. An NCT (NFT, etc.) is an NCA (NFA, etc.) with outputs.

A transducer T is k-valued ($k \geq 1$) if for every u, there are at most k distinct strings v such that (u, v) is in $L(T)$. T is finite-valued if T is k-valued for some k.

A transducer T is k-ambiguous ($k \geq 1$) if T with outputs ignored is a k-ambiguous acceptor. (Again 1-ambiguous is the same as unambiguous).

Theorem 5. *It is decidable, given a 1-ambiguous NPDT T augmented with 1-reversal counters and an integer $k \geq 1$, whether T is k-valued.*

Proof. Consider the case $k = 1$, Given T, we construct an NPDA M with 1-reversal counters, which uses two additional 1-reversal counters $C(1, 2)$ and $C(2, 1)$. M on input x, simulates T suppressing the outputs and accepts x if it finds two outputs y_1 and y_2 such that y_1 and y_2 disagree in some position p and x is accepted by T. (Note that if $|y_1| \neq |y_2|$, they disagree on the last symbol of the longer string.) In order to do this, during the simulation, M uses $C(1, 2)$ and $C(2, 1)$ to record the positions i and j (chosen nondeterministically) in y_1 and y_2, respectively, and the symbols a and b in these positions, such that $i = j$ and $a \neq b$. Clearly, a and b can be remembered in the state. Storing i and j need only increment $C(1, 2)$ and $C(2, 1)$ during the simulation. To check that $i = j$, after the simulation, M decrements $C(1, 2)$ and $C(2, 1)$ simultaneously and verifies that they reach zero at the same time. Note that since T is 1-ambiguous, M's accepting computation on x (except for the outputs) is unique and therefore the procedure just described can be accomplished by M on a single accepting run on the input. Clearly, T is 1-valued if and only if $L(M) = \varnothing$, which is decidable, since emptiness of NPDAs augmented with 1-reversal counters is decidable [10].

The above construction generalizes for any $k \geq 1$. Now M on input x, checks that there are at least $k + 1$ distinct outputs $y_1, \ldots, y_{k+1}$. M uses $k(k + 1)$ additional 1-reversal counters. In the simulation, for $1 \leq i \leq k+1$, M nondeterministically selects k positions $p(i, 1), \ldots, p(i, i-1), p(i, i+1), \ldots, p(i, k+1)$ in output y_i and records these positions in counters $C(i, 1), \ldots, C(i, i-1), C(i, i+1), \ldots, C(i, k+1)$ and the symbols at these positions in the state. At the end of the simulation, M accepts x if for all $1 \leq i, j \leq k + 1$ such that $i \neq j$, the symbol in position $p(i, j)$ is different from the symbol in position $p(j, i)$ and the value of counter $C(i, j)$ is the same as the value of $C(j, i)$. $\square$

The construction in the proof above does not work when T is k-ambiguous for any $k \geq 2$. This is because the computation of T on an input x may not be unique, so it is possible, e.g., that one accepting run on x produces output y_1 and a different accepting run on x produces output y_2. So to determine if $y_1 \neq y_2$, we need to simulate two runs on input x, i.e., M will no longer be one-way. In fact, we can prove:

Theorem 6. *For any $k \geq 1$, it is undecidable, given a $(k + 1)$-ambiguous NCT (resp., 1-reversal NPDT) T, whether T is k-valued.*

Proof. In the last part of the proof of Theorem 2, the NCA M_k constructed is $(k + 1)$-ambiguous, but it is undecidable whether M_k is k-ambiguous. Construct from M_k an NCT T_k which, on input u, simulates M_k and also outputs the sequences of "rules" used during the computation. Then T_k is k-valued if and only if M is k-ambiguous. The proof for NPDT is similar using Theorem 1. $\sqcap$

However, Theorem 5 holds for finite-crossing 2NFTs augmented with 1-reversal counters, even when there is no restriction on ambiguity.

Theorem 7. *It is decidable, given a finite-crossing 2NFT T augmented with 1-reversal counters and an integer $k \geq 1$, whether T is k-valued.*

Proof. Given T, we construct a finite-crossing 2NCM M augmented with 1-reversal counters. Consider the case $k = 1$. M will have two new 1-reversal counters C_1 and C_2. M simulates two accepting runs of T (suppressing the outputs) and uses C_1 and C_2 to check that the runs produce two distinct outputs. For a general k, M simulates $k + 1$ runs of T and checks using $k(k+1)$ 1-reversal counters to check that the runs generate at least $k + 1$ distinct outputs. $\square$

While it is decidable, given an NFT, whether it is finite-valued [16], we have the following open problem:

Open: Is finite-valuedness for 1-ambiguous NFT augmented with 1-reversal counters decidable?

4   Lossiness in Transducers

A transducer T is k-lossy (for a a given $k \geq 1$) if there are at most k distinct inputs that are mapped into the same output. T is finitely-lossy if there is some k such that T is k-lossy. A 1-lossy transducer is also called a lossless transducer.

A transducer T and an acceptor M are of the same type if they have the same infinite memory structure. So, e.g., NPDT and DPDT are of the same type as NPDA and DPDA, NFT and DFT are of the same type as NFA and DFA, etc.

Lossy transducers were studied in [11] as abstract models of communication channels in analyzing lossy rates of these channels. The following result was stated in [11], but because of space limitation, the proof was omitted. We give the proof here for completeness, as we will need this result.

Theorem 8. *The following statements are equivalent, where M and T are one-way acceptor and transducer of the same type:*

1. *It is undecidable, given a nondeterministic acceptor M, whether M is k-ambiguous for a given k (resp., finitely-ambiguous).*
2. *It is undecidable, given a deterministic transducer T, whether T is k-lossy for a given k (resp., finitely-lossy).*

Proof. First we prove that if (1) is undecidable, then (2) is also undecidable. Let Γ be the set of rules of M (i.e., each rule is represented by a symbol). We construct a deterministic transducer of the same type as M whose input alphabet is Γ. Given a string w in Γ^* (thus $w = r_1 \cdots r_n$, where each r_i is a rule), T deterministically simulates M's computation by reading w symbol-by-symbol and executes rule r_i and outputting the input symbol or ε involved in rule r_i and making sure that w is an accepting sequence of computation. It follows that if M is k-ambiguous for a given k (resp., finitely- ambiguous), then T is k-lossy (resp., finitely-lossy).

Now we show that if (2) is undecidable, then (1) is also undecidable. Suppose T is a deterministic transducer with input and output alphabets Σ and Δ, respectively. We construct a nondeterministic acceptor M with input alphabet Δ. M on input w in Δ^*, guesses a string x in Σ^* symbol-by-symbol (without writing them) and simulates T on x and checks that w is the output of T on input x. M accepts if T accepts. Clearly, since T is deterministic, M is k-ambiguous for a given k (resp., finitely- ambiguous) if T is k-lossy (resp., finitely-lossy). $\square$

An NPDT is 1-reversal if its stack only makes one reversal. From Theorems 1, 2, 3, and 8, we get the following corollary (note that DPDT and DCT are the deterministic versions of NPDT and NCT):

Corollary 2. *It is undecidable, given a 1-reversal DPDT (resp., DCT) T, whether it is k-lossy for a given k (resp., finitely-lossy).*

There is a close connection between finite-valuedness and finite-lossiness in one-way nondeterministic transducers. Let T be a one-way nondeterministic transducer of a given type. We can construct another nondeterministic one-way transducer T' of the same type as T such that $L(T') = \{(w, u) : (u, w) \in L(T)\}$. Clearly, T is k-lossy for a given k (resp., finitely-lossy) if and only if T' is k-valued (resp., finite-valued). The converse is also true.

The above relationship does not hold when the transducers are two-way: Theorem 7 shows that it decidable, given a finite-crossing 2NFT T augmented with 1-reversal counters and an integer $k \geq 1$, whether T is k-valued. However, it is undecidable, given a 2DFT which makes only 1-turn on the input tape (left-to-right then right-to-left), whether it is k-lossy for any k [11]. But, since Theorem 7 obviously holds when the input is one-way we have:

Corollary 3. *It is decidable, given an NFT T augmented with 1-reversal counters and an integer $k \geq 1$, whether T is k-lossy.*

The relationship between valuedness and lossiness does not preserve the degree of ambiguity: We have seen in Theorem 5 that it is decidable, given a 1-ambiguous NPDT T augmented with 1-reversal counters and an integer $k \geq 1$, whether T is k-valued. However, from Corollary 2, it is undecidable, given a 1-reversal DPDT (i.e., the stack makes only one reversal), whether it is k-lossy for any k (resp., finitely-lossy).

Finally, we note that it is decidable to determine, given an NFT T, whether it is finite-valued [16]. Hence, we have:

Corollary 4. *It is decidable to determine, given an NFT T, whether T is finitely-lossy.*

5   Finite-Valuedness in Context-Free Transducers

The notions of ambiguous, finite-valuedness, and lossiness can also be defined for context-free grammars (CFGs) with outputs.

A context-free transducer (CFT) T is a CFG with outputs, i.e., the rules are of the form $A \rightarrow (\alpha, y)$, where α is a string of terminals and nonterminals, and y is an output string (possibly ε). We assume that the underlying CFG G of T, i.e., the grammar obtained by deleting the outputs, has no ε-rules (i.e.,no rules of the form $A \rightarrow \varepsilon$) and unit-rules (i.e., no rules of the form $A \rightarrow B$), where A, B are nonterminals). Moreover, we assume that all nonterminals are useful (i.e., reachable from the start nonterminal S and can reach a terminal string).

Throughout the paper, we will only consider *leftmost* derivations in T, i.e., at each step, the leftmost nonterminal is expanded). Thus T generates transductions (u, v) (where u is a terminal string and v is an output string) derived in a sequence of rule applications in a *leftmost* derivation: $(S, \varepsilon) \Rightarrow^+ (u, v)$

A nonterminal A in the underlying CFG of a CFT is self-embedding if there is some leftmost derivation $A \Rightarrow^+ \alpha A \beta$ where α, β are strings of terminals and nonterminals. (Note that $|\alpha\beta| > 0$, since there are no ε-rules and unit-rules.)

A CFT T is k-ambiguous for a given k (resp., finitely-ambiguous) if its underlying CFG is k-ambiguous (resp., finitely-ambiguous).

Lemma 1. *Let T be a finitely-ambiguous CFT with terminal and nonterminal alphabets Σ and N, respectively. Let G be its underlying finitely-ambiguous CFG. Let A be a nonterminal such that $A \Rightarrow^+ \alpha A\beta$, where $\alpha, \beta \in (\Sigma \cup N)^*$. Then this derivation (of $\alpha A\beta$) is unique.*

Proof. Suppose two distinct derivations $A \Rightarrow^+ \alpha A\beta$ exist. Then, since A is a useful nonterminal, there are an exponential (in k) distinct leftmost derivations: $S \Rightarrow^* xAy \Rightarrow^+ x\alpha^k A\beta^k y \Rightarrow^+ xu^k zv^k y$ for some x, u, z, v, y in Σ^*, for any $k \geq 1$. This contradicts the assumption that the CFG G is finitely-ambiguous.   □

Lemma 2. *It is decidable, given a finitely-ambiguous CFT T, whether there exist a nonterminal A and a leftmost derivation $A \Rightarrow^+ \alpha A\beta$ for some $\alpha, \beta \in (\Sigma \cup N)^*$ (note that $\alpha\beta| > 0$), such that there are at least two distinct outputs generated in the derivation.*

Proof. Let A be a nonterminal and $L = \{w \mid w = \alpha A\beta,$ for some $\alpha, \beta \in (\Sigma \cup N)^*$ such that $|\alpha\beta| > 0$, and $A \Rightarrow^+ \alpha A\beta$ produces at least two distinct outputs$\}$. (Thus $L \subseteq (\Sigma \cup N)^*$.)

We construct an NPDA M with 1-reversal counters to accept L. M, when given input w, tries to simulates a leftmost derivation $A \Rightarrow^+ \alpha A\beta$ (which, if it exists, is unique, by Lemma 1) and checks that there are at least two distinct outputs generated in the derivation. Initially, A is the only symbol on the stack. Each derivation step is of the form $B \rightarrow x\varphi$, where x is in Σ^* and φ is in $N(\Sigma \cup N)^* \cup \{\varepsilon\}$. If B is the symbol on the top of the stack, then M simulates this step by checking that the remaining input segment to be read has prefix x (if $x \neq \varepsilon$) and replacing B by φ on the pushdown stack. It uses two 1-reversal counters C_1 and C_2 to check that there is a discrepancy in the outputs corresponding to the derivation $A \Rightarrow^+ \alpha A\beta$. Since the derivation $A \Rightarrow^+ \alpha A\beta$ is unique, this can be done in the same manner as described in the proof of Theorem 5.

At some point in the derivation, M guesses that the stack contains a string of the form $z = \gamma_1 A\gamma_2$, where $\gamma_1, \gamma_2 \in (\Sigma \cup N)^*$. M then pops the stack and checks that the remaining input yet to be read is $\gamma_1 A\gamma_2$ and accepts if there were two distinct outputs generated in the derivation.

It is easily verified that $L(M) = L$. The result follows since the emptiness problem for NPDAs with 1- reversal counters is decidable [10].   □

Lemma 3. *Let T be a finitely-ambiguous CFT and G be its underlying CFG. Suppose for some nonterminal A, there is a leftmost derivation $A \Rightarrow^+ \alpha A\beta$ for some $\alpha, \beta \in (\Sigma \cup N)^*$, with $|\alpha\beta| > 0$ such that there are at least two distinct outputs generated in the derivation. Then T is not finite-valued.*

Proof. Suppose there is a self-embedding nonterminal A and a derivation $A \Rightarrow^+ \alpha A\beta$, where $|\alpha\beta| > 0$ and there are at least two distinct strings y_1 and y_2 that are outputted in the derivation. Since all nonterminals are useful, we have in G, $S \Rightarrow^* wAx \Rightarrow^+ w\alpha A\beta x \Rightarrow^* w\alpha^k A\beta^k x \Rightarrow^+ wu^k zv^k x$ for some terminal strings

w, u, z, v, x, for all $k \geq 1$. Let y_0, y_3, y_4, y_5 be the outputs in the derivations $S \Rightarrow^* wAx$, $\alpha \Rightarrow^* u$, $A \Rightarrow^+ z$, and $\beta \Rightarrow^* v$, respectively. We have two cases:

Case 1. $|y_1| = |y_2|$ but $y_1 \neq y_2$. Then corresponding to the string $wu^k zv^k x$ generated by G, there will be an unbounded number of distinct outputs: $y_0 y_1^k y_3 y_4 y_5, y_0 y_2 y_1^{k-1} y_3 y_4 y_5, y_0 y_2 y_2 y_1^{k-2} y_3 y_4 y_5, \ldots$ It follows that CFT T is not finite-valued.

Case 2. $|y_1| \neq |y_2|$. Assume, e.g., that $|y_1| - |y_2| = p$. Then, again, there will be an unbounded number of distinct outputs: $y_0 y_2^k y_3 y_4 y_5$ (of length $|y_0| + k|y_2| + |y_3| + |y_4| + |y_5|$), $y_0 y_1 y_2^{k-1} y_3 y_4 y_5$ (of length $|y_0| + k|y_2| + p + |y_3| + |y_4| + |y_5|$), $y_0 y_1 y_1 y_2^{k-2} y_3 y_4 y_5$ (of length $|y_0| + k|y_2| + 2p + |y_3| + |y_4| + |y_5|$), ... So again, the CFT T is not finite-valued. $\square$

We now prove the converse of Lemma 3 for linear context-free transducers (LCFTs). A LCFT is a CFT whose underlying grammar is a linear CFG (LCFG). Thus, the rules are of the form $A \to (uBv, y)$ or $A \to (u, y)$, where A, B are nonterminals, u, v are terminal strings with $|uv| > 0$, and y is an output string.

Lemma 4. *Let T be a finitely-ambiguous LCFT and G be its underlying LCFG. Suppose there is no nonterminal A for which there is a leftmost derivation $A \Rightarrow^+ uAv$ for some $u, v \in \Sigma^*$, with $|uv| > 0$ such that there are at least two distinct outputs generated in the derivation. Then T is finite-valued.*

Proof. Let T be m-ambiguous for some $m \geq 1$. We will show that for any string $w \in L(G)$, there are at most md^n distinct outputs where $n = $ number of nonterminals and d is a constant (to be defined later).

Let G ibe the underelying LCFG of T (which is m-ambiguous) Then there are at most m distinct derivations of any string w in $L(G)$. Lct F be one derivation tree of w. By assumption, every loop in the derivation produces only one output.

Now in F, identify the first (from the root) nonterminal A such that $A \Rightarrow^+ uAv$ for some terminal strings u, v and this is the "longest" derivation that A "reaches" A, i.e., after uAv, A cannot "reach" A any more and can only reach a nonterminal different from A. By assumption there is one output generated in this derivation. We consider two cases:

Case 1. Suppose A can reach a self-embedding nonterminal B (which must be different from A). Then we proceed as above, i.e., identify the first nonterminal B such that $B \Rightarrow^+ u'Bv'$ for some terminal strings u', v'. Again there is only one output generated in this derivation.

By iterating the process described above, we will eventually end up with the following:

Case 2. $A \Rightarrow^+ x$, where x is a terminal string and, in this derivation, no self-embedding nonterminal is encountered.

The maximum number of iterations is the number of self -embedding nonterminals in the derivation tree F, which is at most n. Now the loops involving the self-embedding nonterminals do not increase the number of distinct output values. Let d be the maximum number of distinct outputs that can be generated in any derivation of the form $A \Rightarrow^+ xBy$ or of the form $A \Rightarrow^+ x$, and no self-embedding nonterminal is encountered in these derivations. Note that d can effectively be computed. Then the upper bound on the number of distinct outputs that can be generated in derivation F is d^n, independent of the length of the derivation F. Since there at most m distinct derivations, the number of distinct outputs that T can produce is at most md^n, independent of the length of the derivation. It follows that T is finite-valued.   □

Theorem 9. *It is decidable, given a finitely-ambiguous LCFT T, whether it is finite-valued.*

Proof. Let G be the underlying LCFG of T. We determine if there is a self-embedding nonterminal A in G for which there is a derivation $A \Rightarrow^+ uAv$ (note that $|uv| > 0$), and in this derivation T outputs at least two distinct strings y_1 and y_2. By Lemma 2, there is an algorithm for this. The result then follows from Lemmas 3 and 4,   □

When the LCFT is not finitely-ambiguous, Theorem 9 does not hold:

Theorem 10. *It is undecidable, given a LCFT T, whether it is finite-valued.*

Proof. In [17], it was shown that there is a class of LCFGs for which every grammar in the class is either unambiguous or unboundedly ambiguous, but determining which is the case is undecidable. Let G be a LCFG in this class. Number the rules in G and construct a LCFT T which outputs the rule number corresponding to each rule. The result then follows.   □

Lemma 4 does not hold for finitely-ambiguous CFTs. For consider the 1-ambiguous CFT: $S \rightarrow (SA, 0) \mid (a, 0)$, $A \rightarrow (a, \{0, 1\})$, where S, A are nonterminals, a is a terminal symbol, and $0, 1$ are output symbols. This CFT satisfies the hypothesis of Lemma 4, but it is not finite-valued.

But for the case of nonterminal-bounded context-free transducers (NTBCFTs) (i.e., there is an $s \geq 1$ such that every sentential form derivable in the underlying grammar has at most s nonterminals), we can show that Lemma 4 holds. Hence:

Theorem 11. *It is decidable, given a finitely-ambiguous NTBCFT T, whether it is finite-valued.*

In the journal version of this paper, we will show that finite-valuedness for 1-ambiguous CFTs is decidable.

6   Undecidable Problems Concerning 2-Tape NFAs

A n-tape NFA is an NFA with n (one-way) heads operating independently on n input tapes. We assume that n-tape NFAs and n- DFAs have right end markers on the tapes, and acceptance is when all heads eventually reach \$ on their tapes and the machine enters an accepting state. When a head reaches \$, it remains on $. The set of tuples accepted by a multitape NFA (DFA) M is denoted by $L(M)$. Note that the set of tuples (language) accepted by 2-tape NFAs are exactly the set of transductions accepted by NFTs. Multitape finite automata were introduced in the 1960's in the papers [3] and [15], and the relations they define (i.e., accept) are commonly referred to as rational relations. An n-tape NFA is k-ambiguous if every accepted n-tuple is accepted in at most k distinct accepting computations.

It is a known result that equivalence of multitape DFAs is decidable [8] (see also [18] for the complexity of this problem). In fact, this decidability generalizes to 1-ambiguous multitape NFAs [8]. We will prove a contrasting result below.

A multitape DFA is synchronized [2] if at each step during any accepting computation, all heads that have not yet reached $ move synchronously one position to the right. Thus the heads that have not yet reached $ are always aligned.

The proof of the next result uses ideas in [2,12].

Theorem 12. *It is undecidable to determine, given a 2-ambiguous 2-tape NFA M and a synchronized 2-tape DFA M', whether $L(M) = L(M')$ (resp., $L(M) \subseteq L(M')$).*

Proof. We use the undecidability of the halting problem for deterministic TMs on blank tape. Let Γ be the alphabet used to encode the sequence w of instantaneous descriptions (IDs), separated by the separators #'s, that describes the halting computation of the TM on blank tape, if it exists. Let $L = L_1 \cup L_2$, where $L_1 = \{(x,y) \mid x, y \in \Gamma^*, x \neq y\}$ and $L_2 = \{(x,x) \mid x$ is the halting sequence of IDs of the TM on blank tape $\}$. Note that $L_2 \neq \varnothing$ if and only if the TM halts on blank tape.

We construct a 2-ambiguous 2-tape NFA M that accepts L. Given an input (x,y), M nondeterministically guesses whether it is in L_1 or L_2 as follows:

1. If M guesses that (x,y) is in L_1, then M moves both heads to the right simultaneously and accepts if $x \neq y$. Clearly, for this computation, M is deterministic and synchronized.
2. If M guesses that (x,y) is in L_2, then M assumes that the first tape and second tape are identical. (During the computation, if the tapes are different, then it does not matter what M does, since the input would already be accepted when M guesses that the input is in L_1.) M initially moves its second head until it reads the first ID $q_0\#$ without moving the first head. M then checks that the (identical tapes) correspond to the computation of the TM. The first head of M would lag by one ID until the very end stage when the second head reaches an accepting state, after which the first head moves to the right end marker, and M accepts. Again the process is deterministic. The initialization guarantees that the lag is bigger than 0, but upperbounded by some constant.

Clearly, M is 2-ambiguous. Moreover, it is synchronized and accepts only strings in L_1 if and only if the TM does not halt on blank tape.

Now construct a synchronized 2-tape DFA M' to accept L_1. Clearly, $L(M') \subseteq L(M)$. It follows that $L(M) = L(M')$ if and only $L(M) \subseteq L(M')$, and if and only if the TM does not halt on blank tape, which is undecidable. $\square$

7   Conclusion

The emptiness problem for NPDAs with reversal-bounded counters has recently been shown to be NP-complete [7]. Using this result, we can derive lower and upper bounds for many decidable problems discussed in the paper. We intend to do this in the journal version of the paper.

Acknowledgement. I would like to thank K. Wich for pointing out that the undecidability of unbounded ambiguity for linear context-free grammars was shown in his PhD thesis, E. Filiot for bringing reference [7] to my attention, and O. Finkel and B. Ravikumar for comments on this work.

References

1. Culik, K., Karhumaki, J.: The equivalence of finite valued transducers (on HDTOL languages) is decidable. Theoret. Comput. Sci. 47, 71–84 (1986)
2. Eğecioğlu, Ö., Ibarra, O.H., Tran, N.Q.: Multitape NFA: Weak synchronization of the input heads. In: Bieliková, M., Friedrich, G., Gottlob, G., Katzenbeisser, S., Turán, G. (eds.) SOFSEM 2012. LNCS, vol. 7147, pp. 238–250. Springer, Heidelberg (2012)
3. Elgot, C.C., Mezei, J.E.: On relations defined by generalized finite automata. IBM J. Res. 9, 45–65 (1965)
4. Filiot, E., Raskin, J.-F., Reynier, P.-A., Servais, F., Talbot, J.-M.: Properties of visibly pushdown transducers. In: Hliněný, P., Kučera, A. (eds.) MFCS 2010. LNCS, vol. 6281, pp. 355–367. Springer, Heidelberg (2010)
5. Finkel, O.: Personal communication
6. Gurari, E., Ibarra, O.H.: The complexity of decision problems for finite-turn multicounter machines. J. Comput. Syst. Sci. 22, 220–229 (1981)
7. Hague, M., Lin, A.W.: Model checking recursive programs with numeric data types. In: Gopalakrishnan, G., Qadeer, S. (eds.) CAV 2011. LNCS, vol. 6806, pp. 743–759. Springer, Heidelberg (2011)
8. Harju, T., Karhumaki, J.: The equivalence problem of multitape finite automata. Theoret. Comput. Sci. 78(2), 347–355 (1991)
9. Hopcroft, J.E., Ullman, J.D.: Introduction to Automata, Languages and Computation. Addison-Wesley (1978)
10. Ibarra, O.H.: Reversal-bounded multicounter machines and their decision problems. J. of the ACM 25(1), 116–133 (1978)
11. Ibarra, O.H., Cui, C., Dang, Z., Fischer, T.R.: Lossiness of communication channels modeled by transducers. In: Beckmann, A., Csuhaj-Varjú, E., Meer, K. (eds.) CiE 2014. LNCS, vol. 8493, pp. 224–233. Springer, Heidelberg (2014)
12. Ibarra, O.H., Seki, S.: On the open problem of Ginsburg concerning semilinear sets and related problems. Theor. Comput. Sci. 501, 11–19 (2013)
13. Minsky, M.: Recursive unsolvability of Post's problem of Tag and other topics in the theory of Turing machines. Ann. of Math. (74), 437–455 (1961)

14. Sakarovitch, J., de Souza, R.: On the decidability of bounded valuedness for transducers. In: Ochmański, E., Tyszkiewicz, J. (eds.) MFCS 2008. LNCS, vol. 5162, pp. 588–600. Springer, Heidelberg (2008)
15. Schutzenberger, M.P.: A remark on finite transducers. Inform. and Control 4, 185–196 (1961)
16. Weber, A.: On the valuedness of finite transducers. Acta Inform. 27, 749–780 (1990)
17. Wich, K.: Exponential ambiguity of context-free grammars. In: Proc. of 4th Int. Conf. on Developments in Language Theory, pp. 125–138. World Scientific (1999)
18. Worrell, J.: Revisiting the equivalence problem for finite multitape automata. In: Fomin, F.V., Freivalds, R., Kwiatkowska, M., Peleg, D. (eds.) ICALP 2013, Part II. LNCS, vol. 7966, pp. 422–433. Springer, Heidelberg (2013)

Kleene Closure on Regular
and Prefix-Free Languages

Galina Jirásková [1,*], Matúš Palmovský [1], and Juraj Šebej [2,**]

[1] Mathematical Institute, Slovak Academy of Sciences
Grešákova 6, 040 01 Košice, Slovakia
jiraskov@saske.sk, matp93@gmail.com
[2] Institute of Computer Science, Faculty of Science, P.J. Šafárik University
Jesenná 5, 040 01 Košice, Slovakia
juraj.sebej@gmail.com

Abstract. We study the Kleene closure operation on regular and prefix-free languages. Using an alphabet of size $2n$, we get the contiguous range from 1 to $3/4 \cdot 2^n$ of complexities of the Kleene closure of regular languages accepted by minimal n-state deterministic finite automata. In the case of prefix-free languages, the Kleene closure may attain just three possible complexities $n - 2$, $n - 1$, and n.

1   Introduction

Kleene closure is a basic operation on formal languages which is defined as $L^* = \{w \mid w = v_1 v_2 \cdots v_k, k \geq 0, v_i \in L \text{ for all } i\}$. It is known that if a language L is recognized by an n-state deterministic finite automaton (DFA), then the language L^* is recognized by a DFA of at most $3/4 \cdot 2^n$ states [11,14]. The first worst-case example meeting this upper bound was presented already by Maslov in 1970 [11], although he claimed upper bound $3/4 \cdot 2^n - 1$ in his paper. The proof of the fact that his witness meets the bound $3/4 \cdot 2^n$ can be found in [10].

Yu, Zhuang, and Salomaa [14] proved that the size of the minimal DFA for Kleene closure depends on the number of final states of a given DFA, and that the upper bound is $2^{n-1} + 2^{n-1-k}$, where k is the number of final and non-initial states. These upper bounds have been shown to be tight in the binary case for all k with $1 \leq k \leq n - 1$ in [10]. If a regular language L is accepted by a DFA, in which the initial state is the unique final state, then $L^* = L$. Thus the state complexities of a language and of its Kleene closure may be the same; here the state complexity of a regular language L, $\mathrm{sc}(L)$, is the smallest number of states in any DFA recognizing the language L. If a language L over an alphabet Σ contains all the one-symbol strings a with $a \in \Sigma$, then $L^* = \Sigma^*$, so $\mathrm{sc}(L^*) = 1$.

Hence we get the range from 1 to $3/4 \cdot 2^n$ of possible values of the state complexity of the Kleene closure of a regular language with state complexity n,

---

* Research supported by grant APVV-0035-10.
** Research supported by grant VEGA 1/0479/12.

M. Holzer and M. Kutrib (Eds.): CIAA 2014, LNCS 8587, pp. 226–237, 2014.

in which the values $1, n, 3/4 \cdot 2^n$, and $2^{n-1} + 2^{n-1-k}$ with $2 \le k \le n - 1$ are attainable by the state complexity of the Kleene closure of binary languages.

In this paper, we consider the question whether or not the remaining values in this range are attainable. Our motivation comes from the paper by Iwama et al. [5], in which the authors stated the problem of whether there always exists a regular language represented by a minimal n-state nondeterministic finite automaton (NFA) such that the minimal deterministic automaton for the language has α states for all integers n and α with $n \le \alpha \le 2^n$. The problem was solved positively in [9] by using a ternary alphabet. On the other hand, as shown by Geffert in [3], in the unary case, there are a lot of holes in the range from n to $2^{\Theta(\sqrt{n \log n})}$ that cannot be attained by the state complexity of any unary language represented by a minimal n-state NFA. However, no specific holes are known for NFA-to-DFA conversion.

In the case of Kleene closure on unary languages, the holes in the range from 1 to $(n - 1)^2 + 1$ also exists [2]. Moreover, Čevorová in [2] described for every n two specific holes of length n close to the upper bound $(n-1)^2 + 1$. On the other hand, the contiguous range of complexities of Kleene closure from 1 to $3/4 \cdot 2^n$ have been obtained in [8] using an alphabet that grows exponentially with n.

In the first part of our paper, we improve the result from [8] by decreasing the size of the alphabet to $2n$. We show that for all n and α with $1 \le \alpha \le 3/4 \cdot 2^n$, there exists a language L over an alphabet of size $2n$ accepted by a minimal DFA of n states and such that the minimal DFA for L^* has α states.

In the second part of the paper, we study the Kleene closure operation on prefix-free languages. Here the known upper bound is n [4], and we prove that the state complexity of Kleene closure may attain only three values $n-2, n-1, n$.

First, let us recall some basic definitions. For further details and all unexplained notions, the reader may refer to [12,13]. A *nondeterministic finite automaton* (NFA) is a quintuple $A = (Q, \Sigma, \delta, I, F)$, where Q is a finite set of states, Σ is a finite alphabet, $\delta : Q \times \Sigma \to 2^Q$ is the transition function which is extended to the domain $2^Q \times \Sigma^*$ in the natural way, $I \subseteq Q$ is the set of initial states, and $F \subseteq Q$ is the set of final states. The language accepted by A is the set $L(A) = \{w \in \Sigma^* \mid \delta(I, w) \cap F \ne \emptyset\}$. An NFA A is *deterministic* (and complete) if $|I| = 1$ and $|\delta(q, a)| = 1$ for each q in Q and each a in Σ. In such a case, we write $\delta(q, a) = q'$ or simply $q \cdot a = q'$ instead of $\delta(q, a) = \{q'\}$.

The state complexity of a regular language L, $\mathrm{sc}(L)$, is the number of states in the minimal DFA for L. It is well known that a DFA is minimal if all its states are reachable from its initial state, and no two of its states are equivalent.

Every NFA $A = (Q, \Sigma, \delta, I, F)$ can be converted to an equivalent DFA $A' = (2^Q, \Sigma, \cdot, I, F')$, where $R \cdot a = \delta(R, a)$ and $F' = \{R \in 2^Q \mid R \cap F \ne \emptyset\}$. The DFA A' is called the *subset automaton* of the NFA A. The subset automaton need not be minimal since some of its states may be unreachable or equivalent. However, if for each state q of the NFA A, there exists a string w_q that is accepted by A only from the state q, then the subset automaton of the NFA A does not have equivalent states since if two subsets of the subset automaton differ in a state q, then they are distinguishable by w_q.

The *concatenation* of two languages K and L is the language $K \cdot L = \{uv \mid u \in K \text{ and } v \in L\}$. The *Kleene closure* or *star* a language L is the language $L^* = \bigcup_{i \geq 0} L^i$, where $L^0 = \{\varepsilon\}$.

2 Kleene Closure on Regular Languages: Contiguous Range of Complexities for Linear Alphabet

If a language L is accepted by an n-state DFA, then the language L^* is accepted by a DFA of at most $3/4 \cdot 2^n$ states, and this bound is known to be tight in the binary case [10,11,14]. On the other hand, if a language L contains all one-symbol strings, then $L^* = \Sigma^*$. Hence all the possible complexities of the star of a language with state complexity n are in the range from 1 to $3/4 \cdot 2^n$.

It has been shown in [8] that for all n and α with $1 \leq \alpha \leq 3/4 \cdot 2^n$, there exists a regular language L over an alphabet of size 2^n such that $\mathrm{sc}(L) = n$ and $\mathrm{sc}(L^*) = \alpha$. Thus the range of possible complexities for star is contiguous without any holes in it providing that the alphabet contains an exponential number of symbols. The huge alphabet allows us to reach an exponential number of subsets in a subset automaton for star using a new symbol for each subset. The aim of this section is to prove a similar result using an alphabet of size at most $2n$.

To do this, we will describe three constructions. In each of them, we will add a new state and transitions on two new symbols to a given n-state DFA A with $\mathrm{sc}(L(A)^*) = \alpha$. The first construction will produce an $(n + 1)$-state DFA B with $\mathrm{sc}(L(B)^*) = \alpha + 1$. The second and the third construction, will result in $(n + 1)$-state DFAs C and D with $\mathrm{sc}(L(C)^*) = 2\alpha$ and $\mathrm{sc}(L(D)^*) = 2\alpha + 1$, respectively.

Using the induction hypothesis that all the values from n to $3/4 \cdot 2^n - 1$ are attainable as the complexities of the star of minimal n-state DFAs over an alphabet of size $2n$, we will be able to show that all the complexities from $n + 1$ to $3/4 \cdot 2^{n+1} - 1$ are attainable for minimal $(n + 1)$-state DFAs over an alphabet of size $2(n+1)$. The remaining values in the range from 1 to $3/4 \cdot 2^n$ are known to be attained by the complexity of the star of unary or binary languages [2,8,14].

First, let us recall the construction of an NFA A^* for the language L^*. Let a language L be accepted by a DFA $A = (Q, \Sigma, \cdot, s, F)$. To get an NFA A^* for the language L^* from the DFA A, we add a transition on a symbol a from a state q to the initial state s whenever $q \cdot a \in F$. Moreover, if the initial state s is non-final, we add a new initial state q_0. By each symbol a, the state q_0 goes to $\{s \cdot a\}$ if $s \cdot a \notin F$, and it goes to $\{s \cdot a, s\}$ if $s \cdot a \in F$.

Throughout this section, all DFAs will have a final initial state. Therefore, in the construction of the NFA A^*, there is no need to add a new initial state q_0. Let us start with the following example.

Example 1. Consider the 3-state DFA A shown in Fig. 1. The DFA A has two final states 1 and f. Construct an NFA A^* for $L(A)^*$ by adding transitions on a_0, a_1, a_2 from the states $f, 1, 2$, respectively, to the initial state 1. The NFA A^* and the 4-state subset automaton A' of the NFA A^* are shown in Fig. 1 (bottom). Notice that the following conditions are satisfied:

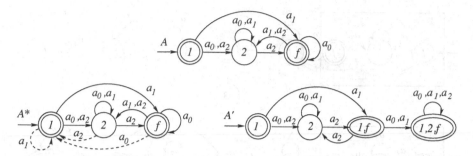

Fig. 1. The DFA A for a language L (top), the NFA A^* for the language L^* (left-bottom), and the subset automaton A' (right-bottom)

(C1) the DFA A accepts the string a_0 only from the state f, the string a_1 only from the state 1, and the string a_2 only from the state 2;

(C2) the initial state of the subset automaton A' is $\{1\}$, the subsets $\{2\}$, $\{1, f\}$, and $\{1, 2, f\}$ are reachable in the subset automaton A', while the empty set is unreachable since A is deterministic.

Since A satisfies (C1), it is minimal. Moreover, it follows from the construction of the NFA A^*, that the NFA A^* also satisfies (C1), that is, the string a_0 is accepted by A^* only from f, and a_i is accepted only from i for $i = 1, 2$. This means that all the states in the subset automaton A' of the NFA A^* are pairwise distinguishable.

Our aim is to get 4-state DFAs B, C, and D whose stars require $4 + 1$, $2 \cdot 4$, and $2 \cdot 4 + 1$ states, respectively.

(1) To get a 4-state DFA B for a language whose star requires 5 states, add a new non-final state 3 to the dfa A, and transitions on two new symbols a_3 and b_3 as follows. The new state 3 goes to itself on old symbols a_0, a_1, and a_2. Next,

$3 \cdot a_3 = f$ and $q \cdot a_3 = 2$ if $q \neq 3$;
$q \cdot b_3 = 3$ for each state q of B.

The initial state of B is the state 1, and final states are 1 and f. Construct an NFA B^* for the star of the language $L(B)$ by adding transitions on a_0, a_1, a_2, a_3 from states $f, 1, 2, 3$, respectively, to the initial state 1. In the subset automaton B' of the NFA B^*, all the states that have been reachable in A' are reachable since the initial state of B' is $\{1\}$, and we do not change the transitions on a_0, a_1, a_2 on the states of A. Moreover, state $\{3\}$ is reached from state $\{1\}$ by b_3. No other state is reachable in B' because A' satisfies (C2). Since B accepts a_0 only from f, and a_i only from i ($1 \leq i \leq 3$), DFA B is minimal, and all the states of the subset automaton B' are pairwise distinguishable. Finally, the subset automaton B' satisfies (C2). Fig. 2 (top) shows the new state and the new transitions in the DFA B, and the reachability of the states in B'.

(2) Now we would like to duplicate the number of states in the subset automaton A', that is, to construct a 4-state DFA C for a language whose star requires $2 \cdot 4$ states. We again add a new non-final state 3 to the DFA A, going to itself

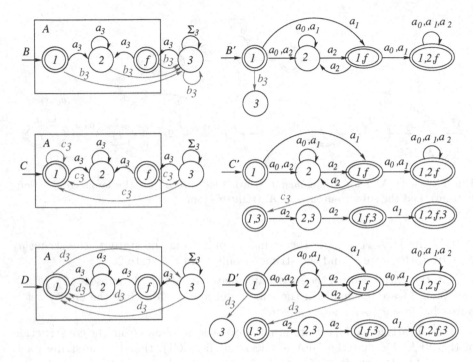

Fig. 2. The new state and the new transitions in the DFAs B, C, D (left) and the reachable states of the subset automata B', C', D' (right)

on each old symbol. This time we define transitions on new symbols a_3, c_3. The transitions on a_3 are the same as in case (1), and the transitions on c_3 are as follows:

$f \cdot c_3 = 3$ and $q \cdot c_3 = 1$ if $q \neq f$.

The new state and new transitions in the DFA C, and the reachability of the states in C' are illustrated in Fig. 2 (middle).

(3) To get a 4-state DFA D for a language whose star requires $2 \cdot 4 + 1$ states, we again add a new state 3, going to itself on old symbols, and transitions on new symbols a_3 and d_3. The transitions on a_3 are as above, and

$1 \cdot d_3 = 3$ and $q \cdot d_3 = 1$ if $q \neq 1$.

The new state and new transitions in the DFA D, and the reachability of the states in D' are illustrated in Fig. 2 (bottom).   □

Let us show that the above described constructions work in the general case.

Lemma 1. *Let* $4 \leq n \leq \alpha \leq 3/4 \cdot 2^n - 1$. *There exists a regular language* L *over an alphabet* Σ_n *with* $|\Sigma_n| \leq 2n$ *such that* $\mathrm{sc}(L) = n$ *and* $\mathrm{sc}(L^*) = \alpha$.

Proof. If a DFA A accepts the empty string, then we can construct an NFA A^* for the language $(L(A))^*$ from the DFA A by adding the transition on a symbol

a from a state q to the initial state of A whenever the state q goes by a in A to a final state.

We prove by induction on n the following claim: For every α with $n \leq \alpha \leq 3/4 \cdot 2^n - 1$, there exists an n-state DFA A over an alphabet Σ_n of size at most $2n$ and such that $\{a_0, a_1, \ldots, a_{n-1}\} \subseteq \Sigma_n$, with the state set $Q = \{1, 2, \ldots, n-1, f\}$, the initial state 1, the set of final states $\{1, f\}$, satisfying the following conditions:

(C1) the string a_0 is accepted by A only from the state f, and the string a_i is accepted only from the state i ($1 \leq i \leq n-1$);

(C2) the subset automaton A' of the NFA A^* has α reachable states, and the states $\{1\}, \{2\}, \{1, f\}, \{1, 2, f\}$ are reachable in A', while the empty set is not reachable.

We first prove the induction step, and then will discuss the basis.

Assume that our claim holds for n and all α with $n \leq \alpha \leq 3/4 \cdot 2^n - 1$, and let us show that then it also holds for $n+1$ and all α with $n+1 \leq \alpha \leq 3/4 \cdot 2^{n+1} - 1$.

To this end, let α be an integer with $n \leq \alpha \leq 3/4 \cdot 2^n - 1$, and let A be an n-state DFA satisfying the induction hypothesis. We will show that we are able to construct $(n+1)$-state DFAs B, C, D from the DFA A by adding a new state n and transitions on two new symbols, so that the star of languages $L(B), L(C), L(D)$ will require $\alpha+1, 2\alpha$, and $2\alpha+1$ states respectively. Moreover, the DFAs B, C, D and the subset automata B', C', D' will satisfy (C1) and (C2).

(1) Let us start with the construction of the DFA B for a language whose star requires $\alpha + 1$ states. We add a new state n to the DFA A, going to itself on each symbol in Σ_n. Next, we add transitions on two new symbols a_n and b_n defined as follows:

$n \cdot a_n = f$ and $q \cdot a_n = 2$ if $q \neq n$;
$q \cdot b_n = n$ for each state q of the DFA B.

Notice that the string a_n is accepted by B only from the state n, while the unique acceptance of the other strings a_i remains the same as in A. Hence B satisfies (C1).

Construct the NFA B^* for the star of the language $L(B)$. In the subset automaton B' of the NFA B^*, the initial state is $\{1\}$, and all the subsets that are reachable in the subset automaton A' are also reachable in B' because we do not change the transitions on the symbols in Σ_n on states of A. Thus B satisfies (C2). Moreover, the subset $\{n\}$ is reachable in B' since $\{1\}$ goes to $\{n\}$ by b_n.

We need to show that no other subset is reachable in B'. To this aim consider the family of $\alpha + 1$ sets

$$\mathcal{R} = \{S \mid S \text{ is reachable in } A'\} \cup \{\{n\}\}.$$

Since the initial state of B' is in $\mathcal{R}$, we only need to show that each set R in $\mathcal{R}$ goes to a set in $\mathcal{R}$ by each symbol in $\Sigma_{n+1} = \Sigma_n \cup \{a_n, b_n\}$. This is straightforward for symbols in Σ_n. Let S be a reachable state of A'. Then S goes to $\{2\}$ by a_n. The set $\{2\}$ is in $\mathcal{R}$ since it is a reachable state of A' by (C2). By b_n, the set S goes to $\{n\}$ which is in $\mathcal{R}$. The set $\{n\}$ goes by b_n to itself, and by a_n to $\{1, f\}$ which is $\mathcal{R}$ by (C2).

Due to unique acceptance of the (one-symbol) strings a_i, automata B and B' do not have equivalent states. Thus B is minimal, and the star of the language $L(B)$ requires exactly $\alpha + 1$ states.

(2) Now we describe the construction of the DFA C for a language whose star requires 2α states. We again add a new state n to the DFA A going to itself on each symbol in Σ_n. We also add the transitions on a_n as in case (1). Thus C will satisfy (C1). Next, we add transitions on new symbol c_n defined as follows:

$f \cdot c_n = n$ and $q \cdot c_n = 1$ if $q \neq f$.

Construct the NFA C^* for $L(C)^*$. In the subset automaton C' of the NFA C^*, all the subsets that are reachable in A' are reachable as well, and so C' satisfies (C2). Moreover, state $\{1, f\}$ goes to $\{1, n\}$ by c_n, and then to $S \cup \{n\}$ for every reachable set of A' because state n goes to itself on every symbol in Σ_n in the DFA C. Thus each set in the family

$$\mathcal{R} = \{S \mid S \text{ is reachable in } A'\} \cup \{S \cup \{n\} \mid S \text{ is reachable in } A'\}$$

is reachable in C'. We need to show that no other set is reachable in C'. To do this, it is enough to show that each set R in $\mathcal{R}$ goes to a set in $\mathcal{R}$ by each symbol in $\Sigma_{n+1} = \Sigma_n \cup \{a_n, c_n\}$. This is straightforward for symbols in Σ_n. Each set S that is reachable in A' goes to $\{2\}$ by a_n. The set $\{2\}$ is in $\mathcal{R}$ by (C2). By c_n, the set S goes either to $\{1\}$ or to $\{1, n\}$. Both these sets are in $\mathcal{R}$. Now consider a set $S \cup \{n\}$ in $\mathcal{R}$; notice that S is non-empty since A is deterministic. By a_n, the set $S \cup \{n\}$ goes to $\{1, 2, f\}$, and by c_n, it goes either to $\{1\}$ or to $\{1, n\}$. All the resulting sets are in $\mathcal{R}$ by (C2). Thus no other set is reachable in C'. Since C satisfies (C1), automata C and C' do not have equivalent states. Hence the DFA C is minimal, and the star of the language $L(C)$ requires exactly 2α states.

(3) To get an $(n+1)$-state DFA D for a language requiring $2\alpha + 1$ states for its star, we add a new state n going to itself on each symbol of Σ_n, and transitions on all two new symbols a_n, d_n. The transitions on a_n are the same as above, so the DFA D satisfies (C1). The transitions on d_n are as follows:

$1 \cdot d_n = n$ and $q \cdot d_n = 1$ if $q \neq 1$.

Construct the NFA D^* for the language $L(D)^*$. Now consider the following family of $2\alpha + 1$ sets

$$\mathcal{R} = \{S \mid S \text{ is reachable in } A'\} \cup \{S \cup \{n\} \mid S \text{ is reachable in } A'\} \cup \{\{n\}\};$$

recall that the empty set is not reachable in A'. Each reachable set in A' is also reachable in the subset automaton D' of the NFA D^*. Next, state $\{1, f\}$, which is reachable in A' by (C2), goes to $\{1, n\}$ by d_n. From $\{1, n\}$, each state $S \cup \{n\}$ with S reachable in A' is reachable in D'. The set $\{n\}$ is reached from $\{1\}$ by d_n. Thus D' satisfies (C2). To prove that no other set is reachable in D', we only need to show that each set R in the family $\mathcal{R}$ goes to a set in $\mathcal{R}$ by each symbol in $\Sigma_n = \Sigma_{n-1} \cup \{a_n, d_n\}$. This is straightforward for symbols in Σ_n. Let S be a reachable state in A'. Then S goes to $\{2\}$ by a_n, and it goes either to $\{n\}$ or to $\{1, n\}$ by d_n. All the resulting sets are in $\mathcal{R}$. Now let $R = S \cup \{n\}$, where S is

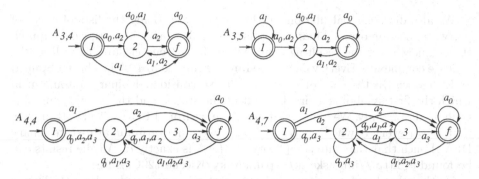

Fig. 3. The basic automata $A_{n,\alpha}$

reachable in A', thus S is non-empty. By a_n, the set R goes to $\{1, 2, f\}$ which is in $\mathcal{R}$. By d_n, it goes either to $\{1\}$ or to $\{n\}$ or to $\{1, n\}$, all of which are in $\mathcal{R}$. The set $\{n\}$ goes to $\{1, f\}$ by a_n, and it goes to $\{1\}$ by d_n. Both the resulting sets are in $\mathcal{R}$. Thus D' has exactly $2\alpha + 1$ reachable states. Since D satisfies (C1), it is minimal, and the star of the language $L(C)$ requires $2\alpha + 1$ states.

Hence if all the values from n to $3/4 \cdot 2^n - 1$ are attainable as the complexity of stars of minimal n-state DFAs, then all the complexities from $2n$ to $3/4 \cdot 2^{n+1} - 1$ are attainable for minimal $(n + 1)$-state DFAs using the second and the third construction. The first construction gives the values from $n + 1$ to $2n - 1$. This completes the induction step.

Now, let us deal with the basis. We describe the basic automata for $n = 3$ and $\alpha = 4, 5$, as well as for $n = 4$ and $\alpha = 4, 7$. All these automata will satisfy conditions (C1) and (C2). From these automata, using our constructions we can get all the DFAs with $n = 4$ and $4 \le \alpha \le 11$, which proves the basis.

The basic automata $A_{n,\alpha}$ for languages $L_{n,\alpha}$ with $sc(L_{n,\alpha}) = n$ and $sc(L_{n,\alpha}^*) = \alpha$ are shown in Fig. 3. All four DFAs satisfy (C1). The complexity of their star, as well as the validity of condition (C2), can be verified using a software, for example, JFLAP [1].   □

The next two propositions recall the known facts that the remaining values in the range from 1 to $3/4 \cdot 2^n$ can be attained by unary or binary languages.

Proposition 1 ([2,8]). *Let $n \ge 2$ and $1 \le \alpha \le n$. There exists a language over an alphabet of size at most two such that $sc(L) = n$ and $sc(L^*) = \alpha$.*   □

Proposition 2 ([10,11,14]). *Let $n \ge 2$. There exists a binary language L such that $sc(L) = n$ and $sc(L^*) = 3/4 \cdot 2^n$.*   □

All the results of this section are summarized in the following theorem.

Theorem 1. *Let $n \ge 2$ and $1 \le \alpha \le 3/4 \cdot 2^n$. There exists a regular language L over an alphabet Σ_n with $|\Sigma_n| \le 2n$ such that $sc(L) = n$ and $sc(L^*) = \alpha$.*   □

We also did some calculations in the binary case. Using the lists of codes of pairwise non-isomorphic binary automata of 2, 3, 4, and 5 states, we computed the frequencies of the resulting complexities for Kleene closure, as well as the average complexity. Every value in the range from 1 to $3/4 \cdot 2^n$ has been obtained at least once. In the case of $n = 6, 7, 8, 9$, we considered binary automata, in which the first symbol is a circular shift of the states, and the second symbol is generated randomly. We found out that all values from 1 to $3/4 \cdot 2^n$ are attainable, that is, for every α with $1 \le \alpha \le 3/4 \cdot 2^n$, we found a minimal n-state binary DFA A such that the state complexity of $L(A)^*$ is exactly α. All the results can be found at http://im.saske.sk/$\sim$ palmovsky/Kleene%20Closure.

On the other hand, the situation is different in the unary case. Here the known upper bound is $(n-1)^2 + 1$, and it is tight [14]. However, as shown in [2], there are at least two gaps of length n in the range from 1 to $(n-1)^2 + 1$ that cannot be attained by the state complexity of the star of any unary language with state complexity n.

3   Kleene Closure on Prefix-Free Languages

If $w = uv$ for a strings u and v, then u is a *prefix* of w. If, moreover, the string v is non-empty, then u is a *proper prefix* of w. A language is *prefix-free* if it does not contain two distinct strings, one of which is a prefix of the other. A DFA is prefix-free if it accepts a prefix-free language. The following characterization of minimal DFAs accepting prefix-free languages is well known.

Proposition 3 ([4]). *Let A be a minimal DFA for a non-empty language L. Then L is prefix-free if and only if A has exactly one final state that goes to the dead state on each symbol of the input alphabet.* □

Using this characterization, a DFA A^* for the star of a prefix-free language L, accepted by an n-state DFA $A = (Q, \Sigma, \cdot, s, \{f\})$, can be constructed as follows. We make the final state f initial, and redirect the transition on each symbol a from the final state f to the state $s \cdot a$. This gives an n-state DFA A^* for the language L^* [4]. The aim of this section is to show that the resulting complexity of L^* may be $n-2, n-1$ or n. Let us start with the following observation.

Lemma 2. *Let A be a minimal prefix-free DFA with the final state f and the dead state d. Let p and q be two distinct states different from d. Then p and q can be distinguished by a string w such that the computations of A on the string w starting in the states p and q do not use any transition from f to d.*

Proof. Let $\cdot$ be the transition function of A. Let a string w be accepted from p and rejected from q. Then the computation on w from p cannot use any transition from f to d, otherwise w would be rejected from p. If the computation on w from q uses a transition from f to d on a symbol a, then w can be factorized as $w = uav$ such that $q \cdot u = f$, $f \cdot a = d$, and $d \cdot v = d$. Hence u is accepted from q. Consider the computation on u from p. Since u is a proper prefix of w, this computation is rejecting, and does not use any transition from f to d. Thus u is the desired string, and the proof is complete. □

Now we are ready to get a lower bound on the state complexity of the star of prefix-free languages.

Lemma 3. *Let L be a prefix-free regular language with $\mathrm{sc}(L) = n$, where $n \geq 3$. Then $n - 2 \leq \mathrm{sc}(L^*) \leq n$.*

Proof. Let $A = (\{1, 2, \ldots, n - 2, f, d\}, \Sigma, \cdot, 1, \{f\})$ be the minimal DFA for a prefix-free language L with the dead state d. Since L is prefix-free, all the transitions from the final state f go to the dead state d.

To get a DFA A^* for the language L^* from the DFA A [4], we make the state f initial, and redirect the transition from f to d on each symbol a in Σ to the state $1 \cdot a$. In the resulting DFA A^*, the states 1 and d may be unreachable, while any other state is reachable since A is minimal.

First, consider the case that the state 1 is reachable in A^*. Then all the reachable states are pairwise distinguishable since d is the only state that does not accept any string, f is the only final state, and the remaining states are distinguishable in the DFA A^* by Lemma 2.

Now, assume that the state 1 is unreachable in A^*, that is, the state 1 does not have any in-transition in the DFA A. It follows that no out-transition from the state 1 can be used to show the distinguishability of the states $2, 3, \ldots, n - 2$ in DFA A, and so these states are distinguishable in the DFA A^* by Lemma 2. This completes the proof. $\qquad\square$

Lemma 4. *Let $n \geq 4$ and $n - 2 \leq \alpha \leq n$. There exists a binary prefix-free language L such that $\mathrm{sc}(L) = n$ and $\mathrm{sc}(L^*) = \alpha$.*

Proof. The minimal DFAs of prefix-free regular languages whose stars meet the complexities $n, n - 1$, and $n - 2$ are shown in Fig. 4 (top, middle, and bottom, respectively). $\qquad\square$

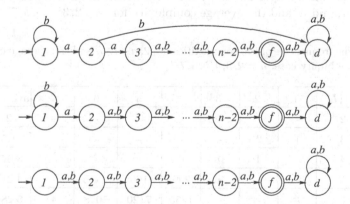

Fig. 4. The DFAs of the prefix-free languages meeting the complexities n (top), $n - 1$ (middle), and $n - 2$ (bottom) for their star

Let us denote by $R_k(n)$ the set of possible complexities of the star of prefix-free languages with state complexity n over a k-letter alphabet, that is,

$$R_k(n) = \{\mathrm{sc}(L^*) \mid L \subseteq \Sigma^*, |\Sigma| = k, L \text{ is prefix-free and } \mathrm{sc}(L) = n\}.$$

Using this notation, we get the following result.

Theorem 2. *Let $R_k(n)$ be the set of possible complexities of the star of prefix-free languages over a k-letter alphabet, as defined above. Then we have*

(a) $R_k(1) = R_k(2) = \{2\}$ where $k \geq 1$;
(b) $R_1(n) = \{n - 2\}$ where $n \geq 3$;
(c) $R_2(3) = \{1, 2\}$ and $R_k(3) = \{1, 2, 3\}$ if $k \geq 3$;
(d) $R_k(n) = \{n - 2, n - 1, n\}$ if $k \geq 2$ and $n \geq 4$.

Proof. (a) The only prefix-free languages with state complexity 1 and 2 are the empty language and the language $\{\varepsilon\}$, respectively. The star of both these languages is $\{\varepsilon\}$.

(b) The only prefix-free unary language with the state complexity n, where $n \geq 3$, is the language $\{a^{n-2}\}$. Its star is the language $(a^{n-2})^*$ with the state complexity $n - 2$.

(c) Using a binary alphabet, we cannot reach the initial and the dead state of a 3-state DFA A from the initial state. Therefore, in the DFA A^*, at most two states are reachable. The languages b^*a and $a + b$ meet the complexities 2 and 1, respectively. The language b^*a over the ternary alphabet $\{a, b, c\}$ meets complexity 3.

(d) The equality is given by Lemma 3 and Lemma 4. $\square$

We also did some computations. Having the lists of binary minimal and pairwise non-isomorphic prefix-free DFAs, we computed the complexities of the stars of the accepted languages. The table below shows the frequencies of complexities $n - 2, n - 1$, and n, and the average complexity for $n = 2, 3, 4, 5, 6, 7$.

Table 1. The frequencies of the complexities $n-2, n-1$, n, and the average complexity of star in the binary case; $n = 2, 3, 4, 5, 6, 7$

$n \backslash \mathrm{sc}(L^*)$	1	2	3	4	5	6	7	average
2	-	1	-	-	-	-	-	2
3	1	2	-	-	-	-	-	1.5
4	-	4	18	6	-	-	-	3.071
5	-	-	56	299	166	-	-	4.211
6	-	-	-	1255	7120	5078	-	5.284
7	-	-	-	-	37 733	222 125	184 182	6.600

4   Conclusions

We investigated the operation of Kleene closure (star) on regular and prefix-free languages. In the case of regular languages, we obtained the contiguous range of complexities from 1 to $3/4 \cdot 2^n$ for an alphabet of size $2n$. We proved that for all n and α with $1 \le \alpha \le 3/4 \cdot 2^n$, there exists a regular language defined over an alphabet of size at most $2n$ with the state complexity n such that the state complexity of its star is α. This improves a similar result from [8] that uses an alphabet that grows exponentially with n.

We did some computations in the binary case, and we obtained a contiguous range of complexities of stars from 1 to $3/4 \cdot 2^n$ for all n with $2 \le n \le 9$. Whether or not this is true for every n remains open.

In the second part of the paper, we examined a similar problem for prefix-free languages. We showed that the state complexity of the star of a prefix-free language with state complexity n may attain just three values $n-2, n-1$, and n.

References

1. JFLAP, http://www.jflap.org/
2. Čevorová, K.: Kleene star on unary regular languages. In: Jurgensen, H., Reis, R. (eds.) DCFS 2013. LNCS, vol. 8031, pp. 277–288. Springer, Heidelberg (2013)
3. Geffert, V.: Magic numbers in the state hierarchy of finite automata. Inform. Comput. 205, 1652–1670 (2007)
4. Han, Y.-S., Salomaa, K., Wood, D.: Operational state complexity of prefix-free regular languages. In: Automata, Formal Languages, and Related Topics, pp. 99–115. Institute of Informatics, University of Szeged (2009)
5. Iwama, K., Kambayashi, Y., Takaki, K.: Tight bounds on the number of states of DFAs that are equivalent to n-state NFAs. Theoret. Comput. Sci. 237, 485–494 (2000)
6. Jirásek, J., Jirásková, G., Szabari, A.: State complexity of concatenation and complementation. Internat. J. Found. Comput. Sci. 16, 511–529 (2005)
7. Jirásková, G.: State complexity of some operations on binary regular languages. Theoret. Comput. Sci. 330, 287–298 (2005)
8. Jirásková, G.: On the state complexity of complements, stars, and reversals of regular languages. In: Ito, M., Toyama, M. (eds.) DLT 2008. LNCS, vol. 5257, pp. 431–442. Springer, Heidelberg (2008)
9. Jirásková, G.: Magic numbers and ternary alphabet. Internat. J. Found. Comput. Sci. 22, 331–344 (2011)
10. Jirásková, G., Palmovský, M.: Kleene closure and state complexity. In: Vinar, T. (ed.) ITAT 2013, vol. 2013, pp. 94–100 (2013), CEUR-WS.org
11. Maslov, A.N.: Estimates of the number of states of finite automata. Soviet Math. Doklady 11, 1373–1375 (1970)
12. Sipser, M.: Introduction to the theory of computation. PWS Publishing Company, Boston (1997)
13. Yu, S.: Regular languages. In: Rozenberg, G., Salomaa, A. (eds.) Handbook of Formal Languages, vol. I, pp. 41–110. Springer, Heidelberg (1997)
14. Yu, S., Zhuang, Q., Salomaa, K.: The state complexity of some basic operations on regular languages. Theoret. Comput. Sci. 125, 315–328 (1994)

Left is Better than Right for Reducing Nondeterminism of NFAs

Sang-Ki Ko and Yo-Sub Han

Department of Computer Science, Yonsei University
50, Yonsei-Ro, Seodaemun-Gu, Seoul 120-749, Korea
{narame7,emmous}@cs.yonsei.ac.kr

Abstract. We study the NFA reductions by invariant equivalences. It is well-known that the NFA minimization problem is PSPACE-complete. Therefore, there have been approaches to reduce the size of NFAs in low polynomial time by computing invariant equivalence and merging the states within same equivalence class. Here we consider the nondeterminism reduction of NFAs by invariant equivalences. We, in particular, show that the left-invariant equivalence is more useful than the right-invariant equivalence for reducing NFA nondeterminism. We also present experimental evidence for showing that NFA reduction by left-invariant equivalence achieves the better reduction of nondeterminism than right-invariant equivalence.

Keywords: Nondeterministic finite automata, Regular expression, NFA reduction, Invariant equivalences.

1 Introduction

Regular expressions are widely used for many applications such as search engine, text editor, programming language, and so on. People often use regular expressions to describe a set of pattern strings for the pattern matching problem.

Once a regular expression is given, then we convert a regular expression into an equivalent nondeterministic finite-state automaton (NFA) by automata constructions such as Thompson construction [21] or the position construction[1] [6,17]. In some cases, the obtained NFA should be converted into a deterministic one by the subset construction. However, the size of the deterministic finite-state automaton (DFA) for the regular expression may be exponential. In addition to that, the problem of minimizing NFAs is PSPACE-complete [14], thus, intractable.

Since DFAs are usually much faster than NFAs, the most of applications prefer DFAs to NFAs. For example, consider the *membership problem* which is the simplest form of pattern matching problem based on FAs. Given an FA of size m and a string of length n, the problem requires $O(n)$ time if the FA is deterministic whereas it takes $O(m^2 n)$ time [7,22] in the worst-case if the FA is nondeterministic.

---

[1] Also known as Glushkov construction.

M. Holzer and M. Kutrib (Eds.): CIAA 2014, LNCS 8587, pp. 238–251, 2014.
© Springer International Publishing Switzerland 2014

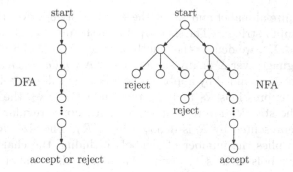

Fig. 1. Difference between deterministic and nondeterministic computations

The real problem is, it is impossible to have small DFAs as NFAs for the same regular languages. It is well known that exponential number of states may be required for an NFA to be represented by a DFA. As an alternative solution, there have been many approaches on NFA reduction techniques for the space-efficient implementations of the applications using regular expressions.

The idea of reducing the size of NFAs by equivalence relations was first proposed by Ilie and Yu [11]. Champarnaud and Coulon [5] modified the idea to use preorders over the set of states instead of equivalences for the better reduction. Later, Ilie et al. [9] showed that it is possible to reduce the size of an NFA with n states and m transitions in $O(m \log n)$ time by equivalences and $O(mn)$ time by preorders. Ilie et al. [10] also showed that the optimal use of equivalences can be computed in polynomial time and the optimal use of preorders is NP-hard.

Here we consider the problem of reducing the nondeterminism of NFAs by using invariant equivalences because the nondeterminism is also a very important factor for the efficient simulation of NFAs. We define the *computation graph* for estimating the nondeterminism of NFAs and investigate several properties. Then, we compare the right- and left-invariant equivalences by reducing NFAs by the equivalences and give experimental results with uniformly generated random regular expressions.

The paper is organized as follows. In Section 2, we shall give some definitions and notations. We introduce the well-known construction of the position automaton from a regular expression in Section 3. We present NFA reduction by invariant equivalences in Section 4 and consider the nondeterminism of NFAs in Section 5. The experimental results are given in Section 6. Section 7 concludes the paper.

2 Preliminaries

Here we briefly recall the basic definitions used throughout the paper. For complete background knowledge in automata theory, the reader may refer to textbooks [7,22].

Let Σ be a finite alphabet and Σ^* be the set of all strings over the alphabet Σ including the empty string λ. The size $|\Sigma|$ of Σ is the number of characters in Σ. For a string $w \in \Sigma^*$, we denote the length of w by $|w|$ and the ith character of w by w_i. A language over Σ is any subset of Σ^*. A *regular expression* over Σ is $\emptyset, \lambda$, or $a \in \Sigma$, or is obtained by applying the following rules finitely many times. For two regular expressions $\mathcal{R}_1$ and $\mathcal{R}_2$, the union $\mathcal{R}_1 + \mathcal{R}_2$, the concatenation $\mathcal{R}_1 \cdot \mathcal{R}_2$, and the star $\mathcal{R}_1^*$ are regular expressions. For a regular expression $\mathcal{R}$, the language represented by $\mathcal{R}$ is denoted by $\mathcal{L}(\mathcal{R})$. The size $|\mathcal{R}|$ of a regular expression $\mathcal{R}$ implies the number of symbols including the characters from Σ and syntactic symbols such as $+, \cdot$, and $*$. We denote the number of occurrences of characters from Σ in $\mathcal{R}$ by $|\mathcal{R}|_\Sigma$.

A *nondeterministic finite-state automaton* (NFA) $\mathcal{A}$ is specified by a 5-tuple $(Q, \Sigma, \delta, s, F)$, where Q is a finite set of states, Σ is an input alphabet, $\delta :$ $Q \times \Sigma \to 2^Q$ is a multi-valued transition function, $s \in Q$ is the initial state and $F \subseteq Q$ is a set of final states.

For a transition $q \in \delta(p, a)$ in $\mathcal{A}$, we say that p has an *out-transition* and q has an *in-transition*. Furthermore, p is a *source state* of q and q is a *target state* of p. The transition function δ can be extended to a function $Q \times \Sigma^* \to 2^Q$ that reflects sequences of inputs. A string w over Σ is accepted by $\mathcal{A}$ if there is a labeled path from s to a state in F such that this path spells out the string w, namely, $\delta(s, w) \cap F \neq \emptyset$. The language $\mathcal{L}(\mathcal{A})$ recognized by $\mathcal{A}$ is the set of all strings that are spelled out by paths from s to a final state in F. Formally we write

$$\mathcal{L}(\mathcal{A}) = \{w \in \Sigma^* \mid \delta(s, w) \cap F \neq \emptyset\}.$$

For a state $q \in Q$, we denote

$$\mathcal{L}_L(A, q) = \{w \in \Sigma^* \mid q \in \delta(s, w)\}, \quad \mathcal{L}_R(A, q) = \{w \in \Sigma^* \mid \delta(q, w) \cap F \neq \emptyset\};$$

when A is understood from the context, we simply write $\mathcal{L}_L(q), \mathcal{L}_R(q)$, respectively.

For a state $q \in Q$ and a string $w \in \Sigma^*$, the *q-computation tree* $T_{A,q,w}$ of $\mathcal{A}$ on w is a labeled tree where the nodes are labeled by elements of $Q \times (\Sigma \cup \{\lambda, \natural\})$, where $\natural \notin \Sigma$. Note that $T_{A,q,\lambda}$ is a single-node tree labeled by (q, λ). Assume that $w = au$, where $a \in \Sigma$, $u \in \Sigma^*$, and $\delta(q, a) = \emptyset$. Then, $T_{A,q,w}$ is again a single-node tree where the only node is labeled by $(q, \natural)$. If $\delta(q, a) = \{p_1, \ldots, p_m\}$, where $m \geq 1$, then $T_{A,q,w}$ is the tree with the root node labeled by (q, a) and the root node has m children where the subtree rooted at the ith child is $T_{A,p_i,u}$ for $i = 1, \ldots, m$. We call the tree $T_{A,s,w}$ the *computation tree* of $\mathcal{A}$ on w and simply denote $T_{A,w}$. If there is an accepting computation for w in the NFA $\mathcal{A}$, $T_{A,w}$ has a leaf labeled by (q, λ), where $q \in F$.

We also define the *computation graph* $G_{A,w}$ of $\mathcal{A}$ on w by merging equivalent subtrees of the computation tree as a single subtree. If $T_{A,w}$ has two computation trees $T_{A,q,v}$ as subtrees, where $w = uv$, $w, u, v \in \Sigma^*$, and $q \in Q$, then we merge the trees into one.

We denote the number of nodes and the number of edges of a computation tree $T_{A,w}$ by $|T_{A,w}|_N$ and $|T_{A,w}|_E$, respectively. We define the size $|T_{A,w}|$ of a computation tree $T_{A,w}$ to be $|T_{A,w}|_N + |T_{A,w}|_E$. Note that the similar notations are defined analogously for the size of computation graph.

3 NFA Constructions from Regular Expressions

We first recall the well-known construction called the *position construction* for obtaining NFAs from regular expressions [6,17]. The automaton obtained from the construction is called the *position automaton* which is also called the *Glushkov automaton*.

Given a regular expression $\mathcal{R}$, we first mark each character of $\mathcal{R}$ with a unique index called the position. From the leftmost character of $\mathcal{R}$, we mark the index of each character with the number from 1 to $|\mathcal{R}|_\Sigma$. The set of indices is called the *positions* of $\mathcal{R}$ and denoted by $\mathsf{pos}(\mathcal{R}) = \{1, 2, \ldots, |\mathcal{R}|_\Sigma\}$. We also denote $\mathsf{pos}_0(\mathcal{R}) = \mathsf{pos}(\mathcal{R}) \cup \{0\}$. We denote the marked regular expression obtained from $\mathcal{R}$ by $\overline{\mathcal{R}}$. Note that $\mathcal{L}(\overline{\mathcal{R}}) \subseteq \overline{A}^*$, where $\overline{A} = \{a_i \mid a \in \Sigma, 1 \le i \le |\mathcal{R}|_\Sigma\}$. For instance, if $\mathcal{R} = abc + d(ef)^*$, then $\overline{\mathcal{R}} = a_1 b_2 c_3 + d_4(e_5 f_6)^*$. For $a \in \Sigma$, $a = \overline{\overline{a}}$.

For a regular expression $\mathcal{R}$, we define first, last, and follow as follows:

$$\mathsf{first}(\mathcal{R}) = \{i \mid a_i w \in L(\overline{\mathcal{R}})\},$$

$$\mathsf{last}(\mathcal{R}) = \{i \mid w a_i \in L(\overline{\mathcal{R}})\},$$

$$\mathsf{follow}(\mathcal{R}, i) = \{j \mid u a_i a_j v \in L(\overline{\mathcal{R}})\}.$$

We extend $\mathsf{follow}(\mathcal{R}, 0) = \mathsf{first}(\mathcal{R})$ and define $\mathsf{last}_0(\mathcal{R})$ to be $\mathsf{last}(\mathcal{R})$ if $\lambda \in L(\mathcal{R})$ and $\mathsf{last}(\mathcal{R}) \cup \{0\}$ otherwise.

Then, the position automaton of $\mathcal{R}$ is defined as follows:

$$\mathcal{A}_{\mathsf{pos}}(\mathcal{R}) = (\mathsf{pos}_0(\mathcal{R}), \Sigma, \delta_{\mathsf{pos}}, 0, \mathsf{last}_0(\mathcal{R})),$$

where

$$\delta_{\mathsf{pos}} = \{(i, a, j) \mid j \in \mathsf{follow}(\mathcal{R}, i), a = \overline{a_j}\}.$$

Notice that the position automaton of $\mathcal{R}$ recognizes the same language with the regular expression $\mathcal{R}$, that is, $\mathcal{L}(\mathcal{R}) = \mathcal{L}(\mathcal{A}_{\mathsf{pos}}(\mathcal{R}))$.

The position automaton has two useful properties as follows.

Property 1. The position automaton for the regular expression $\mathcal{R}$ has always $|\mathcal{R}|_\Sigma + 1$ states.

Proof. Recall that every position automaton satisfies Property 2. The NFA $\mathcal{A}_{\mathsf{pos}}(\mathcal{R})/_{\equiv_L}$ is obtained from $\mathcal{A}_{\mathsf{pos}}(\mathcal{R})$ by merging states if they are in the same left-invariant equivalence class. In other words, any two states q and p are merged if $p \equiv_L q$, thus, $\mathcal{L}_L(p) = \mathcal{L}_L(q)$. This implies that p and q should have in-transitions consuming the same character because otherwise $\mathcal{L}_L(p) \ne \mathcal{L}_L(q)$. Therefore, the merged state by the left-invariant equivalence $\equiv_L$ also has in-transitions labeled by the same character. □

Property 1 guarantees that the position automaton always has smaller number of states than Thompson's automaton [21].

Property 2. All in-transitions for any state of the position automaton are labeled by the same character.

Caron and Ziadi [4] named the second property as the *homogeneous* property. We say that an FA is homogeneous if all in-transitions to a state have the same label. The homogeneous property helps to improve the regular expression search algorithms because we can represent the DFA using $O(2^{|\mathcal{R}|_\Sigma} + |\Sigma|)$ bit-masks of length $|\mathcal{R}_\Sigma|$ instead of $O(2^{|\mathcal{R}|_\Sigma} \cdot |\Sigma|)$ [18,19]. We can compute the position automaton in quadratic time in the size of regular expression using inductive definition of first, last, and follow [3].

Note that there have been proposed two more algorithms for obtaining smaller NFAs than position automata from regular expressions. Antimirov [2] introduced an NFA construction based on partial derivatives called the *partial derivate automaton*. Ilie and Yu [12] constructed an NFA called the *follow automaton* based on the follow relation. It is already proven that a partial derivative automaton and a follow automaton are quotients of the position automaton and always smaller than the position automaton.

4   NFA Reduction by Invariant Equivalences

There have been many results for reducing the size of NFAs by using invariant equivalences [9,11,13]. Here we briefly recall how the reduction works.

Basically, the idea of NFA reduction is from DFA minimization in the sense that we find indistinguishable states and merge them to reduce the size of DFAs. Let $\mathcal{A} = (Q, \Sigma, \delta, s, F)$ be an NFA. For any two states p and q of $\mathcal{A}$, we say that p and q are *distinguishable* if there exists a string w such that $\delta(q, w) \cap F \neq \emptyset$ and $\delta(p, w) \cap F = \emptyset$. Naturally, this leads to the fact that p and q are *indistinguishable* if and only if $\mathcal{L}_R(p) = \mathcal{L}_R(q)$. If $\equiv$ is an equivalence on Q which is right-invariant with respect to $\mathcal{A}$, then $p \equiv q$ implies that p and q are indistinguishable.

The largest right-invariant equivalence relation $\equiv_R$ over Q should satisfy the following properties:

(i) $\equiv_R \cap (F \times (Q - F)) = \emptyset$,
(ii) for any $p, q \in Q, a \in \Sigma$, $p \equiv_R q$ if for all $q' \in \delta(q, a)$, there exists $p' \in \delta(p, a)$ such that $q' \equiv_R p'$.

After computing $\equiv_R$, we can reduce the NFA $\mathcal{A}$ by simply merging all states in the same equivalence class. Given an equivalence $\equiv$ and an NFA $\mathcal{A}$, we denote the NFA obtained after merging the equivalent states by $\mathcal{A}/_\equiv$. For any regular expression $\mathcal{R}$, $\mathcal{A}(\mathcal{R})/_{\equiv_R}$ is always smaller than the partial derivative automaton and the follow automaton since $\equiv_R$ is the largest one among all the right-invariant equivalence relations.

Note that the largest left-invariant equivalence relation $\equiv_L$ can be computed by reversing the given NFA and computing the largest right-invariant equivalence of it. Although the partial derivative automaton [2] and the follow automaton [12] can be obtained by the right-invariant equivalence relations, the left-invariant equivalence has some nice properties.

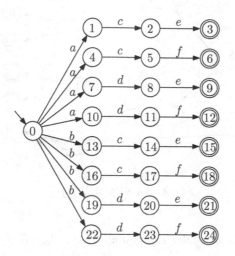

Fig. 2. A position automaton $\mathcal{A}_{pos}(\mathcal{R})$ where $\mathcal{R} = ace + acf + ade + adf + bce + bcf + bde + bdf$

See the position automaton $\mathcal{A}_{pos}(\mathcal{R})$ in Fig. 2 as an inspiring example. Note that this example is already used in the paper by Ilie and Yu [13]. Fig. 3 is an NFA reduced by $\equiv_R$ and Fig. 4 is an NFA reduced by $\equiv_L$.

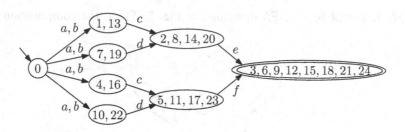

Fig. 3. An NFA $\mathcal{A}_{pos}(\mathcal{R})_{\equiv_R}$

While the number of states is smaller when reduced by $\equiv_R$ than $\equiv_L$, the NFA reduced by $\equiv_L$ is deterministic. Note that the NFA reduction by $\equiv_L$ does not always produce DFAs. However, this example implies that the left-invariant equivalence is useful for reducing nondeterminism of NFAs since the equivalence relation is computed in the same direction as the simulation of NFAs.

We also mention that the NFA reduction by the left-invariant equivalence preserves the homogeneous property which is very useful for the regular expression search algorithms.

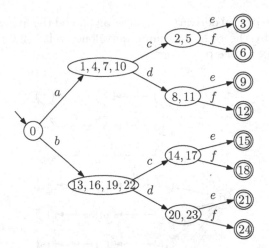

Fig. 4. An NFA $\mathcal{A}_{\mathsf{pos}}(\mathcal{R})_{\equiv_L}$

Lemma 1. *Let $\mathcal{R}$ be a regular expression. Then, $\mathcal{A}_{\mathsf{pos}}(\mathcal{R})/_{\equiv_L}$ is homogeneous.*

5   Nondeterminism of NFAs

Many researchers have studied various measures of nondeterminism in NFAs [8,20]. Here we compare the nondeterminism of NFAs using the size of the computation graph. We give an example for the comparison of the computation tree and the computation graph.

Example 1. Let $\mathcal{A}$ be an NFA described in Fig. 5. Then, the computation tree

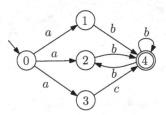

Fig. 5. An NFA $\mathcal{A}$

and the computation graph of $\mathcal{A}$ on the string *abbb* are depicted in Fig. 6.

Now let us discuss the reason why we compare the nondeterminism of NFAs using the computation graph instead of the computation tree. Let $\mathcal{A}$ be an NFA of size m and w be a string of length n. Hromkovič et al. [8] showed that the number of accepting computation can be exponential in the worst-case.

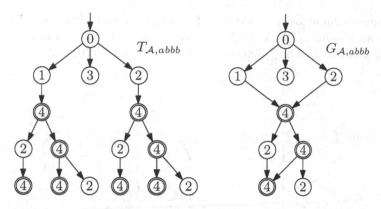

Fig. 6. The computation tree and the computation graph of the NFA $\mathcal{A}$ on the string $aaab$

This follows that the size of the computation tree can be also exponential since a computation tree has corresponding leaves for each accepting computation. This implies that the size of the computation tree can be quite different with the runtime complexity for simulating $\mathcal{A}$ on w, which is $O(m^2 n)$ in the worst-case. On the other hand, the size of computation graph almost coincides with the algorithmic complexity of NFA simulation.

Lemma 2. *Given an NFA $\mathcal{A}$ with m states and a string $w \in \Sigma^*$ of length n, the following statements hold:*

(i) the number of nodes and edges in the computation graph of $\mathcal{A}$ on w are at most $mn + 1$ and $m + m^2(n-1)$, respectively, and

(ii) the number of nodes and edges in the computation tree of $\mathcal{A}$ on w are at most $\frac{m^{n+1}-1}{m-1}$ and $\frac{m^{n+1}-1}{m-1} - 1$, respectively.

Proof. We first prove (i). Since we assume that $\mathcal{A}$ has only one initial state, the simulation of $\mathcal{A}$ starts with one state. After then, $\mathcal{A}$ may simulate all states in the worst-case because of the nondeterminism. Therefore, the number of nodes in the computation graph of an NFA can be $mn + 1$ in the worst-case.

Moreover, the graph can have m edges from the initial node since $\mathcal{A}$ can move to all states by reading a character in the worst-case. Then, since each state of $\mathcal{A}$ can have m options to move by reading a character, the number of edges in the computation graph of $\mathcal{A}$ can be $m + m^2(n-1)$ in the worst-case.

Now we prove (ii). The computation tree of $\mathcal{A}$ on w has one initial node and has m children by reading any character. Then, every node of the computation tree can have m children since there can be up to m transitions for each state and character. Thus, the total number of nodes can be

$$1 + m + m^2 + \cdots + m^{n-1} + m^n = \frac{m^{n+1} - 1}{m - 1}.$$

Note that the number of edges can be $\frac{m^{n+1}-1}{m-1} - 1$ since a tree always has $t - 1$ edges if there are t nodes. $\qquad\square$

We have a simple lower bound NFA for the given upper bounds in Lemma 2, which is an m-state NFA $\mathcal{A} = (Q, \Sigma, \delta, s, F)$, where $Q = F$ and $Q = \delta(q, a)$ for all $q \in Q$ and $a \in \Sigma$. Fig. 7 depicts the lower bound when $m = 2$.

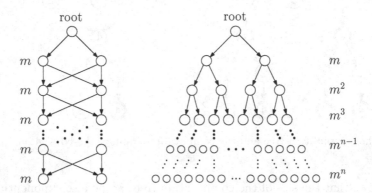

Fig. 7. A lower bound example when $m = 2$ for the upper bound in Lemma 2

We establish the following result as a corollary.

Corollary 1. *Given an NFA $\mathcal{A}$ with m states and a string $w \in \Sigma^*$ of length n, $|G_{\mathcal{A},w}|$ is in $O(m^2 n)$ and $|T_{\mathcal{A},w}|$ is in $O(m^n)$.*

However, the size of the computation graph and tree is linear in the length of the input string for DFAs in the worst-case.

Lemma 3. *Let $\mathcal{A} = (Q, \Sigma, \delta, s, F)$ be a DFA and $w \in \Sigma^*$ be a string. Then, $|T_{\mathcal{A},w}|_N \leq |w| + 1$ and $|T_{\mathcal{A},w}|_E \leq |w|$.*

Proof. Since $\mathcal{A}$ is a DFA, we have only one transition from any state of $\mathcal{A}$ to proceed by reading an input character. Therefore, we always have a sequence of states visited by reading the string w instead of tree structure. If $w = \lambda$, then $|T_{\mathcal{A},w}| = 1$ since $T_{\mathcal{A},\lambda}$ consists of a single node labeled by (s, λ). Otherwise, we have an accepting computation

$$s \to q_1 \to q_2 \to q_3 \to \cdots \to q_{|w|-1} \to q_{|w|}$$

of length $|w| + 1$ in the worst-case, where $\delta(s, w_1) = q_1, \delta(q_i, w_{i+1}) = q_{i+1}$ and $q_i \in Q$ for $i = 1, \ldots, |w|$. Note that the sequence itself is the computation tree in DFAs. If the DFA is incomplete and the computation fails before completed, the length of the computation becomes shorter than $|w| + 1$. □

Corollary 2. *Let $\mathcal{A} = (Q, \Sigma, \delta, s, F)$ be a DFA and $w \in \Sigma^*$ be a string. Then, $|G_{\mathcal{A},w}|_N \leq |w| + 1$ and $|G_{\mathcal{A},w}|_E \leq |w|$.*

Considering that the size of the computation graph represents the computational cost for simulating FAs well, it is evident that DFAs are much better than

NFAs regardless of the size of FAs because they make a deterministic choice at each step.

Here we consider the advantage of reducing NFAs by the left-invariant equivalence. The empirical studies on NFA reductions show that the right-invariant equivalence is more powerful in terms of better reduction on the number of states and transitions. However, it turns out that the left-invariant equivalence can reduce the nondeterminism of NFAs better than the right-invariant.

Lemma 4. *Let $\mathcal{A} = (Q, \Sigma, \delta, s, F)$ be an NFA and $\mathcal{A}'$ be an NFA obtained from $\mathcal{A}$ by merging two equivalent states q and p in Q. Then, there exists a string $w \in \Sigma^*$ such that $|G_{\mathcal{A}',w}|_N < |G_{\mathcal{A},w}|_N$ if and only if $\mathcal{L}_L(p) \cap \mathcal{L}_L(q) \neq \emptyset$.*

Proof.
($\Longleftarrow$) We first prove the statement that if $\mathcal{L}_L(p) \cap \mathcal{L}_L(q) \neq \emptyset$, then there exists a string w such that $|G_{\mathcal{A}',w}|_N < |G_{\mathcal{A},w}|_N$. Since $\mathcal{L}_L(p) \cap \mathcal{L}_L(q) \neq \emptyset$, we say a string $w' \in \mathcal{L}_L(p) \cap \mathcal{L}_L(q)$. Consider the computation graphs $G_{\mathcal{A},w}$ and $G_{\mathcal{A}',w}$, where $w = w'u$ and $w', u \in \Sigma^*$. While the computation of $\mathcal{A}$ on w maintains two states p and q after reading w', the computation of $\mathcal{A}'$ only maintains the merged state. As a result, the number of nodes in the computation graph decrease.
($\Longrightarrow$) We prove that if there exists a string w such that $|G_{\mathcal{A}',w}|_N < |G_{\mathcal{A},w}|_N$, then $\mathcal{L}_L(p) \cap \mathcal{L}_L(q) \neq \emptyset$. The decrease on the number of nodes in the computation graph $G_{\mathcal{A}',w}$ implies that at least one state visited during the simulation of $\mathcal{A}$ on w is merged with the other state. This means that there exists a string w', where $w'u = w$ and $w', u \in \Sigma^*$, which makes $\mathcal{A}$ to visit the merged state. Since the merged states are p and q by assumption, $w' \in \mathcal{L}_L(p) \cap \mathcal{L}_L(q)$. $\square$

From Lemma 4, the following result is immediate.

Theorem 1. *Let $\mathcal{A}$ be an NFA and $\equiv$ be a left-invariant equivalence. If there exist two distinct equivalent states in $\mathcal{A}$, then there exists a string $w \in \Sigma^*$ such that $|G_{\mathcal{A}/\equiv,w}| < |G_{\mathcal{A},w}|$.*

We also observe that NFA reduction by the right-invariant does not guarantee the reduction of nondeterminism.

Corollary 3. *There exist an NFA $\mathcal{A}$ with two distinct equivalent states and a right-invariant equivalence $\equiv$ such that for any string $w \in \Sigma^*$, $|G_{\mathcal{A}/\equiv,w}| < |G_{\mathcal{A},w}|$.*

6   Experimental Results

We present experimental results regarding the NFA reduction by invariant equivalences. Especially, we aim to analyze how the NFA reduction affects the nondeterminism of NFAs. For experiments, we have used uniformly generated random regular expressions by FAdo [1].

FAdo [1] is an ongoing project developed by Almeida et al. that provides a set of formal language manipulation tools. We have used 1,000 uniformly generated regular expression by the FAdo system. Note that the random generation of regular expressions is based on Mairson's work [16] for generating words in a context-free language uniformly. The context-free language used for the random generation of regular expressions is presented by Lee and Shallit [15].

6.1   Size Reduction of NFAs

First we look at the size reduction of NFAs constructed from random regular expressions by invariant equivalences. See Table 1 for the result.

Table 1. The average states/transitions in position automata and reduced NFAs constructed from uniform random regular expressions

$|\mathcal{R}|$	$|\Sigma|$	Number of states/transitions				
		$\mathcal{A}_{\mathsf{pos}}$	$\mathcal{A}_{\mathsf{pos}}/{\equiv_L}$	$\mathcal{A}_{\mathsf{pos}}/{\equiv_R}$	$\mathcal{A}_{\mathsf{pos}}/{\equiv_{LR}}$	$\mathcal{A}_{\mathsf{pos}}/{\equiv_{RL}}$
	2	13.2/21.0	11.6/16.9	9.5/14.0	9.2/13.5	9.0/13.3
20	5	16.3/19.7	15.8/19.0	13.2/16.0	13.0/15.9	13.0/15.8
	10	17.3/19.5	17.1/19.2	14.8/16.9	14.7/16.8	14.7/16.7
	2	29.8/58.3	24.4/41.4	20.2/34.7	19.0/32.4	18.6/31.8
50	5	36.6/51.8	34.7/47.8	28.4/38.5	27.7/38.0	27.6/37.5
	10	40.7/50.1	39.6/48.1	34.1/41.2	33.7/40.8	33.6/40.6
	2	57.9/122.9	45.5/83.8	38.2/70.2	35.5/65.8	34.7/63.1
100	5	71.0/108.4	66.3/97.3	54.7/76.5	53.1/75.3	52.9/73.9
	10	80.0/103.0	77.8/98.9	67.1/81.9	66.1/81.1	65.9/80.4

When we compare the size reduction effects of the right- and left-invariant equivalences, it is obvious that the right-invariant equivalence is superior on average for reducing the size of NFAs from the result.

On the average of all the position automata used in the experiment, the number of states is reduced 8.3% by $\equiv_L$ and 22.7% by $\equiv_R$. The number of transitions is reduced 14.8% by $\equiv_L$ whereas 29.7% is reduced by $\equiv_R$.

We also compare two more reductions, where we reduce the NFAs in both directions. For simplicity, we write $\mathcal{A}_{\mathsf{pos}}/{\equiv_{LR}}$ and $\mathcal{A}_{\mathsf{pos}}/{\equiv_{RL}}$ instead of $(\mathcal{A}_{\mathsf{pos}}/{\equiv_L})/{\equiv_R}$ and $(\mathcal{A}_{\mathsf{pos}}/{\equiv_R})/{\equiv_L}$, respectively. On average, $\equiv_{RL}$ is slightly better in terms of the size reduction of NFAs than $\equiv_{LR}$ since $\equiv_{RL}$ reduces 25.0% of states and 31.6% of transitions whereas $\equiv_{LR}$ reduces 25.5% of states and 32.7% of transitions. However, the difference between the two-way reductions is very small compared to the difference between $\equiv_R$ and $\equiv_L$.

6.2   Reduction of Nondeterminism

For measuring the degree of the practical nondeterminism in NFAs, we use the following definition which can be the measurement of nondeterminism for simulating strings with the NFAs.

Let $\mathcal{A}$ be an NFA and w be a string. Then, we define the *redundancy of simulation* $\mathsf{RS}_{\mathcal{A},w}$ of $\mathcal{A}$ on w as follows:

$$\mathsf{RS}_{\mathcal{A},w} = \frac{|G_{\mathcal{A},w}|_E}{|w|}.$$

Recall that the number of edges in the computation graph $G_{\mathcal{A},w}$ almost coincides with the practical time complexity for simulating w on $\mathcal{A}$. We divide $|G_{\mathcal{A},w}|_E$ by $|w|$ to obtain the redundancy of simulation since $|w|$ is the optimal number of edges in the computation graph for simulating w on any FA $\mathcal{A}$ if $w \in \mathcal{L}(\mathcal{A})$. Therefore, $\mathsf{RS}_{\mathcal{A},w} = 1$ if the given NFA is deterministic or simulates the string w deterministically. We conduct experiments with the randomly generated regular expressions used in the previous experiments.

For generating random strings of the regular expressions, we use a Java library called Xeger[2]. We use random strings from the regular expressions instead of uniformly generated random strings because if the computations fail easily, it is difficult to compare the nondeterminism of NFAs.

Once we choose 1,000 random regular expressions, we convert the regular expressions into position automata and reduce the automata by four different equivalences $\equiv_L$, $\equiv_R$, $\equiv_{LR}$, and $\equiv_{RL}$. Then, we generate 10,000 random strings by Xeger for each regular expression and simulate the five types of automata with the strings.

Table 2 summarizes the result of the experiment. Under the assumption that the redundancy ratio reflects the nondeterminism of NFAs for simulating strings, the best reduction is obtained by $\equiv_{LR}$ in all cases.

Table 2. The average states/transitions in position automata and reduced NFAs constructed from uniform random regular expressions

$|\mathcal{R}|$	$|\Sigma|$	$\mathsf{RS}_{\mathcal{A},w}$				
		$\mathcal{A}_{\mathsf{pos}}$	$\mathcal{A}_{\mathsf{pos}}/{\equiv_L}$	$\mathcal{A}_{\mathsf{pos}}/{\equiv_R}$	$\mathcal{A}_{\mathsf{pos}}/{\equiv_{LR}}$	$\mathcal{A}_{\mathsf{pos}}/{\equiv_{RL}}$
	2	1.625	1.331	1.469	1.296	1.401
20	5	1.103	1.046	1.096	1.046	1.076
	10	1.043	1.009	1.043	1.009	1.024
	2	2.227	1.662	1.946	1.628	1.795
50	5	1.132	1.043	1.122	1.042	1.086
	10	1.062	1.018	1.059	1.018	1.034
	2	2.516	2.149	2.365	2.115	2.314
100	5	1.172	1.066	1.164	1.065	1.112
	10	1.059	1.015	1.057	1.015	1.033

The interesting result is that $\equiv_L$ shows the better reduction than $\equiv_R$ as anticipated in Theorem 1. On average, the redundancy ratio is reduced 12.4%

---

[2] Xeger   generates   a   random   string   from   a   regular   expression. https://code.google.com/p/xeger/

by $\equiv_L$ and 4.8% by $\equiv_R$. This result clearly suggests that the reduction by the left-invariant equivalence is more useful than the right-invariant one to reduce the nondeterminism of NFAs.

One more thing to note is, $\equiv_L$ shows the better result than $\equiv_{RL}$. Recalling that $\equiv_{LR}$ shows better result than $\equiv_R$ in terms of the size reduction of NFAs, this result is noticeable. From the empirical result, $\equiv_{LR}$ can be the best option for reducing the size and the nondeterminism of NFAs at the same time.

7   Conclusions

We have studied the relationship between NFA reductions and nondeterminism of NFAs. The NFA reduction techniques based on the equivalence and preorder relations are well investigated in literature.

Here we have considered the NFA reduction by invariant equivalences. While the most of NFA constructions focus on the right-invariant equivalence for obtaining small NFAs from regular expressions, we have revealed that the reduction by left-invariant equivalence helps to reduce the nondeterminism of NFAs better than the right-invariant equivalence. We have presented empirical results with randomly generated regular expressions.

In future, we aim at comparing the NFA reduction techniques by equivalences and preorders with respect to the nondeterminism of reduced NFAs. Investigating how to optimally reduce the nondeterminism of NFAs is an open problem.

References

1. Almeida, A., Almeida, M., Alves, J., Moreira, N., Reis, R.: FAdo and gUItar: Tools for automata manipulation and visualization. In: Maneth, S. (ed.) CIAA 2009. LNCS, vol. 5642, pp. 65–74. Springer, Heidelberg (2009)
2. Antimirov, V.: Partial derivatives of regular expressions and finite automaton constructions. Theoretical Computer Science 155(2), 291–319 (1996)
3. Brüggemann-Klein, A.: Regular expressions into finite automata. Theoretical Computer Science 120, 197–213 (1993)
4. Caron, P., Ziadi, D.: Characterization of Glushkov automata. Theoretical Computer Science 233(1-2), 75–90 (2000)
5. Champarnaud, J.-M., Coulon, F.: NFA reduction algorithms by means of regular inequalities. Theoretical Computer Science 327(3), 241–253 (2004)
6. Glushkov, V.M.: The Abstract Theory of Automata. Russian Mathematical Surveys 16, 1–53 (1961)
7. Hopcroft, J., Ullman, J.: Introduction to Automata Theory, Languages, and Computation, 2nd edn. Addison-Wesley, Reading (1979)
8. Hromkovič, J., Seibert, S., Karhumäki, J., Klauck, H., Schnitger, G.: Communication complexity method for measuring nondeterminism in finite automata. Information and Computation 172(2), 202–217 (2002)
9. Ilie, L., Navarro, G., Yu, S.: On NFA reductions. In: Karhumäki, J., Maurer, H., Păun, G., Rozenberg, G. (eds.) Theory Is Forever. LNCS, vol. 3113, pp. 112–124. Springer, Heidelberg (2004)

10. Ilie, L., Solis-Oba, R., Yu, S.: Reducing the size of nFAs by using equivalences and preorders. In: Apostolico, A., Crochemore, M., Park, K. (eds.) CPM 2005. LNCS, vol. 3537, pp. 310–321. Springer, Heidelberg (2005)
11. Ilie, L., Yu, S.: Algorithms for computing small NFAs. In: Diks, K., Rytter, W. (eds.) MFCS 2002. LNCS, vol. 2420, pp. 328–340. Springer, Heidelberg (2002)
12. Ilie, L., Yu, S.: Follow automata. Information and Computation 186, 140–162 (2003)
13. Ilie, L., Yu, S.: Reducing NFAs by invariant equivalences. Theoretical Computer Science 306(1-3), 373–390 (2003)
14. Jiang, T., Ravikumar, B.: Minimal NFA problems are hard. SIAM Journal on Computing 22(6), 1117–1141 (1993)
15. Lee, J., Shallit, J.: Enumerating regular expressions and their languages. In: Domaratzki, M., Okhotin, A., Salomaa, K., Yu, S. (eds.) CIAA 2004. LNCS, vol. 3317, pp. 2–22. Springer, Heidelberg (2005)
16. Mairson, H.G.: Generating words in a context-free language uniformly at random. Information Processing Letters 49(2), 95–99 (1994)
17. McNaughton, R., Yamada, H.: Regular expressions and state graphs for automata. IRE Transactions on Electronic Computers 9(1), 39–47 (1960)
18. Navarro, G., Raffinot, M.: Compact DFA representation for fast regular expression search. In: Proceedings of the 5th International Workshop on Algorithm Engineering, pp. 1–12 (2001)
19. Navarro, G., Raffinot, M.: Flexible Pattern Matching in Strings: Practical On-line Search Algorithms for Texts and Biological Sequences. Cambridge University Press, New York (2002)
20. Palioudakis, A., Salomaa, K., Akl, S.G.: Comparisons between measures of nondeterminism on finite automata. In: Jurgensen, H., Reis, R. (eds.) DCFS 2013. LNCS, vol. 8031, pp. 217–228. Springer, Heidelberg (2013)
21. Thompson, K.: Regular expression search algorithm. Communications of the ACM 11(6), 419–422 (1968)
22. Wood, D.: Theory of Computation. Harper & Row (1987)

Analytic Functions Computable
by Finite State Transducers

Petr Kůrka[1] and Tomáš Vávra[2]

[1] Center for Theoretical Study
Academy of Sciences and Charles University in Prague
Jilská 1, 11000 Praha 1, Czechia
[2] Department of Mathematics FNSPE
Czech Technical University in Prague
Trojanova 13, 12000 Praha 2, Czechia

Abstract. We show that the only analytic functions computable by finite state transducers in sofic Möbius number systems are Möbius transformations.

Keywords: exact real algorithms, absorptions, emissions.

1 Introduction

Exact real arithmetical algorithms have been introduced in an unpublished manuscript of Gosper [5] and developed by Vuillemin [16], Potts [14] or Kornerup and Matula [10,9]. These algorithms perform a sequence of **input absorptions** and **output emissions** and update their inner state which may be a $(2 \times 2 \times 2)$-tensor in the case of binary operations like addition or multiplication or a (2×2)-matrix in the case of a Möbius transformation. If the norm of these matrices remains bounded, then the algorithm runs only through a finite number of states and can be therefore computed by a finite state transducer. Delacourt and Kůrka [3] show that this happens if the digits of the number system are represented by modular matrices, i.e., by matrices with integer entries and unit determinant. This generalizes a result of Raney [15] that a Möbius transformation can be computed by a finite state transducer in the number system of continued fractions. Frougny [4] shows that in positional number systems with an irrational Pisot base $\beta > 1$, the addition can be also computed by a finite state transducer.

In the opposite direction, Konečný [8] shows that under certain assumptions, a finite state transducer can compute only Möbius transformations. In the present paper we strenghten and generalize this result and show that if an analytic function is computed by a finite state transducer in a number system with sofic expansion subshift, then this function is a Möbius transformation (Theorem 10). Since modular number systems have some disadvantages (slow convergence), we address the question whether a Möbius transformation can be computed by a finite state transducer also in nonmodular systems which are expansive,

M. Holzer and M. Kutrib (Eds.): CIAA 2014, LNCS 8587, pp. 252–263, 2014.

so that they converge faster. Kůrka and Delacourt [13] show that in the bimodular number system (which extends the binary signed system) the computation of a Möbius transformation has an asymptotically linear time complexity. Although the norm of the state matrices is not bounded, it remains small most of the time. In the present paper we show that this result cannot be improved. For any expansive number systems whose transformations have integer entries and determinant at most 2 there exists a Möbius transformation which cannot be computed by a finite state transducer (Theorem 15).

2   Subshifts

For a finite alphabet A denote by $A^* = \bigcup_{m \geq 0} A^m$ the set of finite words. The length of a word $u = u_0 \ldots u_{m-1} \in A^m$ is $|u| = m$. Denote by $A^{\mathbb{N}}$ the Cantor space of infinite words with the metric

$$d(u, v) = 2^{-k}, \text{ where } k = \min\{i \geq 0 : u_i \neq v_i\}.$$

We say that $v \in A^*$ is a subword of $u \in A^* \cup A^{\mathbb{N}}$ and write $v \sqsubseteq u$, if $v = u_{[i,j)} = u_i \ldots u_{j-1}$ for some $0 \leq i \leq j \leq |u|$. The **shift map** $\sigma : A^{\mathbb{N}} \to A^{\mathbb{N}}$ is defined by $\sigma(u)_i = u_{i+1}$. A **subshift** is a nonempty set $\Sigma \subseteq A^{\mathbb{N}}$ which is closed and σ-invariant, i.e., $\sigma(\Sigma) \subseteq \Sigma$. If $D \subseteq A^*$ then

$$\Sigma_D = \{u \in A^{\mathbb{N}} : \forall v \sqsubseteq u, v \notin D\}$$

is the subshift (provided it is nonempty) with **forbidden words** D. Any subshift can be obtained in this way. A subshift is uniquely determined by its **language** $\mathcal{L}(\Sigma) = \{v \in A^* : \exists u \in \Sigma, v \sqsubseteq u\}$. A nonempty language $L \subseteq A^*$ is **extendable**, if for each word $u \in L$, each subword v of u belongs to L, and there exists a letter $a \in A$ such that $ua \in L$. If Σ is a subshift, then $\mathcal{L}(\Sigma)$ is an extendable language and conversely, for each extendable language $L \subseteq A^*$ there exists a unique subshift $\Sigma \subseteq A^{\mathbb{N}}$ such that $L = \mathcal{L}(\Sigma)$. The **cylinder** of a finite word $u \in \mathcal{L}(\Sigma)$ is the set of infinite words with prefix u: $[u] = \{v \in \Sigma : v_{[0,|u|)} = u\}$.

3   Finite Accepting Automata

We consider finite automata which accept (regular) extendable languages, so the classical definition simplifies: we do not need accepting states (see Kůrka [11]).

Definition 1. *A (deterministic) **finite automaton** over an alphabet A is a triple $\mathcal{A} = (B, \delta, \iota)$, where B is a finite set of states, $\delta : A \times B \to B$ is a partial transition function, and $\iota \in B$ is an initial state.*

A finite automaton determines a labelled graph, whose vertices are states $p \in B$ and whose labelled edges are $p \xrightarrow{a} q$ provided $\delta(a, p) = q$. For each $a \in A$ we have a partial mapping $\delta_a : B \to B$ defined by $\delta_a(p) = \delta(a, p)$ and for each $u \in A^*$ we have a partial mapping $\delta_u : B \to B$ defined by $\delta_u = \delta_{u_{|u|-1}} \circ \cdots \circ \delta_{u_0}$. We write

$\exists \delta_u(p)$ if δ_u is defined on p. For $u \in A^{\mathbb{N}}$ we write $\exists \delta_u(p)$ if $\exists \delta_{u_{[0,n)}}(p)$ for each prefix $u_{[0,n)}$ of u. The **follower set** of a state $p \in B$ is $\mathcal{F}_p = \{u \in A^{\mathbb{N}} : \exists \delta_u(p)\}$.

We assume that every state of $\mathcal{A}$ is accessible from the initial state, i.e., for every $q \in B$ there exists $u \in A^*$ such that $\delta_u(\iota) = q$. The states that are not accessible can be omitted without changing the function of the automaton. The language accepted by $\mathcal{A}$ is $L_{\mathcal{A}} = \{u \in A^* : \exists \delta_u(\iota)\}$, so a word u is accepted iff there exists a path with source ι and label u. We say that $\Sigma \subseteq A^{\mathbb{N}}$ is a **sofic** subshift, if its language is regular iff it is accepted by a finite automaton, i.e., if there exists an automaton $\mathcal{A}$ such that $\Sigma = \mathcal{F}_\iota = \{u \in A^{\mathbb{N}} : \exists \delta_u(\iota)\}$.

4 Möbius Transformations

On the **extended real line** $\overline{\mathbb{R}} = \mathbb{R} \cup \{\infty\}$ we have **homogeneous coordinates** $x = (x_0, x_1) \in \mathbb{R}^2 \setminus \{(0,0)\}$ with equality $x = y$ iff $\det(x, y) = x_0 y_1 - x_1 y_0 = 0$. We regard $x \in \overline{\mathbb{R}}$ as a column vector, and write it usually as $x = \frac{x_0}{x_1}$, for example $\infty = \frac{1}{0}$. A real **Möbius transformation** (MT) is a self-map of $\overline{\mathbb{R}}$ of the form

$$M(x) = \frac{ax + b}{cx + d} = \frac{ax_0 + bx_1}{cx_0 + dx_1},$$

where $a, b, c, d \in \mathbb{R}$ and $\det(M) = ad - bc \neq 0$. If $\det(M) > 0$, we say that M is **increasing**. An MT is determined by a (2×2)-matrix which we write as a pair of fractions of its left and right column $M = (\frac{a}{c}, \frac{b}{d})$. If $m \neq 0$, then $(\frac{ma}{mc}, \frac{mb}{md})$ determines the same transformation as M. Denote by $\mathbb{M}(\mathbb{R})$ the set of real MT and by $\mathbb{M}^+(\mathbb{R})$ the set of increasing MT. The composition of MT corresponds to the product of matrices. The inverse of a transformation is $(\frac{a}{c}, \frac{b}{d})^{-1} = (\frac{d}{-c}, \frac{-b}{a})$. Denote by M^n the n-th iteration of M.

The **stereographic projection** $\mathbf{h}(z) = (iz + 1)/(z + i)$ maps $\overline{\mathbb{R}}$ to the unit circle $\mathbb{T} = \{z \in C : |z| = 1\}$ in the complex plane. For each $M \in \mathbb{M}(\mathbb{R})$ we get a **disc Möbius transformation** $\widehat{M} : \mathbb{T} \to \mathbb{T}$ given by $\widehat{M}(z) = \mathbf{h} \circ M \circ \mathbf{h}^{-1}(z)$. The **circle derivation** of M at $x \in \overline{\mathbb{R}}$ is

$$M^\bullet(x) = |\widehat{M}'(\mathbf{h}(x))| = \frac{|\det(M)| \cdot ||x||^2}{||M(x)||^2},$$

where $||x|| = \sqrt{x_0^2 + x_1^2}$. The **trace** and **norm** of $M = (\frac{a}{c}, \frac{b}{d}) \in \mathbb{M}^+(\mathbb{R})$ are

$$\text{tr}(M) = \frac{|a + d|}{\sqrt{ad - bc}}, \quad ||M|| = \frac{\sqrt{a^2 + b^2 + c^2 + d^2}}{\sqrt{ad - bc}}.$$

We say that $x \in \overline{\mathbb{R}}$ is a **fixed point** of M if $M(x) = x$. If $M = (\frac{a}{c}, \frac{b}{d})$ is not the identity, $M(x) = x$ yields a quadratic equation $bx_0^2 + (d - a)x_0 x_1 - cx_1^2 = 0$ with discriminant $D = (a - d)^2 + 4bc = (a + d)^2 - 4(ad - bc)$, so $D \geq 0$ iff $\text{tr}(M) \geq 2$. If $\text{tr}(M) < 2$, then M has no fixed point and we say that M is **elliptic**. If $\text{tr}(M) = 2$, then M has one fixed point and we say that M is **parabolic**. If $\text{tr}(M) > 2$, then M has two fixed points and we say that M is **hyperbolic**.

Definition 2. *The* **similarity,** **translation** *and* **rotation** *are transformations with matrices*

$$S_r = \begin{pmatrix} r & 0 \\ 0 & 1 \end{pmatrix}, \; T_t = \begin{pmatrix} 1 & t \\ 0 & 1 \end{pmatrix}, \; R_t = \begin{pmatrix} \cos\frac{t}{2} & \sin\frac{t}{2} \\ -\sin\frac{t}{2} & \cos\frac{t}{2} \end{pmatrix}.$$

S_r is a hyperbolic transformation with the fixed points $0, \infty$. T_t is a parabolic transformation with the fixed point ∞, and R_t is an elliptic transformation.

Definition 3. *We say that transformations* $P, Q \in \mathrm{M}^+(\mathbb{R})$ *are* **conjugated** *if there exists a transformation* $M \in \mathrm{M}(\mathbb{R})$ *such that* $Q = M^{-1}PM$.

Conjugated transformations have the same dynamical properties and the same trace. A direct computation shows that $\mathrm{tr}(PQ) = \sum_{i,j} P_{ij}Q_{ji} = \mathrm{tr}(QP)$. It follows that if $Q = M^{-1}PM$, then $\mathrm{tr}(Q) = \mathrm{tr}(PMM^{-1}) = \mathrm{tr}(P)$. If x is a fixed point of P, then $y = M^{-1}x$ is a fixed point of Q and $Q^\bullet(y) = P^\bullet(x)$.

Theorem 4 (Beardon [2]).
1. *Transformations* $P, Q \in \mathrm{M}^+(\mathbb{R})$ *are conjugated iff* $\mathrm{tr}(P) = \mathrm{tr}(Q)$.
2. *Each hyperbolic transformation* P *is conjugated to a similarity with quotient* $0 < r < 1$. P *has an unstable fixed point* $\mathbf{u}(P)$ *and a stable fixed point* $\mathbf{s}(P)$ *such that* $\lim_{n \to \infty} P^n(x) = \mathbf{s}(P)$ *for each* $x \neq \mathbf{u}(P)$.
3. *Each parabolic transformation* P *is conjugated to the translation* $T_1(x) = x + 1$. P *has a unique fixed point* $\mathbf{s}(P)$ *such that* $\lim_{n \to \infty} P^n(x) = \mathbf{s}(P)$ *for each* $x \in \overline{\mathbb{R}}$.
4. *Each elliptic transformation is conjugated to a rotation* R_t *with* $0 < t \leq \pi$.

5 Möbius Number Systems

An **iterative system** over a finite alphabet A is a system of Möbius transformations $F = \{F_a \in \mathrm{M}^+(\mathbb{R}) : a \in A\}$. For each finite word $u \in A^n$, we have the composition $F_u = F_{u_{n-1}} \circ \cdots \circ F_{u_0}$, so $F_{uv}(x) = F_v(F_u(x))$ for any $uv \in A^*$ ($F_\lambda = \mathrm{Id}_{\overline{\mathbb{R}}}$ is the identity). The **convergence space** $\mathbb{X}_F \subseteq A^{\mathbb{N}}$ and the **value function** $\Phi : \mathbb{X}_F \to \overline{\mathbb{R}}$ are defined by

$$\mathbb{X}_F = \{u \in A^{\mathbb{N}} : \lim_{n \to \infty} F^{-1}_{u_{[0,n)}}(i) \in \overline{\mathbb{R}}\}, \quad \Phi(u) = \lim_{n \to \infty} F^{-1}_{u_{[0,n)}}(i).$$

Here i is the imaginary unit. If $u \in \mathbb{X}_F$ then $\Phi(u) = \lim_{n \to \infty} F^{-1}_{u_{[0,n)}}(z)$ for every complex z with positive imaginary part and also for most of the real z. The concept of convergence space is related to the concept of convergence of infinite product of matrices considered in the theory of weighted finite automata (see Culik II et al. [6] or Kari et al [7]).

Proposition 5 (Kůrka [12]). *Let* F *be an iterative system over* A.
1. *For* $v \in A^+$, $u \in A^{\mathbb{N}}$ *we have* $vu \in \mathbb{X}_F$ *iff* $u \in \mathbb{X}_F$, *and then* $\Phi(vu) = F_v^{-1}(\Phi(u))$.

256   P. Kůrka and T. Vávra

2. For $v \in A^+$ we have $v^\infty \in \mathbb{X}_F$ iff F_v is not elliptic. In this case $\Phi(v^\infty) = s(F_v^{-1})$ is the stable fixed point of F_v^{-1}.

Definition 6. We say that (F, Σ) is a **number system** if F is an iterative system and $\Sigma \subseteq \mathbb{X}_F$ is a subshift such that $\Phi : \Sigma \to \overline{\mathbb{R}}$ is continuous and surjective. We say that (F, Σ) is an **expansive number system** if for each $u \in \Sigma$, we have $F_{u_0}^\bullet(\Phi(u)) > 1$. We say that $(F, \Sigma, \mathcal{A})$ is a **sofic number system**, if (F, Σ) is a number system and $\mathcal{A}$ is a finite automaton with $L_{\mathcal{A}} = \mathcal{L}(\Sigma)$.

If (F, Σ) is expansive, then the convergence in $\Phi(u) = \lim_{n \to \infty} F_{u_{[0,n)}}^{-1}(i)$ is geometric. In nonexpansive systems this convergence may be much slower (see Delacourt and Kůrka [13]).

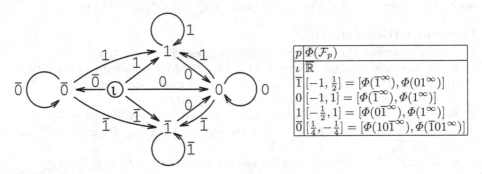

Fig. 1. The accepting automaton of the subshift of the binary signed system with forbidden words $D = \{\overline{1}0, 0\overline{0}, 1\overline{0}, \overline{0}0, 1\overline{1}, \overline{1}1\}$ (left) and Φ-images of the follower sets (right). Here $[\frac{1}{4}, -\frac{1}{4}] = \{x \in \mathbb{R} : x \geq \frac{1}{4} \text{ or } x \leq -\frac{1}{4}\} \cup \{\infty\}$ is an unbounded interval which contains ∞.

Example 1. The binary signed system (F, Σ_D) has alphabet $A = \{\overline{1}, 0, 1, \overline{0}\}$, transformations

$$F_{\overline{1}}(x) = 2x + 1, \ F_0(x) = 2x, \ F_1(x) = 2x - 1, \ F_{\overline{0}}(x) = x/2,$$

and forbidden words $D = \{\overline{1}0, 0\overline{0}, 1\overline{0}, \overline{0}0, 1\overline{1}, \overline{1}1\}$.

The digits $\overline{1}, \overline{0}$ stand for -1 and ∞. A finite word of Σ_D can be written as $\overline{0}^m u$, where $m \geq 0$ and $u \in \{\overline{1}, 0, 1\}^*$. If $|u| = n$ then

$$F_{\overline{0}^m u}^{-1}(x) = 2^m \left(\frac{u_0}{2} + \cdots + \frac{u_{n-1}}{2^n} + \frac{x}{2^n} \right),$$

so for $u \in \{\overline{1}, 0, 1\}^{\mathbb{N}}$ we get

$$\Phi(\overline{0}^m u) = \lim_{n \to \infty} F_{\overline{0}^m u}^{-1}(i) = \sum_{i \geq 0} u_i \cdot 2^{m-i-1}.$$

Thus $\Sigma_D \subseteq \mathbb{X}_F$ and $\Phi : \Sigma_D \to \overline{\mathbb{R}}$ is continuous and surjective. The subshift Σ_D is sofic. Its accepting automaton has states $B = \{\iota, \overline{1}, 0, 1, \overline{0}\}$, initial state ι and transitions which can be seen in Figure 1 left. Computing for each $p \in B$ the minimum and maximum of paths which start at p, we obtain the Φ-images of the follower sets in Figure 1 right.

6   Finite State Transducers

Definition 7. *A finite state transducer over an alphabet A is a quadruple $\mathcal{T} = (B, \delta, \tau, \iota)$, where (B, δ, ι) is a finite automaton over A and $\tau : A \times B \to A^*$ is a partial output function with the same domain as δ.*

For each $u \in A$ we have a partial mapping $\tau_u : B \to A^*$ defined by induction: $\tau_\lambda(p) = \lambda$, $\tau_{ua}(p) = \tau_u(p)\tau(a, \delta_u(p))$ (concatenation). The output mapping works also on infinite words. If u is a prefix of v, then $\tau_u(p)$ is a prefix of $\tau_v(p)$, so for each $p \in B$ and $u \in A^{\mathbb{N}}$ we have $\tau_u(p) \in A^* \cup A^{\mathbb{N}}$. A finite state transducer determines a labelled oriented graph, whose vertices are elements of B. There is an oriented edge $p \xrightarrow{a/v} q$ iff $\delta_a(p) = q$ and $\tau_a(p) = v$. The label of a path is the concatenation of the labels of its edges, so there is a path $p \xrightarrow{u/v} q$ iff $\delta_u(p) = q$ and $\tau_u(p) = v$.

Definition 8. *We say that a finite state transducer $\mathcal{T} = (B, \delta, \tau, \iota)$ computes a real function $G : \overline{\mathbb{R}} \to \overline{\mathbb{R}}$ in a number system (F, Σ) with sofic expansion subshift Σ, if for any $u \in A^{\mathbb{N}}$ we have $\exists \delta_u(\iota)$ iff $u \in \Sigma$ and in this case $\Phi(\tau_u(\iota)) = G(\Phi(u))$.*

Proposition 9. *Assume that a finite state transducer $\mathcal{T}$ computes a real function G in a number system (F, Σ) with sofic expansion subshift. Then for every state $p \in B$ there exists a real function $G_p : \Phi(\mathcal{F}_p) \to \overline{\mathbb{R}}$ such that if $w \in \mathcal{F}_p$ and $\tau_w(p) = z$, then $\Phi(z) = G_p\Phi(w)$. We say that $\mathcal{T}$ computes G_p at the state p. If $u, v \in \mathcal{L}(\Sigma)$, $\delta_u(p) = q$ and $\tau_u(p) = v$ then $G_q = F_v G_p F_u^{-1}$.*

Proof. Assume that $\iota \xrightarrow{u/v} p \xrightarrow{w/z}$ and set $G_p = F_v G F_u^{-1}$. By Proposition 5,

$$G_p\Phi(w) = F_v G F_u^{-1}\Phi(w) = F_v G\Phi(uw) = F_v\Phi(vz) = \Phi(z),$$

so $\mathcal{T}$ computes G_p at p. If $p \xrightarrow{u/v} q \xrightarrow{w/z}$, then

$$F_v G_p F_u^{-1}\Phi(w) = F_v G_p\Phi(uw) = F_v\Phi(vz) = \Phi(z),$$

so $\mathcal{T}$ computes $F_v G_p F_u^{-1}$ at q and must be equal to G_q.

7   Analytic Functions

A real function $G : \overline{\mathbb{R}} \to \overline{\mathbb{R}}$ is **analytic**, if it can be written as a power series $G(x) = \sum_{n \geq 0} a_n(x - w)^n$ in a neighbourhood of every point $w \in \overline{\mathbb{R}}$. For $w = \infty$ this means that the function $G(1/x)$ is analytic at 0. Every rational function, i.e., a ratio of two polynomials is analytic in $\overline{\mathbb{R}}$. The functions e^x, $\sin x$ or $\cos x$ are analytic in $\mathbb{R}$ but not in $\overline{\mathbb{R}}$.

Lemma 1. *Let* $G : \overline{\mathbb{R}} \to \overline{\mathbb{R}}$ *be a nonzero analytic function and let* $F_0, F_1 \in M^+(\mathbb{R})$ *be hyperbolic transformations such that* $F_0 G = G F_1$. *Then* G *is a rational function.*

Proof. Any hyperbolic transformation is conjugated to a similarity $S_r(x) = rx$ with $0 < r < 1$. Thus there exist transformations f_0, f_1 and $0 < r_0, r_1 < 1$ such that $F_0 = f_0 S_{r_0} f_0^{-1}$, $F_1 = f_1 S_{r_1} f_1^{-1}$. For $H = f_0^{-1} G f_1$ we get

$$S_{r_0} H = S_{r_0} f_0^{-1} G f_1 = f_0^{-1} F_0 G f_1 = f_0^{-1} G F_1 f_1 = H f_1^{-1} F_1 f_1 = H S_{r_1}.$$

Since G is analytic, H also is analytic and $H(x) = a_0 + a_1 x + a_2 x^2 + \cdots$ in a neighbourhood of zero, so

$$r_0 a_0 + r_0 a_1 x + r_0 a_2 x^2 + \cdots = a_0 + a_1 r_1 x + a_2 r_1^2 x^2 + \cdots$$

Since $r_0 \neq 0$ we get $a_0 = 0$. If n is the first integer with $a_n \neq 0$, then $r_0 = r_1^n$. For $m > n$ we get $r_1^n a_m = a_m r_1^m$, so $a_m = 0$. Thus $H(x) = a_n x^n$ and therefore $G = f_0 H f_1^{-1}$ is a rational function.

Konečný [8] proves essentially Lemma 1 but makes the assumption that the derivation of G at the fixed point of F_1 is nonzero, i.e., $H'(0) \neq 0$ which implies that H is linear. Without the assumption of analyticity, we would get a much larger class of functions. Given $0 < r_0, r_1 < 1$, let $h : [r_1, 1] \to [r_0, 1]$ be any continuous function with $h(r_1) = r_0$, $h(1) = 1$. Then the function $H : (0, \infty) \to (0, \infty)$ defined by $H(x) = r_0^n \cdot h(r_1^{-n} \cdot x)$ for $r_1^{n+1} \leq x \leq r_1^n$, $n \in \mathbb{Z}$, satisfies $H(r_1 x) = r_0 H(x)$. We can define H similarly on $(-\infty, 0)$, and if we set $H(0) = 0$, $H(\infty) = \infty$, then $H : \overline{\mathbb{R}} \to \overline{\mathbb{R}}$ is continuous but not necessarily analytic or differentiable.

To exclude rational functions of degree $n \geq 2$, we prove Lemma 2. Recall that the degree of a rational function is the maximum of the degree of the numerator and denominator, so rational functions of degree 1 are just Möbius transformations.

Lemma 2. *Let* G *be a rational function of degree* $n \geq 2$, *and let* $F_0, F_1, F_2, F_3 \in M^+(\mathbb{R})$ *be hyperbolic transformations such that* $F_0 G = G F_1$, $F_2 G = G F_3$. *Then* F_2 *has the same fixed points as* F_0 *and* F_3 *has the same fixed points as* F_1.

Proof. By Lemma 1 there exist transformations f_0, f_1 and $0 < r_0, r_1 < 1$ such that $F_0 = f_0 S_{r_0} f_0^{-1}$, $F_1 = f_1 S_{r_1} f_1^{-1}$, and $H = f_0^{-1} G f_1$ is a function of the form $H(x) = p x^n$ with $n \geq 2$. Since $G = f_0 H f_1^{-1}$, we get

$$f_0^{-1} F_2 f_0 H = f_0^{-1} F_2 G f_1 = f_0^{-1} G F_3 f_1 = H f_1^{-1} F_3 f_1.$$

Setting $f_0^{-1} F_2 f_0 = (\frac{a}{c}, \frac{b}{d})$, $f_1^{-1} F_3 f_1 = (\frac{A}{C}, \frac{B}{D})$ we get

$$(a p x^n + b)(C x + D)^n = p(c p x^n + d)(A x + B)^n$$

Comparing the coeficients at x^{2n} and x^{2n-1} we get $aC^n = pcA^n$, $aC^{n-1}D = pcA^{n-1}B$. Thus $pcA^n D = aC^n D = pcA^{n-1}BC$, so $pcA^{n-1}(AD - BC) = 0$ and therefore $cA = 0$ and it follows $aC = 0$. Comparing the coeficients at x and x^0, we get $bCD^{n-1} = pdAB^{n-1}, bD^n = pdB^n$, so $pdAB^{n-1}D = bcD^n = pdCB^n$ and $pdB^{n-1}(AD - BC) = 0$. Thus $dB = 0$ and it follows $bD = 0$. We have therefore proved $cA = aC = dB = bD = 0$. It follows that either $A = D = a = d = 0$ or $B = C = b = c = 0$. In the former case, F_2 and F_3 would be elliptic which is excluded by the assumption. Thus $B = C = b = c = 0$, so both $f_0^{-1}F_2f_0$ and $f_1^{-1}F_3f_1$ have the fixed points 0 and ∞, which are also fixed points of S_{r_0} and S_{r_1}. It follows that F_2 has the same fixed points as F_0 and F_3 has the same fixed points as F_1.

Lemma 3. *Let $G : \overline{\mathbb{R}} \to \overline{\mathbb{R}}$ be an analytic function and let $F_0, F_1 \in \mathrm{M}^+(\overline{\mathbb{R}})$ be parabolic transformations such that $F_0G = GF_1$. Then $G \in \mathrm{M}(\mathbb{R})$ is a MT.*

Proof. A parabolic transformation is conjugated to the translation $T_1(x) = x+1$. Thus there exist transformations f_0, f_1 such that $F_0 = f_0T_1f_0^{-1}$, $F_1 = f_1T_1f_1^{-1}$. For $H = f_0^{-1}Gf_1$ we get $T_1H = HT_1$. The function $H_0(x) = H(x) - x$ is then periodic with period 1, i.e., $H_0(x + 1) = H_0(x)$. Since H_0 is analytic at ∞, it must be zero, otherwise it would not be even continuous at ∞. Thus $H(x) = x$ and G is an MT.

Lemma 4. *Let $G : \overline{\mathbb{R}} \to \overline{\mathbb{R}}$ be an analytic function and let $F_0, F_1 \in \mathrm{M}^+(\overline{\mathbb{R}})$ be transformations such that $F_0G = GF_1$. If one of the F_0, F_1 is hyperbolic and the other is parabolic, then G is the zero function.*

Proof. Let $H = f_0^{-1}Gf_1$ as in the proof of Lemma 3. If $H(x)+1 = H(r_1x)$, where $H(x) = a_0 + a_1x + \cdots$, then we get $a_0 + 1 = a_0$ which is impossible. Suppose $r_0 \cdot H(x) = H(x + 1)$ with $0 < r_0 < 1$. If $H(0) = 0$, then $H(n) = 0$ for all $n \in \mathbb{Z}$ and $H = 0$, since H is continuous at ∞. If $H(0) \neq 0$, then $H(n) = H(0) \cdot r_0^n$, so $\lim_{n \to \infty} H(n) = 0$, $\lim_{n \to -\infty} H(n) = \infty$ which is impossible.

Theorem 10. *Let (F, Σ) be a number system with sofic subshift Σ. If $G : \overline{\mathbb{R}} \to \overline{\mathbb{R}}$ is a nonzero analytic function computed in Σ by a finite state transducer, then $G \in \mathrm{M}(\mathbb{R})$ is a Möbius transformation (the determinant of G may be negative).*

Proof. Let $\iota \xrightarrow{u/v} p \xrightarrow{w/z} p$ be a path in the graph of the transducer. By Proposition 9, $G_p = F_vGF_u^{-1}$ is analytic and $G_pF_w = F_zG_p$. By Proposition 5, F_w, F_z cannot be elliptic and by Lemma 1, 3, 4, G_p must be a rational function, so $G = F_v^{-1}G_pF_u$ is rational too. Assume by contradiction that the degree of G is at least $n \geq 2$. Then all G_p must have degree n and by Lemma 3 and 4, F_u, F_v must be hyperbolic whenever $p \xrightarrow{u/v} p$. Take any infinite path u/v. There exists a state $p \in B$ which occurs infinitely often in this path, so we have words $u^{(i)}$, $v^{(i)}$ such that $u = u^{(0)}u^{(1)}u^{(2)} \cdots$ and

$$\iota \xrightarrow{u^{(0)}/v^{(0)}} p \xrightarrow{u^{(1)}/v^{(1)}} p \xrightarrow{u^{(2)}/v^{(2)}} p \cdots .$$

By Lemma 2, all $F_{u^{(i)}}$ with $i > 0$ have the same fixed points. It follows that $\Phi(u) = F_{u^{(0)}}(s)$, where s is one of the fixed points of $F_{u^{(1)}}$. However the set of such numbers is countable, while the mapping $\Phi : \Sigma \to \overline{\mathbb{R}}$ is assumed to be surjective, so we have a contradiction. Thus $G_p \in \mathrm{M}(\mathbb{R})$ and therefore $G \in \mathrm{M}(\mathbb{R})$.

8 Rational Transformations and Intervals

Denote by $\mathbb{Z}$ the set of integers and by $\overline{\mathbb{Q}} = \{x \in \mathbb{Z}^2 \setminus \{{}^0_0\} : \gcd(x) = 1\}$ the set of (homogeneous coordinates of) rational numbers which we understand as a subset of $\overline{\mathbb{R}}$. Here $\gcd(x)$ is the greatest common divisor of x_0 and x_1. The norm $\|x\| = \sqrt{x_0^2 + x_1^2}$ of $x \in \overline{\mathbb{Q}}$ does not depend on the representation of x. We have the cancellation map $\mathbf{d} : \mathbb{Z}^2 \setminus \{{}^0_0\} \to \overline{\mathbb{Q}}$ given by $\mathbf{d}(x) = \frac{x_0/\gcd(x)}{x_1/\gcd(x)}$. Denote by $\mathbb{Z}^{2\times2}$ the set of 2×2 matrices with integer entries and

$$\mathbb{M}(\mathbb{Z}) = \{M \in \mathbb{Z}^{2\times2} : \gcd(M) = 1,\ \det(M) > 0\}.$$

We say that a Möbius transformation is rational if its matrix belongs to $\mathbb{M}(\mathbb{Z})$.

For $x \in \overline{\mathbb{Q}}$ we distinguish $M \cdot x \in \mathbb{Z}^2$ from $Mx = \mathbf{d}(M \cdot x) \in \overline{\mathbb{Q}}$. For $M = (\frac{a}{c}, \frac{b}{d}) \in \mathbb{Z}^{2\times2}$ denote by $\mathbf{d}(M) = (\frac{a/g}{c/g}, \frac{b/g}{d/g})$, where $g = \gcd(M)$, so we have a cancellation map $\mathbf{d} : \mathbb{Z}^{2\times2} \setminus \{(\frac{0}{0}, \frac{0}{0})\} \to \mathbb{M}(\mathbb{Z})$. We distinguish the matrix multiplication $M \cdot N$ from the multiplication $MN = \mathbf{d}(M \cdot N)$ in $\mathbb{M}(\mathbb{Z})$. The inverse of $M = (\frac{a}{c}, \frac{b}{d}) \in \mathbb{M}(\mathbb{Z})$ is $M^{-1} = (\frac{d}{-c}, \frac{-b}{a})$, so $M \cdot M^{-1} = \det(M) \cdot I$, $MM^{-1} = I$.

Lemma 5. *If $M, N \in \mathbb{M}(\mathbb{Z})$, then $g = \gcd(M \cdot N)$ divides both $\det(M)$ and $\det(N)$.*

Proof. Clearly g divides $M^{-1} \cdot M \cdot N = \det(M) \cdot N$. Since $\gcd(N) = 1$, g divides $\det(M)$. For the similar reason, g divides $\det(N)$.

Definition 11. *A number system (F, Σ) is **rational**, if all its transformations belong to $\mathbb{M}(\mathbb{Z})$. A rational number system is **modular**, if all its transformations have determinant 1.*

Theorem 12 (Delacourt and Kůrka [3]). *If (F, Σ) is a sofic modular number system, then each transformation $M \in \mathbb{M}^+(\mathbb{Z})$ can be computed in (F, Σ) by a finite state transducer.*

Proposition 13. *A modular number system cannot be expansive.*

Proof. Assume by contradiction that a modular system (F, Σ) is expansive and let $u \in \Sigma$ be such that $\Phi(u) = 0$, so $F_{u_0}^\bullet(0) > 1$. If $F_{u_0} = (\frac{a}{c}, \frac{b}{d})$, then $F_{u_0}^\bullet(0) = \frac{1}{b^2+d^2} > 1$, so $b = d = 0$ and therefore $\det(F_{u_0}) = 0$ which is a contradiction.

9 The Binary Signed System

It is well-known that in redundant number systems, the addition can be computed by a finite state transducer (see e.g. Avizienis [1] or Frougny [4]), provided both operands are from a bounded interval. The binary signed system of Example 1 is redundant, since the intervals $V_p = \Phi(\mathcal{F}_p)$ overlap: their interiors cover whole $\overline{\mathbb{R}}$. It is not difficult to show that any linear function $G(x) = rx$, where r is rational, can be computed by a finite state transducer. This is based on the fact that the matrices $F_v \cdot G_p \cdot F_u^{-1}$ have a common factor which can be cancelled:

$$\left(\tfrac{1}{0}, \tfrac{0}{2^m}\right) \cdot \left(\tfrac{p}{0}, \tfrac{0}{q}\right) \cdot \left(\tfrac{2^n}{0}, \tfrac{0}{1}\right) = \left(\tfrac{2^n p}{0}, \tfrac{0}{2^m q}\right),$$

$$\left(\tfrac{2^m}{0}, \tfrac{-b}{1}\right) \cdot \left(\tfrac{p}{0}, \tfrac{0}{q}\right) \cdot \left(\tfrac{1}{0}, \tfrac{a}{2^n}\right) = \left(\tfrac{2^m p}{0}, \tfrac{2^m ap - 2^n bq}{2^n q}\right),$$

On the other hand we have

Proposition 14. *The function $G(x) = x + 1$ is not computable by a finite state transducer in the binary signed system.*

Proof. Assume that $\mathcal{T} = (B, \delta, \tau, \iota)$ computes $G(x) = x + 1$. Since $\tau_{\overline{0}^\infty}(\iota) = \overline{0}^\infty$, there exists $p \in B$ and $r, s \geq 0$, $m, n > 0$ such that $\iota \xrightarrow{\overline{0}^r / \overline{0}^s} p \xrightarrow{\overline{0}^n / \overline{0}^m} p$. However, for $G_p = F_{\overline{0}^s} G F_{\overline{0}^r}^{-1} = \left(\tfrac{2^r}{0}, \tfrac{1}{2^s}\right)$ we get

$$F_{\overline{0}^m} G_p F_{\overline{0}^n}^{-1} = \left(\tfrac{1}{0}, \tfrac{0}{2^m}\right) \cdot \left(\tfrac{2^r}{0}, \tfrac{1}{2^s}\right) \cdot \left(\tfrac{2^n}{0}, \tfrac{0}{1}\right) = \left(\tfrac{2^{r+n}}{0}, \tfrac{1}{2^{m+s}}\right) \neq G_p.$$

and this is a contradiction.

10   Bimodular Systems

We are going to prove another negative result concerning the computation of a Möbious transformations in expansive number systems. We say that a rational number system (F, Σ) is **bimodular**, if $F_a \in \mathbb{M}(\mathbb{Z})$ and $\det(F_a) \leq 2$ for each $a \in A$. Kůrka and Delacourt [13] show that there exists a bimodular number system (which extends the binary signed system) in which the computation of a Möbius transformation has an asymptotically linear time complexity. Although the norm of the state matrices is not bounded, it remains small most of the time. We show that this result cannot be improved. There exist transformations which cannot be computed by a finite state transducer.

Lemma 6. *Assume $F \in \mathbb{M}(\mathbb{Z})$ and $\det(F) \leq 2$.*
1. If $F^\bullet(0) > 1$, then either $F = \left(\tfrac{2}{c}, \tfrac{0}{1}\right)$, $F(0) = 0$, or $F = \left(\tfrac{a}{2}, \tfrac{-1}{0}\right)$, $F(0) = \infty$.
2. If $F^\bullet(\infty) > 1$, then either $F = \left(\tfrac{0}{-1}, \tfrac{2}{d}\right)$, $F(\infty) = 0$, or $F = \left(\tfrac{1}{0}, \tfrac{b}{2}\right)$, $F(\infty) = \infty$.

Proof. Let $F = \left(\tfrac{a}{c}, \tfrac{b}{d}\right)$. If $F^\bullet(0) = \tfrac{\det(F)}{b^2 + d^2} > 1$, then $\det(F) = 2$ since b, d cannot be both zero. Thus $b^2 + d^2 < 2$ and $b, d \in \{-1, 0, 1\}$, so either $F = \left(\tfrac{2}{c}, \tfrac{0}{1}\right)$ or $F = \left(\tfrac{a}{2}, \tfrac{-1}{0}\right)$. If $F^\bullet(\infty) = \tfrac{\det(F)}{a^2 + c^2} > 1$, then $\det(F) = 2$, $a, c \in \{-1, 0, 1\}$ and either $F = \left(\tfrac{1}{0}, \tfrac{b}{2}\right)$, or $F = \left(\tfrac{0}{-1}, \tfrac{2}{d}\right)$.

Theorem 15. *Let (F, Σ) be a rational bimodular system. Then there exists a transformation $G \in \mathbb{M}(\mathbb{Z})$ which cannot be computed by a finite state transducer in (F, Σ).*

Proof. Denote by mod_2 the modulo 2 function. Choose any transformation G such that $G(0) = 0$ and $\mathrm{mod}_2(G) = (\begin{smallmatrix} 0 & 0 \\ 1 & 0 \end{smallmatrix})$, e.g., $G(x) = \frac{2x}{x+2}$. Pick a word $u \in \Sigma$ with $\Phi(u) = 0$ and assume that we have a finite state transducer which computes G on u with the result v, so $\Phi(v) = 0$. The computation of the transducer determines a path whose vertices compute functions $G_{n,m} = F_{v_{[0,m)}} G F_{u_{[0,n)}}^{-1}$ and in each transition we have either $G_{n,m} \xrightarrow{u_n/\lambda} G_{n+1,m}$ or $G_{n,m} \xrightarrow{\lambda/v_n} G_{n,m+1}$. We show by induction that during the process no cancellation ever occurs: either $\det(G_{n+1,m}) = 2 \det(G_{n,m})$ or $\det(G_{n,m+1}) = 2 \det(G_{n,m})$. Denote by $x_n = \Phi(u_{[n,\infty)}) = F_{u_{[0,n)}} \Phi(u) = F_{u_{[0,n)}}(0)$, so $x_0 = 0$ and $y_m = F_{v_{[0,m)}} G \Phi(u) = F_{v_{[0,m)}}(0)$, so $y_0 = 0$. Denote by $H_{n,m} = \mathrm{mod}_2(G_{n,m})$. We show by induction that $x_n, y_m \in \{0, \infty\}$, and $H_{n,m}$ is determined by x_n, y_m by the table

x_n, y_m	$0,0$	$0,\infty$	$\infty,0$	∞,∞
$H_{n,m}$	$(\begin{smallmatrix} 0 & 0 \\ 1 & 0 \end{smallmatrix})$	$(\begin{smallmatrix} 1 & 0 \\ 0 & 0 \end{smallmatrix})$	$(\begin{smallmatrix} 0 & 0 \\ 0 & 1 \end{smallmatrix})$	$(\begin{smallmatrix} 0 & 1 \\ 0 & 0 \end{smallmatrix})$

If $x_n = y_m = 0$, then $F_{u_n}^{\bullet}(0) > 1$ so by Lemma 6 either $x_{n+1} = F_{u_n} F_{u_{[0,n)}}(0) = F_{u_n}(x_n) = 0$ and then $H_{n+1,m} = (\begin{smallmatrix} 0 & 0 \\ 1 & 0 \end{smallmatrix}) \cdot (\begin{smallmatrix} 0 & 0 \\ c & 1 \end{smallmatrix})^{-1} = (\begin{smallmatrix} 0 & 0 \\ 1 & 0 \end{smallmatrix}) \cdot (\begin{smallmatrix} 1 & 0 \\ c & 0 \end{smallmatrix}) = (\begin{smallmatrix} 0 & 0 \\ 1 & 0 \end{smallmatrix})$, or $x_{n+1} = \infty$ and then $H_{n+1,m} = (\begin{smallmatrix} 0 & 0 \\ 1 & 0 \end{smallmatrix}) \cdot (\begin{smallmatrix} a & 0 \\ 1 & 0 \end{smallmatrix})^{-1} = (\begin{smallmatrix} 0 & 0 \\ 1 & 0 \end{smallmatrix}) \cdot (\begin{smallmatrix} 0 & 0 \\ 1 & a \end{smallmatrix}) = (\begin{smallmatrix} 0 & 0 \\ 0 & 1 \end{smallmatrix})$. Similarly $F_{v_m}^{\bullet}(0) > 1$ so by Lemma 6 either $y_{m+1} = 0$ and then $H_{n,m+1} = (\begin{smallmatrix} 0 & 0 \\ c & 1 \end{smallmatrix}) \cdot (\begin{smallmatrix} 0 & 0 \\ 1 & 0 \end{smallmatrix}) = (\begin{smallmatrix} 0 & 0 \\ 1 & 0 \end{smallmatrix})$, or $y_{m+1} = \infty$ and then $H_{n,m+1} = (\begin{smallmatrix} a & 1 \\ 0 & 0 \end{smallmatrix}) \cdot (\begin{smallmatrix} 0 & 0 \\ 1 & 0 \end{smallmatrix}) = (\begin{smallmatrix} 1 & 0 \\ 0 & 0 \end{smallmatrix})$. If $(x_n, y_m) = (0, \infty)$, then either $x_{n+1} = 0$ and $H_{n+1,m} = (\begin{smallmatrix} 1 & 0 \\ 0 & 0 \end{smallmatrix}) \cdot (\begin{smallmatrix} 1 & 0 \\ c & 0 \end{smallmatrix}) = (\begin{smallmatrix} 1 & 0 \\ 0 & 0 \end{smallmatrix})$, or $x_{n+1} = \infty$ and $H_{n+1,m} = (\begin{smallmatrix} 1 & 0 \\ 0 & 0 \end{smallmatrix}) \cdot (\begin{smallmatrix} 0 & 0 \\ 1 & a \end{smallmatrix}) = (\begin{smallmatrix} 0 & 0 \\ 0 & 1 \end{smallmatrix})$, or $y_{m+1} = 0$ and $H_{n,m+1} = (\begin{smallmatrix} 0 & 0 \\ 1 & d \end{smallmatrix}) \cdot (\begin{smallmatrix} 0 & 0 \\ 1 & 0 \end{smallmatrix}) = (\begin{smallmatrix} 0 & 0 \\ 1 & 0 \end{smallmatrix})$, or $y_{m+1} = \infty$ and $H_{n,m+1} = (\begin{smallmatrix} 0 & 0 \\ 0 & 1 \end{smallmatrix}) \cdot (\begin{smallmatrix} 1 & 0 \\ 0 & 0 \end{smallmatrix}) = (\begin{smallmatrix} 1 & 0 \\ 0 & 0 \end{smallmatrix})$. If $(x_n, y_m) = (\infty, 0)$ then either $x_{n+1} = 0$ and $H_{n+1,m} = (\begin{smallmatrix} 0 & 0 \\ 0 & 1 \end{smallmatrix}) \cdot (\begin{smallmatrix} d & 0 \\ 1 & 0 \end{smallmatrix}) = (\begin{smallmatrix} 0 & 0 \\ 1 & 0 \end{smallmatrix})$, or $x_{n+1} = \infty$ and $H_{n+1,m} = (\begin{smallmatrix} 0 & 0 \\ 0 & 1 \end{smallmatrix}) \cdot (\begin{smallmatrix} 0 & 0 \\ 0 & 1 \end{smallmatrix}) = (\begin{smallmatrix} 0 & 0 \\ 0 & 1 \end{smallmatrix})$, or $y_{m+1} = 0$ and $H_{n,m+1} = (\begin{smallmatrix} 0 & 0 \\ c & 1 \end{smallmatrix}) \cdot (\begin{smallmatrix} 0 & 0 \\ 0 & 1 \end{smallmatrix}) = (\begin{smallmatrix} 0 & 0 \\ 0 & 1 \end{smallmatrix})$ or $y_{m+1} = \infty$ and $H_{n,m+1} = (\begin{smallmatrix} a & 1 \\ 0 & 0 \end{smallmatrix}) \cdot (\begin{smallmatrix} 0 & 0 \\ 0 & 1 \end{smallmatrix}) = (\begin{smallmatrix} 0 & 1 \\ 0 & 0 \end{smallmatrix})$. If $(x_n, y_n) = (\infty, \infty)$ then either $x_{n+1} = 0$ and $H_{n+1,m} = (\begin{smallmatrix} 0 & 1 \\ 0 & 0 \end{smallmatrix}) \cdot (\begin{smallmatrix} 1 & 0 \\ d & 0 \end{smallmatrix}) = (\begin{smallmatrix} 1 & 0 \\ 0 & 0 \end{smallmatrix})$, or $x_{n+1} = \infty$ and $H_{n+1,m} = (\begin{smallmatrix} 0 & 1 \\ 0 & 0 \end{smallmatrix}) \cdot (\begin{smallmatrix} 0 & 0 \\ 0 & 1 \end{smallmatrix}) = (\begin{smallmatrix} 0 & 1 \\ 0 & 0 \end{smallmatrix})$, or $y_{m+1} = 0$ and $H_{n,m+1} = (\begin{smallmatrix} 0 & 0 \\ 1 & d \end{smallmatrix}) \cdot (\begin{smallmatrix} 0 & 1 \\ 0 & 0 \end{smallmatrix}) = (\begin{smallmatrix} 0 & 0 \\ 0 & 1 \end{smallmatrix})$, or $y_{m+1} = \infty$ and $H_{n,m+1} = (\begin{smallmatrix} 1 & b \\ 0 & 0 \end{smallmatrix}) \cdot (\begin{smallmatrix} 0 & 1 \\ 0 & 0 \end{smallmatrix}) = (\begin{smallmatrix} 0 & 1 \\ 0 & 0 \end{smallmatrix})$. It follows that in all cases $\det(G_{n,m}) = 2^{n+m} \det(G)$. If $n + m \neq n' + m'$, then $G_{n,m} \neq G_{n',m'}$ and the corresponding states of the transducer must be different. Thus the number of states cannot be finite.

11   Conclusions

We have shown that the only analytical functions in extended real line computable by finite state transducers in Möbius number systems are Möbius transformations. However, many questions remain still open. For example, without the assumption of analyticity, we obtain a wider class of functions. Namely, in the signed binary system from Example 1, the function $f(x) = |x|$ can be also computed by a finite state transducer. Thus the question arises, which continuous, or differentiable (to some degree) functions can be computed in such a way. Similarly, one can also ask which "wild" functions, such as nowhere continuous, or continuous but nowhere differentiable functions, can be computable by finite state transducers in Möbius number systems.

Acknowledgment. The research was supported by the Czech Science Foundation research project GAČR 13-03538S and by the Grant Agency of the Czech Technical University in Prague, grant No. SGS14/205/OHK4/3T/14.

References

1. Avizienis, A.: Signed-digit number representations for fast parallel arithmetic. IRE Transactions on Electronic Computers EC-10 (1961)
2. Beardon, A.F.: The geometry of discrete groups. Springer, Berlin (1995)
3. Delacourt, M., Kůrka, P.: Finite state transducers for modular möbius number systems. In: Rovan, B., Sassone, V., Widmayer, P. (eds.) MFCS 2012. LNCS, vol. 7464, pp. 323–334. Springer, Heidelberg (2012)
4. Frougny, C.: On-line addition in real base. In: Kutyłowski, M., Wierzbicki, T., Pacholski, L. (eds.) MFCS 1999. LNCS, vol. 1672, pp. 1–11. Springer, Heidelberg (1999)
5. Gosper, R.W.: Continued fractions arithmetic (1977) (unpublished manuscript), http://www.tweedledum.com/rwg/cfup.htm
6. Culik II, K., Karhumaki, J.: Finite automata computing real functions. SIAM Journal on Computing 23(4), 789–914 (1994)
7. Kari, J., Kazda, A., Steinby, P.: On continuous weighted finite automata. Linear Algebra and its Applications 436, 1791–1824 (2012)
8. Konečný, M.: Real functions incrementally computable by finite automata. Theoretical Computer Science 315(1), 109–133 (2004)
9. Kornerup, P., Matula, D.W.: Finite precision number systems and arithmetic. Cambridge University Press, Cambridge (2010)
10. Kornerup, P., Matula, D.W.: An algorithm for redundant binary bit-pipelined rational arithmetic. IEEE Transactions on Computers 39(8), 1106–1115 (1990)
11. Kůrka, P.: Topological and symbolic dynamics, volume 11 of Cours spécialisés. Société Mathématique de France, Paris (2003)
12. Kůrka, P.: Möbius number systems with sofic subshifts. Nonlinearity 22, 437–456 (2009)
13. Kůrka, P., Delacourt, M.: The unary arithmetical algorithm in bimodular number systems. In: 2013 IEEE 21st Symposium on Computer Arithmetic ARITH-21, pp. 127–134. IEEE Computer Society (2013)
14. Potts, P.J.: Exact real arithmetic using Möbius transformations. PhD thesis, University of London, Imperial College, London (1998)
15. Raney, G.N.: On contiuned fractions and finite automata. Mathematische Annalen 206, 265–283 (1973)
16. Vuillemin, J.E.: Exact real computer arithmetic with continued fractions. IEEE Transactions on Computers 39(8), 1087–1105 (1990)

Partial Derivative and Position Bisimilarity Automata

Eva Maia, Nelma Moreira, and Rogério Reis

CMUP & DCC, Faculdade de Ciências da Universidade do Porto
Rua do Campo Alegre, 4169-007 Porto, Portugal
{emaia,nam,rvr}@dcc.fc.up.pt

Abstract. Minimization of nondeterministic finite automata (NFA) is a hard problem (PSPACE-complete). Bisimulations are then an attractive alternative for reducing the size of NFAs, as even bisimilarity (the largest bisimulation) is almost linear using the Paige and Tarjan algorithm. NFAs obtained from regular expressions (REs) can have the number of states linear with respect to the size of the REs and conversion methods from REs to equivalent NFAs can produce NFAs without or with transitions labelled with the empty word (ε-NFA). The standard conversion without ε-transitions is the position automaton, $\mathcal{A}_{pos}$. Other conversions, such as partial derivative automata ($\mathcal{A}_{pd}$) or follow automata ($\mathcal{A}_f$), were proven to be quotients of the position automata (by some bisimulations). Recent experimental results suggested that for REs in (normalized) star normal form the position bisimilarity almost coincide with the $\mathcal{A}_{pd}$ automaton. Our goal is to have a better characterization of $\mathcal{A}_{pd}$ automata and their relation with the bisimilarity of the position automata. In this paper, we consider $\mathcal{A}_{pd}$ automata for regular expressions without Kleene star and establish under which conditions they are isomorphic to the bisimilarity of $\mathcal{A}_{pos}$.

1   Introduction

Regular expressions (REs), because of their succinctness and clear syntax, are the common choice to represent regular languages. The minimal deterministic finite automaton (DFA) equivalent to a RE can be exponentially larger than the RE. However, nondeterministic finite automata (NFAs) equivalent to REs can have the number of states linear with respect to (w.r.t) the size of the REs. But, minimization of NFAs is a hard problem (PSPACE-complete). Bisimulations are then an attractive alternative for reducing the size of NFAs, as even bisimilarity (the largest bisimulation) can be computed in almost linear time using the Paige and Tarjan algorithm [20].

Conversion methods from REs to equivalent NFAs can produce NFAs without or with transitions labelled with the empty word (ε-NFA). The standard conversion without ε-transitions is the position automaton ($\mathcal{A}_{pos}$) [12,17]. Other conversions such as partial derivative automata ($\mathcal{A}_{pd}$) [1,18], follow automata

M. Holzer and M. Kutrib (Eds.): CIAA 2014, LNCS 8587, pp. 264–277, 2014.

$(\mathcal{A}_f)$ [14], or the construction by Garcia et al. $(\mathcal{A}_u)$ [11] were proved to be quotients of the position automata, by specific bisimulations[1] [10,14]. When REs are in (normalized) star normal form, i.e. when subexpressions of the star operator do not accept ε, the $\mathcal{A}_{pd}$ automaton is a quotient of the $\mathcal{A}_f$ [8].

The $\mathcal{A}_{pos}$ bisimilarity was studied in [15], and of course it is always not larger than all other quotients. Nevertheless, some experimental results on uniform random generated REs suggested that for REs in (normalized) star normal form the $\mathcal{A}_{pos}$ bisimilarity automata almost coincide with the $\mathcal{A}_{pd}$ automata [13].

Our goal is to have a better characterization of $\mathcal{A}_{pd}$ automata and their relation with the $\mathcal{A}_{pos}$ bisimilarity. All the above mentioned automata ($\mathcal{A}_{pos}$, $\mathcal{A}_{pd}$, $\mathcal{A}_f$, and $\mathcal{A}_u$) can be obtained from a given RE by specific algorithms (without considering the correspondent bisimulation of $\mathcal{A}_{pos}$) in quadratic time. We aim to obtain a similar algorithm that computes, directly from a regular expression, the position bisimilarity automaton.

In this paper, we review the construction of $\mathcal{A}_{pd}$ as a quotient of $\mathcal{A}_{pos}$ and study several of its properties. For regular expressions without Kleene star we characterize the $\mathcal{A}_{pd}$ automata and we prove that the $\mathcal{A}_{pd}$ automaton is isomorphic to the position bisimilarity automaton, under certain conditions. Thus, for these special regular expressions, we conclude that the $\mathcal{A}_{pd}$ is an optimal conversion method. We close considering the difficulties of relating the two automata for general regular expressions.

2   Regular Expressions and Automata

Given an alphabet $\Sigma = \{\sigma_1, \sigma_2, \ldots, \sigma_k\}$ of size k, the set RE of *regular expressions* α over Σ is defined by the following grammar:

$$\alpha := \emptyset \mid \varepsilon \mid \sigma_1 \mid \cdots \mid \sigma_k \mid (\alpha + \alpha) \mid (\alpha \cdot \alpha) \mid (\alpha)^\star, \tag{1}$$

where the symbol $\cdot$ is often omitted. If two regular expressions α and β are syntactically equal, we write $\alpha \equiv \beta$. The *size* of a regular expression α, $|\alpha|$, is its number of symbols, disregarding parenthesis; its *alphabetic size*, $|\alpha|_\Sigma$, is the number of occurrences of letters from Σ; and $|\alpha|_\varepsilon$ denotes the number of occurrences of ε in α. A regular expression α is *linear* if all its letters are distinct.

The language represented by a RE α is denoted by $\mathcal{L}(\alpha)$. Two REs α and β are *equivalent* if $\mathcal{L}(\alpha) = \mathcal{L}(\beta)$, and one writes $\alpha = \beta$. We define $\varepsilon(\alpha) = \varepsilon$ if $\varepsilon \in \mathcal{L}(\alpha)$ and $\varepsilon(\alpha) = \emptyset$, otherwise. We can inductively define $\epsilon(\alpha)$ as follows:

$$\begin{aligned}
\varepsilon(\sigma) = \varepsilon(\emptyset) &= \emptyset & \varepsilon(\alpha + \beta) &= \begin{cases} \varepsilon & \text{if } (\varepsilon(\alpha) = \varepsilon) \vee (\varepsilon(\beta) = \varepsilon) \\ \emptyset & \text{otherwise} \end{cases} \\
\varepsilon(\varepsilon) &= \varepsilon \\
\varepsilon(\alpha^*) &= \varepsilon & \varepsilon(\alpha\beta) &= \begin{cases} \varepsilon & \text{if } (\varepsilon(\alpha) = \varepsilon) \wedge (\varepsilon(\beta) = \varepsilon) \\ \emptyset & \text{otherwise} \end{cases}
\end{aligned}$$

---

[1] Also called right-invariant equivalence relations.

The algebraic structure $(\mathsf{RE}, +, ., \emptyset, \varepsilon)$ constitutes an idempotent semiring, and with the Kleene star operator $\star$, a Kleene algebra. The axioms for the star operator can be defined by the following rules [16]:

$$\varepsilon + \alpha\alpha^\star = \alpha^\star \text{ and } \varepsilon + \alpha^\star\alpha = \alpha^\star,$$

$$\beta + \alpha\gamma \leq \gamma \implies \alpha^\star\beta \leq \gamma \text{ and } \beta + \gamma\alpha \leq \gamma \implies \beta\alpha^\star \leq \gamma,$$

where $\alpha \leq \beta$ means $\alpha + \beta = \beta$. Given a language $L \subseteq \Sigma^\star$ and a word $w \in \Sigma^\star$, the *left-quotient* of L w.r.t. w is the language $w^{-1}L = \{x \mid wx \in L\}$. Brzozowski [6] defined the syntactic notion of derivative of a RE α w.r.t. a word w, $d_w(\alpha)$, such that $\mathcal{L}(d_w(\alpha)) = w^{-1}\mathcal{L}(\alpha)$, and showed that the set of derivatives of a regular expression w.r.t. all words is finite, modulo associativity (A), commutativity (C), and idempotence (I) of $+$ (which we denote by modulo ACI).

In this paper, we only consider REs α *normalized* under the following conditions:

- The expression α is *reduced* according to:
  - the equations $\emptyset + \alpha = \alpha + \emptyset = \alpha$, $\varepsilon.\alpha = \alpha.\varepsilon = \alpha$, $\emptyset.\alpha = \alpha.\emptyset = \emptyset$;
  - and the rule, for all subexpressions β of α, $\beta = \gamma + \varepsilon \implies \varepsilon(\gamma) = \emptyset$.
- The expression α is in *star normal form* (snf) [5], i.e. for all subexpressions $\beta^\star$ of α, $\varepsilon(\beta) = \emptyset$.

Every regular expression can be converted into an equivalent normalized RE in linear time.

A *nondeterministic finite automaton* (NFA) is a five-tuple $A = (Q, \Sigma, \delta, q_0, F)$ where Q is a finite set of states, Σ is a finite alphabet, q_0 in Q is the initial state, $F \subseteq Q$ is the set of final states, and $\delta : Q \times \Sigma \to 2^Q$ is the transition function. This transition function can be extended to words in the natural way. The language accepted by A is $\mathcal{L}(A) = \{w \in \Sigma^\star \mid \delta(q, w) \cap F \neq \emptyset\}$. Two NFAs are *equivalent* if they accept the same language. If two NFAs A and B are isomorphic, we write $A \simeq B$. An NFA is deterministic (DFA) if for all $(q, \sigma) \in Q \times \Sigma$, $|\delta(q, \sigma)| \leq 1$. A DFA is minimal if there is no equivalent DFA with fewer states. Minimal DFAs are unique up to isomorphism.

A binary symmetric and reflexive relation R on Q is a *bisimulation* if $\forall p, q \in Q$ and $\forall \sigma \in \Sigma$ if pRq then

- $p \in F$ if and only if $q \in F$;
- $\forall p' \in \delta(p, \sigma) \ \exists q' \in \delta(q, \sigma)$ such that $p'Rq'$.

The sets of bisimulations on Q are closed under finite union. The largest bisimulation, i.e., the union of all bisimulation relations on Q, is called *bisimilarity* ($\equiv_b$), and it is an equivalence relation. Bisimilarity can be computed in almost linear time using the Paige and Tarjan algorithm [20]. If R is a equivalence bisimulation on Q the *quotient automaton* A/R can be constructed by $A/R = (Q/R, \Sigma, \delta/R, [q_0], F/R)$, where $[q]$ is the equivalence class that contains $q \in Q$; $S/R = \{[q] \mid q \in S\}$, with $S \subseteq Q$; and $\delta/R = \{([p], \sigma, [q]) \mid (p, \sigma, q) \in \delta\}$.

It is easy to see that $\mathcal{L}(^A\!/_R) = \mathcal{L}(A)$. The quotient automaton $^A\!/_{\equiv_b}$ is the minimal automaton among all quotient automata $^A\!/_R$, where R is a bisimulation on Q, and it is unique up to isomorphism. By language abuse, we will call $^A\!/_{\equiv_b}$ the *bisimilarity* of automaton A. If A is a DFA, $^A\!/_{\equiv_b}$ is the minimal DFA equivalent to A.

2.1   Position Automaton

The position automaton was introduced independently by Glushkov [12] and McNaughton and Yamada [17]. The states in the position automaton, equivalent to a regular expression α, correspond to the positions of letters in α plus an additional initial state. Let $\overline{\alpha}$ denote the linear regular expression obtained by marking each letter with its position in α, i.e., $\mathcal{L}(\overline{\alpha}) \in \overline{\Sigma}^*$ where $\overline{\Sigma} = \{\sigma_i \mid \sigma \in \Sigma, 1 \le i \le |\alpha|_\Sigma\}$. For example, the marked version of the regular expression $\tau = (ab^\star + b)^\star a$ is $\overline{\tau} = (a_1 b_2^\star + b_3)^\star a_4$. The same notation is used to remove the markings, i.e., $\overline{\overline{\alpha}} = \alpha$. Let $\mathsf{Pos}(\alpha) = \{1, 2, \ldots, |\alpha|_\Sigma\}$, and $\mathsf{Pos}_0(\alpha) = \mathsf{Pos}(\alpha) \cup \{0\}$.

We consider the construction of the position automaton following Berry and Sethi [3] and Champarnaud and Ziadi [10], i.e. the c-continuation automaton. In this way the relation between $\mathcal{A}_{pos}$ and $\mathcal{A}_{pd}$ is immediate.

If α is linear, for every symbol $\sigma \in \overline{\Sigma}$ and every word $w \in \overline{\Sigma}^*$, $d_{w\sigma}(\alpha)$ is either $\emptyset$ or unique modulo ACI [3]. If $d_{w\sigma}(\alpha)$ is different from $\emptyset$, it is named *c-continuation* of α w.r.t. $\sigma \in \overline{\Sigma}$, and denoted by $c_\sigma(\alpha)$. We define $c_0(\alpha) = d_\varepsilon(\alpha) = \alpha$. This means that we can associate to each position $i \in \mathsf{Pos}_0(\alpha)$, a unique c-continuation. For example, given $\overline{\tau} = (a_1 b_2^\star + b_3)^\star a_4$ we have $c_{a_1}(\overline{\tau}) = b_2^\star \overline{\tau}$, $c_{b_2}(\overline{\tau}) = b_2^\star \overline{\tau}$, $c_{b_3}(\overline{\tau}) = \overline{\tau}$, and $c_{a_4}(\overline{\tau}) = \varepsilon$. The c-continuation automaton for α is $\mathcal{A}_c(\alpha) = (Q_c, \Sigma, \delta_c, q_0, F_c)$ where $Q_c = \{q_0\} \cup \{(i, c_{\sigma_i}(\overline{\alpha})) \mid i \in \mathsf{Pos}(\alpha)\}$, $q_0 = (0, c_0(\overline{\alpha}))$, $F_c = \{(i, c_{\sigma_i}(\overline{\alpha})) \mid \varepsilon(c_{\sigma_i}(\overline{\alpha})) = \varepsilon\}$, $\delta_c = \{((i, c_{\sigma_i}(\overline{\alpha})), b, (j, c_{\sigma_j}(\overline{\alpha}))) \mid \overline{\sigma_j} = b \land d_{\sigma_j}(c_{\sigma_i}(\overline{\alpha})) \ne \emptyset\}$. The $\mathcal{A}_c(\tau)$ is represented in Figure 1.

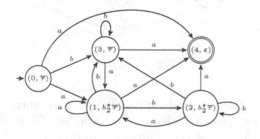

Fig. 1. $\mathcal{A}_c(\tau)$

If we ignore the c-continuations in the label of each state, we obtain the position automaton.

Proposition 1 (Champarnaud & Ziadi). $\forall \alpha \in \mathsf{RE}, \mathcal{A}_{pos}(\alpha) \simeq \mathcal{A}_c(\alpha)$.

2.2 Partial Derivative Automaton

The partial derivative automaton of a regular expression was introduced independently by Mirkin [18] and Antimirov [1]. Champarnaud and Ziadi [9] proved that the two formulations are equivalent. For a RE α and a symbol $\sigma \in \Sigma$, the set of partial derivatives of α w.r.t. σ is defined inductively as follows:

$$\partial_\sigma(\emptyset) = \partial_\sigma(\varepsilon) = \emptyset \qquad\qquad \partial_\sigma(\alpha + \beta) = \partial_\sigma(\alpha) \cup \partial_\sigma(\beta)$$
$$\partial_\sigma(\sigma') = \begin{cases} \{\varepsilon\}, & \text{if } \sigma' = \sigma \\ \emptyset, & \text{otherwise} \end{cases} \qquad \begin{aligned} \partial_\sigma(\alpha\beta) &= \partial_\sigma(\alpha)\beta \cup \varepsilon(\alpha)\partial_\sigma(\beta) \\ \partial_\sigma(\alpha^\star) &= \partial_\sigma(\alpha)\alpha^\star \end{aligned} \tag{2}$$

where for any $S \subseteq \mathsf{RE}$, $\beta \in \mathsf{RE}$, $S\emptyset = \emptyset S = \emptyset$, $S\varepsilon = \varepsilon S = S$, and $S\beta = \{\alpha\beta | \alpha \in S\}$ if $\beta \neq \emptyset$, and $\beta \neq \varepsilon$.

The definition of partial derivative can be extended to sets of regular expressions, words, and languages. Given $\alpha \in \mathsf{RE}$ and $\sigma \in \Sigma$, $\partial_\sigma(S) = \bigcup_{\alpha \in S} \partial_\sigma(\alpha)$ for $S \subseteq \mathsf{RE}$, $\partial_\varepsilon(\alpha) = \alpha$ and $\partial_{w\sigma}(\alpha) = \partial_\sigma(\partial_w(\alpha))$, for any $w \in \Sigma^\star, \sigma \in \Sigma$, and $\partial_L(\alpha) = \bigcup_{w \in L} \partial_w(\alpha)$ for $L \subseteq \Sigma^\star$. We know that $\bigcup_{\tau \in \partial_w(\alpha)} \mathcal{L}(\tau) = w^{-1}\mathcal{L}(\alpha)$. The set of all partial derivatives of α w.r.t. words is denoted by $\mathsf{PD}(\alpha) = \bigcup_{w \in \Sigma^\star} \partial_w(\alpha)$. Note that the set $\mathsf{PD}(\alpha)$ is always finite [1], as opposed to what happens for the Brzozowski derivatives set which is only finite modulo ACI.

The partial derivative automaton is defined by $\mathcal{A}_{pd}(\alpha) = (\mathsf{PD}(\alpha), \Sigma, \delta_{pd}, \alpha, F_{pd})$, where $\delta_{pd} = \{(\tau, \sigma, \tau') \mid \tau \in \mathsf{PD}(\alpha) \text{ and } \tau' \in \partial_\sigma(\tau)\}$ and $F_{pd} = \{\tau \in \mathsf{PD}(\alpha) \mid \varepsilon(\tau) = \varepsilon\}$. Considering $\tau = (ab^\star + b)^\star a$, the Figure 2 shows $\mathcal{A}_{pd}(\tau)$.

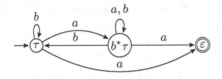

Fig. 2. $\mathcal{A}_{pd}(\tau)$

Note that if α is a linear regular expression, for every word w, $|\partial_w(\alpha)| \leq 1$ and the partial derivative coincide with $d_w(\alpha)$ modulo ACI. Given the c-continuation automaton $\mathcal{A}_c(\alpha)$, let $\equiv_c$ be the bisimulation on Q_c defined by $(i, c_{\sigma_i}(\overline{\alpha})) \equiv_c (j, c_{\sigma_j}(\overline{\alpha}))$ if $\overline{c_{\sigma_i}(\overline{\alpha})} \equiv \overline{c_{\sigma_j}(\overline{\alpha})}$. That the $\mathcal{A}_{pd}$ is isomorphic to the resulting quotient automaton, follows from the proposition below. For our running example, we have $(0, c_\varepsilon) \equiv_c (3, c_{b_3})$ and $(1, c_{a_1}) \equiv_c (2, c_{b_2})$. In Figure 2, we can see the merged states, and that the corresponding REs are unmarked.

Proposition 2 (Champarnaud & Ziadi). $\forall \alpha \in \mathsf{RE}, \mathcal{A}_{pd}(\alpha) \simeq {}^{\mathcal{A}_c(\alpha)}\!/_{\equiv_c}$.

Inductive Characterization of $\mathcal{A}_{pd}$. Mirkin's construction of the $\mathcal{A}_{pd}(\alpha)$ is based on solving a system of equations $\alpha_i = \sigma_1\alpha_{i1} + \ldots + \sigma_k\alpha_{ik} + \varepsilon(\alpha_i)$, with $\alpha_0 \equiv \alpha$ and α_{ij}, $1 \le j \le k$, linear combinations the α_i, $0 \le i \le n$, $n \ge 0$. A solution $\pi(\alpha) = \{\alpha_1, \ldots, \alpha_n\}$ can be obtained inductively on the structure of α as follows:

$$
\begin{aligned}
\pi(\emptyset) &= \emptyset & \pi(\alpha \cup \beta) &= \pi(\alpha) \cup \pi(\beta) \\
\pi(\varepsilon) &= \emptyset & \pi(\alpha\beta) &= \pi(\alpha)\beta \cup \pi(\beta) \\
\pi(\sigma) &= \{\varepsilon\} & \pi(\alpha^\star) &= \pi(\alpha)\alpha^\star.
\end{aligned}
\tag{3}
$$

Champarnaud and Ziadi [9] proved that $\mathsf{PD}(\alpha) = \pi(\alpha) \cup \{\alpha\}$ and that the two constructions led to the same automaton.

As noted by Broda et. al [4], Mirkin's algorithm to compute $\pi(\alpha)$ also provides an inductive definition of the set of transitions of $\mathcal{A}_{pd}(\alpha)$. Let $\varphi(\alpha) = \{(\sigma, \gamma) \mid \gamma \in \partial_\sigma(\alpha), \sigma \in \Sigma\}$ and $\lambda(\alpha) = \{\alpha' \mid \alpha' \in \pi(\alpha), \varepsilon(\alpha') = \varepsilon\}$, where both sets can be inductively defined using (2) and (3). We have, $\delta_{pd} = \{\alpha\} \times \varphi(\alpha) \cup F(\alpha)$ where the result of the $\times$ operation is seen as a set of triples and the set F is defined inductively by:

$$
\begin{aligned}
F(\emptyset) = F(\varepsilon) = F(\sigma) &= \emptyset, \ \sigma \in \Sigma \\
F(\alpha + \beta) &= F(\alpha) \cup F(\beta) \\
F(\alpha\beta) &= F(\alpha)\beta \cup F(\beta) \cup \lambda(\alpha)\beta \times \varphi(\beta) \\
F(\alpha^\star) &= F(\alpha)\alpha^\star \cup (\lambda(\alpha) \times \varphi(\alpha))\alpha^\star.
\end{aligned}
\tag{4}
$$

Then, we can inductively construct the partial derivative automaton of α using the following result.

Proposition 3. *For all $\alpha \in \mathsf{RE}$, and $\lambda'(\alpha) = \lambda(\alpha) \cup \varepsilon(\alpha)\{\alpha\}$,*

$$\mathcal{A}_{pd}(\alpha) = (\pi(\alpha) \cup \{\alpha\}, \Sigma, \{\alpha\} \times \varphi(\alpha) \cup F(\alpha), \alpha, \lambda'(\alpha)),$$

Figure 3 illustrates this inductive construction, where we assume that states are merged whenever they correspond to syntactically equal REs.

A new proof of Proposition 2 can also be given using the function π. Let π' be a function that coincides with π except that $\pi'(\sigma) = \{(\sigma, \varepsilon)\}$ and in the two last rules the regular expression, either β or $\alpha^\star$, is concatenated to the second component of each pair in π'.

Proposition 4. *Let $\alpha \in RE$, $\pi'(\overline{\alpha}) = \{(i, c_{\sigma_i}(\overline{\alpha})) \mid i \in \mathsf{Pos}(\overline{\alpha})\}$.*

By Proposition 4, we can conclude that if we compute $\pi'(\overline{\alpha})$ we obtain exactly[2] the set of states $Q_c \setminus \{(0, c_\varepsilon)\}$ of the c-continuation automaton $\mathcal{A}_c(\alpha)$. Then it is easy to see that $\pi(\alpha)$ is obtained by unmarking the c-continuations and removing the first component of each pair, and thus $Q_c/_{\equiv_c} = \pi(\alpha) \cup \{\alpha\}$. Considering $\overline{\tau} = (a_1 b_2^\star + b_3)^\star a_4$, $\pi'(\overline{\tau}) = \{(a_1, b_2^\star\overline{\tau}), (b_2, b_2^\star\overline{\tau}), (b_3, \overline{\tau}), (b_4, \varepsilon)\}$, which corresponds exactly to the set of states (excluding the initial) of $\mathcal{A}_c(\overline{\tau})$, presented in Figure 1. The set $\pi(\tau)$ is $\{b^\star\tau, \tau, \varepsilon\}$. That the other components are quotients, also follows.

---

[2] Considering, for each position i, the marked letter σ_i.

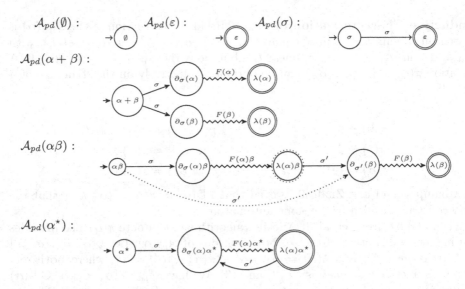

Fig. 3. Inductive construction of $\mathcal{A}_{pd}$. The initial states are final if ε belongs to its language. Note that only if $\varepsilon(\beta) = \varepsilon$ the dotted arrow in $\mathcal{A}_{pd}(\alpha\beta)$ exists and the state $\lambda(\alpha)\beta$ is final.

3 $\mathcal{A}_{pd}$ Characterizations and Bisimilarity

We aim to obtain some characterizations of $\mathcal{A}_{pd}$ automaton and to determine when it coincides with the bisimilarity of the position automaton, i.e. $\mathcal{A}^{pos}/_{\equiv_b}$. We assume that all regular expressions are normalized. This ensures that the $\mathcal{A}_{pd}$ is a quotient of $\mathcal{A}_f$, so the smaller known direct ε-free automaton construction from a regular expression. As we discuss in Subsection 3.4, to solve the problem in the general case it is difficult, mainly because the lack of unique normal forms. Here, we give some partial solutions. First, we consider linear regular expressions and, in Subsection 3.2, we solve the problem for regular expressions representing finite languages.

3.1   Linear Regular Expressions

Given a linear regular expression α, it is obvious that the position automaton $\mathcal{A}_{pos}(\alpha)$ is a DFA. In this case, all positions correspond to distinct letters and transitions from a same state are all distinct. Thus, $\mathcal{A}_{pd}(\alpha)$ is also a DFA.

Proposition 5. *If α is a normalized linear regular expression, $\mathcal{A}_{pd}(\alpha)$ is minimal.*

Proof. By [8, Theorem 2] we know that

$$c_{\sigma_x}(\alpha) \not\equiv c_{\sigma_y}(\alpha) \Leftrightarrow \{\sigma \mid \partial_\sigma(c_{\sigma_x}(\alpha)) \neq \emptyset\} \neq \{\sigma \mid \partial_\sigma(c_{\sigma_y}(\alpha)) \neq \emptyset\}$$

where α is a normalized linear regular expression and σ_x and σ_y are two distinct letters. We want to prove that any two states $c_{\sigma_x}(\alpha)$ and $c_{\sigma_y}(\alpha)$ of $\mathcal{A}_{pd}(\alpha)$ are distinguishable. Consider $\sigma' \in \Sigma$ such that $\sigma' \in \{\sigma \mid \partial_\sigma(c_{\sigma_x}(\alpha)) \neq \emptyset\}$ but $\sigma' \notin \{\sigma \mid \partial_\sigma(c_{\sigma_y}(\alpha)) \neq \emptyset\}$. Then $\delta_{pd}(c_{\sigma_x}(\alpha), \sigma') = c_{\sigma'}(\alpha)$. By construction, we know that $\exists w \in \Sigma^*$ such that $\delta_{pd}(c_{\sigma'}(\alpha), w) \in F_{pd}$. Let $w' = \sigma'w$. Therefore $\delta_{pd}(c_{\sigma_x}(\alpha), w') = \delta_{pd}(c_{\sigma'}(\alpha), w) \in F_{pd}$ and either δ_{pd} is not defined for $(c_{\sigma_y}(\alpha), w')$ or $\delta_{pd}(c_{\sigma_y}(\alpha), w')$ is a non final dead state. Thus, the two states are distinguishable. $\qquad\square$

It follows, from this, that for any linear regular expressions α,

$$\mathcal{A}_{pd}(\alpha) \simeq {}^{\mathcal{A}_{pos}(\alpha)}\!/_{\equiv_b}.$$

3.2   Finite Languages

In this section, we consider normalized regular expressions without the Kleene star operator, i.e. that represent finite languages. These regular expressions are called *finite regular expressions*. The following results characterize NFAs that are $\mathcal{A}_{pd}$ automaton.

Proposition 6. *The $\mathcal{A}_{pd}(\alpha) = (\mathsf{PD}(\alpha), \Sigma, \delta_\alpha, \alpha, F_\alpha)$ automaton of any finite regular expression $\alpha \not\equiv \emptyset$ has the following properties:*

1. *The state ε always exists and it is a final state;*
2. *The state ε is reachable from any other state;*
3. *All other final states, $q \in F_\alpha \setminus \{\varepsilon\}$, are of the form $(\alpha_1 + \varepsilon)\ldots(\alpha_n + \varepsilon)$;*
4. *$|F_\alpha| \leq |\alpha|_\varepsilon + 1$;*
5. *The size of each element of $\mathsf{PD}(\alpha)$ is not greater than $|\alpha|$.*

Proof. We use the inductive construction of $\mathcal{A}_{pd}(\alpha)$.

1. For the base cases this is obviously true. If α is $\gamma + \beta$, then $\pi(\alpha) = \pi(\gamma) \cup \pi(\beta)$. As $\varepsilon \in \pi(\gamma)$ and $\varepsilon \in \pi(\beta)$, by inductive hypothesis, then $\varepsilon \in \pi(\alpha)$. If α is $\gamma\beta$, then $\pi(\alpha) = \pi(\gamma)\beta \cup \pi(\beta)$. As $\varepsilon \in \pi(\beta)$, $\varepsilon \in \pi(\alpha)$.
2. If α is ε or σ it is obviously true. Let α be $\gamma + \beta$. The states of $\mathcal{A}_{pd}(\alpha)$ are $\{\alpha\} \cup \pi(\gamma) \cup \pi(\beta)$. By construction, there exists at least a transition from the state α to a (distinct) state in $\pi(\gamma) \cup \pi(\beta)$. Let α be $\gamma\beta$. The states of $\mathcal{A}_{pd}(\alpha)$ are $\{\alpha\} \cup \pi(\gamma)\beta \cup \pi(\beta)$. For $\beta' \in \{\beta\} \cup \pi(\beta)$, $\exists w_\beta \; \varepsilon \in \partial_{w_\beta}(\beta')$. In the same way, for $\gamma' \in \{\gamma\} \cup \pi(\gamma)$, $\exists w_\gamma \; \varepsilon \in \partial_{w_\gamma}(\gamma')$. Thus, for $\alpha' = \gamma'\beta \in \pi(\gamma)\beta$, we can conclude that $\varepsilon \in \partial_{w_\gamma w_\beta}(\alpha')$. From the state α we can reach the state ε because the transitions leaving it go to states which reach the state ε.
3. It is obvious, because final states must accept ε.
4. For the base cases it is obviously true. Let α be $\gamma + \beta$. We know that $|\alpha|_\varepsilon = |\gamma|_\varepsilon + |\beta|_\varepsilon$, $|F_\alpha| \leq |F_\gamma| + |F_\beta| - 1$, and that $\varepsilon(\alpha) = \varepsilon$ if either $\varepsilon(\gamma)$ or $\varepsilon(\beta)$ are ε. Then $|F_\alpha| \leq |\gamma|_\varepsilon + |\beta|_\varepsilon + 1 \leq |\alpha|_\varepsilon + 1$. If α is $\gamma\beta$ we know also that $|\alpha|_\varepsilon = |\gamma|_\varepsilon + |\beta|_\varepsilon$ and that $\varepsilon(\alpha) = \varepsilon$ if $\varepsilon(\gamma)$ and $\varepsilon(\beta)$ are ε. If $\varepsilon(\beta) = \varepsilon$, then $|F_\alpha| \leq |F_\gamma| + |F_\beta| - 1$. Otherwise, $|F_\alpha| = |F_\beta|$. We have, in the both cases, $|F_\alpha| \leq |\gamma|_\varepsilon + |\beta|_\varepsilon + 1 \leq |\alpha|_\varepsilon + 1$.

5. If α is ε or σ it is obvious that the proposition is true. Let α be $\gamma + \beta$. For all $\alpha_i \in \pi(\alpha) = \pi(\gamma) \cup \pi(\beta)$, $|\alpha_i| \leq |\gamma|$ or $|\alpha_i| \leq |\beta|$, and thus $|\alpha_i| \leq |\alpha|$. If α is $\gamma\beta$, then $\pi(\alpha) = \pi(\gamma)\beta \cup \pi(\beta)$. For $\gamma_i \in \pi(\gamma)$, $|\gamma_i| \leq \gamma$. If $\alpha_i \in \pi(\gamma)\beta$, $\alpha_i = \gamma_i\beta$ and $|\alpha_i| \leq |\gamma| + |\beta| \leq |\alpha|$. If $\alpha_i \in \pi(\beta)$, $|\alpha_i| \leq |\beta| \leq |\alpha|$. □

Caron and Ziadi [7] characterized the position automaton in terms of the properties of the underlying digraph. We consider a similar approach to characterize the $\mathcal{A}_{pd}$ for finite languages. We restrict the analysis to acyclic NFAs. We first observe that $\mathcal{A}_{pos}$ are series-parallel automata [19] which is not the case for all $\mathcal{A}_{pd}$ as can be seen considering $\mathcal{A}_{pd}(a(ac + b) + bc)$.

Let $A = (Q, \Sigma, \delta, q_0, F)$ be an acyclic NFA. A is an *hammock* if it has the following properties. If $|Q| = 1$, A has no transitions. Otherwise, there exists an unique $f \in F$ such that for any state $q \in Q$ one can find a path from q_0 to f going through q. The state q_0 is called the *root* and f the *anti-root*. The *rank* of a state $q \in Q$, named $rk(q)$, is the length of the longest word $w \in \Sigma^*$ such that $\delta(q, w) \in F$. In an hammock, the anti-root has rank 0. Each state q of rank $r \geq 1$, has only transitions for states in smaller ranks and at least one transition for a state in rank $r - 1$.

Proposition 7. *For every finite regular expression α, $\mathcal{A}_{pd}(\alpha)$ is an hammock.*

Proof. If the partial derivative automaton has a unique state then it is the $\mathcal{A}_{pd}(\varepsilon)$ or $\mathcal{A}_{pd}(\emptyset)$ which has no transitions. Otherwise, for all $q \in \mathsf{PD}(\alpha)$ there exists at least one path from $q_0 = \alpha$ to q because $\mathcal{A}_{pd}(\alpha)$ is initially connected; also there exists at least one path from q to ε, the anti-root, by Proposition 6, item 2. □

Proposition 8. *An acyclic NFA $A = (Q, \Sigma, \delta, q_0, F)$ is a partial derivative automaton of some finite regular expression α, if the following conditions holds:*

1. *A is an hammock;*
2. *$\forall q, q' \in Q \; rk(q) = rk(q') \implies \exists \sigma \in \Sigma \; \delta(q, \sigma) \neq \delta(q', \sigma)$.*

Proof. First we give an algorithm that allows to associate to each state of an hammock A a regular expression. Then, we show that if the second condition holds, A is the $\mathcal{A}_{pd}(\alpha)$ where α is the RE associated to the initial state.

We label each state q with a regular expression $RE(q)$, considering the states by increasing rank order. We define for the anti-root f, $RE(f) = \varepsilon$. Suppose that all states of ranks less then n are already labelled. Let $q \in Q$ with $rk(q) = n$. For $\sigma \in \Sigma$, with $\delta(q, \sigma) = \{q_1, \ldots, q_m\}$ and $RE(q_i) = \beta_i$ we construct the regular expression $\sigma(\beta_1 + \cdots + \beta_m)$. Then,

$$RE(q) = \sum_{\sigma \in \Sigma} \sigma(\beta_1 + \cdots + \beta_m)$$

where we omit all $\sigma \in \Sigma$ such that $\delta(q, \sigma) = \emptyset$. We have, $RE(q_0) = \alpha$

To show that if A satisfies condition 2. then $A \simeq \mathcal{A}_{pd}(\alpha)$, we need to prove that $RE(q) \not\equiv RE(q')$ for all $q, q' \in Q$ with $q \neq q'$. We prove by induction on the rank. For rank 0, it is obvious. Suppose that all states with rank $m < n$

are labelled by different regular expressions. Let $q \in Q$, with $rk(q) = n$. We must prove that $RE(q) \not\equiv RE(q')$ for all q' with $rk(q') \leq n$. Suppose that $rk(q) = rk(q')$, $RE(q) = \sigma_1(\alpha_1 + \cdots + \alpha_n) + \cdots + \sigma_i(\beta_1 + \cdots + \beta_m)$, and $RE(q') = \sigma'_1(\alpha'_1 + \cdots + \alpha'_{n'}) + \cdots + \sigma'_j(\beta'_1 + \cdots + \beta'_{m'})$. We know that $\exists \sigma \delta(q, \sigma) \neq \delta(q', \sigma)$. Suppose that $\sigma = \sigma_1 = \sigma'_1$. Then we know that $\exists t, t'$ $\alpha_t \neq \alpha'_{t'}$, thus $RE(q) \not\equiv RE(q')$. If $rk(q) > rk(q')$, then there exists a $w \in \Sigma^*$ with $|w| = n$ such that $\delta(q, w) \cap F \neq \emptyset$ and $\delta(q', w) \cap F = \emptyset$. Thus $RE(q) \not\equiv RE(q')$. $\square$

3.3 Comparing $\mathcal{A}_{pd}$ and $\mathcal{A}_{pos}/_{\equiv_b}$

As we already mentioned, there are many (normalized) regular expressions α for which $\mathcal{A}_{pd}(\alpha) \simeq \mathcal{A}_{pos}(\alpha)/_{\equiv_b}$. But, even for REs representing finite languages that is not always true. Taking, for example, $\tau_1 = a(a+b)c + b(ac+bc) + a(c+c)$, we have $\mathsf{PD}(\tau_1) = \{\tau_1, ac + bc, (a + b)c, c + c, c, \varepsilon\}$, $F_{pd} = \{\varepsilon\}$, $\delta_{pd}(\tau_1, a) = \{(a + b)c, c + c\}$, $\delta_{pd}(\tau_1, b) = \{ac + bc\}$, $\delta_{pd}(ac + bc, a) = \delta_{pd}(ac + bc, b) = \delta_{pd}((a + b)c, a) = \delta_{pd}((a + b)c, b) = \{c\}$ and $\delta_{pd}(c + c, c) = \delta_{pd}(c, c) = \{\varepsilon\}$. One can see that $c \equiv_b (c + c)$ and $(ac + bc) \equiv_b (a + b)c$. Thus, $\mathcal{A}_{pos}(\tau_1)/_{\equiv_b}$ has two states less than $\mathcal{A}_{pd}(\tau_1)$. The states that are bisimilar are equivalent modulo the $+$ idempotence and left-distributivity. It is also easy to see that two states are bisimilar if they are equivalent modulo $+$ associativity or $+$ commutativity.

Considering an order $<$ on Σ and that $\cdot < +$, we can extend $<$ to REs. Then, the following rewriting system is confluent and terminating:

$$\alpha + (\beta + \gamma) \to (\alpha + \beta) + \gamma \qquad \text{(+ \textbf{Associativity})}$$
$$\alpha + \beta \to \beta + \alpha \qquad \text{if } \beta < \alpha \ \text{(+ \textbf{Commutativity})}$$
$$\alpha + \alpha \to \alpha \qquad \text{(+ \textbf{Idempotence})}$$
$$(\alpha\beta)\gamma \to \alpha(\beta\gamma) \qquad \text{(. \textbf{Associativity})}$$
$$(\alpha + \gamma)\beta \to \alpha\beta + \gamma\beta \qquad \text{(\textbf{Left distributivity}).}$$

A (normalized) regular expression α that can not be rewritten anymore by this system is called an *irreducible regular expression modulo* ACIAL.

Remark 9. An irreducible regular expression modulo ACIAL α is of the form:

$$w_1 + \ldots + w_n + w'_1\alpha_1 + \ldots + w'_m\alpha_m \tag{5}$$

where w_i, w'_j are words for $1 \leq i \leq n$, $1 \leq j \leq m$, and α_j are expressions of the same form of α, for $1 \leq j \leq m$. For for each normalized RE without the Kleene star operator, there exits a unique normal form.

For example, considering $a < b < c$, the normal form for the RE τ_1 given above is $\tau_2 = ac + a(ac + bc) + b(ac + bc)$ and $\mathcal{A}_{pd}(\tau_2) \simeq \mathcal{A}_{pos}(\tau_2)/_{\equiv_b}$. As we will see next, for normal forms this isomorphism always holds.

The following lemmas are needed to prove the main result.

Lemma 10. *For $\sigma \in \Sigma$, the function ∂_σ is closed modulo* ACIAL.

Proof. We know that α has the form $w_1 + \ldots + w_n + w'_1\alpha_1 + \ldots + w'_i\alpha_m$, where $w_i = \sigma v_i, v_i \in \Sigma^\star$, $w'_j = \sigma v'_j, v'_j \in \Sigma^\star$, $i \in \{1, \cdots, n\}$, $j \in \{1, \cdots, m\}$. Thus, $\forall \sigma \in \Sigma \; \partial_\sigma(\alpha) = \partial_\sigma(w_1) \cup \cdots \cup \partial_\sigma(w_n) \cup \partial_\sigma(w'_1)\alpha_1 \cup \cdots \cup \partial_\sigma(w'_i)\alpha_m$, where $\partial_\sigma(w_i) = v_i$ and $\partial_\sigma(w'_j)\alpha_j = v'_j\alpha_j$. Then it is obvious that the both possible results are irreducible modulo ACIAL. Thus the proposition holds. $\qquad\square$

Lemma 11. *For $w, w' \in \Sigma^\star$,*

1. *$(\forall \sigma \in \Sigma) \; |\partial_\sigma(w)| \leq 1$.*
2. *$w \neq w' \implies (\forall \sigma \in \Sigma) \; \partial_\sigma(w) \neq \partial_\sigma(w') \vee \partial_\sigma(w) = \partial_\sigma(w') = \emptyset$.*
3. *$(\forall \sigma \in \Sigma)\partial_\sigma(w\alpha) = \partial_\sigma(w)\alpha = \{w'\alpha\}$, if $w = \sigma w'$.*

Proof. 1. Let $w = \sigma w'$. Then $\partial_\sigma(w) = \partial_\sigma(\sigma w') = \{w'\}$. For $\sigma \neq \sigma', \partial_{\sigma'}(w) = \emptyset$.
2. We need to consider three cases:
   (a) if $\sigma \notin \mathsf{First}(w)$ and $\sigma \notin \mathsf{First}(w')$ then $\partial_\sigma(w) = \emptyset$ and $\partial_\sigma(w') = \emptyset$.
   (b) if $\sigma \in \mathsf{First}(w)$ and $\sigma \notin \mathsf{First}(w')$ then $\partial_\sigma(w) \neq \emptyset$ and $\partial_\sigma(w') = \emptyset$.
   (c) if $\sigma \in \mathsf{First}(w)$, $\sigma \in \mathsf{First}(w')$ and $w = \sigma v$, $w' = \sigma v'$ then $v \neq v'$. As $\partial_\sigma(w) = v$ and $\partial_\sigma(w') = v'$ then $\partial_\sigma(w) \neq \partial_\sigma(w')$.
3. Let $w = \sigma w'$. Then $\partial_\sigma(w\alpha) = \partial_\sigma(w)\alpha = \partial_\sigma(\sigma w')\alpha = \{w'\alpha\}$. For $\sigma \neq \sigma'$, $\partial_{\sigma'}(w\alpha) = \emptyset$. $\qquad\square$

Proposition 12. *Given α and β irreducible finite regular expressions modulo ACIAL,*
$$\alpha \not\equiv \beta \implies \exists \sigma \in \Sigma \; \partial_\sigma(\alpha) \neq \partial_\sigma(\beta).$$

Proof. Let $\alpha \not\equiv \beta$. We know that $\alpha = w_1 + \cdots + w_n + w'_1\alpha_1 + \cdots + w'_m\alpha_m$ and $\beta = x_1 + \cdots + x_{n'} + x'_1\beta_1 + \cdots + x'_{m'}\beta_{m'}$. The sets of partial derivatives of α and β w.r.t a $\sigma \in \Sigma$ can be written as:

$$\partial_\sigma(\alpha) = A \cup \partial_\sigma(w_{i_1}) \cup \cdots \cup \partial_\sigma(w_{i_j}) \cup \partial_\sigma(w'_{l_1})\alpha_{l_1} \cup \cdots \cup \partial_\sigma(w'_{l_t})\alpha_{l_t},$$
$$\partial_\sigma(\beta) = A \cup \partial_\sigma(x_{i'_1}) \cup \cdots \cup \partial_\sigma(x_{i'_{j'}}) \cup \partial_\sigma(x'_{l'_1})\beta_{l'_1} \cup \cdots \cup \partial_\sigma(x'_{l'_{t'}})\beta_{l'_{t'}},$$

where the set A contains all partial derivatives φ such that $\varphi \in \partial_\sigma(\gamma)$ if, and only if, γ is a common summand of α and β, i.e. if $\gamma \equiv w_i \equiv x_j$ or $\gamma \equiv w'_l\alpha_l \equiv x'_k\beta_k$ for some i, j, l, and k. Without loss of generality, consider the following three cases:

1. If $i_1 \neq 0$ and $i'_1 \neq 0$, we know that for $k \in \{i'_1, \ldots, i'_{j'}\}$, $w_{i_1} \neq x_k$ and, by Lemma 11, $\partial_\sigma(w_{i_1}) \neq \partial_\sigma(x_k)$. And also, by Lemma 11, $\partial_\sigma(w_{i_1}) \neq \partial_\sigma(x'_k)\beta_k$, for $k \in \{l'_1, \ldots, l'_{t'}\}$. Thus, $\partial_\sigma(w_i) \cap \partial_\sigma(\beta) = \emptyset$.
2. If $i_1 \neq 0$ and $i'_j = 0$, this case corresponds to the second part of the previous one.
3. If $i_j = i'_{j'} = 0$, for $k \in \{l'_1, \ldots, l'_{t'}\}$, we have $w'_{l_1}\alpha_{l_1} \neq x'_k\alpha_k$ and then either $w'_{l_1} \neq x'_k$ or $\alpha_{l_1} \neq \beta_k$. If $w'_{l_1} \neq x'_k$ then $\partial_\sigma(w'_{l_1}) \neq \partial_\sigma(x'_k)$ and thus $\partial_\sigma(w'_{l_1})\alpha_{l_1} \neq \partial_\sigma(x'_k)\alpha_k$. If $\alpha_{l_1} \neq \beta_k$ it is obvious that $\partial_\sigma(w'_l)\alpha_l \neq \partial_\sigma(x'_k)\alpha_k$. Thus, $\partial_\sigma(w'_{l_1})\alpha_{l_1} \cap \partial_\sigma(\beta) = \emptyset$. $\qquad\square$

Theorem 13. *Let α be a irreducible finite regular expression modulo ACIAL. Then, $\mathcal{A}_{pd}(\alpha) \simeq {}^{\mathcal{A}_{pos}(\alpha)}\!/_{\equiv_b}$.*

Proof. Let $\mathcal{A}_{pd}(\alpha) = (\mathsf{PD}(\alpha), \Sigma, \delta_{pd}, \alpha, F_{pd})$. We want to prove that no pair of states of $\mathcal{A}_{pd}(\alpha)$ is bisimilar. As in Proposition 8, we proceed by induction on the rank of the states. The only state in rank 0 is ε, for which the proposition is obvious. Suppose that all pair of states with rank $m < n$ are not bisimilar. Let $\gamma, \beta \in \mathsf{PD}(\alpha)$ with $n = rk(\gamma) \geq rk(\beta)$. Then, there exists $\gamma' \in \partial_\sigma(\gamma)$ that is distinct of every $\beta' \in \partial_\sigma(\beta)$, by Proposition 12. Because $rk(\beta') < n$ and $rk(\gamma') < n$, by inductive hypothesis, $\gamma' \not\equiv_b \beta'$. Thus $\gamma \not\equiv_b \beta$.   □

Despite $\mathcal{A}_{pd}(\alpha) \simeq \mathcal{A}_{pos}(\alpha)/_{\equiv_b}$, for irreducible REs modulo ACIAL, these NFAs are not necessarily minimal. For example, if $\tau_3 = ba(a + b) + c(aa + ab)$, both NFAs have seven states and a minimal equivalent NFA has four states.

Finally, note that for general regular expressions representing finite languages, $\mathcal{A}_{pos}(\alpha)/_{\equiv_b}$ can be arbitrarily more succinct than $\mathcal{A}_{pd}$. For example, considering the family of REs

$$\alpha_n = aa_1 + a(a_1 + a_2) + a(a_1 + a_2 + a_3) + \ldots + a(a_1 + a_2 + \ldots + a_n)$$

the $\mathcal{A}_{pd}(\alpha_n)$ has $n + 2$ states and $\mathcal{A}_{pos}(\alpha)/_{\equiv_b}$ has three states independently of n.

3.4   General Regular Languages

If we consider general regular expressions with the Kleene star operator, it is easy to find REs α such that $\mathcal{A}_{pd}(\alpha) \not\simeq \mathcal{A}_{pos}(\alpha)/_{\equiv_b}$. This is true even if $\mathcal{A}_{pos}(\alpha)$ is a DFA, i.e. if α is one-unambiguous [5]. For example, for $\alpha = aa^\star + b(\varepsilon + aa^\star)$ the $\mathcal{A}_{pd}(\alpha)$ has one more state than $\mathcal{A}_{pos}(\alpha)/_{\equiv_b}$. Ilie and Yu [15] presented a family of REs

$$\alpha_n = (a + b + \varepsilon)(a + b + \varepsilon) \ldots (a + b + \varepsilon)(a + b)^\star,$$

where $(a + b + \varepsilon)$ is repeated n times, for which $\mathcal{A}_{pd}(\alpha_n)$ has $n + 1$ states and $\mathcal{A}_{pos}(\alpha_n)/_{\equiv_b}$ has one state independently of n. Considering $n = 3$ the $\mathcal{A}_{pd}(\alpha_3)$ are represented in Figure 4.

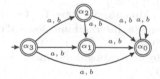

Fig. 4. $\mathcal{A}_{pd}((a + b + \varepsilon)(a + b + \varepsilon)(a + b + \varepsilon)(a + b)^\star)$

In concurrency theory, the characterization of regular expressions for which equivalent NFAs are bisimilar has been extensively studied. Baeten et. al [2] defined a normal form that corresponds to the normal form (5), in the finite case.

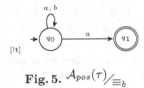

Fig. 5. $\mathcal{A}_{pos}(\tau)/_{\equiv_b}$

For regular expressions with Kleene star operator the normal form defined by those authors is neither irreducible nor unique. In that case, we can find regular expressions α in normal form such that $\mathcal{A}_{pd}(\alpha) \not\simeq \mathcal{A}_{pos}(\alpha)/_{\equiv_b}$. For example, for $\tau = (ab^\star + b)^\star$ the $\mathcal{A}_{pd}(\tau)$ has three states, as seen before in Figure 2, and $\mathcal{A}_{pos}(\tau)/_{\equiv_b}$ has two states, as shown in Figure 5. Other example is $\tau_4 = a(\varepsilon + aa^\star) + ba^\star$, where $|\mathsf{PD}(\tau_4)| = 3$, and in $\mathcal{A}_{pos}(\tau_4)/_{\equiv_b}$ a state is saved because $(\varepsilon + aa^\star) \equiv_b a^\star$. This corresponds to an instance of one of the axioms of Kleene algebra (for the star operator).

As no confluent or even terminating rewrite system modulo these axioms is known, for general REs it will be difficult to obtain a characterization similar to the one of Theorem 13.

Acknowledgements. This work was partially funded by the European Regional Development Fund through the programme COMPETE and by the Portuguese Government through the FCT under project PEst-C/MAT/UI0144/2013 and project FCOMP-01-0124-FEDER-020486.

Eva Maia was also funded by FCT grant SFRH/BD/78392/2011.

The authors would like to thank the valuable remarks and corrections suggested by the anonymous referees.

References

1. Antimirov, V.M.: Partial derivatives of regular expressions and finite automaton constructions. Theor. Comput. Sci. 155(2), 291–319 (1996)
2. Baeten, J.C.M., Corradini, F., Grabmayer, C.A.: A characterization of regular expressions under bisimulation. J. ACM 54(2) (April 2007)
3. Berry, G., Sethi, R.: From regular expressions to deterministic automata. Theor. Comput. Sci. 48(1), 117–126 (1986)
4. Broda, S., Machiavelo, A., Moreira, N., Reis, R.: On the average size of Glushkov and partial derivative automata. International Journal of Foundations of Computer Science 23(5), 969–984 (2012)
5. Brüggemann-Klein, A.: Regular expressions into finite automata. Theoret. Comput. Sci. 48, 197–213 (1993)
6. Brzozowski, J.A.: Derivatives of regular expressions. J. ACM 11(4), 481–494 (1964)
7. Caron, P., Ziadi, D.: Characterization of Glushkov automata. Theoret. Comput. Sci. 233(1-2), 75–90 (2000)
8. Champarnaud, J.M., Ouardi, F., Ziadi, D.: Follow automaton versus equation automaton. In: Ilie, L., Wotschke, D. (eds.) DCFS. vol. Report No. 619, pp. 145–153. Department of Computer Science, The University of Western Ontario, Canada (2004)

9. Champarnaud, J.M., Ziadi, D.: From Mirkin's prebases to Antimirov's word partial derivatives. Fundam. Inform. 45(3), 195–205 (2001)
10. Champarnaud, J.M., Ziadi, D.: Canonical derivatives, partial derivatives and finite automaton constructions. Theor. Comput. Sci. 289(1), 137–163 (2002)
11. García, P., López, D., Ruiz, J., Alvarez, G.I.: From regular expressions to smaller nfas. Theor. Comput. Sci. 412(41), 5802–5807 (2011)
12. Glushkov, V.M.: The abstract theory of automata. Russian Mathematical Surveys 16(5), 1–53 (1961)
13. Gouveia, H., Moreira, N., Reis, R.: Small nfas from regular expressions: Some experimental results. CoRR abs/1009.3599 (2010)
14. Ilie, L., Yu, S.: Follow automata. Inf. Comput. 186(1), 140–162 (2003)
15. Ilie, L., Yu, S.: Reducing nfas by invariant equivalences. Theor. Comput. Sci. 306(1-3), 373–390 (2003)
16. Kozen, D.C.: Automata and Computability. Springer (1997)
17. McNaughton, R., Yamada, H.: Regular expressions and state graphs for automata. IEEE Transactions on Electronic Computers 9, 39–47 (1960)
18. Mirkin, B.: An algorithm for constructing a base in a language of regular expressions. Engineering Cybernetics 5, 110–116 (1966)
19. Moreira, N., Reis, R.: Series-parallel automata and short regular expressions. Fundam. Inform. 91(3-4), 611–629 (2009)
20. Paige, R., Tarjan, R.E.: Three partition refinement algorithms. SIAM J. Comput. 16(6), 973–989 (1987)

The Power of Regularity-Preserving Multi Bottom-up Tree Transducers

Andreas Maletti*

Institute of Computer Science, Universität Leipzig
Augustusplatz 10–11, 04109 Leipzig, Germany
maletti@informatik.uni-leipzig.de

Abstract. The expressive power of regularity-preserving multi bottom-up tree transducers (MBOT) is investigated. These MBOT have very attractive theoretical and algorithmic properties. However, their expressive power is not well understood. It is proved that despite the restriction their power still exceeds that of composition chains of linear extended top-down tree transducers with regular look-ahead (XTOP^R), which are a natural super-class of STSG. In particular, topicalization can be modeled by such MBOT, whereas composition chains of XTOP^R cannot implement it. However, the inverse of topicalization cannot be implemented by any MBOT. An interesting, promising, and widely applicable proof technique is used to prove those statements.

1 Introduction

Statistical machine translation [15] deals with the automatic translation of natural language texts. A central component of each statistical machine translation model is the *translation model*, which is the model that actually performs the translation. Various other models support the translation (such as language models), but the type of transformations computable by the system is essentially determined by the translation model. Various different translation models are currently in use: (i) Phrase-based [23] systems essentially use a finite-state transducer [13]. (ii) Hierarchical phrase-based systems [4] use a synchronous context-free grammar (SCFG), and (iii) syntax-based systems use a form of synchronous tree grammar such as synchronous tree substitution grammars (STSG) [5], synchronous tree-adjoining grammars (STAG) [24], or synchronous tree-sequence substitution grammars (STSSG) [25]. In this contribution, we will focus on syntax-based systems. Since machine translation systems are trained on large data, the used translation model must meet two contradictory goals. Its expressive power should be large in order to be able to model all typical phenomena that occur in translation. On the other hand, the model should have nice algorithmic properties and important operations should have low computational complexity. The mentioned models cover a wide spectrum along this axis with SCFG and STSG

* The author was financially supported of the German Research Foundation (DFG) grant MA / 4959 / 1-1.

M. Holzer and M. Kutrib (Eds.): CIAA 2014, LNCS 8587, pp. 278–289, 2014.
© Springer International Publishing Switzerland 2014

as the weakest models with the best parsing complexities. It is thus essential for the evaluation of a translation model to accurately determine its expressive power and the complexities of its principal operations [14].

A relatively recent proposal for another translation model suggests the multi bottom-up tree transducer (MBOT) [18,19]. It can be understood as an extension of STSG that allows discontinuity on the output side or as a restriction of STSSG that disallows discontinuity on the input side. MBOT are thus a natural (half-way) model in between STSG and STSSG. In addition, [18,19] demonstrated that MBOT have very good theoretical and algorithmic properties in comparison to both STSG and STSSG. They have been implemented [2] in the machine translation framework MOSES [16] and were successfully evaluated in an English-to-German translation task, in which they significantly outperformed the STSG baseline. However, MBOT also have a feature called *finite copying* [7], which on the positive side yields that the output string language can be a multiple context-free language (or equivalently a linear context-free rewriting system language) [12]. Since this class of languages is much more powerful than context-free languages, its algorithmic properties are not as nice as those of the regular tree languages [10,11], which can be used to represent the parse trees of context-free grammars. It is not clear whether this added complexity is necessary to model common discontinuities like topicalization.

In this contribution we demonstrate that the regularity-preserving MBOT (i.e., those whose output is always a regular tree language) retain the power to compute discontinuities such as topicalization. Moreover, these MBOT remain more powerful than arbitrary composition chains [22] of STSG. In particular, no chain of STSG can implement topicalization. However, whereas STSG can trivially be inverted, neither MBOT nor regularity-preserving MBOT can be inverted in general. In fact, we show that the inverse of topicalization cannot be implemented by any MBOT, which confirms the bottom-up nature of MBOT. Overall, these results allow us to relate the expressive power of regularity-preserving MBOT to the other classes (see Figure 6). Secondly, we want to promote the use of explicit links as a tool for analyses. Links naturally record which parts of the input and output tree have to develop synchronously in a derivation step. However, once expanded, the "used" links are typically dropped [3]. Here we retain all links in a special component of the sentential form in the spirit of [20,9]. We investigate the properties of these links and then use them to prove our main results. With the links the proofs split into a standard technical part that establishes certain mandatory links [9] and a rather straightforward high-level argumentation that refutes that the obtained link ensemble is well-formed. We believe that this proof method holds much potential and can successfully be applied to many additional setups.

2   Notation

We write $\mathbb{N}$ for the set of all nonnegative integers (including 0). Given a relation $R \subseteq S_1 \times S_2$ and $S \subseteq S_1$, we let $R(S) = \{s_2 \mid \exists s_1 \in S : (s_1, s_2) \in R\}$

and $R^{-1} = \{(s_2, s_1) \mid (s_1, s_2) \in R\}$ be the elements of S_2 related to elements of S (via R) and the inverse relation of R, respectively. Instead of $R(\{s\})$ with $s \in S_1$ we also write $R(s)$. The composition of two relations $R_1 \subseteq S_1 \times S_2$ and $R_2 \subseteq S_2 \times S_3$ is the relation $R_1 \,;\, R_2 \subseteq S_1 \times S_3$ given by

$$R_1 \,;\, R_2 = \{(s_1, s_3) \mid R_1(s_1) \cap R_2^{-1}(s_3) \neq \emptyset\} \ .$$

As usual, S^* denotes the set of all (finite) words over a set S with the empty word ε. We simply write $v.w$ or vw for the concatenation of the words $v, w \in S^*$, and the length of a word $w \in S^*$ is $|w|$. Given languages $L, L' \subseteq S^*$, we let $L \cdot L' = \{v.w \mid v \in L, w \in L'\}$. An alphabet Σ is a nonempty and finite set of symbols. Given an alphabet Σ and a set S, the set $T_\Sigma(S)$ of Σ-trees indexed by S is the smallest set such that $S \subseteq T_\Sigma(S)$ and $\sigma(t_1, \ldots, t_k) \in T_\Sigma(S)$ for all $k \in \mathbb{N}$, $\sigma \in \Sigma$, and $t_1, \ldots, t_k \in T_\Sigma(S)$. We write T_Σ for $T_\Sigma(\emptyset)$.

Whenever we need to address specific parts of a tree, we use positions. Each position is a word of $\mathbb{N}^*$. The root of a tree has position ε and the position $i.p$ with $i \in \mathbb{N}$ and $p \in \mathbb{N}^*$ addresses the position p in the ith direct child of the root. The set $\mathrm{pos}(t)$ denotes the set of all positions in a tree $t \in T_\Sigma(S)$. We note that positions are totally ordered by the lexicographic order $\sqsubseteq$ on $\mathbb{N}^*$ and partially ordered by the prefix order $\leq$ on $\mathbb{N}^*$. The total order $\sqsubseteq$ allows us to turn a finite set $P \subseteq \mathbb{N}^*$ into a vector $\boldsymbol{P}$ by letting $\boldsymbol{P} = (w_1, \ldots, w_m)$ if $P = \{w_1, \ldots, w_m\}$ with $w_1 \sqsubset \cdots \sqsubset w_m$. Given a tree $t \in T_\Sigma(S)$ its size $|t|$ is the number of its nodes (i.e., $|t| = |\mathrm{pos}(t)|$), and its height $\mathrm{ht}(t)$ coincides with the length of the longest position (i.e., $\mathrm{ht}(t) = \max_{w \in \mathrm{pos}(t)} |w|$).

We conclude this section with some essential operations on trees. To this end, let $t, u \in T_\Sigma(S)$ be trees and $w \in \mathrm{pos}(t)$ be a position in t. The label of t at w is $t(w)$. For every $s \in S$, we let $\mathrm{pos}_s(t) = \{w \in \mathrm{pos}(t) \mid t(w) = s\}$ be those positions in t that are labeled by s. The tree $t \in T_\Sigma(S)$ is linear if $|\mathrm{pos}_s(t)| \leq 1$ for every $s \in S$. We let $\mathrm{idx}(t) = \{s \in S \mid \mathrm{pos}_s(t) \neq \emptyset\}$. Finally, the expression $t[u]_w$ denotes the tree that is obtained from t by replacing the subtree at w by u. We also extend this notation to sequences $\boldsymbol{u} = (u_1, \ldots, u_m)$ of trees and positions $\boldsymbol{w} = (w_1, \ldots, w_m)$ of t that are pairwise incomparable with respect to $\leq$. Thus, $t[\boldsymbol{u}]_{\boldsymbol{w}}$ denotes the tree obtained from t by replacing the subtree at w_i by u_i for all $1 \leq i \leq m$. Formally, $t[\boldsymbol{u}]_{\boldsymbol{w}} = (\cdots (t[u_1]_{w_1}) \cdots)[u_m]_{w_m}$.

3 Formal Models

The main transformational grammar formalism under discussion is the multi bottom-up tree transducer (MBOT), which was introduced by [17,1]. An English theoretical treatment can be found in [6]. In general, MBOT are synchronous grammars [3] with potentially discontinuous output sides, which makes them more powerful than the commonly used STSG [5]. Thus, each rule of an MBOT specifies potentially several parts of the output tree. We essentially recall the definition of [20].

Definition 1. *A* multi bottom-up tree transducer *(for short:* MBOT*) is a tuple* $G = (N, \Sigma, S, P)$, *where*

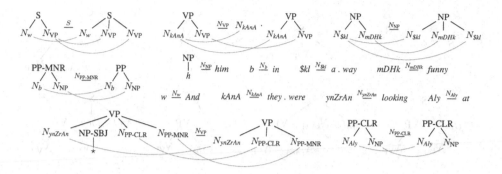

Fig. 1. Example productions

- N *is its finite set of* nonterminals,
- Σ *is its alphabet of* input *and* output symbols,
- $S \in N$ *is its* initial *nonterminal, and*
- $P \subseteq T_\Sigma(N) \times N \times T_\Sigma(N)^*$ *is its finite set of* productions *such that* $\ell \notin N$, ℓ *is linear, and* $\bigcup_{1 \leq i \leq |r|} \mathrm{idx}(r_i) \subseteq \mathrm{idx}(\ell)$ *for every* $\langle \ell, n, r \rangle \in P$.

If all productions $\langle \ell, n, r \rangle \in P$ *obey* $|r| \leq 1$, *then* G *is a (linear) extended top-down tree transducer with regular look-ahead* (XTOP$^\mathrm{R}$) *[21], and if they even fulfill* $|r| = 1$, *then it is a (linear)* nondeleting extended top-down tree transducer (n-XTOP).

In comparison to [20] we added the requirement of $\ell \notin N$, which could be called *input ε-freeness*. To avoid repetition, we henceforth let $G = (N, \Sigma, S, P)$ be an arbitrary MBOT. As usual, ℓ and r of a production $\langle \ell, n, r \rangle \in P$ are called *left-* and *right-hand side*, respectively. We also write $\ell \xrightarrow{n} r$ instead of $\langle \ell, n, r \rangle$. The productions of our running example MBOT are displayed in Figure 1. The initial nonterminal is S, and we omit an explicit representation of the set N of nonterminals (containing the various slanted N_x and S) and the set Σ of symbols because they can be deduced from the productions. For completeness' sake, the leftmost production on the first line in Figure 1 can be written as

$$\langle \mathrm{S}(N_w, N_{\mathrm{VP}}), S, \mathrm{S}(N_w, N_{\mathrm{VP}}, N_{\mathrm{VP}}) \rangle .$$

In contrast to [19,2], which present the semantics of MBOT using a bottom-up process based on pre-translations, we present a top-down semantics in the style of [20] here. As usual [3], the top-down semantics requires us to keep track of the positions that are supposed to develop synchronously in the input and output. Such related positions are called *linked positions*, and the additional data structure recording the linked positions is called the *link structure*. Although the link structure might at first be seen as an overhead (since it is not required for the bottom-up semantics), it will be an essential tool later on. In fact, all our later arguments will be based on the link structure, so we explicitly want to promote the use of link structures and an investigation into their detailed properties.

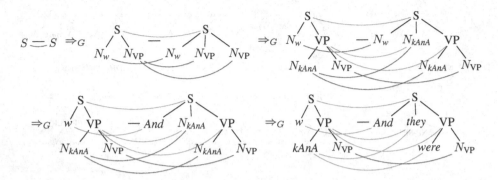

Fig. 2. Partial derivation using the productions of Figure 1. The active links are clearly marked, whereas disabled links are light.

We start with the introduction of the link structure resulting from a single production. In fact, the link structure of a production is implicit because we assume that an occurrence of a nonterminal n in the left-hand side is linked to all its occurrences in the right-hand side. We usually depict these links as (light) splines in graphical illustrations of productions (see Figure 1). However, once we move to the derivation process, an explicit representation of these links is required to keep track of synchronously developing nonterminals.

Definition 2. *Given $\ell \xrightarrow{n'} r \in P$ and positions v and $\boldsymbol{w} = (w_1, \ldots, w_m)$ with $m = |\boldsymbol{r}|$ to which the production should be applied, we define the* link structure $\text{links}_{v,\boldsymbol{w}}(\ell \xrightarrow{n'} r)$ *by* $\bigcup_{n \in N, 1 \leq i \leq m}(\{v\} \cdot \text{pos}_n(\ell)) \times (\{w_i\} \cdot \text{pos}_n(r_i))$.

In other words, besides linking occurrences of the same nonterminal as already mentioned, we prefix the positions by the corresponding position given as a parameter. These argument positions will hold the positions to which the production shall be applied to. Now we are ready to present the semantics. Simply said, we select an input position, its actively linked output positions, and a production that has the right number of right-hand sides. Then we disable the selected links, substitute the production components into the corresponding selected positions, and add the link structure of the production to the set of active links. Formally, a sentential form is simply a tuple $\langle t, A, D, u \rangle$ consisting of an input and an output tree $t, u \in T_\Sigma(N)$ and two sets of links $A, D \subseteq \text{pos}(t) \times \text{pos}(u)$ containing *active* and *disabled* links, respectively. We let $\mathcal{SF}(G)$ be the set of all sentential forms, and $\mathcal{D}(G)$ is the set $\{\langle t, D, u \rangle \mid \langle t, \emptyset, D, u \rangle \in \mathcal{SF}(G), t, u \in T_\Sigma\}$ of all potential dependencies for nonterminal-free input and output trees. In graphical representations we only present the input and output trees and illustrate the links of A and D as clear and light splines, respectively.

Definition 3. *We write $\langle t, A, D, u \rangle \Rightarrow_G \langle t', A', D', u' \rangle$ for two sentential forms $\langle t, A, D, u \rangle, \langle t', A', D', u' \rangle \in \mathcal{SF}(G)$, if there exist a nonterminal $n \in N$, an input* *position* $v \in pos_n(t)$ *labeled by* n, *actively linked output positions* $A(v)$, *and a production* $\ell \overset{n}{-} r \in P$ *such that*

- $|r| = |A(v)|$ *and* $w = A(v)$,
- $t' = t[\ell]_v$ *and* $u' = u[r]_w$, *and*
- $A' = (A \setminus L) \cup links_{v,w}(\ell \overset{n}{-} r)$ *and* $D' = D \cup L$ *with* $L = \{(v, w) \mid w \in A(v)\}$.

As usual, $\Rightarrow_G^*$ *is the reflexive and transitive closure of* $\Rightarrow_G$. *The* MBOT G *computes the dependencies* $dep(G) \subseteq \mathcal{D}(G)$ *given by*

$$\{\langle t, D, u \rangle \in \mathcal{D}(G) \mid \langle S, \{(\varepsilon, \varepsilon)\}, \emptyset, S \rangle \Rightarrow_G^* \langle t, \emptyset, D, u \rangle\} \ .$$

Finally, the MBOT G *computes the relation* $G \subseteq T_\Sigma \times T_\Sigma$, *which is given by* $G = \{\langle t, u \rangle \mid \langle t, D, u \rangle \in dep(G)\}$.

Note that disabled links are often not preserved in the sentential forms in the literature [3], but we want to investigate and reason about those links as in [20,9], so we preserve them. The first steps of a derivation using the productions of Figure 1 are presented in Figure 2.

In the remaining part of this section, we recall the notion of regular tree languages [10,11] and some properties on dependencies [20,9]. Any subset $L \subseteq T_\Sigma$ is a *tree language*, and a tree language $L \subseteq T_\Sigma$ is *regular* [6] if and only if there exists an MBOT $G = (N, \Sigma, S, P)$ such that $L = G^{-1}(T_\Sigma)$ (i.e., L is the domain of G). A relation $R \subseteq T_\Sigma \times T_\Sigma$ *preserves regularity* if $R(L)$ is regular for every regular tree language $L \subseteq T_\Sigma$.

Next, we recall the properties on dependencies of [20,9]. We only define them for the input side, but assume that they are also defined (in the same manner) for the output side.

Definition 4. *A dependency* $\langle t, D, u \rangle \in \mathcal{D}(G)$ *is*

- *input hierarchical if* $w_2 \not< w_1$ *and there exists* $(v_1, w_1') \in D$ *with* $w_1' \leq w_2$ *for all* $(v_1, w_1), (v_2, w_2) \in D$ *with* $v_1 < v_2$,
- *strictly input hierarchical if (i)* $v_1 < v_2$ *implies* $w_1 \leq w_2$ *and (ii)* $v_1 = v_2$ *implies* $w_1 \leq w_2$ *or* $w_2 \leq w_1$ *for all* $(v_1, w_1), (v_2, w_2) \in D$,
- *input link-distance bounded by* $b \in \mathbb{N}$ *if for all links* $(v_1, w_1), (v_1v', w_2) \in D$ *with* $|v'| > b$ *there exists* $(v_1v, w_3) \in D$ *such that* $v < v'$ *and* $1 \leq |v| \leq b$,
- *strict input link-distance bounded by* b *if for all positions* $v_1, v_1v' \in pos(t)$ *with* $|v'| > b$ *there exists* $(v_1v, w_3) \in D$ *such that* $v < v'$ *and* $1 \leq |v| \leq b$.

The set $dep(G)$ *has those properties if each dependency* $\langle t, D, u \rangle \in dep(G)$ *has them.*

We also say that $dep(G)$ is *input link-distance bounded* if there exists an integer $b \in \mathbb{N}$ such that it is input link-distance bounded by b. We summarize the known properties in Table 1.

4   Main Results

In this contribution, we want to investigate the expressive power of regularity-preserving MBOT, which constitute the class of all MBOT whose computed relation preserves regularity. This class has very nice (algorithmic) properties

Table 1. Summary of the properties of the dependencies dep(G) for grammars G belonging to the various grammar formalisms [20,9]

Model \ Property	hierarchical		link-distance bounded		regular		
	input	output	input	output	domain	range	pres.
n-XTOP	strictly	strictly	strictly	strictly	✓	✓	✓
XTOP^R	strictly	strictly	✓	strictly	✓	✓	✓
MBOT	✓	strictly	✓	strictly	✓	✗	✗
reg.-pres. MBOT	✓	strictly	✓	strictly	✓	✓	✓

(see Table 1). It was already argued by [19] that regularity should be preserved by any grammar formalism (used in syntax-based machine translation) in order to obtain an efficient representation of the output tree language. In fact, several (syntactic) ways to obtain regularity preserving MBOT are discussed there, but these all yield subclasses of the class of all regularity-preserving MBOT. On the other hand, XTOP^R and n-XTOP, which are both slightly more powerful than the commonly used STSG [5] but strictly less powerful than regularity-preserving MBOT, are not closed under composition [21], but always preserve regularity. Consequently, [22] consider the efficient evaluation of (composition) chains of n-XTOP, and their approach can easily be extended to XTOP^R. Obviously, every chain of XTOP^R can be simulated by a regularity-preserving MBOT because each individual XTOP^R can be simulated and MBOT are closed under composition [6]. However, the exact relation between these two classes remained open. This question is interesting because it solves whether the (non-copying) features of MBOT (such as discontinuity) can be achieved by chains of XTOP^R. In particular, it settles the question whether chains of XTOP^R can handle discontinuities, which, in general, cannot be handled by a single XTOP^R.

The author assumes that the question remained open because both possible answers require deep insight. If the classes coincide, then we should be able to simulate each regularity-preserving MBOT by a chain of XTOP^R, which is complicated due to the fact that "regularity-preserving" is a semantic property on the computed relation. Such a construction would (most likely) shed light on the exact (syntactic) consequences of the restriction to regularity-preserving MBOT. On the other hand, if regularity-preserving MBOT are more powerful than chains of XTOP^R (which we prove in this contribution), then we need to exhibit a relation that cannot be computed by any chain of XTOP^R, which requires deep insight into the relations computable by chains of XTOP^R. Fortunately, there was recent progress in the latter area. In [8] it was shown that a chain of 3 XTOP^R can simulate any chain of XTOP^R. Together with the linking technique of [20,9], this will allow us to present a regularity-preserving MBOT that cannot be simulated by any chain of XTOP^R. The counterexample is even linguistically motivated in the sense that it abstractly represents *topicalization* (see Figure 3), which is a common form of discontinuity.

Example 5. Let Tpc $= (N, \Sigma, S, P)$ be the MBOT with $N = \{S, T, T', T'', U\}$, symbols $\Sigma = \{\sigma, \delta, \gamma, \alpha\}$, and the productions P illustrated in Figure 3. It is

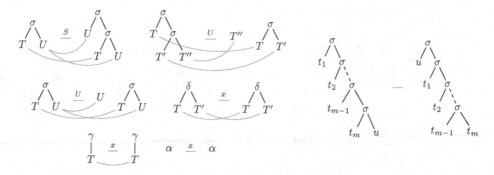

Fig. 3. Productions of the counterexample MBOT Tpc with $x \in \{T, T', T''\}$ and relation (topicalization) computed by it for all $m \in \mathbb{N}$ and arbitrary trees $u, t_1, \ldots, t_m$, which can contain binary δ-symbols, unary γ-symbols, and nullary α-symbols

clearly regularity-preserving because it is straightforward to develop two n-XTOP G_1 and G_2 that compute transformations similar to topicalization (see Figure 3), but just preserving u and just preserving $t_1, \ldots, t_m$, respectively. Thus, the language Tpc(L) for a regular tree language L is obtained as $G_1(L) \cap G_2(L)$. Since n-XTOP preserve regularity [21], $G_1(L)$ and $G_2(L)$ are regular tree languages, and regular tree languages are closed under intersection [10,11]. The relation computed by Tpc is depicted in Figure 3.

Theorem 6. *The relation Tpc cannot be computed by any chain of* XTOP^R.

Proof (Sketch). We already remarked that 3 XTOP^R suffice to simulate any chain of XTOP^R according to [8]. Consequently, in order to derive a contradiction we assume that there exist 3 XTOP^R G_1, G_2, G_3 such that Tpc $= G_1; G_2; G_3$. We know that dep(G_1), dep(G_2), dep(G_3) are input and output link-distance bounded (see Table 1), so let $b \in \mathbb{N}$ be such that all link-distances (for all 3 XTOP^R) are bounded by b. Using an involved technical argumentation based on the link properties and size arguments [9] (using only the symbols γ and α for the trees $u, t_1, \ldots, t_m$), we can deduce the existence of the light dependencies depicted in Figure 4 (for the input and output tree and two intermediate trees without the clearly marked links), in which $m \gg b^3$. Consequently, the ellipsis (clearly marked dots) in the output tree (last tree in Figure 4) hides at least b^2 links that point to this part of the output because there must be a link every b positions by the link-distance bound. Let $(v''_1, w''_1), \ldots, (v''_{m''}, w''_{m''})$ with $m'' \gg b^2$ be those links such that $w''_1 < \cdots < w''_{m''}$. These links are marked with (1) in Figure 4. Clearly, $w''_{m''}$ dominates (via $\leq$) the positions of the subtrees t_{m-1} and t_m, but it does not dominate that of the subtree u. The input positions of those links, which point to positions inside the third tree in Figure 4, automatically fulfill $v''_1 \leq \cdots \leq v''_{m''}$ since dep(G_3) is strictly output hierarchical. A straightforward induction can be used to show that (for any XTOP^R) all links sharing the same input positions must be incomparable with respect to the prefix order $\leq$ [9], which uses the restriction that $\ell \notin N$ for each production $\langle \ell, n, r \rangle \in P$. Consequently, $v''_1 < \cdots < v''_{m''}$.

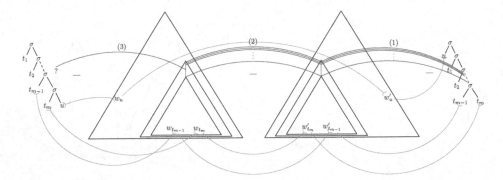

Fig. 4. Illustration of the dependencies discussed in the proof of Theorem 6. Inverted arrow heads indicate that the link points to a position below the one indicated by the spline. The links relating the roots of the trees are omitted.

Similarly, we can conclude $v''_{m''} < w'_{t_{m-1}}$, $v''_{m''} < w'_{t_m}$, and $v''_1 \not\leq w'_u$, where the last statement uses that $\mathrm{dep}(G_3)$ is strictly input hierarchical. Repeating essentially the same arguments for $\mathrm{dep}(G_2)$, we obtain links $(v'_1, w'_1), \ldots, (v'_{m'}, w'_{m'})$ with $m' \gg b$ such that $v''_1 \leq w'_1 < \cdots < w'_{m'} \leq v''_{m''}$ and $v'_1 < \cdots < v'_{m'}$. These links are labeled by (2) in Figure 4. Moreover, $v'_{m'} < w_{t_{m-1}}$, $v'_{m'} < w_{t_m}$, and $v'_1 \not\leq w_u$. Using the arguments once more for $\mathrm{dep}(G_1)$, we obtain a link (v, w) such that $v'_1 \leq w \leq v'_{m'}$. This final link is marked (3) in Figure 4. Moreover, we have $v < v_{t_{m-1}}$ and $v < v_{t_m}$, but $v \not\leq v_u$. However, such a position does not exist in the input tree, which completes the desired contradiction.   □

It is noteworthy that the proof can be achieved using high-level arguments based on the links and their properties. In fact, the whole proof is rather straightforward once the basic links (light in Figure 4) are established using [9]. Arguably, the omitted part of the proof that establishes those links is quite technical and involved (using size arguments and thus the particular shape of the trees $u, t_1, \ldots, t_m$), but it can be reused in similar setups as it generally establishes links in the presented way between identical subtrees (for which infinitely many trees are possible). The proof nicely demonstrates that the difficult argumentation via two unknown intermediate trees reduces to (relatively) simple arguments with the help of the links. The author believes that the links will provide a powerful and versatile tool in the future and have been neglected for too long. From Theorem 6 it follows that (some) topicalizations cannot be computed by any chain of XTOP^R (or any chain of n-XTOP), and since Tpc is computed by a regularity-preserving MBOT, we can conclude that regularity-preserving MBOT are strictly more powerful than chains of XTOP^R.

Corollary 7. *Regularity-preserving MBOT are strictly more powerful than composition chains of* XTOP^R *(and composition chains of n-XTOP).*

Our next result will limit the expressive power of MBOT. Using the linking technique [9] once more (this time for MBOT), we will prove that the inverse

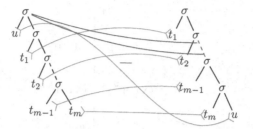

Fig. 5. Illustration of the dependencies discussed in the proof of Theorem 8. Inverted arrow heads indicate that the link points to a position below the one indicated by the spline. The links relating the roots of the trees are omitted.

relation Tpc^{-1} cannot be computed by any MBOT. This confirms the bottom-up nature of the device. It can "grab" deeply nested subtrees and transport them towards the root, but it cannot achieve the converse.

Theorem 8. *The relation Tpc^{-1} cannot be computed by any (composition chain of) MBOT.*

Proof (Sketch). Since MBOT are closed under composition [6], we need to consider only a single MBOT. In order to derive a contradiction, let $G = (N, \Sigma, S, P)$ be an MBOT such that $G = \text{Tpc}^{-1}$. As usual, we know that $\text{dep}(G)$ is input and output link-distance bounded (see Table 1), so let $b \in \mathbb{N}$ be a suitable bound. Moreover, let $a > |\boldsymbol{r}|$ for all productions $\langle \ell, n, \boldsymbol{r} \rangle \in P$. Hence a is an upper bound for the length of the right-hand sides. Finally, let $k > \max(a, b)$ be our main constant.

Again we need to use a (different, but similar) involved technical argumentation [9] based on the link properties and size and height arguments (that uses the symbols δ, γ, and α for the subtrees $u, t_1, \ldots, t_m$) to deduce the existence of the light dependencies shown in Figure 5, in which $m \gg 2k$. Consequently, the ellipsis (clearly marked dots) in the output tree hides at least 2 links that point to this part of the output because there must be a link every b positions by the link-distance bound. Let $(v, w), (v', w')$ be those links such that $w < w'$. These links are clearly indicated in Figure 5.

Clearly, w' dominates the positions of the subtrees t_m and u. Since $\text{dep}(G)$ is strictly output hierarchical (see Table 1), we obtain that (i) $v \leq v'$ and (ii) v' dominates the input positions of the (light) links pointing into the subtrees t_m and u. Obviously, the root ε is the only suitable position, so $v = v' = \varepsilon$ as indicated in Figure 5. Another straightforward induction can be used to show that (for any MBOT) all links sharing the same input positions must be incomparable with respect to the prefix order $\leq$ [9], which uses the restriction that $\ell \notin N$ for each production $\langle \ell, n, \boldsymbol{r} \rangle \in P$. However, (ε, w) and (ε, w') are two links with the same source and comparable target because $w < w'$, so we derived the desired contradiction.   □

Again we note that the proof could be straightforwardly achieved using high-level arguments on the links and their interrelation after establishing the basic links (light in Figure 5). Then the link-distance can be used to conclude the existence of links and their input and output target can be related to existing links

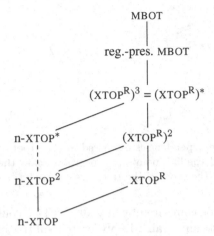

Fig. 6. HASSE diagram for the discussed classes, where C^* is the composition closure of class C and the dashed line just indicates that all powers in between form a chain and are thus strictly contained as well

using the hierarchical properties. In this way, we could in both cases derive a contradiction in rather straightforward ways, which would not have been possible without the links. Typically, such (negative) statements are proved using the fooling technique (see [1] or [21] for examples), which requires a rather detailed case analysis of all possible intermediate trees and applied productions, which then individually have to be contradicted. In a scenario with 2 unknown intermediate trees such an approach becomes (nearly) impossible to handle. Thus, we strongly want to promote the use of links and their interrelations in the analysis of translation models.

Theorem 8 yields that regularity-preserving MBOT are not closed under inversion. In other words, there are regularity-preserving MBOT G (such as Tpc), whose inverted computed relation G^{-1} cannot be computed by any MBOT.

Corollary 9. *Regularity-preserving* MBOT *(and general* MBOT*) are not closed under inversion.*

Let us now collect these results together with some minor consequences in a HASSE diagram (see Figure 6).

References

1. Arnold, A., Dauchet, M.: Morphismes et bimorphismes d'arbres. Theoret. Comput. Sci. 20(1), 33–93 (1982)
2. Braune, F., Seemann, N., Quernheim, D., Maletti, A.: Shallow local multi bottom-up tree transducers in statistical machine translation. In: Proc. 51st ACL, pp. 811–821. Association for Computational Linguistics (2013)
3. Chiang, D.: An introduction to synchronous grammars. In: Proc. 44th ACL. Association for Computational Linguistics, part of a tutorial given with Kevin Knight (2006)

4. Chiang, D.: Hierarchical phrase-based translation. Comput. Linguist. 33(2), 201–228 (2007)
5. Eisner, J.: Learning non-isomorphic tree mappings for machine translation. In: Proc. 41st ACL, pp. 205–208. Association for Computational Linguistics (2003)
6. Engelfriet, J., Lilin, E., Maletti, A.: Composition and decomposition of extended multi bottom-up tree transducers. Acta Inform. 46(8), 561–590 (2009)
7. Engelfriet, J., Rozenberg, G., Slutzki, G.: Tree transducers, L systems, and two-way machines. J. Comput. System Sci. 20(2), 150–202 (1980)
8. Fülöp, Z., Maletti, A.: Composition closure of ε-free linear extended top-down tree transducers. In: Béal, M.-P., Carton, O. (eds.) DLT 2013. LNCS, vol. 7907, pp. 239–251. Springer, Heidelberg (2013)
9. Fülöp, Z., Maletti, A.: Linking theorems for tree transducers (2014) (manuscript)
10. Gécseg, F., Steinby, M.: Tree Automata. Akadémiai Kiadó, Budapest (1984)
11. Gécseg, F., Steinby, M.: Tree languages. In: Rozenberg, G., Salomaa, A. (eds.) Handbook of Formal Languages, vol. 3, ch. 1, pp. 1–68. Springer (1997)
12. Gildea, D.: On the string translations produced by multi bottom-up tree transducers. Comput. Linguist. 38(3), 673–693 (2012)
13. de Gispert, A., Iglesias, G., Blackwood, G.W., Banga, E.R., Byrne, W.: Hierarchical phrase-based translation with weighted finite-state transducers and shallow-n grammars. Comput. Linguist. 36(3), 505–533 (2010)
14. Knight, K., Graehl, J.: An overview of probabilistic tree transducers for natural language processing. In: Gelbukh, A. (ed.) CICLing 2005. LNCS, vol. 3406, pp. 1–24. Springer, Heidelberg (2005)
15. Koehn, P.: Statistical Machine Translation. Cambridge University Press (2010)
16. Koehn, P., Hoang, H., Birch, A., Callison-Burch, C., Federico, M., Bertoldi, N., Cowan, B., Shen, W., Moran, C., Zens, R., Dyer, C., Bojar, O., Constantin, A., Herbst, E.: Moses: Open source toolkit for statistical machine translation. In: Proc. 45th ACL, pp. 177–180. Association for Computational Linguistics (2007)
17. Lilin, E.: Propriétés de clôture d'une extension de transducteurs d'arbres déterministes. In: Astesiano, E., Böhm, C. (eds.) CAAP 1981. LNCS, vol. 112, pp. 280–289. Springer, Heidelberg (1981)
18. Maletti, A.: Why synchronous tree substitution grammars? In: Proc. HLT-NAACL 2010, pp. 876–884. Association for Computational Linguistics (2010)
19. Maletti, A.: How to train your multi bottom-up tree transducer. In: Proc. 49th ACL, pp. 825–834. Association for Computational Linguistics (2011)
20. Maletti, A.: Tree transformations and dependencies. In: Kanazawa, M., Kornai, A., Kracht, M., Seki, H. (eds.) MOL 12. LNCS (LNAI), vol. 6878, pp. 1–20. Springer, Heidelberg (2011)
21. Maletti, A., Graehl, J., Hopkins, M., Knight, K.: The power of extended top-down tree transducers. SIAM J. Comput. 39(2), 410–430 (2009)
22. May, J., Knight, K., Vogler, H.: Efficient inference through cascades of weighted tree transducers. In: Proc. 48th ACL, pp. 1058–1066. Association for Computational Linguistics (2010)
23. Och, F.J., Ney, H.: The alignment template approach to statistical machine translation. Comput. Linguist. 30(4), 417–449 (2004)
24. Shieber, S.M., Schabes, Y.: Synchronous tree-adjoining grammars. In: Proc. 13th CoLing, vol. 3, pp. 253–258 (1990)
25. Zhang, M., Jiang, H., Aw, A., Li, H., Tan, C.L., Li, S.: A tree sequence alignment-based tree-to-tree translation model. In: Proc. 46th ACL, pp. 559–567. Association for Computational Linguistics (2008)

Pushdown Machines for Weighted Context-Free Tree Translation

Johannes Osterholzer

Faculty of Computer Science
Technische Universität Dresden
01062 Dresden, Germany
johannes.osterholzer@tu-dresden.de

Abstract. In this paper, we consider weighted synchronous context-free tree grammars and identify a certain syntactic restriction of these grammars. We suggest a new weighted tree transducer formalism and prove that the transformations of the restricted grammars are precisely those of the linear and nondeleting instances of these transducers.

1 Introduction

Synchronous context-free grammars (or: syntax-directed translation schemata) were introduced in the context of compiler construction in the late 1960s [12]. They define string transductions by the simultaneous derivation of an input and an output word. In contrast, modern systems for machine translation of natural language employ weighted tree transformations to account for the grammatical structure of the input sentence and the ambiguity inherent in spoken language (cf. the survey in [10]). Such transformations may be computed by weighted synchronous tree substitution grammars [17], or by weighted linear and nondeleting extended top-down tree transducers [7]. The former derive input and output trees simultaneously, while the latter model has unidirectional semantics: it derives the output from its input tree. The advantage of unidirectional semantics is that it may serve as a starting point to define weighted tree transformations which are conditional probability distributions [3].

Synchronous context-free tree grammars (scftg) have been proposed as a generalization of synchronous tree substitution grammars that may allow modelling even more linguistic phenomena [13]. In this work, we consider weighted scftg (wscftg), and investigate *simple* wscftg (s-wscftg), a syntactic restriction of wscftg (Section 3). In Section 4 we introduce a formalism with unidirectional derivation semantics, called weighted pushdown extended top-down tree transducer (wpxtop). It is a weighted extended top-down tree transducer [7] whose finite state control is equipped with a tree pushdown [9,6]. We devise certain normal forms for s-wscftg as well as for linear and nondeleting wpxtop. Section 5 contains the main result: we prove that the transformations of linear and nondeleting wpxtop are exactly those of s-wscftg. This proof relies on a close correspondence between the normal forms of the respective formalisms.

M. Holzer and M. Kutrib (Eds.): CIAA 2014, LNCS 8587, pp. 290–303, 2014.

Let us turn back to string grammars: In [12, Thms. 2 and 3], the subclass of *simple* synchronous context-free grammars was identified and characterized by pushdown transducers. Thus, this work is also a generalization of a classical result from formal languages to weighted tree languages.

2 Preliminaries

We denote the nonnegative integers by $\mathbb{N}$, and $\{1,\ldots,n\}$ by $[n]$ for every $n \in \mathbb{N}$. A set Σ equipped with a function $\mathrm{rk}_\Sigma\colon \Sigma \to \mathbb{N}$ is a *ranked set*, its elements are *symbols*. A *ranked alphabet* is a finite ranked set. Let Σ be a ranked set. When Σ is obvious, we write rk instead of rk_Σ. Let $k \in \mathbb{N}$. We set $\Sigma^{(k)} = \mathrm{rk}_\Sigma^{-1}(k)$. Let U be a set and Λ denote $\Sigma \cup U \cup C$, where C is made up of the three symbols '(', ')', and ','. If $\Omega \subseteq \Sigma^{(k)}$ and $T_1, \ldots, T_k \subseteq \Lambda^*$, then define $\Omega(T_1,\ldots,T_k) = \{\sigma(t_1,\ldots,t_k) \mid \sigma \in \Omega, t_i \in T_i, i \in [k]\}$. The set $T_\Sigma(U)$ of *trees (over Σ indexed by U)* is the smallest set $T \subseteq \Lambda^*$ such that $U \subseteq T$ and for every $k \in \mathbb{N}$, $\Sigma^{(k)}(T,\ldots,T) \subseteq T$. A tree $\alpha()$ is abbreviated by α, and $T_\Sigma(\emptyset)$ by T_Σ. Let $t \in T_\Sigma(U)$. The set of *positions (Gorn addresses) of t* is denoted by $\mathrm{pos}(t) \subseteq \mathbb{N}^*$. We denote the *lexicographic order* on $\mathbb{N}^*$ by $\leq_\mathrm{lex}$, the *label* of t at its position w by $t(w)$, and the *subtree* of t at w by $t|_w$. For any $\Delta \subseteq \Sigma \cup U$, let $\mathrm{pos}_\Delta(t) = \{w \in \mathrm{pos}(t) \mid t(w) \in \Delta\}$. Given $k \in \mathbb{N}$, $t_1, \ldots, t_k \in T_\Sigma(U)$, and pairwise different $u_1,\ldots, u_k \in U$, denote by $t[u_1/t_1,\ldots, u_k/t_k]$ the result of *substituting* every occurrence of u_i in t with the tree t_i, where $i \in [k]$. For $V \subseteq U$, t is *linear* (resp. *nondeleting*) *in V* if every $v \in V$ occurs at most (resp. at least) once in t. The set of all trees in $T_\Sigma(U)$ which are linear and nondeleting in U is denoted by $T_\Sigma^\mathrm{ln}(U)$. In the following, let X (resp. Y) denote the sets of *variables* $\{x_1, x_2, \ldots\}$ (resp. $\{y_1, y_2, \ldots\}$). Let $k \in \mathbb{N}$. We set $X_k = \{x_i \mid i \in [k]\}$ and $Y_k = \{y_i \mid i \in [k]\}$. Let $t_1, \ldots, t_k \in T_\Sigma$. We write $t[t_1,\ldots,t_k]$ for $t[x_1/t_1,\ldots,x_k/t_k]$ if $t \in T_\Sigma(X_k)$, and for $t[y_1/t_1,\ldots,y_k/t_k]$ if $t \in T_\Sigma(Y_k)$. A tree $\xi \in T_\Sigma^\mathrm{ln}(X_k)$ is a *(k-)context (over Σ)* if the variables occur in the order $x_1, \ldots, x_k$ within the word $\xi \in \Lambda^*$, when read from left to right. The set of all k-contexts over Σ is denoted by $C_\Sigma(X_k)$. Define $\mathrm{dec}_k(t) = \{(\xi, t_1\ldots t_k) \mid \xi \in T_\Sigma^\mathrm{ln}(X_k), t_1,\ldots,t_k \in T_\Sigma(U), t = \xi[t_1,\ldots,t_k]\}$. Let $t \in T_\Sigma$, and $T \subseteq T_\Sigma$. Then $\mathrm{lin}_T(t)$ denotes the *linearization of t with respect to T*, i.e. the tuple $(\xi, t_1\ldots t_n)$ such that there is an $n \in \mathbb{N}$ with *(i)* $\xi \in C_\Sigma(X_n)$, $t_1, \ldots, t_n \in T$; *(ii)* $t = \xi[t_1,\ldots,t_n]$; and *(iii)* $|\mathrm{pos}_\Sigma(\xi)|$ is minimal with respect to *(i)* and *(ii)*. Unless stated otherwise, Σ, Γ, and N denote arbitrary ranked alphabets.

A tuple $(S, +, \cdot, 0, 1)$ is a *semiring* if $(S, +, 0)$ is a commutative monoid, $(S, \cdot, 1)$ is a monoid, multiplication distributes over addition from the left and from the right, and 0 is annihilating with respect to multiplication. Following convention, such a semiring is referred to by its carrier set S. We call S *complete* if it is equipped with an infinitary sum $\sum$ that maps every indexed family of elements of S into S, where $\sum$ must extend $+$ and satisfy infinitary associativity, commutativity, and distributivity laws [4]. The *Boolean semiring* $(\mathbb{B}, \vee, \wedge, 0, 1, \bigvee)$, with $\mathbb{B} = \{0, 1\}$, and the *semiring of nonnegative real numbers* $(\mathbb{R}_{\geq 0}^\infty, +, \cdot, 0, 1, \sum)$ are two examples of complete semirings. In the sequel of this work, let S denote an arbitrary complete semiring $(S, +, \cdot, 0, 1, \sum)$. A mapping $L \colon T_\Sigma \to S$ is a *weighted tree language*, and a mapping $\tau \colon T_\Sigma \times T_\Sigma \to S$ a *weighted tree transformation*. Refer to [8] for a survey of weighted tree languages and transformations. Let U, U' be sets. We identify any function $f \colon U \to \mathbb{B}$ with the subset $f^{-1}(1)$ of U, and write $U \subseteq_{\text{fin}} U'$ if U is a finite subset of U'.

3   Weighted Synchronous Context-Free Tree Grammar

Context-free tree grammars (cftg) [14] generalize context-free grammars to trees. Their sentential forms are trees that consist of terminal and nonterminal symbols from Σ, resp. N. Both may appear at any position of the tree. In the application of a production, a nonterminal A is substituted with its right-hand side, a tree over $N \cup \Sigma$ with variables, which are, in turn, substituted with the subtrees of A.

The semantics of *synchronous* cftg, in particular, is based on the concept of synchronized trees. We say that two trees ξ and ζ over $N \cup \Sigma$ are *synchronized* if there is a one-to-one relation between the occurrences of nonterminals in ξ and in ζ. Two nonterminals are *linked* if they are thus related. The relation is specified implicitly, by equipping the occurrences of nonterminals in ξ and in ζ with *indices* that are unique in the respective tree. Nonterminal occurrences in ξ and ζ are linked iff both are equipped with the same index. Since the actual values of the indices are irrelevant, synchronized trees which are identical up to renaming of indices are identified in the following formalization.

For each $M \subseteq \mathbb{N}$, let $N^{\boxed{M}}$ denote the ranked set $\{A^{\boxed{i}} \mid A \in N, i \in M\}$, where $\operatorname{rk}(A^{\boxed{i}}) = \operatorname{rk}(A)$ for every $i \in M$. In the sequel, let M, M_1, M_2, etc., denote arbitrary finite subsets of $\mathbb{N}$. For every set U, define $\mathcal{I}^{\boxed{M}}(N, \Sigma, U)$ to be the set of all $\xi \in T_{N^{\boxed{M}} \cup \Sigma}(U)$ such that for every $i \in M$, there is exactly one position $w \in \operatorname{pos}(\xi)$ with $\xi(w) = A^{\boxed{i}}$, for some $A \in N$. In this situation, $\operatorname{nt}^{(i)}(\xi)$ denotes A and $\operatorname{pos}^{(i)}(\xi)$ denotes w. The set $\left(\mathcal{I}^{\boxed{M}}(N, \Sigma, U)\right)^2$ is denoted by $\mathcal{S}^{\boxed{M}}(N, \Sigma, U)$. Abbreviate $\bigcup_{M \subseteq_{\text{fin}} \mathbb{N}} \mathcal{I}^{\boxed{M}}(N, \Sigma, U)$ by $\mathcal{I}(N, \Sigma, U)$, and $\bigcup_{M \subseteq_{\text{fin}} \mathbb{N}} \mathcal{S}^{\boxed{M}}(N, \Sigma, U)$ by $\mathcal{S}(N, \Sigma, U)$.

Let $(\xi_1, \xi_2) \in \mathcal{S}^{\boxed{M_1}}(N, \Sigma, U)$, and $(\zeta_1, \zeta_2) \in \mathcal{S}^{\boxed{M_2}}(N, \Sigma, U)$. We say that (ξ_1, ξ_2) and (ζ_1, ζ_2) are *identical up to renaming of indices*, denoted by $(\xi_1, \xi_2) \equiv (\zeta_1, \zeta_2)$, if there is a function $\rho \colon M_1 \to M_2$ such that for every $j \in \{1, 2\}$, ζ_j is the result of replacing every $A^{\boxed{i}} \in N^{\boxed{M_1}}$ in ξ_j by $A^{\boxed{\rho(i)}}$. The relation $\equiv$ is an equivalence on $\mathcal{S}(N, \Sigma, U)$. The equivalence class of (ξ_1, ξ_2) with respect to $\equiv$ is denoted by $[\xi_1, \xi_2]$ and the factor set $\mathcal{S}(N, \Sigma, U)/_{\equiv}$ by $\mathcal{S}_\equiv(N, \Sigma, U)$. We call (ζ_1, ζ_2) *fresh for* (ξ_1, ξ_2) if $M_1 \cap M_2 = \emptyset$. When considering the equivalence classes $[\xi_1, \xi_2]$ and $[\zeta_1, \zeta_2]$, their representatives (ξ_1, ξ_2) and (ζ_1, ζ_2) can always be chosen fresh for each other. We omit U from the above sets' identifiers whenever $U = \emptyset$.

A wscftg can now be understood as a context-free tree grammar such that the right-hand sides of its productions are (equivalence classes of) pairs of synchronized trees. Formally, a *weighted synchronous context-free tree grammar (wscftg)* [13] is a tuple $\mathcal{G} = (N, S, \Sigma, Z, P, wt)$ such that N is a ranked alphabet of *nonterminal symbols*, Σ is a ranked alphabet of *terminal symbols*, $Z \in N^{(0)}$ is the *initial nonterminal*, P is a finite set of *productions* p, each of the form

$$A(x_1, \ldots, x_k) \to [\xi, \zeta],$$ (1)

where $k \in \mathbb{N}$, $A \in N^{(k)}$, and furthermore, *(i)* $(\xi, \zeta) \in \mathcal{S}^{\boxed{M}}(N, \Sigma, X_k)$ for some $M \subseteq_{\mathrm{fin}} \mathbb{N}$; *(ii)* $\mathrm{nt}^{(i)}(\xi) = \mathrm{nt}^{(i)}(\zeta)$ for every $i \in M$; *(iii)* ξ is a k-context, while ζ is linear and nondeleting in X_k; and finally, *(iv)* $wt: P \to S \setminus \{0\}$ is a mapping that assigns to every production its *weight.* In a production of the form (1), the tree ξ is referred to as its *input component,* and ζ as its *output component.* Note that this definition of wscftg is more restricted than the one in [13], as linked nonterminals must be identical, and therefore of the same rank. However, for this work we will adhere to the definition above, as the pushdown machine characterization in Section 5 does not apply to the more general definition. If all nonterminals of $\mathcal{G}$ are nullary, i.e. $N = N^{(0)}$, then $\mathcal{G}$ is a *weighted synchronous regular tree grammar (wsrtg).*

The set $\mathcal{S}_{\equiv}(N, \Sigma)$ is the set of *sentential forms* of $\mathcal{G}$, denoted by $\mathrm{SF}(\mathcal{G})$. Given a production $p \in P$ as in (1), we define the binary relation $\Rightarrow_{\mathcal{G}}^p$ on $\mathrm{SF}(\mathcal{G})$. Let $[\eta_1, \eta_2], [\kappa_1, \kappa_2] \in \mathrm{SF}(\mathcal{G})$, where (ξ, ζ) is fresh for (η_1, η_2). Then $[\eta_1, \eta_2] \Rightarrow_{\mathcal{G}}^p [\kappa_1, \kappa_2]$ if there are $i \in \mathbb{N}$, $\varphi_0, \psi_0 \in \mathcal{I}(N, \Sigma, X_1)$ both linear and nondeleting in X_1, and $\varphi_1, \ldots, \varphi_k, \psi_1, \ldots, \psi_k \in \mathcal{I}(N, \Sigma)$ such that

$$\eta_1 = \varphi_0 \big[A^{\boxed{i}}(\varphi_1, \ldots, \varphi_k) \big], \qquad \eta_2 = \psi_0 \big[A^{\boxed{i}}(\psi_1, \ldots, \psi_k) \big],$$

$$\kappa_1 = \varphi_0 \big[\, \xi \, [\varphi_1, \ldots, \varphi_k] \big], \quad \text{and} \quad \kappa_2 = \psi_0 \big[\, \zeta \, [\psi_1, \ldots, \psi_k] \big].$$

Note that $\Rightarrow_{\mathcal{G}}^p$ is indeed well-defined, because ξ and ζ are chosen fresh, and hence the relation $\equiv$ is compatible with substitution. Refer to Fig. 1 for an example of a production and its application to a sentential form.

The weight of a tuple of trees (s, t) under the transformation of $\mathcal{G}$ is defined as the sum of the weights of all its derivations, and the weight of such a derivation is the product of the weights of the contained productions. However, we must take care not to sum up unappropriately often – therefore the sum is restricted to derivations that are in a certain sense *leftmost.* Formally, the relation $\Rightarrow_{\mathcal{G},\mathrm{LO}}^p$ on $\mathrm{SF}(\mathcal{G})$ is defined just like $\Rightarrow_{\mathcal{G}}^p$, only with the additional requirement that the nonterminal $A^{\boxed{i}}$ occurs *leftmost-outermost* in the tree η_1. That is, we require that for every $w \in \mathrm{pos}(\eta_1)$, if $w <_{\mathrm{lex}} \mathrm{pos}^{(i)}(\eta_1)$, then $\eta_1(w) \notin N^{\boxed{\mathbb{N}}}$. Let $m \in \mathbb{N}$, $\pi_0, \pi_m \in \mathrm{SF}(\mathcal{G})$, and $d = p_1 \ldots p_m \in P^m$. We write $\pi_0 \Rightarrow_{\mathcal{G},\mathrm{LO}}^d \pi_m$ if there are $\pi_1, \ldots, \pi_{m-1} \in \mathrm{SF}(\mathcal{G})$ such that $\pi_{i-1} \Rightarrow_{\mathcal{G},\mathrm{LO}}^{p_i} \pi_i$ for every $i \in [m]$. The mapping wt is extended to P^* by $wt(p_1 \ldots p_m) = \prod_{i=1}^m wt(p_i)$. Let, for every $m \in \mathbb{N}$, $\pi \in \mathrm{SF}(\mathcal{G})$, and $s, t \in T_\Sigma$,

$$[\![\mathcal{G}]\!]_\pi^{(m)}(s, t) = \sum_{\substack{d \in P^m, \\ \pi \Rightarrow_{\mathcal{G},\mathrm{LO}}^d [s,t]}} wt(d) \quad \text{and} \quad [\![\mathcal{G}]\!](s, t) = \sum_{m \in \mathbb{N}} [\![\mathcal{G}]\!]_{[Z^{\boxed{1}}, Z^{\boxed{1}}]}^{(m)}(s, t).$$

The latter weighted tree transformation $[\![\mathcal{G}]\!]: T_\Sigma \times T_\Sigma \to S$ is the *transformation of* $\mathcal{G}$. Two wscftg $\mathcal{G}_1$ and $\mathcal{G}_2$ are *equivalent* if $[\![\mathcal{G}_1]\!] = [\![\mathcal{G}_2]\!]$.

Next, we will introduce simple wscftg. Intuitively, we demand that for every production of the grammar the call structure in its input component is the same as in its output component. The call structure of a tree ξ comprises the

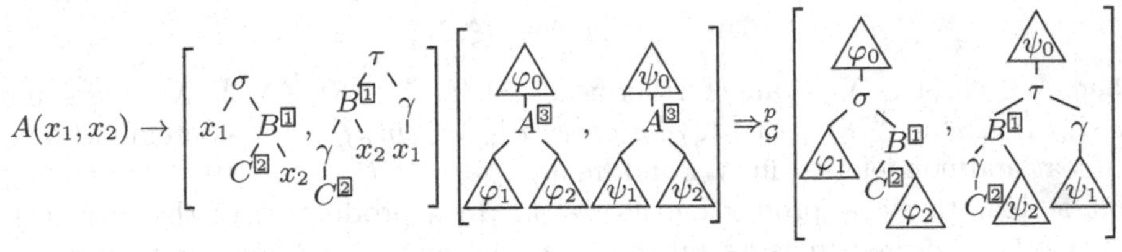

$$A(x_1, x_2) \rightarrow \begin{bmatrix} x_1 & \overset{\sigma}{\underset{C^{\boxed{2}}}{\overset{B^{\boxed{1}}}{\diagdown}}} x_2 & , & \overset{\tau}{\underset{C^{\boxed{2}}}{\overset{B^{\boxed{1}}}{\diagup}}} \gamma \\ & & & x_2 x_1 \end{bmatrix} \begin{bmatrix} \overset{\varphi_0}{\triangle} \\ \triangle A^{\boxed{3}} \triangle \\ \overset{}{\triangle}_{\varphi_1} \overset{}{\triangle}_{\varphi_2} & , & \overset{\psi_0}{\triangle} \\ \triangle A^{\boxed{3}} \triangle \\ \overset{}{\triangle}_{\psi_1} \overset{}{\triangle}_{\psi_2} \end{bmatrix} \Rightarrow_{\mathcal{G}}^p \begin{bmatrix} \overset{\varphi_0}{\triangle} \\ \overset{\sigma}{\underset{\triangle_{\varphi_1}}{\diagup}} \overset{B^{\boxed{1}}}{\underset{C^{\boxed{2}}\triangle_{\psi_2}}{\diagdown}} \\ \triangle_{\varphi_2} & , & \overset{\psi_0}{\triangle} \\ \overset{\tau}{\underset{C^{\boxed{2}}\triangle_{\psi_2}}{\overset{B^{\boxed{1}}}{\diagup}}} \gamma \\ & & \triangle_{\psi_1} \end{bmatrix}$$

Fig. 1. A wscftg production p and its application to a sentential form

entirety of the successor relations between the nonterminals and the variables in ξ. Formally, a wscftg $\mathcal{G}$ is called a *simple wscftg (s-wscftg)* if for every production in P of the form (1), for every $i \in M$, and for every $j \in [\mathrm{rk}(\mathrm{nt}^{(i)}(\xi))]$, the sets $\{\xi(wjv) \in N^{\boxtimes} \cup X \mid w = \mathrm{pos}^{(i)}(\xi), v \in \mathrm{pos}(\xi|_{wj})\}$ and $\{\zeta(wjv) \in N^{\boxtimes} \cup X \mid w = \mathrm{pos}^{(i)}(\zeta), v \in \mathrm{pos}(\zeta|_{wj})\}$ are equal. For example, the production p in Fig. 1 is the production of a s-wscftg, as in both components, C (resp. x_2) is in the first (resp. second) subtree of B, while x_1 has no nonterminal as an ancestor. But p would not belong to a s-wscftg if x_1 and x_2 (or $B^{\boxed{1}}$ and $C^{\boxed{2}}$) were exchanged with each other in the input component. Note that every wsrtg is a s-wscftg. Also note that there are wscftg whose transformations can not be defined by s-wscftg, cf. [1, Thm. 2(b)].

Let us introduce a certain normal form for s-wscftg that is closely related to the Chomsky normal form for context-free grammars. A s-wscftg $\mathcal{G}$ is said to be in *normal form* if every production p in P is of either of the following two forms: *(i)* $A(x_1, \ldots, x_k) \rightarrow [\xi, \xi]$, for some $\xi \in T_{N^{\boxtimes}}(X_k)$; we then call p a *nonterminal production*, and denote the set of all such productions of $\mathcal{G}$ by P_{NT}; or *(ii)* $A(x_1, \ldots, x_k) \rightarrow [\xi, \zeta]$, for $\xi, \zeta \in T_\Sigma(X_k)$; then p is a *terminal production*, and the set of all such productions of $\mathcal{G}$ is denoted by P_{T}.

Lemma 1. *For every s-wscftg there is an equivalent s-wscftg in normal form.*

Proof. Assume that $\mathcal{G}$ is a s-wscftg, and that $p \in P$ is of the form (1), but neither terminal nor nonterminal. Then there must be an occurrence of a terminal symbol in ξ or in ζ. W.l.o.g., let it occur in ξ at position u. Distinguish the following two cases: *(i)* There are positions on the path from the root of ξ to u that are labeled by a nonterminal. Then let v' be the one such position of maximal length (as an element of $\mathbb{N}^*$) and let $j \in \mathbb{N}$ be such that $v'j$ is a prefix of u. By definition of $\mathcal{G}$, there must be a position $w' \in \mathrm{pos}(\zeta)$ with $\zeta(w') = \xi(v')$. Set $v = v'j$ and $w = w'j$. Note that since $\mathcal{G}$ is simple, the set of nonterminals and variables that occur in $\xi|_v$ must be equal to the respective set in $\zeta|_w$. *(ii)* There are no positions on the path from the root of ξ to u that are labeled by nonterminals. In this case, we set $v = w = \varepsilon$. This concludes the case analysis.

Now let $(\varphi, \xi_1 \ldots \xi_{k_1}) = \mathrm{lin}_{\Xi}(\xi|_v)$ and $(\tilde{\psi}, \zeta_1 \ldots \zeta_{k_2}) = \mathrm{lin}_{\Xi}(\zeta|_w)$, where $\Xi = X \cup \{\xi \in T_{N^{\boxtimes} \cup \Sigma}(X) \mid \xi(\varepsilon) \in N^{\boxtimes}\}$. As $\mathcal{G}$ is simple, $k_1 = k_2$ (denoted by k') and there must be a permutation $\omega : [k'] \rightarrow [k']$ such that for every $j \in [k']$,

$\xi_j(\varepsilon) = \zeta_{\omega(j)}(\varepsilon)$. Let $\psi = \tilde{\psi}[x_1/x_{\omega^{-1}(1)}, \ldots, x_{k'}/x_{\omega^{-1}(k')}]$. Replace the subtrees $\xi|_v$ and $\zeta|_w$ in p by $B^{\boxed{l}}(\xi_1, \ldots, \xi_{k'})$, resp. $B^{\boxed{l}}(\zeta_{\omega(1)}, \ldots, \zeta_{\omega(k')})$, where B is a new nonterminal and l a fresh index, and call this production p'. Moreover, let p'' be the production $B(x_1, \ldots, x_{k'}) \to [\varphi, \psi]$. Construct the wscftg $\mathcal{G}' = (N', S, \Sigma, Z, P', wt')$ with $N' = N \cup \{B\}$, $P' = P \setminus \{p\} \cup \{p', p''\}$, and $wt'(p') = wt(p)$, $wt'(p'') = 1$, while $wt'|_{P \cap P'} = wt|_{P \cap P'}$. By inspection of the definition, $\mathcal{G}'$ is also simple. One can show that there is a weight-preserving bijection between successful leftmost-outermost derivations in $\mathcal{G}$ and $\mathcal{G}'$, and therefore $[\![\mathcal{G}]\!] = [\![\mathcal{G}']\!]$. Obviously, this construction can only be applied a finite number of times, and the resulting grammar is an equivalent s-wscftg in normal form. $\square$

As an example for the above, let p be the production $A(x_1, x_2, x_3) \to [\xi, \zeta]$, with

$$[\xi, \zeta] = \left[\sigma\big(x_1, B^{\boxed{1}}(\sigma(C^{\boxed{2}}(x_2), D^{\boxed{3}}(x_3)))\big), \tau\big(B^{\boxed{1}}(\sigma(D^{\boxed{3}}(x_3), C^{\boxed{2}}(x_2))), x_1\big)\right].$$

In the construction's first iteration the terminal position $u = \varepsilon$ in ξ is chosen, thus case (ii) applies and $v = w = \varepsilon$. Hence p is replaced by a production $p' = (A(x_1, x_2, x_3) \to [\xi', \zeta']$, where $[\xi', \zeta']$ is

$$\left[E^{\boxed{4}}\big(x_1, B^{\boxed{1}}(\sigma(C^{\boxed{2}}(x_2), D^{\boxed{3}}(x_3)))\big), E^{\boxed{4}}\big(x_1, B^{\boxed{1}}(\sigma(D^{\boxed{3}}(x_3), C^{\boxed{2}}(x_2)))\big)\right],$$

and by p'' of the form $E(x_1, x_2) \to [\sigma(x_1, x_2), \tau(x_2, x_1)]$. In the second iteration let $u = 21$ in ξ', this yields case (i). Then also $v = w = 21$, and p' is replaced by the production $A(x_1, x_2, x_3) \to [\xi'', \zeta'']$, with $[\xi'', \zeta'']$ equal to

$$\left[E^{\boxed{4}}\big(x_1, B^{\boxed{1}}(F^{\boxed{5}}(C^{\boxed{2}}(x_2), D^{\boxed{3}}(x_3)))\big), E^{\boxed{4}}\big(x_1, B^{\boxed{1}}(F^{\boxed{5}}(C^{\boxed{2}}(x_2), D^{\boxed{3}}(x_3)))\big)\right],$$

and by $F(x_1, x_2) \to [\sigma(x_1, x_2), \sigma(x_2, x_1)]$. All introduced productions satisfy the conditions of the normal form, so the construction may terminate at this point.

4   Weighted Pushdown Extended Tree Transducers

In contrast to the productions of wscftg, the rules of wpxtop are asymmetric, and permit a state-based rewriting of input into output trees. Just as for extended tree transducers, every rule allows matching the current input tree with a context of arbitrary height, and their right-hand sides are output trees at whose frontiers the rewriting process may continue on the remaining subtrees of the input. Unlike the former however, the derivations of wpxtop are controlled by tree pushdowns [9,6]. Thus, a rule can additionally inspect the top symbol of the current tree pushdown, and push further symbols that control the remaining derivation. In the example derivation step in Fig. 2, the context c has already been produced as output, while the input tree $\sigma(s_1, \sigma(s_2, \alpha))$ must yet be rewritten. Since the tree pushdown is $\gamma(\kappa_1, \kappa_2)$, the rule r can be applied, producing some output. The transduction must continue on the remaining inputs s_1 and s_2, controlled by the pushdowns κ_1 and $\gamma(\alpha, \kappa_2)$.

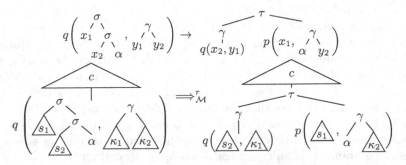

Fig. 2. A wpxtop rule r and its application

A *weighted pushdown extended top-down tree transducer (wpxtop)* is a tuple $\mathcal{M} = (Q, S, \Sigma, \Gamma, q_0, \gamma_0, R, wt)$ where Q is a dyadic ranked alphabet of *states*, i.e. $Q = Q^{(2)}$, Σ and Γ are ranked alphabets of, resp., *terminal* and *pushdown symbols*, $q_0 \in Q$ is the *initial state*, $\gamma_0 \in \Gamma^{(0)}$ is the *initial pushdown symbol*, and R is a finite set of *rules* r of the form

$$q\big(\xi, \gamma(y_1, \ldots, y_k)\big) \to \zeta, \tag{2}$$

for some $q \in Q$, $k, n \in \mathbb{N}$, $\gamma \in \Gamma^{(k)}$, $\xi \in C_\Sigma(X_n)$, and $\zeta \in T_\Sigma(Q(X_n, T_\Gamma(Y_k)))$. Finally, $wt \colon R \to S \setminus \{0\}$ assigns to every rule its *weight*. We call a wpxtop $\mathcal{M}$ *linear and nondeleting (ln-wpxtop)* if for every rule from R of the form (2), ζ is linear and nondeleting in $X_n \cup Y_k$. It is *one-state* if $|Q| = 1$.

Consider a wpxtop $\mathcal{M}$ as above. The set of *derivation forms* of $\mathcal{M}$, denoted by $\mathrm{DF}(\mathcal{M})$, is the set $T_\Sigma(Q(T_\Sigma, T_\Gamma))$. For every $r \in R$ of the form (2), define the binary relation $\Rightarrow^r_\mathcal{M}$ on $\mathrm{DF}(\mathcal{M})$ as follows. Given $s, t \in \mathrm{DF}(\mathcal{M})$, let $s \Rightarrow^r_\mathcal{M} t$ if there are $c \in T_\Sigma(Q(T_\Sigma, T_\Gamma) \cup X_1)$, $s_1, \ldots, s_n \in T_\Sigma$, and $\kappa_1, \ldots, \kappa_k \in T_\Gamma$ such that x_1 appears exactly once in c,

$$s = c\big[q(\xi[s_1, \ldots, s_n], \gamma(\kappa_1, \ldots, \kappa_k))\big], \text{ and}$$
$$t = c\big[\zeta[x_1/s_1, \ldots, x_n/s_n, y_1/\kappa_1, \ldots, y_k/\kappa_k]\big].$$

Again, care must be taken that there is no superfluous summation of the weight of what is essentially the same derivation. Therefore, the *leftmost rewrite relation* $\Rightarrow^r_{\mathcal{M},\mathrm{L}}$ is defined just like $\Rightarrow^r_\mathcal{M}$, but the rule r must be applied to the leftmost state that occurs in s. Formally, we add the restriction that for every $v, w \in \mathrm{pos}(c)$, if $c(w) = x_1$ and $v \leq_{\mathrm{lex}} w$, then $c(v) \notin Q$. Given a sequence $d = r_1 \ldots r_m \in R^m$, $m \in \mathbb{N}$, we write $s \Rightarrow^d_{\mathcal{M},\mathrm{L}} t$ if there are $\xi_0, \ldots, \xi_m \in \mathrm{DF}(\mathcal{M})$ such that $s = \xi_0$, $t = \xi_m$ and $\xi_{i-1} \Rightarrow^{r_i}_{\mathcal{M},\mathrm{L}} \xi_i$ for every $i \in [m]$. The mapping wt is extended to R^* by setting $wt(r_1 \ldots r_m) = \prod_{i=1}^m wt(r_i)$. For every $m \in \mathbb{N}$, $\kappa \in T_\Gamma$, and $s, t \in T_\Sigma$,

$$\llbracket \mathcal{M} \rrbracket^{q,(m)}_\kappa(s, t) = \sum_{\substack{d \in R^m \\ q(s,\kappa) \Rightarrow^d_{\mathcal{M},\mathrm{L}} t}} wt(d), \quad \text{and} \quad \llbracket \mathcal{M} \rrbracket(s, t) = \sum_{m \in \mathbb{N}} \llbracket \mathcal{M} \rrbracket^{q_0,(m)}_{\gamma_0}(s, t).$$

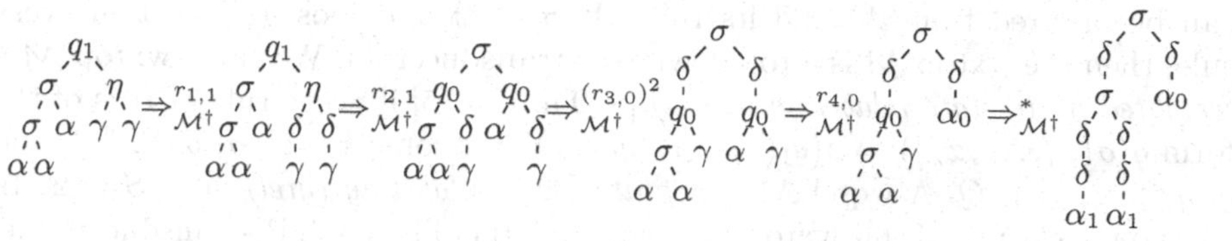

Fig. 3. Derivation of the example wpxtop $\mathcal{M}^\dagger$

We refer to $[\![\mathcal{M}]\!]\colon T_\Sigma \times T_\Sigma \to S$ as the *transformation* of $\mathcal{M}$, and say that two wpxtop $\mathcal{M}$ and $\mathcal{M}'$ are *equivalent* if $[\![\mathcal{M}]\!] = [\![\mathcal{M}']\!]$. From now on, let $\mathcal{M}$ denote the wpxtop $(Q, S, \Sigma, \Gamma, q_0, \gamma_0, R, wt)$, unless stated otherwise.

As an example for the derivation semantics of wpxtops, consider $\mathcal{M}^\dagger = (Q, \mathbb{B}, \Sigma, \Gamma, q_0, \gamma, R, wt)$ with $Q = \{q_0, q_1\}$, $\Sigma^{(2)} = \{\sigma\}$, $\Sigma^{(1)} = \{\delta\}$, and $\Sigma^{(0)} = \{\alpha, \alpha_0, \alpha_1\}$, while $\Gamma^{(2)} = \{\eta\}$, $\Gamma^{(1)} = \{\delta\}$, $\Gamma^{(0)} = \{\gamma\}$, and $\Sigma^{(k)} = \Gamma^{(k)} = \emptyset$ for every $k \geq 3$. Let R be given by the eight rules

$$r_{1,i} = \qquad q_i(x_1, \eta(y_1, y_2)) \to q_i\big(x_1, \eta(\delta(y_1), \delta(y_2))\big),$$
$$r_{2,i} = q_i\big(\sigma(x_1, x_2), \eta(y_1, y_2)\big) \to \sigma\big(q_{(1-i)}(x_1, y_1), q_{(1-i)}(x_2, y_2)\big),$$
$$r_{3,i} = \qquad q_i(x_1, \delta(y_1)) \to \delta\big(q_i(x_1, y_1)\big), \text{ and}$$
$$r_{4,i} = \qquad q_i(\alpha, \gamma) \to \alpha_i,$$

where $i \in \{0, 1\}$. Compare Fig. 3 for a derivation of $\mathcal{M}^\dagger$. The finite state control of $\mathcal{M}^\dagger$ allows rewriting α alternatingly into the symbols α_0 and α_1, while its tree pushdown permits producing an equal unbounded number of symbols δ in independent subtrees. We note that the latter feature of the transformation can not be achieved by any (input-linear) extended tree transducer [7]. This can be proved by a pumping argument.

The following parallel derivation lemma will be used implicitly in many subsequent proofs. It allows the decomposition of a nonempty derivation into the application of a rule and a number of independent subderivations.

Lemma 2. *Let* $k, m \in \mathbb{N}$, $q \in Q$, $s, t \in T_\Sigma$, *and* $\kappa, \kappa_1, \ldots, \kappa_k \in T_\Gamma$ *with* $\kappa = \gamma(\kappa_1, \ldots, \kappa_k)$ *for some* $\gamma \in \Gamma^{(k)}$. *Then*

$$[\![\mathcal{M}]\!]_\kappa^{q,(m+1)}(s, t) = \sum_{\substack{r=(q(\xi, \gamma(y_1, \ldots, y_k)) \to \zeta) \in R, \\ \mathrm{lin}_{Q(X, T_\Gamma(Y))}(\zeta) = (\hat\zeta, q_1(x_{i_1}, \eta_1) \ldots q_l(x_{i_l}, \eta_l)), \\ (\xi, s_1 \ldots s_n) \in \mathrm{dec}_n(s), (\hat\zeta, t_1 \ldots t_l) \in \mathrm{dec}_l(t), \\ m_1, \ldots, m_l \in \mathbb{N}, \sum_{j=1}^l m_j = m}} wt(r) \cdot \prod_{j=1}^{l} [\![\mathcal{M}]\!]_{\eta_j[\kappa_1, \ldots, \kappa_k]}^{q_j, (m_j)}(s_{i_j}, t_j).$$

The wpxtop is related to the following formalisms. If $|\mathrm{pos}_\Sigma(\xi)| \leq 1$ for every rule of the form (2) and $S = \mathbb{B}$, then $\mathcal{M}$ is a *top-down pushdown tree transducer* [18], and if also $\mathrm{lin}_{Q(X, T_\Gamma(Y))}(\zeta) = (\hat\zeta, q_1(x_1, \eta_1) \ldots q_n(x_1, \eta_n))$ with $\hat\zeta = \xi$, then $\mathcal{M}$ is a *pushdown tree automaton* [9]. If $\Gamma = \{\gamma_0\}$, then $\mathcal{M}$ is a *weighted extended top-down tree transducer (wxtop)* [7]. Obviously, in this case the pushdown symbols can be omitted from $\mathcal{M}$ and its rules. If $S = \mathbb{B}$ and $|\text{pos}_\Sigma(\xi)| = 1$ in every rule, then the wxtop $\mathcal{M}$ is a top-down tree transducer [5]. We call a wxtop $\mathcal{M}$ a *weighted finite-state relabeling (wqrel)* [8, Thm. 5.15] if every rule in R is of the form $q(\sigma(x_1, \ldots, x_k)) \to \delta(q_1(x_1), \ldots, q_k(x_k))$ for some $k \in \mathbb{N}$, σ, $\delta \in \Sigma^{(k)}$, and q, q_1, ..., $q_k \in Q$. A wqrel $\mathcal{M}$ is a *finite-state relabeling (qrel)* [5] if $S = \mathbb{B}$. In the nomenclature of [6], wxtop are weighted RT(TR$_{\text{fin}}$ × TP)-transducers, i.e. regular tree grammars equipped with a variant of the tree storage type TR that allows finite lookahead and decomposition, and with a tree pushdown storage type TP.

In the following lemma, we prove the existence of a one-state normal form for ln-wpxtop. The reader's proof idea might be to modify the proof that every creative dendrolanguage can be generated by a one-state creative dendrogrammar [15, Thm. 7], in which the state behaviour of a creative dendrogrammar is encoded into its pushdown symbols. However, this construction does not preserve the properties of linearity and nondeletion. Nevertheless, a different encoding of state transitions on the pushdown can be found which leaves these properties intact. In this encoding scheme, we replace a pushdown symbol $\gamma \in \Gamma^{(k)}$ with $(q, \gamma, q_1 \ldots q_k) \in \Gamma'^{(k)}$, where q_1, ..., q_k and q are states of the original transducer $\mathcal{M}$. This new pushdown symbol has the intended meaning that, if $\mathcal{M}$ processes γ with state q, then it will eventually process the i-th successor symbol γ_i of γ with state q_i, where $i \in [k]$. Of course, γ_i should then again be substituted with a symbol of the form (q_i, γ_i, w_i) for some $w_i \in Q^*$ – i.e., the structure of the constructed tree pushdown must reflect possible state transitions of $\mathcal{M}$. Thus, the described method generalizes the construction of one-state pushdown string automata, cf. e.g. [11, Lect. 25], to linear and nondeleting trees.

Lemma 3. *For every ln-wpxtop $\mathcal{M}$ there is an equivalent one-state ln-wpxtop $\mathcal{M}'$.*

Proof. The new pushdown alphabet is given by $\Gamma'^{(k)} = \{(q, \gamma, q_1 \ldots q_k) \mid \gamma \in \Gamma^{(k)}, q, q_1, \ldots, q_k \in Q\}$ for every $k \in \mathbb{N}$. For every $q \in Q$, $k \in \mathbb{N}$, p_1, ..., $p_k \in Q$, define the qrel $B^q_{p_1 \ldots p_k} = (Q, \mathbb{B}, \Gamma \cup \Gamma' \cup Y_k, q, R_B)$, where R_B contains the rules

$$p_i(y_i) \to y_i \quad \text{and} \quad s(\gamma(x_1, \ldots, x_n)) \to (s, \gamma, s_1 \cdots s_n)(s_1(x_1), \ldots, s_n(x_n))$$

for every $i \in [k]$, $n \in \mathbb{N}$, $\gamma \in \Gamma^{(n)}$ and s, s_1, ..., $s_n \in Q$. Note that the elements of Y_k are nullary terminal symbols of $B^q_{p_1 \ldots p_k}$. The application of $B^q_{p_1 \ldots p_k}$ to a tree pushdown $\eta \in T_\Gamma(Y_k)$ will encode possible state transitions of $\mathcal{M}$ into η, which start out at the root with q and reach y_i in state p_i.

Construct the wpxtop $\mathcal{M}' = (Q', S, \Sigma, \Gamma', \gamma'_0, R', wt')$ with $Q' = \{\star\}$, and $\gamma'_0 = (q_0, \gamma_0, \varepsilon)$. For every rule $r \in R$ of the form (2), where $\text{lin}_{Q(X, T_\Gamma(Y))}(\zeta) = (\hat{\zeta}, q_1(x_{i_1}, \eta_1) \ldots q_n(x_{i_n}, \eta_n))$, and every p_1, ..., $p_k \in Q$, add all r' of the form

$$\star(\xi, (q, \gamma, p_1 \ldots p_k)(y_1, \ldots, y_k)) \to \hat{\zeta}[\star(x_{i_1}, \eta'_1), \ldots, \star(x_{i_n}, \eta'_n)]$$

to R', where $\eta'_j \in [\![B^{q_j}_{p_1 \ldots p_k}]\!](\eta_j)$ for $j \in [n]$, and $wt'(r') = wt(r)$. As the construction modifies no variables from X and Y, $\mathcal{M}'$ is linear and nondeleting.

To prove equivalence of $\mathcal{M}$ and $\mathcal{M}'$, the following auxiliary statement is necessary. Assume $n \in \mathbb{N}$, sets V_i and mappings $f_i \colon V_i \to S$ for every $i \in [n]$. Then

$$\sum\Big(\prod_{i=1}^{n} f_i(v_i) \ \Big| \ v_1 \in V_1, \ldots, v_n \in V_n\Big) = \prod_{i=1}^{n} \sum\Big(f_i(v) \ \Big| \ v \in V_i\Big). \qquad (3)$$

This follows from the distributivity of S. Now, we can prove the following proposition by complete induction: for every $m \in \mathbb{N}$, $q \in Q$, $s, t \in T_\Sigma$, and $\kappa \in T_\Gamma$,

$$\sum\Big(\llbracket\mathcal{M}'\rrbracket_{\kappa'}^{*,(m)}(s,t) \ \Big| \ \kappa' \in \llbracket B_\varepsilon^q\rrbracket(\kappa)\Big) = \llbracket\mathcal{M}\rrbracket_{\kappa}^{q,(m)}(s,t). \qquad (4)$$

For $m = 0$, obviously both sides of the equation are 0, so consider $m+1$. We abbreviate lists of subterms like $a_1, \ldots, a_n$ or $a_1 \ldots a_n$ by $a_{1,n}$, the set $Q(X, T_\Gamma(Y))$ by U, and $Q'(X, T_{\Gamma'}(Y))$ by U'. Assume that $\kappa = \gamma(\kappa_1, \ldots, \kappa_k)$ for $k \in \mathbb{N}$, $\gamma \in \Gamma^{(k)}$, $\kappa_1, \ldots, \kappa_k \in T_\Gamma$. Then

$$\sum\Big(\llbracket\mathcal{M}'\rrbracket_{\kappa'}^{*,(m+1)}(s,t) \ \Big| \ \kappa' \in \llbracket B_\varepsilon^q\rrbracket(\kappa)\Big)$$

$$= \sum\Big(wt'(r') \cdot \prod_{j=1}^{n} \llbracket\mathcal{M}'\rrbracket_{\eta_j'[\kappa_{1,k}']}^{*,(m_j)}(s_{i_j}, t_j)$$
$$\Big| \ p_{1,k} \in Q, \ r' = (\star(\xi, (q, \gamma, p_{1,k})(y_{1,k})) \to \zeta) \in R',$$
$$\lin_{U'}(\zeta) = (\hat\zeta, \star(x_{i_1}, \eta_1') \ldots \star(x_{i_n}, \eta_n')), \ \kappa_u' \in \llbracket B_\varepsilon^{p_u}\rrbracket(\kappa_u), \ u \in [k],$$
$$(\xi, s_{1,n}) \in dec_n(s), \ (\hat\zeta, t_{1,n}) \in dec_n(t), \ m_{1,n} \in \mathbb{N}, \ \textstyle\sum_{j=1}^{n} m_j = m\Big)$$

$$= \sum\Big(wt(r) \cdot \prod_{j=1}^{n} \sum(\llbracket\mathcal{M}'\rrbracket_{\hat\eta_j'[\kappa_{u_1, u_\nu}']}^{*,(m_j)}(s_{i_j}, t_j) \ | \ \lin_Y(\eta_j) = (\hat\eta_j, y_{u_1, u_l}),$$
$$p_{u_1, u_l} \in Q,$$
$$\kappa_{u_\nu}' \in \llbracket B_\varepsilon^{p_{u_\nu}}\rrbracket(\kappa_{u_\nu}), \nu \in [l],$$
$$\eta_j' \in \llbracket B_{p_{u_1, u_l}}^{q_j}\rrbracket(\hat\eta_j[y_{1,l}]))$$
$$\Big| \ r = (q(\xi, \gamma(y_{1,k})) \to \zeta) \in R, \ \lin_U(\zeta) = (\hat\zeta, q_1(x_{i_1}, \eta_1) \ldots q_n(x_{i_n}, \eta_n)),$$
$$(\xi, s_{1,n}) \in dec_n(s), \ (\hat\zeta, t_{1,n}) \in dec_n(t), \ m_{1,n} \in \mathbb{N}, \ \textstyle\sum_{j=1}^{n} m_j = m\Big)$$

$$= \sum\Big(wt(r) \cdot \prod_{j=1}^{n} \sum(\llbracket\mathcal{M}'\rrbracket_\theta^{*,(m_j)}(s_{i_j}, t_j) \ | \ \theta \in \llbracket B_\varepsilon^{q_j}\rrbracket(\eta_j[\kappa_{1,k}]))$$
$$\Big| \ r = (q(\xi, \gamma(y_{1,k})) \to \zeta) \in R, \ \lin_U(\zeta) = (\hat\zeta, q_1(x_{i_1}, \eta_1) \ldots q_n(x_{i_n}, \eta_n)),$$
$$(\xi, s_{1,n}) \in dec_n(s), \ (\hat\zeta, t_{1,n}) \in dec_n(t), \ m_{1,n} \in \mathbb{N}, \ \textstyle\sum_{j=1}^{n} m_j = m\Big)$$

$$\overset{\text{(IH)}}{=} \sum\Big(wt(r) \cdot \prod_{j=1}^{n} \llbracket\mathcal{M}\rrbracket_{\eta_j[\kappa_{1,k}]}^{q_j,(m_j)}(s_{i_j}, t_j)$$
$$\Big| \ r = (q(\xi, \gamma(y_{1,k})) \to \zeta) \in R, \ \lin_U(\zeta) = (\hat\zeta, q_1(x_{i_1}, \eta_1) \ldots q_n(x_{i_n}, \eta_n)),$$
$$(\xi, s_{1,n}) \in dec_n(s), \ (\hat\zeta, t_{1,n}) \in dec_n(t), \ m_{1,n} \in \mathbb{N}, \ \textstyle\sum_{j=1}^{n} m_j = m\Big)$$

$$= \llbracket\mathcal{M}\rrbracket_\kappa^{q,(m+1)}(s,t).$$

First of all, let us explain why the second equation holds. Obviously, the function

$$f_j := \lambda p_1 \dots p_k, \eta_1' \dots \eta_n', \kappa_1' \dots \kappa_k'. [\![\mathcal{M}']\!]_{\eta_j'[\kappa_{1,k}']}^{*,(m_j)}(s_{i_j}, t_j)$$

only depends on the argument η_j', and on those p_u and κ_u' such that y_u occurs in η_j'. Because $\mathcal{M}'$ is linear and nondeleting, we can partition $\{p_1, \dots, p_k\}$ and $\{\kappa_1', \dots, \kappa_k'\}$ into disjoint sets P_j and K_j of those states and tree pushdowns that f_j depends on. Hence we may apply (3) and swap the sum and the product. In the following equation, the relabelings of η_j and of the parameters κ_{u_ν} are combined into a relabeling of $\hat{\eta}_j[\kappa_{u_1}, \dots, \kappa_{u_l}] = \eta_j[\kappa_1, \dots, \kappa_k]$. This is possible because of [8, Lem. 5.9], since qrels are also bottom-up weighted tree transducers. Now, for every $s, t \in T_\Sigma$,

$$[\![\mathcal{M}]\!](s,t) = \sum_{m \in \mathbb{N}} [\![\mathcal{M}]\!]_{\gamma_0}^{q_0,(m)}(s,t) = \sum_{m \in \mathbb{N}} [\![\mathcal{M}']\!]_{(q_0,\gamma_0,\varepsilon)}^{*,(m)}(s,t) = [\![\mathcal{M}']\!](s,t)\,,$$

by (4), because $[\![B_\varepsilon^{q_0}]\!](\gamma_0) = \{(q_0, \gamma_0, \varepsilon)\}$. □

In the normal form for wpxtop we introduce in the following, the rules of the transducer are partitioned into two kinds: *(i)* rules which do neither consume any input nor produce any output, but may, however, push to the pushdown storage, and *(ii)* rules which can consume input and produce output, but may only pop the root of the pushdown store and push no further symbols. Formally, a wpxtop $\mathcal{M}$ is said to be in *index normal form* if every rule $r \in R$ is of one of the following forms: *(i)* $q(x_1, \gamma(y_1, \dots, y_k)) \to q(x_1, \eta)$ for some $q \in Q$, $k \in \mathbb{N}$, and $\eta \in T_\Gamma(Y_k)$; then r is an *index-creating rule*, the set of all such $r \in R$ is denoted R_{IC}; *(ii)* $q(\xi, \gamma(y_1, \dots, y_k)) \to \zeta[q_1(x_1, y_1), \dots, q_k(x_k, y_k)]$ for some $\zeta \in T_\Sigma^{\mathrm{ln}}(Y_k)$ and $q_1, \dots, q_k \in Q$; then r is an *index-erasing rule*, and the set of all such $r \in R$ is denoted R_{IE}.[1]

Lemma 4. *For every ln-wpxtop $\mathcal{M}$, there is an equivalent ln-wpxtop $\mathcal{M}'$ in index normal form. If $\mathcal{M}$ is one-state, then so is $\mathcal{M}'$.*

Proof. Assume that, w.l.o.g., $\Gamma \cap R = \emptyset$. We construe the finite set R as a ranked alphabet with $\mathrm{rk}(r) = n$ for each $r \in R$ as in (2). Construct the ln-wpxtop $\mathcal{M}' = (Q, S, \Sigma, \Gamma', q_0, \gamma_0, R', wt')$, where $\Gamma' = \Gamma \cup R$, and for every rule r from R of the form (2) with $\mathrm{lin}_{Q(X,T_\Gamma(Y))}(\zeta) = (\hat{\zeta}, q_1(x_{i_1}, \eta_{i_1}) \dots q_n(x_{i_n}, \eta_{i_n}))$, the two rules

$$r' = (q(x_1, \gamma(y_1, \dots, y_k)) \to q(x_1, r(\eta_1, \dots, \eta_n)))\,,$$

with $wt'(r') = wt(r)$, and

$$r'' = (q(\xi, r(y_1, \dots, y_n)) \to \hat{\zeta}[q_1(x_{i_1}, y_{i_1}), \dots, q_n(x_{i_n}, y_{i_n})])\,,$$

---

[1] The names of these kinds of rules are not related to the indices from Section 3, but in analogy to the two kinds of productions of creative dendrogrammars [15].

with $wt'(r'') = 1$, are inserted into R'. One can show by complete induction that for every $m \in \mathbb{N}$, $q \in Q$, $s, t \in T_\Sigma$, and $\kappa \in T_\Gamma$,

$$\llbracket \mathcal{M} \rrbracket_\kappa^{q,(m)}(s,t) = \llbracket \mathcal{M}' \rrbracket_\kappa^{q,(2m)}(s,t). \tag{5}$$

Then, for every $s, t \in T_\Sigma$,

$$\llbracket \mathcal{M} \rrbracket(s,t) = \sum_{m \in \mathbb{N}} \llbracket \mathcal{M} \rrbracket_{\gamma_0}^{q_0,(m)}(s,t) = \sum_{m \in \mathbb{N}} \llbracket \mathcal{M}' \rrbracket_{\gamma_0}^{q_0,(2m)}(s,t) = \llbracket \mathcal{M}' \rrbracket(s,t).$$

The second equality holds with (5) because, obviously, there are no derivations d of odd length such that $q_0(s, \gamma_0) \Rightarrow_{\mathcal{M}',\mathrm{L}}^d t$. $\qquad\square$

5 Main Result

Now, we can exploit the similarity between s-wscftg in normal form and one-state wpxtop in index normal form to obtain the proof that both formalisms are equivalent. We require a further concept, however. Given $\xi \in \mathcal{I}(N, \Sigma, X)$, define $\xi{\downarrow}$ as the tree in $T_{N \cup \Sigma}(X)$ which results from ξ by replacing every nonterminal $A^{\boxed{?}}$ in ξ by A.

Definition 1. *Let $\mathcal{G} = (N, S, \Sigma, Z, P, wt_P)$ be a s-wscftg in normal form and $\mathcal{M} = (\{\star\}, S, \Sigma, \Gamma, \star, \gamma_0, R, wt_R)$ be a one-state ln-wpxtop in index normal form. We say that $\mathcal{G}$ and $\mathcal{M}$ are related if $N = \Gamma$, $\gamma_0 = Z$, and (i) the production $p = (A(x_1, \ldots, x_k) \to [\xi, \xi])$ is in P_{NT} iff the rule $r = (\star(x_1, A(y_1, \ldots, y_k)) \to \star(x_1, \xi{\downarrow}[y_1, \ldots, y_k]))$ is in R_{IC}, with $wt_P(p) = wt_R(r)$; and (ii) the production $p' = (A(x_1, \ldots, x_k) \to [\xi, \zeta])$ is in P_{T} iff the rule $r' = (\star(\xi, A(y_1, \ldots, y_k)) \to \zeta[\star(x_1, y_1), \ldots, \star(x_k, y_k)])$ is in R_{IE}, where $wt_P(p') = wt_R(r')$.*

Lemma 5. *Let $\mathcal{G}$ and $\mathcal{M}$ from Def. 1 be related. Then $\llbracket \mathcal{G} \rrbracket = \llbracket \mathcal{M} \rrbracket$.*

Proof. We prove the equation $\llbracket \mathcal{G} \rrbracket_{[\eta,\eta]}^{(m)} = \llbracket \mathcal{M} \rrbracket_{\eta{\downarrow}}^{\star,(m)}$ by complete induction for every $m \in \mathbb{N}$, $\eta \in T_{N\boxed{\mathbb{N}}}$, and $s, t \in T_\Sigma$. The case $m = 0$ is trivial. So let, w.l.o.g, $\eta = A^{\boxed{1}}(\eta_1, \ldots, \eta_k)$ for some $k \in \mathbb{N}$, $A \in N^{(k)}$ and $\eta_1, \ldots, \eta_k \in T_{N\boxed{\mathbb{N}}}$, then

$$\llbracket \mathcal{G} \rrbracket_{[\eta,\eta]}^{(m+1)}(s,t) = \sum \left(wt(d) \,\middle|\, d \in P^{m+1}, [\eta,\eta] \Rightarrow_{\mathcal{G},\mathrm{LO}}^d [s,t] \right)$$

$$= \sum \left(wt_P(p) \cdot \llbracket \mathcal{G} \rrbracket_{[\xi[\eta_{1,k}], \xi[\eta_{1,k}]]}^{(m)}(s,t) \,\middle|\, p = (A(x_{1,k}) \to [\xi,\xi]) \in P_{\mathrm{NT}}, \xi \text{ is fresh} \right)$$

$$+ \sum \left(wt_P(p) \cdot \prod_{j=1}^{k} \llbracket \mathcal{G} \rrbracket_{[\eta_j,\eta_j]}^{(m_j)}(s_j,t_j) \,\middle|\, p = (A(x_{1,k}) \to [\xi,\zeta]) \in P_{\mathrm{T}}, \right.$$
$$\left. (\xi, s_{1,k}) \in \mathrm{dec}_k(s), (\zeta, t_{1,k}) \in \mathrm{dec}_k(t), \right.$$
$$\left. m_1, \ldots, m_k \in \mathbb{N}, \textstyle\sum_{j=1}^{k} m_j = m \right)$$

$$\overset{\text{(IH)}}{=} \sum \left(wt_R(r) \cdot [\![\mathcal{M}]\!]_{\xi[\eta_{1,k}]\downarrow}^{\star,(m)}(s,t) \;\Big|\; r = \star(x_1, A(y_{1,k})) \to \star(x_1, \xi\downarrow[y_{1,k}]) \in R_{\text{IC}} \right)$$

$$+ \sum \left(wt_R(r) \cdot \prod_{j=1}^{k} [\![\mathcal{M}]\!]_{\eta_j\downarrow}^{\star,(m_j)}(s_j,t_j) \;\Big|\; \zeta \in T_\Sigma^{\ln}(X_k), \right.$$
$$r = \star(\xi, A(y_{1,k})) \to \zeta[\star(x_1,y_1),\ldots,\star(x_k,y_k)] \in R_{\text{IE}},$$
$$(\xi, s_{1,k}) \in \text{dec}_k(s),\; (\zeta, t_{1,k}) \in \text{dec}_k(t),$$
$$\left. m_1,\ldots,m_k \in \mathbb{N},\; \textstyle\sum_{j=1}^{k} m_j = m \right)$$

$$= \sum \left(wt(d) \;\Big|\; d \in R^{m+1},\; \star(\eta\downarrow, s) \Rightarrow_{\mathcal{M},\text{L}}^{d} t \right) = [\![\mathcal{M}]\!]_{\eta\downarrow}^{\star,(m+1)}(s,t).$$

In the second step, we have abbreviated the restriction that (ξ,ξ) is fresh for $(\eta_1,\eta_1), \ldots, (\eta_k,\eta_k)$ by "ξ is fresh". Moreover, we again abbreviated sequences like $a_1, \ldots, a_n$ or $a_1 \ldots a_n$ by $a_{1,n}$. Now, for every $s, t \in T_\Sigma$,

$$[\![\mathcal{G}]\!](s,t) = \sum_{m \in \mathbb{N}} [\![\mathcal{G}]\!]_{[Z\boxed{0}, Z\boxed{1}]}^{(m)}(s,t) = \sum_{m \in \mathbb{N}} [\![\mathcal{M}]\!]_{\gamma_0}^{\star,(m)}(s,t) = [\![\mathcal{M}]\!](s,t),$$

concluding the proof. $\qquad\square$

Theorem 1. *The classes of transformations of s-wscftg and ln-wpxtop are equal.*

Proof. Let $\mathcal{G}$ be a s-wscftg. By Lemma 1, we can assume that $\mathcal{G}$ is in normal form. But then, a related ln-wpxtop $\mathcal{M}$ can be constructed according to Definition 1, and by Lemma 5, $[\![\mathcal{G}]\!] = [\![\mathcal{M}]\!]$. Conversely, let $\mathcal{M}$ be a ln-wpxtop. By Lemmas 3 and 4, $\mathcal{M}$ can be assumed to be one-state and in index normal form. So there is a related s-wscftg $\mathcal{G}$ by Definition 1, and $[\![\mathcal{M}]\!] = [\![\mathcal{G}]\!]$ by Lemma 5. $\qquad\square$

6 Conclusion

In this work, we proved the equivalence of linear and nondeleting weighted pushdown extended tree transducers and simple weighted synchronous context-free tree grammars, and thus generalized [12, Thms. 2 and 3] to weighted tree transformations.

We conclude with the claim that the characterization in [1, Thm. 1] can also be generalized to weighted tree transformations. That is, the class of transformations of s-wscftg is exactly the composition of the classes ln-HOM^{-1}, ln-WSCFT, and ln-HOM, that contain, respectively, the transformations of inverse linear and nondeleting tree homomorphisms [7, p. 170], linear and nondeleting weighted context-free tree grammars (ln-wcftg) [2], and linear and nondeleting tree homomorphisms. In fact, the construction for direction $\subseteq$ can be read off directly from the one-state wpxtop $\mathcal{M}$ in index-normal form that is equivalent to a s-wscftg $\mathcal{G}$ by Lemma 5: its index-creating rules determine the productions of a ln-wcftg $\mathcal{G}'$ over the terminal alphabet Γ, and its index-erasing rules the values of the homomorphisms h_1 and h_2. For the other direction $\supseteq$, one may assume the given ln-wcftg $\mathcal{G}$ to be in normal form [16,2]. Then the nonterminal productions of $\mathcal{G}$ directly determine the nonterminal productions of the constructed s-wscftg $\mathcal{G}'$, while the terminal productions of $\mathcal{G}'$ are the result of applying the supposed tree homomorphisms h_1 and h_2 to the right-hand sides of the terminal productions of $\mathcal{G}$.

Contrasted to the bimorphism characterization of wscftg over $\mathbb{B}$ in [13], the above characterization gives an idea of the restricted power of s-wscftg in comparison to wscftg.

References

1. Aho, A.V., Ullman, J.D.: Properties of Syntax Directed Translations. J. Comput. System Sci. 3(3), 319–334 (1969)
2. Bozapalidis, S.: Context-Free Series on Trees. Inform. and Comput. 169(2), 186–229 (2001)
3. Büchse, M., Maletti, A., Vogler, H.: Unidirectional Derivation Semantics for Synchronous Tree-Adjoining Grammars. In: Yen, H.-C., Ibarra, O.H. (eds.) DLT 2012. LNCS, vol. 7410, pp. 368–379. Springer, Heidelberg (2012)
4. Droste, M., Kuich, W.: Semirings and Formal Power Series. In: Kuich, W., Vogler, H., Droste, M. (eds.) Handbook of Weighted Automata, ch. 1, pp. 3–28. Springer (2009)
5. Engelfriet, J.: Bottom-Up and Top-Down Tree Transformations – A Comparison. Math. Systems Theory 9(2), 198–231 (1975)
6. Engelfriet, J., Vogler, H.: Pushdown Machines for the Macro Tree Transducer. Theoret. Comput. Sci. 42(3), 251–368 (1986)
7. Fülöp, Z., Maletti, A., Vogler, H.: Weighted Extended Tree Transducers. Fund. Inform. 111, 163–202 (2011)
8. Fülöp, Z., Vogler, H.: Weighted Tree Automata and Tree Transducers. In: Droste, M., Kuich, W., Vogler, H. (eds.) Handbook of Weighted Automata, ch. 9, pp. 313–404. Springer (2009)
9. Guessarian, I.: Pushdown Tree Automata. Math. Systems Theory 16(1), 237–263 (1983)
10. Knight, K., Graehl, J.: An Overview of Probabilistic Tree Transducers for Natural Language Processing. In: Gelbukh, A. (ed.) CICLing 2005. LNCS, vol. 3406, pp. 1–24. Springer, Heidelberg (2005)
11. Kozen, D.: Automata and Computability. Springer (1997)
12. Lewis, P.M., Stearns, R.E.: Syntax-Directed Transduction. J. ACM 18(3), 465–488 (1968)
13. Nederhof, M.J., Vogler, H.: Synchronous Context-Free Tree Grammars. In: Proc. 11th Int. Workshop Tree Adjoining Grammars and Related Formalisms, pp. 55–63 (2012)
14. Rounds, W.C.: Context-Free Grammars on Trees. In: Proc. 1st ACM Symp. Theory of Comput., pp. 143–148 (1969)
15. Rounds, W.C.: Mappings and Grammars on Trees. Theory Comput. Syst. 4(3), 257–287 (1970)
16. Rounds, W.C.: Tree-Oriented Proofs of Some Theorems on Context-Free and Indexed Languages. In: Proc. 2nd ACM Symp. Theory of Comput., pp. 109–116 (1970)
17. Shieber, S.M.: Synchronous Grammars as Tree Transducers. In: Proc. 7th Int. Workshop Tree Adjoining Grammars and Related Formalisms, pp. 88–95 (2004)
18. Yamasaki, K.: Fundamental Properties of Pushdown Tree Transducers. IEICE Trans. Inf. & Syst. E76-D(10), 1234–1242 (1993)

Weighted Variable Automata over Infinite Alphabets

Maria Pittou and George Rahonis

Department of Mathematics
Aristotle University of Thessaloniki
54124, Thessaloniki, Greece
{mpittou,grahonis}@math.auth.gr

Abstract. We introduce weighted variable automata over infinite alphabets and commutative and idempotent semirings. We prove that the class of their behaviors is closed under sum, and under scalar, Hadamard, Cauchy, and shuffle product, as well as star operation. Furthermore, we consider rational series over infinite alphabets and we state a Kleene-Schützenberger theorem.

Keywords: Infinite alphabets, weighted variable automata, semirings.

1 Introduction

The concept of finite automata with infinite input alphabets is of increasing research interest in the last years. These models are motivated by real practical applications especially data bases, system verification, and web services. Several such automata models have been investigated, namely *register* (cf. [8,12,14]), *pebble* (cf. [12,13]), *data automata* (cf. [2]), *P automata over infinite alphabets* [4] as well as variants of them. Unfortunately, most of these devices are quite complicated according to implementation and application. In [7], the authors considered the model of *variable finite automata* with infinite input alphabets. The main advantage of this model is the simplicity of its definition and operation. More precisely, it is based on an underlying finite automaton with finite input alphabet which consists of a constant subalphabet of the infinite alphabet, and variable symbols of two types, the *bounded* variable symbols and one *free* variable symbol. The variable automaton recognizes a language in the following way. Firstly, it computes the language of the underlying automaton. Then, it substitutes the variable symbols with letters from the infinite alphabet. For these substitutions concrete requirements are imposed. It was shown that variable finite automata have nice properties. In [11] (cf. also [10]), the model of variable automata was extended to the setup of trees over infinite ranked alphabets.

All the aforementioned types of automata refer to qualitative characteristics of the systems applied to. On the other hand, it is well-known that current practical applications require also quantitative features and analysis. Usually, when automata are involved in the investigation, the quantitative analysis is

M. Holzer and M. Kutrib (Eds.): CIAA 2014, LNCS 8587, pp. 304–317, 2014.

achieved by weighted automata models (cf. [5]). According to the authors' best knowledge, a quantitative counterpart for automata over infinite alphabets does not exist. Recently, in [3] the authors considered quantitative infinite alphabets to model controlled variables for a controller synthesis from incompatible situations. It is the scope of this paper to introduce weighted automata over semirings consuming letters from an infinite alphabet. For this, we use the model of [7] since in the weighted setup it also has a simple definition and implementation. Therefore, we consider *weighted variable automata* over commutative and idempotent semirings and infinite alphabets, and investigate several closure properties of the class of their behaviors. For our proofs, we mainly use the techniques which were developed in [11] for variable tree automata over infinite ranked alphabets. Furthermore, we introduce rational expressions over infinite alphabets using the same idea as for our automata models. This enables us to state a Kleene-Schützenberger theorem for the class of series over infinite alphabets obtained as a consequence of the corresponding seminal result for series over finite alphabets. A similar approach for defining regular expressions over infinite alphabets has been followed in [1,9]. Finally, an application of our results, using the Boolean semiring, derives new results and a Kleene theorem for the class of languages accepted by variable finite automata of [7].

The structure of the paper is as follows. Besides this Introduction, the paper contains 5 sections. In Section 2 we present the preliminary notions used in the sequel. In Section 3 we introduce the model of the weighted variable automaton and in Section 4 we prove that the class of series accepted by these automata is closed under sum, scalar product, Hadamard product, Cauchy product, star operation and shuffle product. In Section 5 we deal with rational series over infinite alphabets and we state a Kleene-Schützenberger result. In Section 6 we apply our results to weighted variable automata over the Boolean semiring, and thus we obtain new results for the class of recognizable languages accepted by variable finite automata. Finally, in the Conclusion, we refer to open problems and future research.

2   Preliminaries

Let Σ be an alphabet, i.e., a nonempty (potentially infinite) set. As usually, we denote by Σ^* the set of all finite words over Σ and $\Sigma^+ = \Sigma^* \setminus \{\varepsilon\}$, where ε is the empty word. A subset $L \subseteq \Sigma^*$ is a language over Σ. A word $w = \sigma_0 \ldots \sigma_{n-1}$, where $\sigma_0, \ldots, \sigma_{n-1} \in \Sigma$ $(n \geq 1)$, is written also as $w = w(0) \ldots w(n-1)$ where $w(i) = \sigma_i$ for every $0 \leq i \leq n-1$. If S is a set, then $\mathcal{P}(S)$ will denote the powerset of S.

A *monoid* $(K, \cdot, 1)$ is a nonempty set K which is equipped with an associative operation $\cdot$ and a unit element 1 such that $1 \cdot k = k \cdot 1 = k$ for every $k \in K$. A monoid is called commutative if $\cdot$ is commutative. A *semiring* $(K, +, \cdot, 0, 1)$ is an algebraic structure such that $(K, +, 0)$ is a commutative monoid, $(K, \cdot, 1)$ is a monoid, $0 \neq 1$, $\cdot$ is both left- and right-distributive over $+$, and $0 \cdot k = k \cdot 0 = 0$ for every $k \in K$. If no confusion arises, we shall denote the semiring simply by K and the · operation simply by concatenation. The semiring K is called *commutative* if the monoid $(K, \cdot, 1)$ is commutative. Moreover, K is called *additively idempotent* (or simply *idempotent*), if $1 + 1 = 1$ which in turn implies that $k + k = k$ for every $k \in K$.

Example 1. The following structures constitute semirings.

- The semiring of *natural numbers* $(\mathbb{N}, +, \cdot, 0, 1)$,
- the *Boolean semiring* $\mathbb{B} = (\{0, 1\}, +, \cdot, 0, 1)$,
- the *tropical* or *min-plus semiring* $(\mathbb{R}_+ \cup \{\infty\}, \min, +, \infty, 0)$ where $\mathbb{R}_+ = \{r \in \mathbb{R} \mid r \geq 0\}$,
- the *arctical* or *max-plus semiring* $(\mathbb{R}_+ \cup \{-\infty\}, \max, +, -\infty, 0)$,
- the *Viterbi semiring* $([0, 1], \max, \cdot, 0, 1)$,
- every bounded distributive lattice with the operations supremum and infimum, and especially the *fuzzy semiring* $F = ([0, 1], \max, \min, 0, 1)$.

All the previous semirings, except the first one, are idempotent and commutative.

Let Σ be an alphabet and K a semiring. A *formal series* (or simply *series*) *over Σ and K* is a mapping $s : \Sigma^* \to K$. For every $w \in \Sigma^*$ we write (s, w) for the value $s(w)$ and refer to it as the *coefficient of s on w*. The *support of s* is the set $supp(s) = \{w \in \Sigma^* \mid (s, w) \neq 0\}$. A series with finite support is called a *polynomial*. The *constant series* $\tilde{k}$ ($k \in K$) is defined, for every $w \in \Sigma^*$, by $\left(\tilde{k}, w\right) = k$. Moreover, for every $w \in \Sigma^*$, we denote by $\overline{w}$ the series determined, for every $u \in \Sigma^*$, by $(\overline{w}, u) = 1$ if $u = w$ and 0, otherwise. The class of all series over Σ and K is denoted as usual by $K \langle\langle \Sigma^* \rangle\rangle$, and the class of polynomials over Σ and K by $K \langle \Sigma^* \rangle$.

Let $s, r \in K \langle\langle \Sigma^* \rangle\rangle$ and $k \in K$. The *sum* $s + r$, the *scalar products* ks and sk as well as the *Hadamard product* $s \odot r$ are defined elementwise by $(s + r, w) = (s, w) + (r, w)$, $(ks, w) = k \cdot (s, w)$, $(sk, w) = (s, w) \cdot k$, and $(s \odot r, w) = (s, w) \cdot (r, w)$, respectively, for every $w \in \Sigma^*$. It is well-known that the structures $\left(K \langle\langle \Sigma^* \rangle\rangle, +, \odot, \tilde{0}, \tilde{1}\right)$ and $\left(K \langle \Sigma^* \rangle, +, \odot, \tilde{0}, \tilde{1}\right)$ are semirings, which moreover are commutative (resp. idempotent) whenever K is commutative (resp. idempotent).

The *Cauchy product of r and s* is the series $r \cdot s \in K \langle\langle \Sigma^* \rangle\rangle$ defined for every $w \in \Sigma^*$ by

$$(r \cdot s, w) = \sum \{(r, u) \cdot (s, v) \mid u, v \in \Sigma^*, w = uv\}.$$

The *nth-iteration* $r^n \in K \langle\langle \Sigma^* \rangle\rangle$ ($n \geq 0$) of a series $r \in K \langle\langle \Sigma^* \rangle\rangle$ is defined inductively by

$r^0 = \tilde{\varepsilon}$ and $r^{n+1} = r \cdot r^n$ for $n \geq 0$.

Then, we have $(r^n, w) = \sum \left\{ \prod_{1 \leq i \leq n} (r, u_i) \mid u_i \in \Sigma^*, w = u_1 \ldots u_n \right\}$ for every $w \in \Sigma^*$. A series $r \in K \langle\langle \Sigma^* \rangle\rangle$ is called *proper* whenever $(r, \varepsilon) = 0$. If r is proper, then for every $w \in \Sigma^*$ and $n > |w|$ we have $(r^n, w) = 0$. The *star* $r^* \in K \langle\langle \Sigma^* \rangle\rangle$ of a *proper* series $r \in K \langle\langle \Sigma^* \rangle\rangle$ is defined by $r^* = \sum_{n \geq 0} r^n$. Thus, for every $w \in \Sigma^*$ we have $(r^*, w) = \sum_{0 \leq n \leq |w|} (r^n, w)$.

Finally, the *shuffle product of r and s* is the series $r \shuffle s \in K \langle\langle \Sigma^* \rangle\rangle$ defined for every $w \in \Sigma^*$ by

$$(r \shuffle s, w) = \sum \{ (r, u) \cdot (s, v) \mid u, v \in \Sigma^*, w \in u \shuffle v \}$$

where $u \shuffle v$ denotes the shuffle product of u and v.

Next we turn to weighted automata. For this we assume the alphabet Σ to be finite. A *weighted automaton over Σ and K* is a quadruple $A = (Q, in, wt, ter)$ where Q is the *finite state set*, $in : Q \to K$ is the *initial distribution*, $wt : Q \times \Sigma \times Q \to K$ is a mapping assigning *weights* to the transitions of the automaton, and $ter : Q \to K$ is the *final (or terminal) distribution*.

Let $w = w(0) \ldots w(n-1) \in \Sigma^*$. A *path of A over w* is a sequence of transitions $P_w := ((q_i, w(i), q_{i+1}))_{0 \le i \le n-1}$. The *weight* of P_w is given by the value

$$weight(P_w) = in(q_0) \cdot \prod_{0 \le i \le n-1} wt\,((q_i, w(i), q_{i+1})) \cdot ter(q_n).$$

The *behavior of A* is the series $\|A\| : \Sigma^* \to K$ whose coefficients are given by

$$(\|A\|, w) = \sum_{P_w} weight(P_w)$$

for every $w \in \Sigma^*$.

A series $s \in K \langle\langle \Sigma^* \rangle\rangle$ is called recognizable if $s = \|A\|$ for some weighted automaton A over Σ and K. As usual we denote by $Rec(K, \Sigma)$ the class of recognizable series over Σ and K. Two weighted automata $A = (Q, in, wt, ter)$ and $A' = (Q', in', wt', ter')$ over Σ and K are called *equivalent* if $\|A\| = \|A'\|$.

Finally, a weighted automaton $A = (Q, in, wt, ter)$ over Σ and K is called *normalized* if there exist two states $q_{in}, q_{ter} \in Q$, $q_{in} \ne q_{ter}$, such that:

- $in\,(q) = 1$ if $q = q_{in}$, and 0 otherwise,
- $ter\,(q) = 1$ if $q = q_{ter}$, and 0 otherwise, and
- $wt\,((q, \sigma, q_{in})) = wt\,((q_{ter}, \sigma, q)) = 0$

for every $q \in Q, \sigma \in \Sigma$. We shall denote a normalized weighted automaton $A = (Q, in, wt, ter)$ simply by $A = (Q, q_{in}, wt, q_{ter})$. The next result has been proved by several authors, cf. for instance Chapter 3 in [5].

Proposition 1. *Let $A = (Q, in, wt, ter)$ be a weighted automaton over Σ and K. We can effectively construct a normalized weighted automaton A' such that $(\|A'\|, w) = (\|A\|, w)$ for every $w \in \Sigma^+$ and $(\|A'\|, \varepsilon) = 0$.*

3   Weighted Variable Automata

In this section, we introduce the concept of weighted variable automata. Moreover, we present preliminary results, needed in Section 4, for the proof of the closure properties of the behaviors of our models.

Let Σ, Σ' be (infinite) alphabets. A *relabeling from Σ to Σ'* is a mapping $h : \Sigma \to \mathcal{P}(\Sigma')$. Next let $\Gamma \subseteq \Sigma$ be a finite subalphabet of Σ, Z a finite set whose elements are called *bounded variables* and y an element which is called a *free variable*. We assume that the sets Σ, Z, and $\{y\}$ are pairwise disjoint. A relabeling h from $\Gamma \cup Z \cup \{y\}$ to Σ is called *valid* if

(i) it is the identity on Γ,[1]
(ii) $card(h(z)) = 1$ for every $z \in Z$,
(iii) h is injective on Z and $\Gamma \cap h(Z) = \emptyset$, and
(iv) $h(y) = \Sigma \setminus (\Gamma \cup h(Z))$.

The above definition means that the application of h on a word w over $\Gamma \cup Z \cup \{y\}$ assigns to every occurrence of a symbol $z \in Z$ in w the same symbol from Σ, but it is possible to assign different symbols from Σ to different occurrences of y in w. This justifies the names bounded and free for the set of variables Z and the variable y, respectively. It should be clear that a valid relabeling from $\Gamma \cup Z \cup \{y\}$ to Σ is well-defined if it is defined only on Z satisfying conditions (ii) and (iii). We shall denote by $VR(\Gamma \cup Z \cup \{y\}, \Sigma)$ the set of all valid relabelings from $\Gamma \cup Z \cup \{y\}$ to Σ, and simply by $VR(\Gamma \cup Z \cup \{y\})$ if the alphabet Σ is understood.

We set $\Delta = \Gamma \cup Z \cup \{y\}$ and let $w \in \Sigma^*$. The *preimage of w over Δ* is the set $preim_\Delta(w) = \{u \in \Delta^* \mid \text{there exists } h \in VR(\Delta) \text{ such that } u \in h^{-1}(w)\}$.

Now we are ready to introduce our weighted variable automata.

Definition 1. *A* weighted variable automaton *(wva for short) over Σ and K is a pair $\mathcal{A} = \langle \Sigma, A \rangle$ where Σ is an infinite alphabet and $A = (Q, in, wt, ter)$ is a weighted automaton over Γ_A and K. The input alphabet Γ_A of A is defined by $\Gamma_A = \Sigma_A \cup Z \cup \{y\}$, where $\Sigma_A \subseteq \Sigma$ is a finite subalphabet, Z is a finite alphabet of* bounded variables, *and y is a* free variable.

The *behavior* of $\mathcal{A}$ is the series $\|\mathcal{A}\| : \Sigma^* \to K$ whose coefficients are determined by

$$(\|\mathcal{A}\|, w) = \sum_{u \in preim_{\Gamma_A}(w)} (\|\mathcal{A}\|, u)$$

for every $w \in \Sigma^*$. Clearly, the above sum is finite and thus $(\|\mathcal{A}\|, w)$ is well-defined for every $w \in \Sigma^*$.

Two wva $\mathcal{A}$ and $\mathcal{A}'$ over Σ and K are called *equivalent* whenever $\|\mathcal{A}\| = \|\mathcal{A}'\|$.

A series r over Σ and K is called *v-recognizable* if there exists a wva $\mathcal{A}$ such that $r = \|\mathcal{A}\|$. We shall denote by $VRec(K, \Sigma)$ the class of v-recognizable series over Σ and K. It should be clear that every weighted automaton A over a finite subalphabet $\Sigma' \subseteq \Sigma$ and K can be considered as a wva such that its transitions labelled by variables carry the weight 0. Therefore, we get the next result, where the strictness of the inclusion trivially holds by the definition of wva.

Proposition 2. $\displaystyle\bigcup_{\text{finite } \Sigma' \subseteq \Sigma} Rec(K, \Sigma') \subsetneq VRec(K, \Sigma).$

---

[1] Abusing notation we identify $\{\sigma\}$ with σ, for every $\sigma \in \Gamma$.

Throughout the paper Σ will denote an infinite alphabet and K an idempotent and commutative semiring.

In the sequel, we will call a wva $\mathcal{A} = \langle \Sigma, A \rangle$ over Σ and K, simply a wva.

Definition 2. *A wva $\mathcal{A} = \langle \Sigma, A \rangle$ is called* normalized *if A is normalized.*

Proposition 3. *Let $\mathcal{A} = \langle \Sigma, A \rangle$ be a wva. We can effectively construct a normalized wva $\mathcal{A}'$ such that $(\|\mathcal{A}'\|, w) = (\|\mathcal{A}\|, w)$ for every $w \in \Sigma^+$ and $(\|\mathcal{A}'\|, \varepsilon) = 0$.*

Next, we wish to investigate closure properties of the class $VRec(K, \Sigma)$. For this, we cannot apply the well-known constructions from classical weighted automata theory. For instance, let $\mathcal{A} = \langle \Sigma, A \rangle$ be a normalized wva, where $A = (\{q_{in}, q, q_{ter}\}, q_{in}, wt_A, q_{ter})$, $\Gamma_A = \{a\} \cup \{z\} \cup \{y\}$ and transitions with nonzero weights given by $wt_A((q_{in}, a, q)) = wt_A((q, z, q_{ter})) = 1$. Consider also the normalized wva $\mathcal{A}' = \langle \Sigma, A' \rangle$ where $A' = (\{q'_{in}, q'_{ter}\}, q'_{in}, wt_{A'}, q'_{ter})$, $\Gamma_{A'} = \{a'\} \cup \{z'\} \cup \{y'\}$ and $wt_{A'}((q'_{in}, a', q'_{ter})) = wt_{A'}((q'_{in}, y', q'_{ter})) = 1$. Moreover, let us assume that $a \neq a'$. Clearly, $(\|\mathcal{A}\|, aa') = 1$ and $(\|\mathcal{A}'\|, a') = 1$. Nevertheless, if we consider the disjoint union of A and A', say the weighted automaton B, then $a, a' \in \Gamma_B$ which implies that we cannot apply a valid relabeling assigning the letter a' to z. This in turn, implies that the word aa' does not belong to the support of the wva derived by the weighted automaton B. Furthermore, another problem of this construction is the choice of the free variable among y and y' which moreover causes new inconsistencies. Similar, even more complex, situations arise for the constructions of wva proving closure under further properties like Hadamard, Cauchy, and shuffle product. The subsequent material is needed for our investigation for the closure properties of $VRec(K, \Sigma)$.

Let $\mathcal{A} = \langle \Sigma, A \rangle$ be a wva where $A = (Q, in, wt, ter)$ with $\Gamma_A = \Sigma_A \cup Z \cup \{y\}$, and $\Sigma' \subseteq \Sigma$ a finite alphabet such that $\Sigma' \setminus \Sigma_A \neq \emptyset$. We define on $VR(\Gamma_A)$ the relation $\equiv_{\Sigma'}$ determined for every $h_1, h_2 \in VR(\Gamma_A)$ by

$$h_1 \equiv_{\Sigma'} h_2 \quad \text{iff} \quad h_1(\sigma) \cap \Sigma' = h_2(\sigma) \cap \Sigma' \text{ for every } \sigma \in Z \cup \{y\}.$$

It should be clear that $\equiv_{\Sigma'}$ is an equivalence relation. Moreover, since $Z \cup \{y\}$ and Σ' are finite, the index of $\equiv_{\Sigma'}$ is finite. Let V be a set of representatives of $VR(\Gamma_A) / \equiv_{\Sigma'}$. For every $h \in V$, we let $Z_h = \{z \in Z \mid h(z) \in \Sigma'\}$ and $\Gamma_h = \Sigma_A \cup \Sigma' \cup (Z \setminus Z_h) \cup \{y\}$, and we consider the weighted automaton $A_h = (Q_h, in_h, wt_h, ter_h)$ over Γ_h and K, where $Q_h = \{q_h \mid q \in Q\}$ is a copy of Q, $in_h(q_h) = in(q)$ and $ter_h(q_h) = ter(q)$ for every $q_h \in Q_h$. The weight assignment mapping wt_h is defined as follows. For every $q_h, q'_h \in Q_h, \sigma \in \Gamma_h$, we let

$$wt_h((q_h, \sigma, q'_h)) = \begin{cases} wt((q, \sigma, q')) & \text{if } \sigma \in \Sigma_A \cup (Z \setminus Z_h) \cup \{y\} \\ wt((q, z, q')) & \text{if } \sigma = h(z) \text{ and } z \in Z_h \\ wt((q, y, q')) & \text{if } \sigma \in h(y) \cap \Sigma' \\ 0 & \text{otherwise.} \end{cases}$$

Without any loss, we assume that the sets Q_h are pairwise disjoint. We let $Q_V = \bigcup_{h \in V} Q_h$, $\Gamma_V = \Sigma_A \cup \Sigma' \cup Z \cup \{y\}$, and consider the wva $\mathcal{A}_{(\Sigma', V)} = \langle \Sigma, A_{(\Sigma', V)} \rangle$

over Σ and K, where $A_{(\Sigma',V)} = (Q_V, in_V, wt_V, ter_V)$ is a weighted automaton with input alphabet Γ_V. Its initial and final distribution are defined, respectively, by $in_V(q) = in_h(q)$, $ter_V(q) = ter_h(q)$ for every $q \in Q_h$, $h \in V$. The weight assignment mapping $wt_V : Q_V \times \Gamma_V \times Q_V \to K$ is given by

$$wt_V((q,\sigma,q')) = \begin{cases} wt_h((q,\sigma,q')) & \text{if } q,q' \in Q_h \text{ for some } h \in V \\ 0 & \text{otherwise} \end{cases}$$

for every $q, q' \in Q_V, \sigma \in \Gamma_V$.

Since the weighted automaton $A_{(\Sigma',V)}$ is the disjoint union of A_h, $h \in V$, we get that $\|A_{(\Sigma',V)}\| = \sum_{h \in V} \|A_h\|$. Therefore, for every $w \in \Sigma^*$, we have

$$\left(\|A_{(\Sigma',V)}\|, w\right) = \sum_{u \in preim_{\Gamma_V}(w)} \left(\|A_{(\Sigma',V)}\|, u\right) = \sum_{h \in V} \sum_{u \in preim_{\Gamma_h}(w)} \left(\|A_h\|, u\right).$$

The next result is crucial for the proofs of the closure properties of the class $VRec(K, \Sigma)$.

Lemma 1. $\|A\| = \|A_{(\Sigma',V)}\|$.

Proof. Let $w = w(0) \ldots w(n-1) \in \Sigma^*$. Consider a word $u = u(0) \ldots u(n-1) \in preim_{\Gamma_A}(w)$ and a valid relabeling $h \in VR(\Gamma_A)$ with $w \in h(u)$. We define the word $u' = u'(0) \ldots u'(n-1) \in \Gamma_V^*$ as follows.

$$u'(i) = \begin{cases} u(i) & \text{if } (u(i) \in \Sigma_A \cup Z \setminus Z_h) \text{ or } (u(i) = y \text{ and } w(i) \notin \Sigma' \setminus \Sigma_A) \\ w(i) & \text{if } (u(i) \in Z_h) \text{ or } (u(i) = y \text{ and } w(i) \in \Sigma' \setminus \Sigma_A) \end{cases}$$

for every $0 \le i \le n-1$.

We consider the set of valid relabelings $V' \subseteq V$ as follows: $g \in V'$ implies that $g(z) = h(z)$ for every $z \in Z_h \cap \{u(i) \mid 0 \le i \le n-1\}$ and $g(y) \cap \Sigma' = h(y) \cap \Sigma'$ whenever $u(i) = y$ and $w(i) \in \Sigma'$ for some $0 \le i \le n-1$. Let $P_u^{(A)}$ be a path of A over u. Then, by construction of $A_{(\Sigma',V)}$, for every $g \in V'$, there exists a path $P_{u'}^{(A_g)}$ of A_g over u' with $weight\left(P_{u'}^{(A_g)}\right) = weight\left(P_u^{(A)}\right)$. Clearly, there are $r = card(V')$ such paths and since K is idempotent, we get $\sum_{g \in V'} weight\left(P_{u'}^{(A_g)}\right) = weight\left(P_u^{(A)}\right)$. On the other hand, for every $g \in V \setminus V'$ and path $P_{u'}^{(A_g)}$ of A_g, we have $weight\left(P_{u'}^{(A_g)}\right) = 0$. Therefore, we obtain

$$\sum_{P_u^{(A)}} weight\left(P_u^{(A)}\right) = \sum_{g \in V} \sum_{P_{u'}^{(A_g)}} weight\left(P_{u'}^{(A_g)}\right).$$

We define the valid relabeling $h' \in VR(\Gamma_V)$ by $h'(z) = h(z)$ for every $z \in Z \setminus Z_h$, and we let, nondeterministically, $h'(z) \in \Sigma \setminus (\Sigma_A \cup \Sigma' \cup h(Z \setminus Z_h) \cup \{w(i) \mid 0 \le i \le n-1 \text{ and } w(i) \in h(y)\})$ for every $z \in Z_h$. Then we have $w \in h'(u')$ which implies that $u' \in preim_{\Gamma_V}(w)$.

Conversely, let $u' = u'(0) \ldots u'(n-1) \in preim_{\Gamma_V}(w)$. Hence, there is a valid relabeling $h' \in VR(\Gamma_V)$ such that $w \in h'(u')$. By construction of $A_{(\Sigma',V)}$, there is a valid relabeling h from Γ_A to Σ and a word $u = u(0) \ldots u(n-1) \in \Gamma_A^*$ such that

$$u(i) = \begin{cases} u'(i) & \text{if } u'(i) \in \Sigma_A \cup Z \setminus Z_h \\ z & \text{if } u'(i) = h(z) \text{ and } z \in Z_h \\ y & \text{if } u'(i) \in (h(y) \cap \Sigma') \cup \{y\} \end{cases}$$

for every $0 \leq i \leq n-1$. Keeping the previous notations, for every $g \in V'$, there is a path $P_{u'}^{(A_g)}$ of the weighted automaton A_g over u'. By construction of $A_{(\Sigma',V)}$, all such paths $P_{u'}^{(A_g)}$ $(g \in V')$ have the same weight and there exist $r = card(V')$ such paths. Furthermore, for every $g \in V'$ and $P_{u'}^{(A_g)}$ there is a path $P_u^{(A)}$ of A over u with $weight\left(P_u^{(A)}\right) = weight\left(P_{u'}^{(A_g)}\right)$, and since K is idempotent we get $weight\left(P_u^{(A)}\right) = \sum_{g \in V'} weight\left(P_{u'}^{(A_g)}\right)$. On the other hand, for every $g \in V \setminus V'$ and path $P_{u'}^{(A_g)}$ of A_g, we have that $weight\left(P_{u'}^{(A_g)}\right) = 0$. Therefore $\sum_{g \in V} \sum_{P_{u'}^{(A_g)}} weight\left(P_{u'}^{(A_g)}\right) = \sum_{P_u^{(A)}} weight\left(P_u^{(A)}\right)$. We consider the relabeling h'' from Γ_A to Σ defined in the following way. It is the identity on Σ_A, $h''(z) = h'(z)$ for every $z \in Z \setminus Z_h$, $h''(z) = h(z)$ for every $z \in Z_h$, and $h''(y) = h'(y) \cup ((h(y) \cap \Sigma') \setminus h(Z_h))$ (in fact $(h(y) \cap \Sigma') \cap h(Z_h) = \emptyset$ since h is a valid relabeling on Γ_A). Trivially h'' is a valid relabeling and $w \in h''(u)$ which implies that $u \in preim_{\Gamma_A}(w)$.

We conclude that for every $w \in \Sigma^*$ we have

$$\left(\left\|A_{(\Sigma',V)}\right\|, w\right) = \sum_{u' \in preim_{\Gamma_V}(w)} \left(\left\|A_{(\Sigma',V)}\right\|, u'\right)$$

$$= \sum_{u' \in preim_{\Gamma_V}(w)} \sum_{g \in V} \left(\left\|A_g\right\|, u'\right)$$

$$= \sum_{u' \in preim_{\Gamma_V}(w)} \sum_{g \in V} \sum_{P_{u'}^{(A_g)}} weight\left(P_{u'}^{(A_g)}\right)$$

$$= \sum_{u \in preim_{\Gamma_A}(w)} \sum_{P_u^{(A)}} weight\left(P_u^{(A)}\right)$$

$$= \sum_{u \in preim_{\Gamma_A}(w)} \left(\left\|A\right\|, u\right)$$

$$= \left(\left\|A\right\|, w\right)$$

and our proof is completed.

4   Closure Properties of the Class $VRec\,(K, \Sigma)$.

In this section, we state the closure of the class of v-recognizable series over the infinite alphabet Σ and the semiring K, under sum, and under scalar, Hadamard, Cauchy and shuffle product, as well as star operation. Due to space limitations we present only the proof for the closure under the shuffle product.

Proposition 4. *The class $VRec\,(K, \Sigma)$ is closed under sum.*

Proposition 5. *The class $VRec\,(K, \Sigma)$ is closed under the scalar products.*

Proposition 6. *The class $VRec\,(K, \Sigma)$ is closed under Hadamard product.*

Proposition 7. *The class $VRec\,(K, \Sigma)$ is closed under Cauchy product.*

Proposition 8. *The class $VRec\,(K, \Sigma)$ is closed under the star operation applied to proper series.*

Proposition 9. *The class $VRec\,(K, \Sigma)$ is closed under the shuffle product.*

Proof. Let $r^{(i)} \in VRec\,(K, \Sigma)$ with $i = 1, 2$. We consider the proper series $r'^{(i)}$ $(i = 1, 2)$ over Σ and K defined, for every $w \in \Sigma^*$, by $(r'^{(i)}, w) = (r^{(i)}, w)$ if $w \in \Sigma^+$, and 0 otherwise.

Then $r^{(1)} \sqcup\!\sqcup r^{(2)} = r'^{(1)} \sqcup\!\sqcup r'^{(2)} + (r^{(1)}, \varepsilon)\, r^{(2)} + r^{(1)} \, (r^{(2)}, \varepsilon) + (r^{(1)}, \varepsilon)\,(r^{(2)}, \varepsilon)\, \bar{\varepsilon}$ and by Propositions 2, 5, and 4, it suffices to show that $r'^{(1)} \sqcup\!\sqcup r'^{(2)} \in VRec\,(K, \Sigma)$. By Proposition 3, there are normalized wva $\mathcal{A}^{(i)} = \langle \Sigma, A^{(i)} \rangle$ with $A^{(i)} = \left(Q^{(i)}, q_{in}^{(i)}, wt^{(i)}, q_{ter}^{(i)}\right)$ over $\Gamma^{(i)} = \Sigma^{(i)} \cup Z^{(i)} \cup \{y^{(i)}\}$ and K, accepting respectively $r'^{(i)}$, with $i = 1, 2$. Without any loss, we assume that $Q^{(1)} \cap Q^{(2)} = \emptyset$ and $\left(Z^{(1)} \cup \{y^{(1)}\}\right) \cap \left(Z^{(2)} \cup \{y^{(2)}\}\right) = \emptyset$. We consider the wva $\mathcal{A}^{(1)}_{\left(\Sigma^{(2)}, V_1\right)} = \left\langle \Sigma, A^{(1)}_{\left(\Sigma^{(2)}, V_1\right)} \right\rangle$ and $\mathcal{A}^{(2)}_{\left(\Sigma^{(1)}, V_2\right)} = \left\langle \Sigma, A^{(2)}_{\left(\Sigma^{(1)}, V_2\right)} \right\rangle$ determined by the procedure before Lemma 1. By Proposition 3 and Lemma 1 these wva can be also assumed to be normalized hence, let $A^{(1)}_{\left(\Sigma^{(2)}, V_1\right)} = \left(Q^{(1)}_{V_1}, q^{(1)}_{inv_1}, wt^{(1)}_{V_1}, q^{(1)}_{terv_1}\right)$ over $\Gamma^{(1)} \cup \Sigma^{(2)}$ and $A^{(2)}_{\left(\Sigma^{(1)}, V_2\right)} = \left(Q^{(2)}_{V_2}, q^{(2)}_{inv_2}, wt^{(2)}_{V_2}, q^{(2)}_{terv_2}\right)$ over $\Gamma^{(2)} \cup \Sigma^{(1)}$. Moreover, without any loss, we assume that $Q^{(1)}_{V_1} \cap Q^{(2)}_{V_2} = \emptyset$. We let $y = (y^{(1)}, y^{(2)})$ and consider the set $H = \left(Z^{(1)} \cup \{y^{(1)}\}\right) \times \left(Z^{(2)} \cup \{y^{(2)}\}\right) \setminus \{y\}$ and a maximal subset $G \subseteq H \cup Z^{(1)} \cup Z^{(2)}$ satisfying the following condition: every element of $Z^{(1)}$ (resp. of $Z^{(2)}$) occurs either in at most one pair of H as a left (resp. as a right) coordinate, or as a single element of G. Assume that $G_1, \ldots, G_m$ is an enumeration of all such sets. We let $Q = Q^{(1)}_{V_1} \times Q^{(2)}_{V_2}$, $\Gamma_{G_j} = \Sigma^{(1)} \cup \Sigma^{(2)} \cup G_j \cup \{y\}$, for every $1 \leq j \leq m$, and consider the normalized wva $\mathcal{A}_{G_j} = \langle \Sigma, A_{G_j} \rangle$ over Σ and K with $A_{G_j} = \left(Q, \left(q^{(1)}_{inv_1}, q^{(2)}_{inv_2}\right), wt_{G_j}, \left(q^{(1)}_{terv_1}, q^{(2)}_{terv_2}\right)\right)$ over Γ_{G_j}, where the weight assignment mapping wt_{G_j} is defined for every $1 \leq j \leq m$ as follows.

$$wt_{G_j}\left(\left(\left(q^{(1)}, q^{(2)}\right), \sigma, \left(q'^{(1)}, q'^{(2)}\right)\right)\right) =$$

$$\begin{cases} wt_{V_1}^{(1)}\left(\left(q^{(1)}, \sigma, q'^{(1)}\right)\right) & \text{if } q^{(2)} = q'^{(2)} \text{ and } \sigma \in \Sigma^{(1)} \cup \Sigma^{(2)} \cup \left(Z^{(1)} \cap G_j\right) \\ wt_{V_2}^{(2)}\left(\left(q^{(2)}, \sigma, q'^{(2)}\right)\right) & \text{if } q^{(1)} = q'^{(1)} \text{ and } \sigma \in \Sigma^{(1)} \cup \Sigma^{(2)} \cup \left(Z^{(2)} \cap G_j\right) \\ wt_{V_1}^{(1)}\left(\left(q^{(1)}, x^{(1)}, q'^{(1)}\right)\right) & \text{if } q^{(2)} = q'^{(2)} \text{ and } \sigma = \left(x^{(1)}, x^{(2)}\right) \in G_j \cup \{y\} \\ wt_{V_2}^{(2)}\left(\left(q^{(2)}, x^{(2)}, q'^{(2)}\right)\right) & \text{if } q^{(1)} = q'^{(1)} \text{ and } \sigma = \left(x^{(1)}, x^{(2)}\right) \in G_j \cup \{y\} \\ 0 & \text{otherwise} \end{cases}$$

for every $\left(q^{(1)}, q^{(2)}\right), \left(q'^{(1)}, q'^{(2)}\right) \in Q, \sigma \in \Gamma_{G_j}$.

Next, we show that $\left\|\mathcal{A}_{(\Sigma^{(2)}, V_1)}^{(1)}\right\| \sqcup \left\|\mathcal{A}_{(\Sigma^{(1)}, V_2)}^{(2)}\right\| = \sum\limits_{1 \le j \le m} \|\mathcal{A}_{G_j}\|$.

For this let $w, w_1 = w_1(0) \ldots w_1(n_1 - 1), w_2 = w_2(0) \ldots w_2(n_2 - 1) \in \Sigma^+$ such that $w \in w_1 \sqcup w_2$, and $u_1 \in preim_{\Gamma^{(1)} \cup \Sigma^{(2)}}(w_1), u_2 \in preim_{\Gamma^{(2)} \cup \Sigma^{(1)}}(w_2)$. Hence, there exist valid relabelings $h^{(1)} \in VR\left(\Gamma^{(1)} \cup \Sigma^{(2)}\right)$ and $h^{(2)} \in VR(\Gamma^{(2)} \cup \Sigma^{(1)})$ such that $w_1 \in h^{(1)}(u_1), w_2 \in h^{(2)}(u_2)$. We consider a path $P_{u_1} : \left(q_{inv_1}^{(1)}, u_1(0), q_1^{(1)}\right) \ldots \left(q_{n_1-1}^{(1)}, u_1(n_1 - 1), q_{ter_{V_1}}^{(1)}\right)$ of $\mathcal{A}_{(\Sigma^{(2)}, V_1)}^{(1)}$ over u_1 and a path $P_{u_2} : \left(q_{inv_2}^{(2)}, u_2(0), q_1^{(2)}\right) \ldots \left(q_{n_2-1}^{(2)}, u_2(n_2 - 1), q_{ter_{V_2}}^{(2)}\right)$ of $\mathcal{A}_{(\Sigma^{(1)}, V_2)}^{(2)}$ over u_2. We distinguish the following cases.

- The sets $\{w_1(0), \ldots, w_1(n_1 - 1)\} \cap \left(\Sigma \setminus \left(\Sigma^{(1)} \cup \Sigma^{(2)}\right)\right)$ and $\{w_2(0), \ldots, w_2(n_2 - 1)\} \cap \left(\Sigma \setminus \left(\Sigma^{(1)} \cup \Sigma^{(2)}\right)\right)$ are disjoint. Then, if $weight(P_{u_1}) \ne 0 \ne weight(P_{u_2})$, by the definition of the list $G_1, \ldots, G_m$, there is a set $J \subseteq \{1, \ldots, m\}$ such that for every $j \in J$ there is a path $P_u^{(G_j)}$ of $\mathcal{A}_{G_j}$ over u, for $u \in (u_1 \sqcup u_2) \cap preim_{\Gamma_{G_j}}(w)$ with $weight\left(P_u^{(G_j)}\right) = weight(P_{u_1}) weight(P_{u_2})$. Since K is idempotent it holds $\sum\limits_{j \in J} weight\left(P_u^{(G_j)}\right) = weight(P_{u_1}) weight(P_{u_2})$ and thus $\sum\limits_{1 \le j \le m} weight\left(P_u^{(G_j)}\right) = weight(P_{u_1}) weight(P_{u_2})$.
- We assume that $\left(\{w_1(0), \ldots, w_1(n_1 - 1)\} \cap \left(\Sigma \setminus \left(\Sigma^{(1)} \cup \Sigma^{(2)}\right)\right)\right) \cap \left(\{w_2(0), \ldots, w_2(n_2 - 1)\} \cap \left(\Sigma \setminus \left(\Sigma^{(1)} \cup \Sigma^{(2)}\right)\right)\right) \ne \emptyset$. Moreover, for simplicity, we assume that the two sets have only one common letter σ, and let $0 \le l_1 < \ldots < l_r \le n_1 - 1$ and $0 \le g_1 < \ldots < g_s \le n_2 - 1$ be the positions in w_1, w_2 respectively, such that $w_1(l_1) = \ldots = w_1(l_r) = w_2(g_1) = \ldots = w_2(g_s) = \sigma$. Since $u_1 \in preim_{\Gamma^{(1)} \cup \Sigma^{(2)}}(w_1)$ and $u_2 \in preim_{\Gamma^{(2)} \cup \Sigma^{(1)}}(w_2)$ we get that $u_1(l_1) = \ldots = u_1(l_r) = x^{(1)}$ and $u_2(g_1) = \ldots = u_2(g_s) = x^{(2)}$ for some $x^{(1)} \in Z^{(1)} \cup \{y^{(1)}\}$ and $x^{(2)} \in Z^{(2)} \cup \{y^{(2)}\}$. If $weight(P_{u_1}) \ne 0 \ne weight(P_{u_2})$, by the definition of the list $G_1, \ldots, G_m$, there is a set $J \subseteq \{1, \ldots, m\}$ such that for every $j \in J$ there is a path $P_{u'}^{(G_j)}$ of $\mathcal{A}_{G_j}$ over u', where u' is obtained by u by replacing $x^{(1)}$ (resp. $x^{(2)}$) in u_1 (resp. u_2) at the positions $l_1, \ldots, l_r$ (resp. $g_1, \ldots, g_s$) by the pair $\left(x^{(1)}, x^{(2)}\right)$, and from

the remaining letters we replace every occurrence of $y^{(1)}$ and $y^{(2)}$ with y, for $u \in u_1 \sqcup u_2$. Again, we have $weight\left(P_{u'}^{(G_j)}\right) = weight(P_{u_1})weight(P_{u_2})$ and hence, $\sum_{1 \le j \le m} weight\left(P_{u'}^{(G_j)}\right) = weight(P_{u_1})weight(P_{u_2})$. On the other hand, it is trivially shown that $u' \in preim_{\Gamma_{G_j}}(w)$.

Conversely, keeping the previous notations, for every $w \in \Sigma^+, u' \in preim_{\Gamma_{G_j}}(w)$ for some $1 \le j \le m$, there are $u_1 \in preim_{\Gamma^{(1)} \cup \Sigma^{(2)}}(w_1), u_2 \in preim_{\Gamma^{(2)} \cup \Sigma^{(1)}}(w_2)$ with $w \in w_1 \sqcup w_2$, such that for every path $P_{u'}^{(G_j)}$ of A_{G_j} over u', there are paths P_{u_1} of $A_{(\Sigma^{(2)},V_1)}^{(1)}$ over u_1 and P_{u_2} of $A_{(\Sigma^{(1)},V_2)}^{(2)}$ over u_2, with $weight\left(P_{u'}^{(G_j)}\right) = weight(P_{u_1})weight(P_{u_2})$. With the same argument as above, we get $\sum_{1 \le j \le m} weight\left(P_{u'}^{(G_j)}\right) = weight(P_{u_1})weight(P_{u_2})$.

Now, for every $w \in \Sigma^+$, we get

$$\left(\left\|A_{(\Sigma^{(2)},V_1)}^{(1)}\right\| \sqcup \left\|A_{(\Sigma^{(1)},V_2)}^{(2)}\right\|, w\right)$$

$$= \sum_{\substack{w_1,w_2 \in \Sigma^+ \\ w \in w_1 \sqcup w_2}} \left(\left\|A_{(\Sigma^{(2)},V_1)}^{(1)}\right\|, w_1\right)\left(\left\|A_{(\Sigma^{(1)},V_2)}^{(2)}\right\|, w_2\right)$$

$$= \sum_{\substack{w_1,w_2 \in \Sigma^+ \\ w \in w_1 \sqcup w_2}} \sum_{u_1 \in preim_{\Gamma^{(1)} \cup \Sigma^{(2)}}(w_1)} \left(\left\|A_{(\Sigma^{(2)},V_1)}^{(1)}\right\|, u_1\right)$$

$$\sum_{u_2 \in preim_{\Gamma^{(2)} \cup \Sigma^{(1)}}(w_2)} \left(\left\|A_{(\Sigma^{(1)},V_2)}^{(2)}\right\|, u_2\right)$$

$$= \sum_{\substack{w_1,w_2 \in \Sigma^+ \\ w \in w_1 \sqcup w_2}} \sum_{u_1 \in preim_{\Gamma^{(1)} \cup \Sigma^{(2)}}(w_1)} \sum_{P_{u_1}} weight(P_{u_1})$$

$$\sum_{u_2 \in preim_{\Gamma^{(2)} \cup \Sigma^{(1)}}(w_2)} \sum_{P_{u_2}} weight(P_{u_2})$$

$$= \sum_{\substack{w_1,w_2 \in \Sigma^+ \\ w \in w_1 \sqcup w_2}} \sum_{u_1 \in preim_{\Gamma^{(1)} \cup \Sigma^{(2)}}(w_1)} \sum_{u_2 \in preim_{\Gamma^{(2)} \cup \Sigma^{(1)}}(w_2)} \sum_{P_{u_1}} \sum_{P_{u_2}} weight(P_{u_1})weight(P_{u_2})$$

$$= \sum_{1 \le j \le m} \sum_{u \in preim_{\Gamma_{G_j}}(w)} \sum_{P_u^{(G_j)}} weight\left(P_u^{(G_j)}\right)$$

$$= \left(\sum_{1 \le j \le m} \|A_{G_j}\|, w\right)$$

which implies that $\left\|\mathcal{A}^{(1)}_{(\Sigma^{(2)}, V_1)}\right\| \sqcup \left\|\mathcal{A}^{(2)}_{(\Sigma^{(1)}, V_2)}\right\| = \sum_{1 \leq j \leq m} \|\mathcal{A}_{G_j}\|$, i.e., $r'^{(1)} \sqcup$

$r'^{(2)} = \sum_{1 \leq j \leq m} \|\mathcal{A}_{G_j}\|$. Therefore, by Proposition 4, we conclude that $r'^{(1)} \sqcup r'^{(2)} \in$
$VRec(K, \Sigma)$, as required.

5 Rational Series over Infinite Alphabets

In this section, we deal with the notion of rational series over the infinite alphabet Σ and the semiring K. In fact, we intend to prove a Kleene-Schützenberger type result for v-recognizable series over Σ and K. For this, we define the notion of rationality for series over Σ in the same way we did it for v-recognizable series. Firstly, we recall the concept of rational series over finite alphabets. Let $\Gamma \subseteq \Sigma$ be a finite subalphabet of Σ, Z a finite set of bounded variables, y a free variable and assume that the sets Σ, Z, and $\{y\}$ are pairwise disjoint. The class $Rat(K, \Delta)$ of rational series over $\Delta = \Gamma \cup Z \cup \{y\}$ and K is the least class of series containing the polynomials over Δ and K and being closed under sum, Cauchy product, and star operation applied to proper series.

Definition 3. *A series s over Σ and K is called v-rational if there is a finite alphabet $\Gamma \subseteq \Sigma$ and a rational series s' over $\Delta = \Gamma \cup Z \cup \{y\}$ and K such that*

$$(s, w) = \sum_{u \in preim_\Delta(w)} (s', u)$$

for every $w \in \Sigma^*$.

We shall denote by $VRat(K, \Sigma)$ the class of v-rational series over Σ and K.

One could think of alternative definitions, more precisely, by defining rational series over the infinite alphabet Σ in the same way we do it for rational series over finite alphabets. It is not difficult to see that such a consideration should not derive an expressively equivalent notion to wva. Consider for instance the normalized wva $\mathcal{A} = \langle \Sigma, A \rangle$ where $A = (\{q_{in}, q_{ter}\}, q_{in}, wt, q_{ter})$ with $\Sigma_A = \{a\}$ and $Z = \{z\}$. The only non-zero assignment of wt is given by $wt((q_{in}, z, q_{ter})) = k \neq 0$. Then trivially, $\|\mathcal{A}\| = \sum_{a' \in \Sigma \setminus \{a\}} ka'$ and it is not difficult to see that this series is not rational in the sense of rational series over finite alphabets. Even if we should consider our rational series to contain, by definition, series of the above form, then still this is not sufficient. For instance let us consider the normalized wva $\mathcal{B} = \langle \Sigma, B \rangle$ where $B = (\{p_{in}, p, p_{ter}\}, p_{in}, wt, p_{ter})$ with $\Sigma_B = \{b\}$, $Z = \{z, z'\}$ and non-zero weights $wt((p_{in}, z, p)) = k$, $wt((p, z', p_{ter})) = k'$. Then it is easily obtained that

$$\|\mathcal{B}\| = \sum_{\substack{a, a' \in \Sigma \setminus \{b\} \\ a \neq a'}} kk'aa'.$$

On the other hand, the Cauchy product of the series $\sum_{a\in\Sigma\backslash\{b\}} ka \sum_{a'\in\Sigma\backslash\{b\}} k'a'$ clearly differs from $\|\mathcal{B}\|$. Next, we state our Kleene-Schützenberger type theorem for series over Σ and K.

Theorem 1. $VRec(K, \Sigma) = VRat(K, \Sigma)$.

6 Application to Variable Finite Automata

In this section, we derive new results for the class of languages accepted by variable finite automata (vfa for short) over the infinite alphabet Σ (cf. [7]). Let Z be a finite set of bounded variables and y a free variable. Then, a variable finite automaton over Σ is a pair $\mathcal{A} = \langle \Sigma, A \rangle$ where $A = (Q, \Gamma_A, I, E, F)$ is a finite automaton with input alphabet $\Gamma_A = \Sigma_A \cup Z \cup \{y\}$ ($\Sigma_A \subseteq \Sigma$ is a finite alphabet). The language of $\mathcal{A}$ is defined by

$$L(\mathcal{A}) = \bigcup_{\substack{u\in L(A) \\ h\in VR(\Gamma_A)}} h(u).$$

Then the vfa $\mathcal{A} = \langle \Sigma, A \rangle$ can be considered, in the obvious way, as a wva $\mathcal{A}'$ over the Boolean semiring $\mathbb{B}$. Moreover, it holds $w \in L(\mathcal{A})$ iff $(\|\mathcal{A}'\|, w) = 1$ for every $w \in \Sigma^*$.

Now, we introduce rational languages over Σ. More precisely, a language L over Σ is *rational* if there is a finite alphabet $\Gamma \subseteq \Sigma$ and a rational language L' over $\Delta = \Gamma \cup Z \cup \{y\}$ such that

$$L = \bigcup_{\substack{u\in L' \\ h\in VR(\Delta)}} h(u).$$

Clearly, for every v-rational series $s \in VRat(\mathbb{B}, \Sigma)$ its support $supp(s)$ is a rational language over Σ and vice-versa. Now, a straightforward application of the results of the previous sections, derives the following corollaries.

Corollary 1. *The class of recognizable languages over Σ is closed under union, intersection, concatenation, Kleene star, and shuffle product.*[2]

Corollary 2 (Kleene). *A language over Σ is recognizable iff it is rational.*

Conclusion

We introduced weighted variable automata over an infinite alphabet Σ and a commutative and idempotent semiring K. Our model is the extension of variable finite automata of [7], in the quantitative setup. We proved the closure of the

---

[2] The closure under union and intersection has been also proved in [7].

class of the behaviors of wva under sum, and under scalar, Cauchy, Hadamard, and shuffle product, as well as star operation. We considered v-rational series over Σ and K and showed a Kleene-Schützenberger theorem. The idempotency property of the semiring K is crucial for our proofs. Therefore, it should be interesting to state our results by relaxing this property for K. Moreover, it should be interesting to study the concept of wva over more general structures than semirings, that are currently used in practical applications, for instance valuation monoids [6]. In [10,11] the authors considered and studied variable tree automata over infinite ranked alphabets. We intend to extend and investigate this model to the quantitative setup.

References

1. Barceló, P., Reutter, J., Libkin, L.: Parametrized regular expressions and their languages. Theoret. Comput. Sci. 474, 21–45 (2013)
2. Bojańczyk, M., David, C., Muscholl, A., Schwentick, T., Segoufin, L.: Two-variable logic on data words. ACM Trans. Comput. Log. 12(4), 27 (2011)
3. Černý, P., Gopi, S., Henzinger, T.A., Radhakrishna, A., Totla, N.: Synthesis from incompatible specifications. In: Proceedings of EMSOFT 2012, pp. 53–62. ACM Press (2012)
4. Dassow, J., Vaszil, G.: P finite automata and regular languages over countably infinite alphabets. In: Hoogeboom, H.J., Păun, G., Rozenberg, G., Salomaa, A. (eds.) WMC 2006. LNCS, vol. 4361, pp. 367–381. Springer, Heidelberg (2006)
5. Droste, M., Kuich, W., Vogler, H. (eds.): Handbook of Weighted Automata. EATCS Monographs in Theoretical Computer Science. Springer (2009)
6. Droste, M., Meinecke, I.: Weighted automata and weighted MSO logics for average and long-time behaviors. Inform. and Comput. 220–221, 44–59 (2012)
7. Grumberg, O., Kupferman, O., Sheinvald, S.: Variable automata over infinite alphabets. In: Dediu, A.-H., Fernau, H., Martín-Vide, C. (eds.) LATA 2010. LNCS, vol. 6031, pp. 561–572. Springer, Heidelberg (2010)
8. Kaminski, M., Francez, N.: Finite-memory automata. Theoret. Comput. Sci. 134, 329–363 (1994)
9. Kaminski, M., Tan, T.: Regular expressions for languages over infinite alphabets. Fund. Inform. 69, 301–318 (2006)
10. Mens, I.-.E.: Tree automata over infinite ranked alphabets. Master thesis, Thessaloniki (2011),
   http://invenio.lib.auth.gr/record/128884/files/GRI-2012-8361.pdf
11. Mens, I.-.E., Rahonis, G.: Variable tree automata over infinite ranked alphabets. In: Winkler, F. (ed.) CAI 2011. LNCS, vol. 6742, pp. 247–260. Springer, Heidelberg (2011)
12. Neven, F., Schwentick, T., Vianu, V.: Towards regular languages over infinite alphabets. In: Sgall, J., Pultr, A., Kolman, P. (eds.) MFCS 2001. LNCS, vol. 2136, pp. 560–572. Springer, Heidelberg (2001)
13. Neven, F., Schwentick, T., Vianu, V.: Finite state machines for strings over infinite alphabets. ACM Trans. Comput. Log. 5, 403–435 (2004)
14. Shemesh, Y., Francez, N.: Finite-state unification automata and relational languages. Inform. and Comput. 114, 192–213 (1994)

Implications of Quantum Automata
for Contextuality[*]

Jibran Rashid[1,**] and Abuzer Yakaryılmaz[2,3,***]

[1] Facoltà di Informatica, Università della Svizzera Italiana, Via G. Buffi 13,
6900, Lugano, Switzerland
[2] University of Latvia, Faculty of Computing, Raina bulv. 19, Rīga, 1586, Latvia
[3] National Laboratory for Scientific Computing, Petrópolis, RJ, 25651-075, Brazil
jibran.rashid@usi.ch, abuzer@lncc.br

Abstract. We construct zero-error quantum finite automata (QFAs) for
promise problems which cannot be solved by bounded-error probabilistic
finite automata (PFAs). Here is a summary of our results:

1. There is a promise problem solvable by an exact two-way QFA in
   exponential expected time, but not by any bounded-error sublogarithmic space probabilistic Turing machines.
2. There is a promise problem solvable by a Las Vegas realtime QFA,
   but not by any bounded-error realtime PFA. The same problem can
   be solvable by an exact two-way QFA in linear expected time but
   not by any exact two-way PFA.
3. There is a family of promise problems such that each promise problem can be solvable by a two-state exact realtime QFAs, but, there
   is no such bound on the number of states of realtime bounded-error
   PFAs solving the members of this family.

Our results imply that there exist zero-error quantum computational devices with a *single qubit* of memory that cannot be simulated by any
finite memory classical computational model. This provides a computational perspective on results regarding ontological theories of quantum mechanics [20,28]. As a consequence we find that classical automata
based simulation models [24,6] are not sufficiently powerful to simulate
quantum contextuality. We conclude by highlighting the interplay between results from automata models and their application to developing
a general framework for quantum contextuality.

1 Preliminaries

Observables act as windows through which quantum physics allows us to
extract classical information about quantum entities. More precisely, a quantum
observable refers to a Hermitian operator H, whose eigenvalues correspond to

---

[*] See http://arxiv.org/abs/1404.2761 for the full paper [32].
[**] Jibran Rashid was supported by QSIT Director's Reserve Project.
[***] Abuzer Yakaryılmaz was partially supported by CAPES, ERC Advanced Grant
MQC, and FP7 FET project QALGO.

M. Holzer and M. Kutrib (Eds.): CIAA 2014, LNCS 8587, pp. 318–331, 2014.
© Springer International Publishing Switzerland 2014

the classical values a quantum system can take when it is measured. A pair of observables A and B are compatible if they commute, i.e., $AB - BA = 0$, and they are said to be incompatible otherwise. Intuitively, if the operators commute then it is possible to simultaneously measure them such that the obtained measurement results co-exist at the same time.

Assume we are given the observables for a quantum system of dimension greater than two. We choose to assign values to these observables corresponding to results obtainable if we were to measure the underlying quantum system. **Quantum contextuality** refers to the fact that for sets of commuting observables, there always exists at least one set for which the actual quantum outcomes would contradict our pre-assigned list. In other words, there is no classical hidden variable model which produces the same predictions as quantum physics. This is what Kochen and Specker proved in their seminal 1967 result [25]. We now present a specific example due to Peres and Mermin to make things more concrete [31,26].

Consider the 3×3 grid G, as depicted in Figure 1. The task is to assign entries $A_i \in \{-1, +1\}$ for each cell in the grid such that the parity, i.e., the product of the entries in each row and column is "+1" except for the third column which is required to have parity "−1". Let Ri be the parity for row i and Cj be the parity for column j. The fact that no such assignment exists for the square can be verified by noting that $\prod_{i=1}^{3} Ri = 1$ while $\prod_{j=1}^{3} Cj = -1$.

	C1	C2	C3	
R1	A_1	A_2	A_3	1
R2	A_4	A_5	A_6	1
R3	A_7	A_8	A_9	1
	1	1	−1	?

$Z \otimes \mathbb{I}$	$\mathbb{I} \otimes Z$	$Z \otimes Z$
$\mathbb{I} \otimes X$	$X \otimes \mathbb{I}$	$X \otimes X$
$Z \otimes X$	$X \otimes Z$	$Y \otimes Y$

Fig. 1. The Peres–Mermin magic square on the left. Each entry in the right square gives the measurements performed by the players to generate the corresponding output bit for the Peres–Mermin square. Here X, Y, and Z are the Pauli spin matrices while $\mathbb{I}$ is the identity operator.

Consider Alice and Bob who are presented with the 3×3 grid. After determining a common strategy, the players are spatially separated and the game proceeds as follows. Alice receives input i and Bob receives input j, each chosen uniformly random from the set $\{1, 2, 3\}$. They are required to output cell entries corresponding to row i and column j, respectively, such that the parity requirement is satisfied and furthermore the common cell in their output is consistent, i.e., both of them assign it with the same value.

Even though no classical strategy allows the players to win the magic square game with certainty, if the players share a pair of Bell states given by

$$|\psi\rangle = \frac{1}{\sqrt{2}} (|00\rangle + |11\rangle)^{\otimes 2},$$

then performing the measurements given in Figure 1 result in correlations that always satisfy the magic square requirements. This does not correspond to a fixed assignment to the square, just that in each independent run of the game, Alice and Bob are able to generate output that satisfies the requirements imposed on the rows and columns.

The recent works have focused on developing a general framework for contextuality based on generating a hypergraph for a given contextuality scenario and studying its combinatorial properties [11,2]. Even though graph theoretic structures are appropriate for modelling contextuality, they lack the computational perspective that emerges by modelling the computational procedures that generate contextuality scenarios. Quantum automata provide exactly such a framework. As a direct consequence of such considerations, we find that separations between classical and quantum finite automata imply that no amount of finite memory is in general sufficient to simulate quantum behaviour. Similar results have also been obtained by Hardy [20] and Montina [28].

Kleinmann et al. [24] and Blasiak [6] have suggested a classical simulation of Peres-Mermin magic square using classical memory. Cabello and Joosten [10] have shown that the amount of memory required to simulate the measurement results of the generalized Peres-Mermin square increasingly violate the Holevo bound. Cabello [8] proposed the *principle of bounded memory* which states that the memory a finite physical system can keep is bounded. On the other hand, Cabello [7] has also shown that the memory required to produce quantum predictions grows at least exponentially with the number of qubits n.

We show that a stronger statement follows from our results on the separations between quantum and classical finite automata. More specifically, there exist promise problems that quantum automata equipped with a *single qubit* can solve with zero-error while no classical finite memory model can solve these problems with bounded error. In contrast, the exponential separation obtained by Cabello [7] requires a quantum system of size n. The hidden variable model for a single qubit due to Bell [5] does not apply since there are only finite bits available for the classical simulation.

We assume the reader is familiar with the basics of quantum computation [30] and the basic models in automata theory [36].

1.1   Promise Problems

We denote input alphabet by Σ, which does not include ¢ (the left end-marker) and \$ (the right end-marker), and $\tilde{\Sigma} = \Sigma \cup \{¢, \$\}$. A promise problem is a pair

$P = (P_{yes}, P_{no})$, where $P_{yes}, P_{no} \subseteq \Sigma^*$ and $P_{yes} \cap P_{no} = \emptyset$ [35]. P is said to be solved by a machine $\mathcal{M}$ with error bound $\epsilon \in (0, \frac{1}{2})$ if any member of P_{yes} is accepted with a probability at least $1 - \epsilon$ and any member of P_{no} is rejected by $\mathcal{M}$ with a probability at least $1 - \epsilon$. P is said to be solved by $\mathcal{M}$ with bounded-error if it is solved by $\mathcal{M}$ with an error bound. If $\epsilon = 0$, then it is said that the problem is solved by $\mathcal{M}$ *exactly*. A special case of bounded-error is one-sided bounded-error where either all members of P_{yes} are accepted with probability 1 or all members of P_{no} are rejected with probability 1. $\mathcal{M}$ is said to be Las Vegas with a success probability $p \in (0, 1]$ [22] if

- $\mathcal{M}$ has the ability of giving three answers (instead of two): "accept", "reject", or "don't know";
- for a member of P_{yes}, $\mathcal{M}$ gives the decision of "acceptance" with a probability at least p and gives the decision of "don't know" with the remaining probability; and,
- for a member of P_{no}, $\mathcal{M}$ gives the decision of "rejection" with a probability at least p and gives the decision of "don't know" with the remaining probability.

If P satisfies $P_{yes} \cup P_{no} = \Sigma^*$ and it is *solvable* by $\mathcal{M}$, then it is conventional said that P_{yes} is *recognized* by $\mathcal{M}$.

1.2   Quantum Automata

A two-way finite automaton with quantum and classical states (2QCFA) [3] is a two-way deterministic finite automaton augmented with a fixed-size quantum register. All automata models in this paper have a single-head read-only tape on which the given input string is placed between left and right end-markers. The head never moves beyond the end-markers. The input head can move to the left, move to the right, or stay on the same square. This property is denoted as "two-way". If the input head is not allowed to move to left, then it is called "one-way". As a further restriction, if the input head is allowed to stay on the same square only for a fixed-number of steps, then it is called "realtime". Formally, a 2QCFA[1] is

$$\mathcal{M} = (S, Q, \Sigma, \delta, s_1, q_1, s_a, s_r),$$

where S and Q are the set of classical and quantum states, respectively; $s_1 \in S$ and $q_1 \in Q$ are the initial classical and quantum states, respectively; $s_a \in S$ and $s_r \in S$ ($s_a \neq s_r$) are the accepting and rejecting states, respectively; and δ is the transition function composed by two sub-elements δ_q and δ_c that govern the quantum part and classical part of the machine, respectively. Suppose that $\mathcal{M}$ is in state $s \in S$ and the symbol under the input head is $\sigma \in \tilde{\Sigma}$. In each step, first the quantum part and then the classical part is processed in the following manner:

---

[1] Here, we define a slightly different model than the original one, but, they can simulate each other exactly.

- $\delta_q(s, \sigma)$ determines either a unitary operator, say $\mathsf{U}_{s,\sigma}$, or a projective operator, say $\mathsf{P}_{s,\sigma} = \{\mathsf{P}_{s,\sigma,1}, \ldots, \mathsf{P}_{s,\sigma,k}\}$ for some $k > 0$, and then it is applied to the quantum register. Formally, in the former case,

$$\delta_q(s, \sigma, |\psi\rangle) \to (i = 1, \mathsf{U}_{s,\sigma}|\psi\rangle),$$

where we fix $i = $ "1" if a unitary operator is applied and $\mathsf{U}_{s,\sigma}|\psi\rangle$ is the evolved state. In the latter case,

$$\delta_q(s, \sigma, |\psi\rangle) \to \left\{ \left(i, \frac{|\psi_i\rangle}{\sqrt{\langle\psi_i|\psi_i\rangle}} \right) \middle| \; |\psi_i\rangle = \mathsf{P}_{s,\sigma,i}|\psi\rangle, \langle\psi_i|\psi_i\rangle \neq 0, \text{ and } 1 \leq i \leq k. \right\},$$

where i is the measurement result and $\frac{|\psi_i\rangle}{\sqrt{\langle\psi_i|\psi_i\rangle}}$ is the post-measurement state. Note that only a single outcome ($i \in \{1, \ldots, k\}$) can be observed in the case of a projective measurement.

- After the quantum phase, the machine evolves classically. Formally,

$$\delta(s, \sigma, i) \to (s', d),$$

where i is the measurement outcome of quantum phase, s' is the new classical state, and $d \in \{\leftarrow, \downarrow, \rightarrow\}$ represents the update of the input head.

Note that, for Las Vegas algorithms, we need to define another halting state called s_d corresponding to answer "don't know".

The computation of $\mathcal{M}$ on a given input string w starts in the initial configuration, where the head is on the first symbol of $\tilde{w} = \text{¢}w\$$, the classical state is s_1, and the quantum state is $|q_1\rangle$. The computation is terminated and the input is accepted (resp., rejected) if $\mathcal{M}$ enters to state s_a (resp., s_r).

A two-way automaton is called sweeping if the input head is allowed to change its direction only on the end-markers [34,23]. A very restricted version of sweeping automaton called restarting realtime automaton runs a realtime algorithm in an infinite loop, [37], i.e. if the computation is not terminated on the right end-marker, the same realtime algorithm is executed again. A 2QCFA restricted to a realtime head (no restarting) is denoted by rtQCFA. Formally defined in [39], on each tape square a rtQCFA applies an unitary operator followed by a projective measurement, and then evolves its classical part.[2] The most known restricted realtime QFA model is the Moore-Cruthcfield quantum finite automaton (MCQFA) [29]. It consists of only quantum states and a single unitary operator determined by the scanned symbol is applied on each tape square. A projective measurement is applied at the end of computation. A probabilistic or quantum automaton is called rational or algebraic if all the probabilities or amplitudes associated with transitions are restricted to rational or algebraic numbers, respectively.

---

[2] This definition is sufficient to obtain the most general realtime quantum finite automaton [21,38]. Moreover, allowing more than one quantum or classical transition on the same tape square does not increase the computational power of rtQCFAs. Note that, realtime head must be classical.

2   Quantum Automata for Promise Problems

In this section, we present some promise problems solvable by QFAs *without error* but not solvable by their *bounded-error* probabilistic counterparts. At the end, we will also show that the family of promise problem, which was shown to be solvable by a family of exact rtQFAs (MCQFAs) having only two states [4], cannot be solvable by a family of bounded-error probabilistic finite automata (PFAs) having a fixed number of states.

2.1   Exact Rational (Sweeping) 2QCFA Algorithm

2QCFAs can recognize the language palindromes, i.e., $PAL = \{w \mid w \in \{a, b\}^*$ and $w = w^r\}$, where w^r is the string w reversed and $EQ = \{a^n b^n \mid n > 0\}$ for any one-sided error bound [3,37]. In the case of one-sided error, one decision is always reliable. We use this fact to develop quantum automata for solving promise problems inherited from PAL and EQ. We know that PAL cannot be recognized by bounded-error PTMs using sublogarithmic space [15,17] and EQ can be recognized by bounded-error $o(\log \log n)$-space PTMs only in super-polynomial expected time [19,14]. We take into consideration these facts when formulating our promise problems so that the impossibility results for bounded-error probabilistic algorithms are still applicable for our constructions. (We refer the reader to the full paper [32] for the results based on EQ.)

Our first promise problem is as follows:

$$PromisePAL = (PromisePAL_{yes}, PromisePAL_{no}), \text{ where}$$

– $PromisePAL_{yes} = \{ucv \mid u, v \in \{a, b\}^*, |u| = |v|, u \in PAL, \text{ and } v \notin PAL\}$ and
– $PromisePAL_{no} = \{ucv \mid u, v \in \{a, b\}^*, |u| = |v|, u \notin PAL, \text{ and } v \in PAL\}$.

Each of the two 2QCFA algorithms given for PAL in [3] and [37] have zero-error when they reject. That is, for a given $\epsilon \in (0, \frac{1}{2})$, there exists a 2QCFA $\mathcal{M}_\epsilon$ which always accepts every string $w \in PAL$ and every $w \notin PAL$ is accepted with probability at most ϵ and it is rejected with probability at least $1 - \epsilon$. So, if $\mathcal{M}_\epsilon$ rejects an input, we can be certain that the input is not contained in PAL.

We can design an exact 2QCFA, say $\mathcal{EXACT}_{PAL}$, for $PromisePAL$ based on $\mathcal{M}_\epsilon$ as follows: Let string $w \in PromisePAL$, i.e., $w = ucv$ such that $u, v \in \{a, b\}^*$ and $|u| = |v|$. On input string w, $\mathcal{EXACT}_{PAL}$ proceeds in an infinite loop as follows,

– the computation splits into two branches on the left end-marker with probabilities $\frac{16}{25}$ and $\frac{9}{25}$, respectively, by applying one of the rational unitary operators U_a and U_b [3], given by

$$U_a = \frac{1}{5}\begin{pmatrix} 4 & 3 & 0 \\ -3 & 4 & 0 \\ 0 & 0 & 5 \end{pmatrix} \text{ and } U_b = \frac{1}{5}\begin{pmatrix} 4 & 0 & 3 \\ 0 & 5 & 0 \\ -3 & 0 & 4 \end{pmatrix},$$

to a qubit, i.e. if the quantum state is $(1 \quad 0 \quad 0)^T$ and U_a is applied, then the first and the second states are observed with probability $\frac{16}{25}$ and $\frac{9}{25}$, respectively, after a measurement on the computational basis. This is followed by a measurement in the computational basis;

- in the 1st branch, $\mathcal{EXACT}_{\mathcal{PAL}}$ executes $\mathcal{M}_\epsilon$ on v and *accepts w if $\mathcal{M}_\epsilon$ rejects v*;
- in the 2nd branch, $\mathcal{EXACT}_{\mathcal{PAL}}$ executes $\mathcal{M}_\epsilon$ on u and *rejects w if $\mathcal{M}_\epsilon$ rejects u*; and,
- the computation continues, otherwise.

Note that only a single decision is given in each branch: In the 1st branch, the members of PromisePAL$_{\text{yes}}$ are accepted with a probability at least $1 - \epsilon$ and no decision is given on the members of PromisePAL$_{\text{no}}$. In the 2nd branch, no decision is given on the members of PromisePAL$_{\text{yes}}$ and the members of PromisePAL$_{\text{no}}$ are rejected with a probability at least $1 - \epsilon$. Thus, in a single round, the members of PromisePAL$_{\text{yes}}$ are accepted with a probability at least $\frac{16}{25}(1 - \epsilon)$ and the members of PromisePAL$_{\text{no}}$ are rejected with a probability at least $\frac{9}{25}(1-\epsilon)$. Thus, $\mathcal{EXACT}_{\mathcal{PAL}}$ separates PromisePAL$_{\text{yes}}$ and PromisePAL$_{\text{no}}$ exactly by making an expected linear number of calls to $\mathcal{M}_\epsilon$. This establishes Theorem 1 while the fact that sublogarithmic space PTMs cannot solve PromisePAL is established in Theorem 2.

Theorem 1. PromisePAL *can be solved by an exact rational sweeping 2QCFA in exponential expected time.*

Theorem 2. *Bounded-error sublogarithmic space 2PTMs cannot solve* PromisePAL.

The scheme given above can be easily generalized to many other cases. The size of the quantum register, the type of the head, and the type of the transitions are determined by $\mathcal{M}_\epsilon$. Specifically, (i) if $\mathcal{M}_\epsilon$ is restarting (sweeping), then $\mathcal{EXACT}_{\mathcal{PAL}}$ is restarting (sweeping), too, or (ii) if $\mathcal{M}_\epsilon$ has only rational (algebraic) amplitudes, then $\mathcal{EXACT}_{\mathcal{PAL}}$ has rational (algebraic) amplitudes.

The 2QCFA algorithm for PAL given by Ambainis and Watrous [3] is rational and sweeping. The one given by Yakaryılmaz and Say in [37] is restarting but uses algebraic numbers. Both of them run in expected exponential time. In the next section we present a new promise problem that uses the former algorithm to obtain an exact rational restarting rtQCFA. Currently, we do not know how to obtain a similar result based on the latter model except by utilizing superoperators.

2.2   Exact Rational Restarting rtQCFA Algorithm

In this section, we define a promise problem (a modified version of PromisePAL) solvable by an exact rational restarting rtQCFA but not by any sublogarithmic space PTMs: PromiseTWINPAL = (PromiseTWINPAL$_{\text{yes}}$, PromiseTWINPAL$_{\text{no}}$), where

- PromiseTWINPAL$_{\text{yes}} = \{ucucvcv \mid u, v \in \{a,b\}^+, |u| = |v|, u \in \text{PAL}, \text{and } v \notin \text{PAL}\}$, and
- PromiseTWINPAL$_{\text{no}} = \{ucucvcv \mid u, v \in \{a,b\}^+, |u| = |v|, u \notin \text{PAL}, \text{and } v \in \text{PAL}\}$.

Theorem 3. *There is an exact rational restarting rtQCFA that solves* PromiseTWINPAL *in exponential expected time.*

Theorem 4. PromiseTWINPAL *cannot be solved by any bounded-error $o(\log n)$-space PTM.*

2.3   Las Vegas Rational rtQCFA Algorithm

In this section, we present another promise problem solvable by Las Vegas rtQCFAs or linear-time exact 2QCFAs but not by any bounded-error realtime PFA (rtPFA). Since an exact 2PFA can be simulated by a realtime deterministic finite automaton (rtDFA) [18], exact two-way PFAs (2PFAs) also cannot solve the new promise problem. The new promise problem is given by: EXPPromiseTWINPAL = (EXPPromiseTWINPAL$_{\text{yes}}$, EXPPromiseTWINPAL$_{\text{no}}$), where

- EXPPromiseTWINPAL$_{\text{yes}}$ = $\{(ucucvcvc)^t | u, v \in \{a, b\}^+, |u| = |v|, u \in \text{PAL}, v \notin$ PAL, and $t \geq 25^{|u|}\}$, and
- EXPPromiseTWINPAL$_{\text{no}}$ = $\{(ucucvcvc)^t | u, v \in \{a, b\}^+, |u| = |v|, u \notin \text{PAL}, v \in$ PAL, and $t \geq 25^{|u|}\}$.

Theorem 5. EXPPromiseTWINPAL *is solved by a Las Vegas rational rtQCFA or by an exact rational restarting rtQCFA in linear expected time.*

Theorem 6. *There is no bounded-error rtPFA that solves* EXPPromiseTWINPAL*.*

2.4   Succinctness of Realtime QFAs

For a given positive integer k, EVENODD^k = (EVENODD$^k_{\text{yes}}$, EVENODD$^k_{\text{no}}$) is a promise problem [4] such that

- EVENODD$^k_{\text{yes}}$ = $\{a^{i2^k} \mid i$ is a nonnegative even integer$\}$, and
- EVENODD$^k_{\text{no}}$ = $\{a^{i2^k} \mid i$ is a nonnegative odd integer$\}$.

Ambainis and Yakaryılmaz [4] showed that EVENODD^k can be solved by a 2-state MCQFA exactly, but, the corresponding probabilistic automaton needs at least 2^{k+1} states.[3] We show in Theorem 7 that allowing errors in the output does not help in decreasing the space requirement.

Theorem 7. *Bounded-error rtPFAs need at least 2^{k+1} states to solve* EVENODD^k*.*

3   Noncontextual Inequalities from Automata

We begin by reformulating the Peres-Mermin game in terms of inequalities. Let $\langle A_i A_j A_k \rangle$ be the expected parity of the corresponding entries of the square. We associate with each strategy, a value of the game $\langle \chi \rangle$, which is given by

$$\langle \chi \rangle = \langle A_1 A_2 A_3 \rangle + \langle A_4 A_5 A_6 \rangle + \langle A_7 A_8 A_9 \rangle \\ + \langle A_1 A_4 A_7 \rangle + \langle A_2 A_5 A_8 \rangle - \langle A_3 A_6 A_9 \rangle. \tag{1}$$

---

[3] Some new classical results on EVENODD^k were given in [18] and [1].

The classical bound is $\langle \chi \rangle \leqslant 4$, while the quantum bound is given by $\langle \chi \rangle \leqslant 6$, since there exists a perfect quantum strategy which assigns '+1' to the first five terms and '−1' to the last term in Equation 1. We can construct similar inequalities for the promise problems defined in this paper. The general idea behind the inequalities is to construct a game based on quantum and classical automata separations. Assume Bob is restricted to either N bits of classical memory or N quantum bits and Alice has the task of verifying what type of memory is available to Bob. She can query Bob multiple times on a pre-selected problem that is known to both of them. Conditioned on the classical memory requirement for the problem the idea then is for Alice to iteratively query Bob on input strings of increasing length. Eventually Bob's classical memory becomes insufficient to correctly answer the query and his best response is a random guess.

The problems PAL and PromisePAL can be solved in log space. As a consequence, Alice requires an exponential number of queries in N before Bob's memory is exhausted. The classical exponential memory requirement for EVENODD^k means that number of queries need only be logarithmic before a violation is observed. On the other hand, EVENODD^k is not a single problem but a family of promise problems and the classical memory requirement is for rtPFAs. For PromisePAL we obtain zero error for the quantum strategy while for the classical strategy bounded-error is not possible for 2PFAs.

We base the inequality we present on EVENODD^k. The arguments carry over to PAL and PromisePAL as well. On a given query Bob receives as input an integer k and a unary string $w = a^l$ that is promised to be from either EVENODD$^k_{yes}$ or EVENODD$^k_{no}$, i.e., $l = i2^k$. The task for Bob is to determine the membership of string w, i.e., whether i is even or odd. So, he outputs "+1" if i is even and "−1" otherwise. The identification can be made for any k if Bob has unbounded memory. If Bob is restricted to have memory 2^{n+1} then the identification can still be made perfectly for all integer inputs to Bob with $k \leqslant n$. It becomes impossible to perform this identification perfectly when $k > n$. In this case the amount of memory available to Bob is not sufficient to determine classically the membership of the input string w.

In the quantum case, a perfect strategy exists for all k, if Bob is allowed access to a single qubit $|\psi\rangle$. The state is initialized to $|0\rangle$ and for each "a" in the string w Bob applies the rotation $U_a = \begin{pmatrix} \cos\theta & -\sin\theta \\ \sin\theta & \cos\theta \end{pmatrix}$, with $\theta = \frac{\pi}{2^{k+1}}$. Bob measures in the computational basis once the input is processed and realizes that i is odd if he obtains $|1\rangle$ or $-|1\rangle$ and i is even if the result is $|0\rangle$ or $-|0\rangle$. This procedure guarantees that Bob will always correctly identify the string w.

Assume that the amount of memory available to Bob is N but we do not know N. Let $\langle A^k_y \rangle$ be the expected value of Bob's output when the input is k and i is even. Similarly, $\langle A^k_n \rangle$ represents the expected value for input k and i odd. One way to verify whether Bob is quantum or classically memory bound is to initially query for a choice of k and then sequentially query for increasing size of k. For each choice of k we choose i to be odd or even with a uniform random distribution. We define V to measure how successful Bob is in correctly identifying the string w. Performing the procedure Q times gives us a value

$$V = \sum_{j=1}^{Q} \left(\langle A_y^{k=4j} \rangle - \langle A_n^{k=4j} \rangle \right), \tag{2}$$

where we have chosen to increase the input k by multiples of 4 at each iteration. If $k \leqslant \log N - 1$, then for both the classical and quantum case Bob can achieve $V = Q$. For $k > \log N - 1$, since there is no perfect strategy in the classical case we have $V < Q$, while the quantum strategy still achieve $V = Q$. The classical value can be made much tighter since the optimal classical strategy for $k > \log N - 1$ is just a random guess. We have shown in Section 2.4 that allowing error classically does not help in terms of reducing the memory requirement, i.e. bounded-error rtPFAs need at least 2^{k+1} states to solve EVENODD^k. This implies that the classical value is bounded by $\frac{\log N - 1}{4}$. Similar inequalities may be derived for PAL and PromisePAL and they are summarized in Tables 1 and 2.

Table 1. For both PAL and PromisePAL no 2PFA exists that solves the problem with bounded error. For the family of $\{$EVENODD$^k \mid k > 0\}$, there is no bound on the number of states for rtPFAs that solve the members of this family with bounded error. Given an input string of size n and classical memory N, the table gives the memory requirement for solving the specific instance and the value attained for the non-contextual inequality.

Problem	Type	Classical Memory	Inequality Value
PAL	language recognition	$\log n$	$\frac{2^N}{4}$
EVENODD^k	family of promise problems	2^{k+1}	$\frac{\log N - 1}{4}$
PromisePAL	promise problem	$\log n$	$\frac{2^N}{4}$

Table 2. The weakest known quantum models that solve the given problems and the associated error in the solution. The value attained for the inequality is related to the number of runs Q of the game.

Problem	Quantum Model	Quantum Memory	Quantum Value
PAL	2QCFA	qubit	$Q - \delta$
EVENODD^k	Real-time	qubit	Q
PromisePAL	2QCFA	qubit	Q

It may be possible to improve these inequalities by finding other problems for which we obtain a similar separation as PromisePAL but with an exponential classical memory requirement and a polynomial time quantum automata. There is a key distinction between the inequalities obtained from quantum contextuality, such as Equation 1 for the Peres–Mermin game and Equation 2 for quantum automata. While the terms in quantum contextuality inequality depend on multiple combinations of compatible measurements, the automata inequalities only have a single measurement in each term. In this sense, the automata inequalities do no represent true quantum contextuality. As noted by Terry Rudolph [33], the quantum automata discussed here can be thought of as a classical computation with a vector rotating in a sphere. The catch is that such classical computation could not be done in a fault tolerant method while the quantum schemes admit fault tolerance.

Another possibility is to consider defining the notion of Quantum Contextual Promise Problems (QCPP), where rather than having a single output corresponding to each term in Equation 2, the automaton for the problem outputs multiple bits similar to each term in Equation 1. This would allow for the classification of quantum contextuality within the language hierarchy.

4   Discussion

Perhaps the most alluring charm of quantum automata separations is the possibility they offer of constructing a computational device that could solve a problem which no classic device with finite memory could. Rather than just ruling out hidden variables with an exponential size increase, these computational devices could be used in principle to rule out arbitrary size hidden variables by increasing the problem input size. The thorny issue though is a trade-off between the memory utilized and the amount of precision required in the interactions with the quantum memory. This precision requirement appears either in the form of the matrix entries for the unitaries, as in the case of EVENODD^k, *or* in the form of the ability to resolve two states that may be arbitrarily close to each other, as in PAL.

Any experimental setup is always restricted by some level of precision, if not by technological limitations then by more fundamental restrictions such as the uncertainty principle. The question that we are inevitably led to consider following this line of reasoning is whether it is possible to retain the quantum advantage in the automata model while still requiring finite precision in our interactions with memory.

The intersection of ideas from classical simulation of contextuality and automata theory leads us to the notion of *Finite Precision Quantum Automata* (FPQA). In addition to the usual automata requirements, a FPQA satisfies the additional constraints that for any two unique unitaries U_i and U_j applied during the computation we have $|U_i - U_j| > \epsilon$ and for any two different states $|\psi\rangle$ and $|\phi\rangle$ obtained during the computation we have $|\langle\psi|\phi\rangle|^2 > \delta$.

It is not clear whether we can construct a FPQA that still manages to provide a computational advantage over classical automata. Meyer [27] has argued that the Kochen-Specker theorem [25] does not hold when only finite precision measurements are available. Clifton and Kent [13] have generalized the arguments of Meyer for POVM's. On ther other hand, Mermin and Cabello [12] have indepedently argued that such nullification theorems do not hold. Recently Cabello and Cunha [9] have proposed a two-qutrit contextuality test, claiming it to be free of the finite precision loophole. These tests though do admit a finite memory simulation model. Constructing a FPQA that yields a separation over the classical models would not admit a finite memory simulation model and consequently does provide a stronger separation.

Even if the FPQA model does turn out to be equivalent to the classical model in terms of the class of problems it solves, it does not take away from the succinctness advantage of the quantum model. In the previous section we argued that quantum automata separations serve as witnesses for distinguishing between genuinely quantum and space bounded classical players. We can flip the reasoning around and observe that simulating quantum contextuality is an inherently classical memory intensive task. This difficulty can be used to construct classical Proofs of Space as identified by Dziembowski et al. [16]. The idea is to establish that Bob has access to a certain amount of memory. Bob is asked to simulate an appropriately chosen quantum contextuality scenario. This would require exponential memory on Bob's side while the verifier could directly check that Bob's output satisfies the required quantum correlations. Note also that by definition, pre-computation does not allow Bob to simulate quantum contextuality.

Acknowledgements. We thank anonymous referees for their very helpful comments.

References

1. Ablayev, F., Gainutdinova, A., Khadiev, K., Yakaryılmaz, A.: Very narrow quantum OBDDs and width hierarchies for classical OBDDs. In: DCFS (to appear 2014)
2. Acín, A., Fritz, T., Leverrier, A., Sainz, A.B.: A combinatorial approach to nonlocality and contextuality. Technical report (2014),
   http://arxiv.org/abs/1212.4084
3. Ambainis, A., Watrous, J.: Two–way finite automata with quantum and classical states. Theoretical Computer Science 287(1), 299–311 (2002)
4. Ambainis, A., Yakaryılmaz, A.: Superiority of exact quantum automata for promise problems. Information Processing Letters 112(7), 289–291 (2012)
5. Bell, J.S.: On the Einstein–Podolsky–Rosen paradox. Physics 1(3), 195–200 (1964)
6. Blasiak, P.: Quantum cube: A toy model of a qubit. Physics Letters A 377, 847–850 (2013)
7. Cabello, A.: The contextual computer. A Computable Universe 31, 595–604 (2012)
8. Cabello, A.: The role of bounded memory in the foundations of quantum mechanics. Foundations of Physics 42(1), 68–79 (2012)
9. Cabello, A., Cunha, M.T.: Proposal of a two-qutrit contextuality test free of the finite precision and compatibility loopholes. Physical Review Letters 106, 190401 (2011)
10. Cabello, A., Joosten, J.J.: Hidden variables simulating quantum contextuality increasingly violate the Holevo bound. In: Calude, C.S., Kari, J., Petre, I., Rozenberg, G. (eds.) UC 2011. LNCS, vol. 6714, pp. 64–76. Springer, Heidelberg (2011)

11. Cabello, A., Severini, S., Winter, A.: Graph-theoretic approach to quantum correlations. Physical Review Letters 112, 40401 (2014)
12. Cabello, A.C.: Finite-precision measurement does not nullify the Kochen–Specker theorem. Physical Review Letters 65, 52101 (2002)
13. Clifton, R., Kent, A.: Simulating quantum mechanics by non-contextual hidden variables. Proceedings of the Royal Society A 456, 2101–2114 (2000)
14. Dwork, C., Stockmeyer, L.: A time complexity gap for two-way probabilistic finite-state automata. SIAM Journal on Computing 19(6), 1011–1123 (1990)
15. Dwork, C., Stockmeyer, L.: Finite state verifiers I: The power of interaction. Journal of the ACM 39(4), 800–828 (1992)
16. Dziembowski, S., Faust, S., Kolmogorov, V., Pietrzak, K.: Proofs of space. Cryptology ePrint Archive, Report 2013/796 (2013), http://eprint.iacr.org/
17. Freivalds, R., Karpinski, M.: Lower space bounds for randomized computation. In: Shamir, E., Abiteboul, S. (eds.) ICALP 1994. LNCS, vol. 820, pp. 580–592. Springer, Heidelberg (1994)
18. Geffert, V., Yakaryılmaz, A.: Classical automata on promise problems. In: DCFS, (to appear 2014)
19. Greenberg, A.G., Weiss, A.: A lower bound for probabilistic algorithms for finite state machines. Journal of Computer and System Sciences 33(1), 88–105 (1986)
20. Hardy, L.: Quantum ontological excess baggage. Studies in History and Philosophy of Science Part B: Studies in History and Philosophy of Modern Physics 35(2), 267–276 (2004)
21. Hirvensalo, M.: Quantum automata with open time evolution. International Journal of Natural Computing Research 1(1), 70–85 (2010)
22. Hromkovic, J., Schnitger, G.: On the power of Las Vegas for one-way communication complexity, OBDDs, and finite automata. Information and Computation 169(2), 284–296 (2001)
23. Kapoutsis, C.A., Královic, R., Mömke, T.: Size complexity of rotating and sweeping automata. Journal of Computer and System Sciences 78(2), 537–558 (2012)
24. Kleinmann, M., Gühne, O., Portillo, J.R., Larsson, J.-Å., Cabello, A.: Memory cost of quantum contextuality. New Journal of Physics 13(11), 113011 (2011)
25. Kochen, S., Specker, E.P.: The problem of hidden variables in quantum mechanics. JMM 17, 59–87 (1967)
26. Mermin, N.D.: Simple unified form for the major no-hidden-variable theorems. Physical Review Letters 65(27) (1990)
27. Meyer, D.A.: Finite precision measurement nullifies the Kochen–Specker theorem. Physical Review Letters 83, 3751–3754 (1999)
28. Montina, A.: Exponential complexity and ontological theories of quantum mechanics. Physical Review A 77, 022104 (2008)
29. Moore, C., Crutchfield, J.P.: Quantum automata and quantum grammars. Theoretical Computer Science 237(1-2), 275–306 (2000)
30. Nielsen, M.A., Chuang, I.L.: Quantum Computation and Quantum Information. Cambridge University Press (2000)
31. Peres, A.: Incompatible results of quantum measurements. Pyhsics Letter A 151(3-4), 107–108 (1990)
32. Rashid, J., Yakaryılmaz, A.: Implications of quantum automata for contextuality. Technical Report arXiv:1404.2761. Cornell University Library (2014)
33. Rudolph, T.: Private communication (February 2014)
34. Sipser, M.: Lower bounds on the size of sweeping automata. Journal of Computer and System Sciences 21(2), 195–202 (1980)

35. Watrous, J.: Quantum computational complexity. In: Meyers, R.A. (ed.) Encyclopedia of Complexity and Systems Science, pp. 7174–7201. Springer (2009)
36. Yakaryılmaz, A.: Classical and Quantum Computation with Small Space Bounds. PhD thesis. Boğaziçi University, arXiv:1102.0378 (2011)
37. Yakaryılmaz, A., Say, A.C.C.: Succinctness of two-way probabilistic and quantum finite automata. Discrete Mathematics and Theoretical Computer Science 12(2), 19–40 (2010)
38. Yakaryılmaz, A., Say, A.C.C.: Unbounded-error quantum computation with small space bounds. Information and Computation 279(6), 873–892 (2011)
39. Zheng, S., Qiu, D., Li, L., Gruska, J.: One-way finite automata with quantum and classical states. In: Bordihn, H., Kutrib, M., Truthe, B. (eds.) Dassow Festschrift 2012. LNCS, vol. 7300, pp. 273–290. Springer, Heidelberg (2012)

Pairwise Rational Kernels Obtained by Automaton Operations

Abiel Roche-Lima[1], Michael Domaratzki[1], and Brian Fristensky[2]

[1] Department of Computer Science,
University of Manitoba, Winnipeg, Manitoba, R3T 2N2, Canada
[2] Department of Plant Science
University of Manitoba, Winnipeg, Manitoba, R3T 2N2, Canada

Abstract. Pairwise Rational Kernels (PRKs) are the combination of pairwise kernels, which handle similarities between two pairs of entities, and rational kernels, which are based on finite-state transducer for manipulating sequence data. PRKs have been already used in bioinformatics problems, such as metabolic network prediction, to reduce computational costs in terms of storage and processing.

In this paper, we propose new Pairwise Rational Kernels based on automaton and transducer operations. In this case, we define new operations over pairs of automata to obtain new rational kernels. We develop experiments using these new PRKs to predict metabolic networks. As a result, we obtain better accuracy and execution times when we compare them with previous kernels.

Keywords: Automata Operations, Rational Kernels, Pairwise Rational Kernels, Metabolic Networks.

1 Introduction

Pairwise kernels are measures of similarities between two pairs of entities obtained by converting relations between single entities into a relation between pairs of entities [1]. Pairwise kernel methods have been applied to bioinformatics problems such as prediction of Protein-Protein Interaction (PPI) networks [1, 2], metabolic networks [3] and drug-target interaction networks [4]. Other contexts, such as social networks [5] and semantic relationships [6], have also been benefited for using pairwise kernels.

In our context, we aim to use pairwise kernels to measure similarities between sequence data. Sequence data can also be represented by automata and transducers. Automata and transducers have been used in applications regarding natural language [7, 8], image processing [9] and bioinformatics [10–12]. Finite-state transducers are the theoretical fundamentals for rational kernels which represent similarity measures between sequences or automata [13].

Pairwise rational kernels (PRKs) have already been defined as the combination of pairwise kernels and rational kernels to manipulate sequence data [14]. First, the sequence data are represented as automata, later the rational kernels are obtained from each automaton and, finally, the pairwise operations are computed.

M. Holzer and M. Kutrib (Eds.): CIAA 2014, LNCS 8587, pp. 332–345, 2014.

In this paper, we propose a new family of pairwise rational kernels, but changing the order of the operations. Firstly, we also represent the sequences as automata, later we make pairwise combination using automata operations, and finally, the rational kernels are obtained based on the resulting automata. Some of the used operations are sum, product and compisition. We define three new PRKs based on the pairwise kernel combinations. Furthermore, we propose a general algorithm to use with the new kernels to be combined with Support Vector Machines (SVMs) methods. We ran experiments using these kernels to predict metabolic networks. As a result, performance and accuracy values were improved in comparison with previous PRKs [14].

2   Related Works

In bioinformatics, transducers provide a bridge through biological analysis of indel processes represented as graphs and finite-state machine design [12]. They contribute as a framework for studying mutation rates [15], phylogenetic and multiple alignments [11] and protein classification and essentiality problems [16, 17].

Several kernel methods have been developed to be applied to bioinformatics problems, such as locality-improved kernels [18], remote homology detection [19], mismatch kernels [20], convolution kernels [21] and path kernels [22]. These sequence kernels can be conveniently represented by weighted finite-state transducers, called rational kernels [23].

Other bioinformatics applications have involved rational kernels, for example to determine essentiality in protein sequences [17]. Allauzen et al. [17] designed two sequence kernels (called general domain-based kernels) which are instances of rational kernels (based on finite-state transducers) to predict protein essentiality. In this research, we define new kernels by extending the theory of measuring similarities between sets of sequences represented by automata applied to metabolic network prediction.

Pairwise rational kernels have been recently used in combination with SVM algorithms to manipulate large amount of sequence data [14]. The main goal is to decrease the computational costs in terms of processing and storage by taken advantage of the transducer representations and algorithms. Previous PRKs firstly obtain the rational kernels and later compute the pairwise operations. In this research, we propose a new family of PRKs that compute the pairwise operations first and then the rational kernels.

3   Background

Before to introduce the new PRKs family, several terms and notations are defined in this section based on Cortes et al. [13] .

3.1   Automata, Finite-State Transducers

An automaton M is a 5-tuple $(\Sigma, Q, I, F, \delta)$ [24] where Σ is the input alphabet set, Q is the state set, $I \subseteq Q$ is the subset of initial states, $F \subseteq Q$ is the subset of final states, and $\delta \subseteq Q \times (\Sigma \cup \{\epsilon\}) \times Q$ is the transition set, where ϵ is the empty sequence.

Similarly, a finite-state transducer is an automaton where an output label is included in each transition in addition to the input label. Based on the above definition, a finite-state transducer is a 6-tuple $(\Sigma, \Delta, Q, I, F, \delta)$ [23], where the new term Δ is the output alphabet set and the transition set δ is now $\delta \subseteq Q \times (\Sigma \cup \{\epsilon\}) \times (\Delta \cup \{\epsilon\}) \times Q$.

In addition, automata and finite-state transducers can be weighted, where each transition is labelled with a weight. Thus, a Weighted Automaton (WA) is a 7-tuple $(\Sigma, Q, I, F, \delta, \lambda, \rho)$ and a Weighted Finite-State Transducer (WFST) is a 8-tuple $(\Sigma, \Delta, Q, I, F, \delta, \lambda, \rho)$ [23], where the new terms $\lambda : I \to \mathbb{R}$ is the initial weight function and $\rho : F \to \mathbb{R}$ is the final weight function.

Given the Weighted Finite-State Transducer T and $(x, y) \in \Sigma^* \times \Delta^*$, $T(x, y)$ is computed by adding the weights of all the paths with x and y as input and output labels, respectively. There are several operations defined on WFST and WA, such as sum + (union) and $product$ · (concatenation) [25]. Given two WFST T_1 and T_2, the sum is defined as $(T_1 + T_2)(x, y) = T_1(x, y) + T_2(x, y)$ and the product is $(T_1 \cdot T_2)(x, y) = \sum_{x_1 x_2 = x, y_1 y_2 = y} T_1(x_1, y_1) \cdot T_2(x_2, y_2), \forall (x, y) \in \Sigma^* \times \Delta^*$.

3.2   Kernels and Rational Kernels

Kernel Methods are used in supervised learning approaches to find and study general types of relations in general types of data, e.g., SVM and kernel combinations [26]. In SVM methods a non-linear function Φ maps each point of the input space X to a high-dimensional feature space F. A function $k : X \times X \to \mathbb{R}$, called a kernel, is defined and for two values x and y, $k(x, y)$ coincides with the dot product of their images $\Phi(x)$ and $\Phi(y)$ in a feature space [27], i.e., $k(x, y) = \Phi(x) \cdot \Phi(y)$.

In order to manipulate sequence kernels, FSTs provide a simple representation and efficient algorithms such as composition and shortest-distance [23]. Kernels based on Finite-State Transducers are called Rational Kernels and are effective for analyzing sequences with variable lengths [13].

As a formal definition, given $X \subseteq \Sigma^*$ and Σ^* the set of all finite sequences over Σ, a kernel function $k : X \times X \to \mathbb{R}$ is rational if there exists a weighted transducer U such that $k(x, y) = U(x, y)$ for all sequences x and y. In other words, given two strings x and y and a rational kernel defined by U, $U(x, y)$ is the similarity measure between x and y [13].

3.3 n-gram Kernel as a Rational Kernel

Similarity measures between two biological sequences can be usually defined as a function of the number of subsequences of some type that they share (e.g., mismatch [20], gappy n-gram kernels [28] and n-gram [23]). A sequence kernel can be the sum of the product of the counts of these common subsequences. In this work, we use the n-gram kernel as a rational kernel. It is defined as:

$$k_n(x, y) = \sum_{|z|=n} c_x(z)c_y(z) \tag{1}$$

In this expression, $c_x(z)$ represents how many times z occurs in x.

In [17], new rational kernels have been developed based on Formula (1) to obtain domain-based kernels, which measure similarities between sequences represented by automata. They construct k_n using the weighted transducer $U = T_n \circ T_n^{-1}$, where $T_n(x, z) = c_x(z)$, for all $x, z \in X$ with $|z| = n$, then:

$$k_n(M_x, M_y) = \sum_{x,y \in X} (M_x \circ T_n \circ T_n^{-1} \circ M_y)(x, y) = \sum_{|z|=n} c_{M_x}(z)c_{M_y}(z) \tag{2}$$

where M_x, M_y are the weighted automata for x, y, respectively. M_x and M_y are obtained using the linear finite automata representing x and y augmenting each transition with the input label and by setting all transitions and final weights to one.

3.4   Pairwise Rational Kernels

In previous research, four different pairwise rational kernels were defined, i.e., Direct-Sum Pairwise Rational Kernel, Tensor-Product Pairwise Rational Kernel, Metric Learning Pairwise Rational Kernel and Cartesian Pairwise Rational Kernel [14]. They were obtained based on the state-of-the-art pairwise kernel combinations [29, 30]. Direct-Sum Pairwise Kernels and Tensor Product Pairwise are defined respectively as:

$$K_{PRKDS}((x_1, y_1), (x_2, y_2)) = U(x_1, x_2) + U(y_1, y_2) + U(y_1, x_2) + U(x_1, y_2) \tag{3}$$

$$K_{PRKT}((x_1, y_1), (x_2, y_2)) = U(x_1, x_2) * U(y_1, y_2) + U(x_1, y_2) * U(y_1, x_2) \tag{4}$$

where K is a function $K : (X \times X) \times (X \times X) \to \mathbb{R}$, $X \subseteq \Sigma^*$ and U is a transducer.

As we mentioned, each x_i is converted into automata A_{x_i}, the U transducers are obtained using the automata [17], and later the pairwise combinations are computed [29]. In this paper, we focus on these two types of PRKs, because we aim to obtain similar pairwise combinations. In this case, we use operations over the automata A_{x_i} to obtain final automata that represent the pairwise combinations (i.e., sum operator for Direct Sum kernel and product operator for Tensor Product kernel). Later, we obtain the rational kernels U over the final automata.

4   New Pairwise Rational Kernels as Automaton Operations

4.1   General Definitions

We begin by defining some notations, their algorithms and implementations in order to develop the new PRKs. Firstly, we define V_Σ alphabet as:

Definition 1. *Given the alphabet Σ, the new alphabet V_Σ is defined as* $V_\Sigma = \{\binom{a}{b} : a, b \in \Sigma\} \cup \{\binom{a}{\epsilon}, \binom{\epsilon}{b} : a, b \in \Sigma\}$, *where ϵ is the empty character.*

Example 1. Given the alphabet $\Sigma = \{A, G, C, T\}$ and the nucleotide sequences $x = AGGCCCGTA$, $y = CCCGTA$, then
$$\binom{x}{y} = \binom{A}{C}\binom{G}{C}\binom{G}{C}\binom{C}{G}\binom{C}{T}\binom{C}{A}\binom{G}{\epsilon}\binom{T}{\epsilon}\binom{A}{\epsilon}.$$
   In this example, a new sequence on the alphabet V_Σ is created. Each symbol from the nucleotide sequence x goes to the top and each symbol from y goes to the bottom. When the nucleotide sequences have different lengths, the ϵ symbol is used. E.g., the first symbol from x is the base A and the first symbol from y is the base C, the new symbol of the alphabet V_Σ is $\binom{A}{C}$. The lengths of x and y are different, x is longer than y. Then, the last few symbols in V_Σ have the corresponding symbols from x and the empty symbol (ϵ) in the bottom (i.e., $\binom{G}{\epsilon}\binom{T}{\epsilon}\binom{A}{\epsilon}$).

The following operations over the alphabet V_Σ are stated:

Definition 2. *Given $\binom{x_1}{y_1}, \binom{x_2}{y_2} \in V_\Sigma^*$, then*

- $[\uparrow\uparrow]$ *is the Top-Top Operator where* $\binom{x_1}{y_1}[\uparrow\uparrow]\binom{x_2}{y_2} = \binom{x_1}{x_2}$
- $[\downarrow\downarrow]$ *is the Bottom-Bottom Operator where* $\binom{x_1}{y_1}[\downarrow\downarrow]\binom{x_2}{y_2} = \binom{y_1}{y_2}$
- $[\uparrow\downarrow]$ *is the Top-Bottom Operator where* $\binom{x_1}{y_1}[\uparrow\downarrow]\binom{x_2}{y_2} = \binom{x_1}{y_2}$
- $[\downarrow\uparrow]$ *is the Bottom-Top Operator where* $\binom{x_1}{y_1}[\downarrow\uparrow]\binom{x_2}{y_2} = \binom{y_1}{x_2}$

for all operations $\diamond \in [\uparrow\uparrow], [\downarrow\downarrow], [\uparrow\downarrow], [\downarrow\uparrow]$ and all languages $L_1, L_2 \subseteq V_\Sigma^*$ we have $L_1 \diamond L_2 = \bigcup\limits_{x \in L_1, y \in L_2} x \diamond y$, *based on [31].*

Example 2. Given the nucleotide sequences $x_1 = ACCTG$, $y_1 = AGCT$, $x_2 = TGAC$, $y_2 = CTAG$ and their respective sequences in V_Σ^*
$$\binom{x_1}{y_1} = \binom{A}{A}\binom{C}{G}\binom{C}{C}\binom{T}{T}\binom{G}{\epsilon} \text{ and } \binom{x_2}{y_2} = \binom{T}{C}\binom{G}{T}\binom{A}{A}\binom{C}{G},$$
the *Top-Top operation* is
$$\binom{x_1}{y_1}[\uparrow\uparrow]\binom{x_2}{y_2} = \binom{x_1}{x_2}, \text{ where } \binom{x_1}{x_2} = \binom{A}{T}\binom{C}{G}\binom{C}{A}\binom{T}{C}\binom{G}{\epsilon}.$$
   The *Top-Top operation* takes the two sequences in the top and creates a new sequence in the alphabet V_Σ^*. In this example, the two top sequences are x_1 and x_2, then the new sequence $\binom{x_1}{x_2} \in V_\Sigma^*$ is created. Equivalently, the results for *Bottom-Bottom operation* is
$$\binom{x_1}{y_1}[\downarrow\downarrow]\binom{x_2}{y_2} = \binom{y_1}{y_2} = \binom{A}{C}\binom{G}{T}\binom{C}{A}\binom{T}{G}$$

Fig. 1 shows automata as a result of some of the operations described above. At the beginning, the nucleotide sequences are represented as automata. Part (a) and part (b) show the automata for the *Top-Top* and *Bottom-Bottom* operations, respectively.

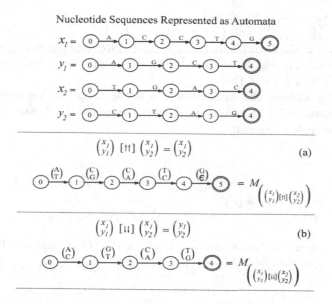

Fig. 1. Example of automata as a result of Top-Top and Bottom-Bottom operations. First, the given nucleotide sequences from Example 2 are represented as automata. (a) Automaton as a result of *Top-Top operation*. (b) Automaton as a result of *Bottom-Bottom operation*.

4.2 Automata Operations

Now, we define a set of automata operators used to create the new family of Pairwise Rational Kernels.

Definition 3. *Let $M_{\binom{x}{y}}$ represent the trivial automaton of $\binom{x}{y}$, for all $\binom{x}{y} \in V_\Sigma^*$. The following pairwise automata operators are defined:*

- *Direct-Sum-Left Pairwise Automata Operator ([⇇]) where*
$$M_{\binom{x_1}{y_1}}[\Leftcolon]M_{\binom{x_2}{y_2}} = M_{\left(\binom{x_1}{y_1}[\uparrow\uparrow]\binom{x_2}{y_2}\right)} \oplus M_{\left(\binom{x_1}{y_1}[\downarrow\downarrow]\binom{x_2}{y_2}\right)}$$
- *Direct-Sum-Right Pairwise Automata Operator ([⇉]) where*
$$M_{\binom{x_1}{y_1}}[\Rightcolon]M_{\binom{x_2}{y_2}} = M_{\left(\binom{x_1}{y_1}[\downarrow\uparrow]\binom{x_2}{y_2}\right)} \oplus M_{\left(\binom{x_1}{y_1}[\downarrow\uparrow]\binom{x_2}{y_2}\right)}$$
- *Tensor-Product-Left Pairwise Automata Operator ([⇐]) where*
$$M_{\binom{x_1}{y_1}}[\Leftarrow]M_{\binom{x_2}{y_2}} = M_{\left(\binom{x_1}{y_1}[\uparrow\uparrow]\binom{x_2}{y_2}\right)} \otimes M_{\left(\binom{x_1}{y_1}[\downarrow\downarrow]\binom{x_2}{y_2}\right)}$$
- *Tensor-Product-Right Pairwise Automata Operator ([⇒]) where*
$$M_{\binom{x_1}{y_1}}[\Rightarrow]M_{\binom{x_2}{y_2}} = M_{\left(\binom{x_1}{y_1}[\uparrow\downarrow]\binom{x_2}{y_2}\right)} \otimes M_{\left(\binom{x_1}{y_1}[\downarrow\uparrow]\binom{x_2}{y_2}\right)}$$

In this context, we have used operations over automata (i.e., $\oplus$ and $\otimes$) that we defined in Section 3.1 (i.e., *sum* + union and *product* · concatenation). We have changed the symbols just to make a difference over automata as graphical representation, but basically there are the same operations.

Example 3. Given the sequences in V_Σ^*
$\binom{x_1}{y_1} = \binom{A}{A}\binom{C}{G}\binom{C}{C}\binom{T}{T}\binom{G}{\epsilon}$ and $\binom{x_2}{y_2} = \binom{T}{C}\binom{G}{T}\binom{A}{A}\binom{C}{G}$,
from Example 2, the *Direct-Sum-Left Pairwise Automata Operator* produces the automaton shows in Fig. 2, as a result of the *sum* operation over the automata obtained in Fig. 1 (a) and (b).

Similarly, Fig. 3 shows the automaton as a result of the *product* operation over the same automata in Fig. 1 (a) and (b), which represent the *Tensor-Product-Left Pairwise Automata Operator*.

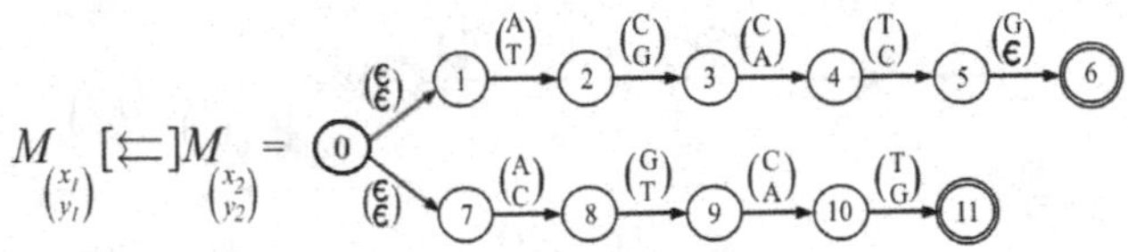

Fig. 2. Example of an Automaton as a result of the Direct-Sum-Left Pairwise Automata Operation (i.e., $M_{\binom{x_1}{y_1}}[\Leftarrow]M_{\binom{x_2}{y_2}} = M_{\left(\binom{x_1}{y_1}[\sqcap]\binom{x_2}{y_2}\right)} \oplus M_{\left(\binom{x_1}{y_1}[\sqcup]\binom{x_2}{y_2}\right)}$)

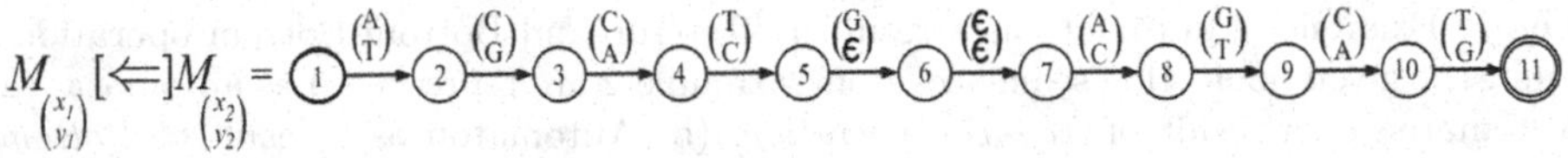

Fig. 3. Example of an Automaton as a result of the Tensor-Product-Left Pairwise Automata Operation (i.e., $M_{\binom{x_1}{y_1}}[\Leftarrow]M_{\binom{x_2}{y_2}} = M_{\left(\binom{x_1}{y_1}[\sqcap]\binom{x_2}{y_2}\right)} \otimes M_{\left(\binom{x_1}{y_1}[\sqcup]\binom{x_2}{y_2}\right)}$).

In this section, we define operations over automata that equivalently make pairwise relations over data (i.e., sequence data). For example, the Direct Sum Learning Pairwise Kernel, described in Section 3.4 Formula 3, is obtained by the sum operation of simple rational kernel represented by the transducer U, computed for a pairs of data. Similarly, the Tensor Learning Pairwise Kernel, Section 3.4 Formula 4, is the combination of product and sum operations over the simple rational kernel, represented by U.

In this research, we are using sequence data. We have represented them in a new alphabet, V_Σ, as pairs of sequences. We converted these pairs of data in automata, over the new alphabet. Then, pairwise operations over automata can be made first, yielding final automata as a result. Then, rational kernels can be obtained from these final automata. Based on this concept, we define a new family of kernels in the next section.

4.3   New Pairwise Rational Kernels

We introduce then the new Pairwise Rational Kernels using automata operations.

Definition 4. *Given $X \subseteq \Sigma^*$ and a transducer U, then a function $K : (X \times X) \times (X \times X) \to \mathbb{R}$ is defined as*

- **Direct-Way Pairwise Rational Kernel on Automaton Operations** $(K_{PRKDW-AO})$:
  $K((x_1, y_1), (x_2, y_2)) = U(I_1, I_2)$, *where* $I_1 = M_{\binom{x_1}{y_1}}$ *and* $I_2 = M_{\binom{x_2}{y_2}}$.
- **Direct-Sum Pairwise Rational Kernel on Automaton Operations** $(K_{PRKDS-AO})$:
  $K((x_1, y_1), (x_2, y_2)) = U(I_3, I_4)$, *where* $I_3 = M_{\binom{x_1}{y_1}}[\leftleftarrows]M_{\binom{x_2}{y_2}}$ *and* $I_4 = M_{\binom{x_1}{y_1}}[\rightrightarrows]M_{\binom{x_2}{y_2}}$.
- **Tensor Pairwise Rational Kernel on Automaton Operations** $(K_{PRKTP-AO})$:
  $K((x_1, y_1), (x_2, y_2)) = U(I_5, I_6)$, *where* $I_5 = M_{\binom{x_1}{y_1}}[\Leftarrow]M_{\binom{x_2}{y_2}}$ *and* $I_6 = M_{\binom{x_1}{y_1}}[\Rightarrow]M_{\binom{x_2}{y_2}}$.

The first PRK, i.e., $K_{PRKDW-AO}$, is obtained directly from the automata that represent the pairs (x_1, y_1) and (x_2, y_2) in the new alphabet V_{Σ}^* (i.e., $\binom{x_1}{y_1}$ and $\binom{x_2}{y_2}$). The other two PRKs represent the Direct-Sum $(K_{PRKDS-AO})$ and Tensor Product $(K_{PRKTP-AO})$ pairwise rational kernels. These new PRKs differ from previous PRK definitions in Section 3.4 in the way the data are represented and kernels are computed. Our motivations to develop these new kernels are based on using automata operations to optimize the performance of kernel computations.

4.4   Algorithms

We use a general algorithm, Algorithm 1, to compute the Pairwise Rational Kernels defined above. This algorithm is based on the basic idea of the Algorithm described by Cortes et al. [13]. The input of the algorithm is the pairs $(x_1, y_1), (x_2, y_2)$. In the first step, the I_i automata are computed using Definition 3, then the transducer composition and shortest-distance algorithm is used to finally obtain the Pairwise Rational Kernel values.

Algorithm 1. *Pairwise Rational Kernel Computation based on Automaton Operation*

INPUT: pairs $(x_1, y_1), (x_2, y_2)$ and WFST U

(i) Compute the operations in definition (4.4):

$$I_1 = M_{\binom{x_1}{y_1}} \qquad\qquad I_2 = M_{\binom{x_2}{y_2}},$$
$$I_3 = M_{\binom{x_1}{y_1}}[\Leftarrowtail]M_{\binom{x_2}{y_2}} \quad I_4 = M_{\binom{x_1}{y_1}}[\rightrightarrows]M_{\binom{x_2}{y_2}},$$
$$I_5 = M_{\binom{x_1}{y_1}}[\Leftarrow]M_{\binom{x_2}{y_2}} \quad I_6 = M_{\binom{x_1}{y_1}}[\Rightarrow]M_{\binom{x_2}{y_2}}.$$

(ii) use transducer composition to compute:

$$N_1 = I_1 \circ U \circ I_2$$
$$N_2 = I_3 \circ U \circ I_4$$
$$N_3 = I_5 \circ U \circ I_6$$

(iii) use a shortest-distance algorithm algorithm to compute the sum of the weights of all paths of N_1, N_2, N_3, and finally

$$K_{PRKDW}((x_1, y_1), (x_2, y_2)) = N_1$$
$$K_{PRKDS}((x_1, y_1), (x_2, y_2)) = N_2$$
$$K_{PRKT}((x_1, y_1), (x_2, y_2)) = N_3$$

RESULTS: values of $K((x_1, y_1), (x_2, y_2))$

---

5   Methods

In this section we describe experiments to predict metabolic networks using SVM-based algorithms in the training process, based on Pairwise Rational Kernels Automata Operations.

5.1   Dataset and Kernels

We used the data set of metabolic pathways of the yeast *Saccharomyces cerevisiae* [1, 17]. The data were taken from the KEGG pathway database [32]. The metabolic pathways were converted to a graph where enzymes are nodes and enzyme-enzyme relations are edges. In this dataset, enzymes are considered globular proteins produce by specific gene that have been previously identified. Enzyme-enzyme relations mean that two genes produce enzymes that catalyse successive reactions [3]. We used as data the nucleotide sequences of the genes.

We implemented the algorithm described in Section 4.4 using Open Finite-State Transducer (OpenFST) [33] and OpenKernels [34] libraries. We use the n-gram kernel described in Section 3.3. In this case, U was defined using a 3-gram ($n = 3$), where $U = T_3 \circ T_3^{-1}$. For example, the Direct-Sum Pairwise Rational Kernel Automata Operation was computed as $K_{PRKDS} = U(I_3, I_4) = \sum_{|z|=3} c_{I_3}(z)c_{I_4}(z)$, where I_3 and I_4 were obtained using Definition 4.

Yu et al. [35] have verified that sequence kernels are not good enough to predict interactions, however they are computationally inexpensive [14]. As recommended by [14, 35], we combine n-gram kernels with other kernels (i.e., PFAM [1] and Phylogenetic [3] kernels) that include evolutionary information to improve accuracy, as well we balance the positive and negative samples.

The Phylogenetic (PHY) kernel was obtained based on a Gaussian RBF kernel that associated each gene to 145 organisms, describing if the gene is present or not present in each organism [36]. The PFAM [3] kernel was computed based on a set of statistics E-values of the Hidden Markov Model (HMM) associated to PFAM database [37]. In bothe cases, we used available data from Ben-Hur et al., [1] and Allauzen et al., [17].

The experiments were separated in three different groups: Exp I included the new Pairwise Rational Kernels (K_{PRKDW}, K_{PRKDS} and K_{PRKTP}), Exp II considered the combination of the new PRKs with the Phylogenetic kernel ($K_{PRKDW+PHY}$, $K_{PRKDS+PHY}$ and $K_{PRKTP+PHY}$), Exp III included the new PRKs with the PFAM kernel ($K_{PRKDW+PFAM}$, $K_{PRKDS+PFAM}$ and $K_{PRKTP+PFAM}$).

5.2   Learning Procedure

We used Support Vector Machines with the new family of Pairwise Rational Kernels to predict the enzyme-enzyme reactions in the metabolic network. To balance our dataset, the program BRS-noint was executed to select non-interacting pairs [35]. We measured the accuracy using the Area Under Curve of Receiver Operating Characteristic (AUC ROC) [38], which defines a function of the rates of true-positives (predicted enzyme pairs are present in the dataset) and false-positives (predicted enzyme pairs are absent in the dataset). We denoted the accuracy as AUC values.

As supervised network inference methods require the knowledge of part of the network in the training process, we used a stratified cross-validation procedure with 10-fold cross-validation. In addition, execution times during the fold cross-validation process were also collected. The execution times included the automaton operations. All the experiments were executed on a PC intel i7CORE, 8MB RAM. We ran them ten times and calculated the average of AUC and processing time values.

We applied the McNemar's non-parametric test to compare the performance of the SVM classification method using the new PRKs with the previous PRKs obtained by [14]. McNemar's tests [39] have been recently used by [40] to prove significant statistical differences between classification methods. We used this test to determine statistical improvement during the prediction of metabolic networks, when the different Pairwise Rational Kernels were used with SVM-based algorithms.

McNemar's test defines the parameters N_{fs} and N_{sf} of two algorithms (Algorithm A and Algorithm B). N_{fs} is the number of times Algorithm A failed and Algorithm B succeeded, and N_{sf} is the number of times Algorithm A succeeded and Algorithm B failed. Then, a z score is calculated as:

$$z = \frac{(|N_{sf} - N_{fs}| - 1)}{\sqrt{(N_{sf} + N_{fs})}} \qquad (5)$$

When z is equal to 0, the two algorithms have similar performance. On the contrary, when z is greater than zero, the algorithm performance differ significantly, based on the confidence levels described by [40]. Additionally, if N_{fs} is larger than N_{sf} then Algorithm B performs better than Algorithm A. During our experiments, we collected data to compute the z score for all kernel combinations.

6   Results

Table 1 shows the SVM performance and execution times grouped by the pairwise rational kernels automata operations mentioned above (Exp I) and their combinations with Phylogenetic (Exp II) and PFAM (Exp III) kernels.

Table 1. Average AUC values, processing times and z score (McNemar's test) for Pairwise Rational Kernels using automaton operations

Exp	Kernel	AUC	Time (sec)	z-score
I	PRK-Direct-Way ($K_{PRKDW-3gram}$)	0.532	9.60	—
	PRK-Direct-Sum ($K_{PRKDS-3gram}$)	0.526	11.4	4.51
	PRK-Tensor-Product ($K_{PRKTP-3gram}$)	0.648	12.0	6.23
II	PRK-Direct-Way + PHY kernel ($K_{PRKDW-3gram+PHY}$)	0.674	129.8	—
	PRK-Direct-Sum + PHY kernel ($K_{PRKDS-3gram+PHY}$)	0.533	133.6	7.51
	PRK-Tensor-Product + PHY kernel ($K_{PRKTP-3gram+PHY}$)	0.789	131.8	6.25
III	PRK-Direct-Way + PFAM kernel($K_{PRKDW-3gram+PFAM}$)	0.771	129.9	—
	PRK-Direct-Sum + PFAM kernel ($K_{PRKDS-3gram+PFAM}$)	0.538	133.4	5.81
	PRK-Tensor-Product + PFAM kernel ($K_{PRKTP-3gram+PFAM}$)	0.877	132.0	6.22

When only n-gram kernels were used (Exp I), the best accuracy value was obtained with the Tensor-Product Pairwise Rational Kernel ($K_{PRKTP-3gram}$), as has also occurred in other proposals, such as [14]. However, the fastest execution time was obtained with the PRK-Direct-Way ($K_{PRKDW-3gram}$), which has been defined and used in this research for first time. Unfortunately, using only n-gram kernels yield low accuracy values (AUC). These results coincided with Yu et al. [35] and Roche-Lima et al. [14], which recommended using other kernels with evolutionary information (i.e., PHY and PFAM kernels) to improve the predictor accuracy. Thus, we also obtained the results in Exp II and Exp III with Phylogenetic and PFAM kernels, respectively. The accuracy values were improved in all cases, while maintaining adequate processing times. The best accuracy value was AUC=0.877, corresponding to *PRK-Tensor-Product + PFAM* ($K_{PRKTP-3gram+PFAM}$) kernel (Exp III). Furthermore, this was the highest accuracy and best processing time obtained with all the PRKs included in this research and reported by [14]. We noted that the increase in processing time in Exp II and Exp III was due to using kernels which were not automata-based. This indicated the efficiency of using auotmaton operations to obtain the kernels in this application.

To compare the results using the PRKs based on Automata Operations with the performance of PRKs described by [14], we used McNemar's test to determine statistical significance. Since the PRK-Direct-Way kernels were defined for first time in this research, we did not include them in this analysis. Thus, we compared the SVM methods using the n-gram Direct-Sum Pairwise Rational Kernels obtained here (i.e., $K_{PRKDS-3gram}$, $K_{PRKDS-3gram+PHY}$ and $K_{PRKDS-3gram+PFAM}$ kernels) with the similar kernels used by [14]. Based on the N_{fs} and N_{sf} values, the z-scores were computed, given in Table 1. In all cases, the new Direct-Sum Pairwise Rational Kernels, described in this paper, performed better. These z-score also proved that the difference was statistically significant with a confidence level of 99% (based on One-Tailed Prediction Confidence Level described by [40]). Likewise, we compared the Tensor-Product Rational Kernels from both papers (i.e., $K_{PRKTP-3gram}$, $K_{PRKTP-3gram+PHY}$ and $K_{PRKTP-3gram+PFAM}$). In this case, all the collected values also yielded statistical significant differences of the performances in favour of the new kernels obtained in this research with confidentiality levels of 99% (see z-scores in Table 1).

7   Conclusion

We introduced new pairwise rational kernels based on automata operations. Using these new kernels, we improved the performance of SVM-based methods to predict metabolic network, decreasing the computational costs in term of processing time.

The new PRKs measured similarities between two pairs of sequences using language and automaton representations. Initially, we defined new notations and operations, and later we described the new kernels based on transducers combined with automata. We ran several experiments using the new Pairwise Rational Kernels for metabolic network prediction methods. As a result, we obtained better accuracy and execution times than other previous works [14]. We have also proved that the performance improvements, using these new PRKs, are statistically significant with high level of confidentiality.

As future work, new pairwise rational kernels may be developed based on other previous pairwise kernel such as metric learning and Cartesian kernels [29, 30], as well as other automaton operations.

Acknowledgments. This work is funded by Natural Sciences and Engineering Research Council of Canada (NSERC) and Microbial Genomics for Biofuels and Co-Products from Biorefining Processes (MGCB2) project.

References

1. Ben-Hur, A., Noble, W.S.: Kernel methods for predicting protein–protein interactions. Bioinformatics 21(suppl. 1), i38–i46 (2005)
2. Tsuda, K., Noble, W.S.: Learning kernels from biological networks by maximizing entropy. Bioinformatics 20(suppl. 1), i326–i333 (2004)
3. Yamanishi, Y.: Supervised inference of metabolic networks from the integration of genomic data and chemical information. In: Elements of Computational Systems Biology, pp. 189–212. Wiley (2010)

4. Yamanishi, Y., Araki, M., Gutteridge, A., Honda, W., Kanehisa, M.: Prediction of drug–target interaction networks from the integration of chemical and genomic spaces. Bioinformatics 24(13), i232–i240 (2008)
5. O'Madadhain, J., Hutchins, J., Smyth, P.: Prediction and ranking algorithms for event-based network data. ACM SIGKDD Explorations Newsletter 7(2), 23–30 (2005)
6. Taskar, B., Wong, M.F., Abbeel, P., Koller, D.: Link prediction in relational data. In: Advances in Neural Information Processing Systems (2003)
7. Lothaire, M.: Applied Combinatorics on Words. Cambridge University Press (2005)
8. Allauzen, C., Mohri, M., Riley, M.: Statistical modeling for unit selection in speech synthesis. In: Proceedings of the 42nd Annual Meeting on Association for Computational Linguistics, ACL 2004. Association for Computational Linguistics, Stroudsburg (2004)
9. Albert, J., Kari, J.: Digital image compression. In: Handbook of weighted automata, EATCS Monographs on Theoretical Computer Science. Springer (2009)
10. Holmes, I.: Using guide trees to construct multiple-sequence evolutionary hmms. Bioinformatics 19, i147–i157 (2003)
11. Westesson, O., Lunter, G., Paten, B., Holmes, I.: Phylogenetic automata, pruning, and multiple alignment. arXiv preprint arXiv:1103.4347 (2011)
12. Bradley, R.K., Holmes, I.: Transducers: An emerging probabilistic framework for modeling indels on trees. Bioinformatics 23(23), 3258–3262 (2007)
13. Cortes, C., Mohri, M.: Learning with weighted transducers. In: Proceedings of the 2009 Conference on Finite-State Methods and Natural Language Processing: Post-Proceedings of the 7th International Workshop FSMNLP 2008, pp. 14–22. IOS Press, Amsterdam (2009)
14. Roche-Lima, A., Domaratzki, M., Fristensky, B.: Metabolic network prediction through pairwise rational kernels. Submitted BMC Bioinformatics (April 2014)
15. Kosiol, C., Holmes, I., Goldman, N.: An empirical codon model for protein sequence evolution. Molecular Biology and Evolution 24(7), 1464–1479 (2007)
16. Roche-Lima, A., Oncina, J.: Bioinformatics applied to genetic study of rumen microorganism. In: 2nd Conference of IT in Agriculture Scienc, Havana, Cuba (November 2007)
17. Allauzen, C., Mohri, M., Talwalkar, A.: Sequence kernels for predicting protein essentiality. In: Proceedings of the 25th International Conference on Machine Learning, ICML 2008, pp. 9–16. ACM, New York (2008)
18. Zien, A., Ratsch, G., Mika, S., Schalkopf, B., Lengauer, T., Macller, K.R.: Engineering support vector machine kernels that recognize translation initiation sites. Bioinformatics (Oxford, England) 16(9), 799–807 (2000)
19. Kuang, R., Ie, E., Wang, K., Wang, K., Siddiqi, M., Freund, Y., Leslie, C.: Profile-based string kernels for remote homology detection and motif extraction. Journal of Bioinformatics and Computational Biology 3(3), 527 (2005)
20. Leslie, C.S., Eskin, E., Cohen, A., Weston, J., Noble, W.S.: Mismatch string kernels for discriminative protein classification. Bioinformatics (Oxford, England) 20(4), 467–476 (2004)
21. Haussler, D.: Convolution kernels on discrete structures. Technical Report UCSCCRL-99-10. University of California at Santa Cruz. (1999)
22. Takimoto, E., Warmuth, M.: Path kernels and multiplicative updates. Journal of Machine Learning Research 4(5), 773–818 (2004)
23. Cortes, C., Haffner, P., Mohri, M.: Rational kernels: Theory and algorithms. J. Mach. Learn. Res. 5, 1035–1062 (2004)

24. Rabin, M.O., Scott, D.: Finite automata and their decision problems. IBM Journal of Research and Development 3(2), 114–125 (1959)
25. Mohri, M.: Weighted automata algorithms. In: Handbook of weighted automata, pp. 213–254. Springer (2009)
26. Lee, K.H., Lee, D., Lee, K., Kim, D.W.: Possibilistic support vector machines. Pattern Recognition 38(8), 1325–1327 (2005)
27. Moreau, Y.: Kernel methods for genomic data fusion. In: Sixth International Workshop on Machine Learning in Systems Biology (MLSB 2012), Basel, Switzerland (September 2012)
28. Lodhi, H., Saunders, C., Shawe-Taylor, J., Cristianini, N., Watkins, C.: Text classification using string kernels. The Journal of Machine Learning Research 2, 419–444 (2002)
29. Brunner, C., Fischer, A., Luig, K., Thies, T.: Pairwise support vector machines and their application to large scale problems. Journal of Machine Learning Research 13, 2279–2292 (2012)
30. Kashima, H., Oyama, S., Yamanishi, Y., Tsuda, K.: Cartesian kernel: An efficient alternative to the pairwise kernel. IEICE TRANSACTIONS on Information and Systems 93(10), 2672–2679 (2010)
31. Kari, L.: On language equations with invertible operations. Theoretical Computer Science 132(1), 129–150 (1994)
32. Kanehisa, M., Araki, M., Goto, S., Hattori, M., Hirakawa, M., Itoh, M., Katayama, T., Kawashima, S., Okuda, S., Tokimatsu, T., et al.: KEGG for linking genomes to life and the environment. Nucleic Acids Research 36(suppl. 1), D480–D484 (2008)
33. Allauzen, C., Riley, M., Schalkwyk, J., Skut, W., Mohri, M.: OpenFST: A general and efficient weighted finite-state transducer library. In: Holub, J., Žd'árek, J. (eds.) CIAA 2007. LNCS, vol. 4783, pp. 11–23. Springer, Heidelberg (2007)
34. Allauzen, C., Mohri, M.: Openkernel library (2012)
35. Yu, J., Guo, M., Needham, C.J., Huang, Y., Cai, L., Westhead, D.R.: Simple sequence-based kernels do not predict protein–protein interactions. Bioinformatics 26(20), 2610–2614 (2010)
36. Yamanishi, Y., Vert, J.P., Kanehisa, M.: Supervised enzyme network inference from the integration of genomic data and chemical information. Bioinformatics 21(suppl. 1), i468–i477 (2005)
37. Gomez, S.M., Noble, W.S., Rzhetsky, A.: Learning to predict protein–protein interactions from protein sequences. Bioinformatics 19(15), 1875–1881 (2003)
38. Gribskov, M., Robinson, N.L.: Use of receiver operating characteristic (ROC) analysis to evaluate sequence matching. Computers & Chemistry 20(1), 25–33 (1996)
39. McNemar, Q.: Note on the sampling error of the difference between correlated proportions or percentages. Psychometrika 12(2), 153–157 (1947)
40. Bostanci, B., Bostanci, E.: An evaluation of classification algorithms using McNemar's test. In: Bansal, J.C., Singh, P.K., Deep, K., Pant, M., Nagar, A.K. (eds.) Proceedings of Seventh International Conference on Bio-Inspired Computing: Theories and Applications (BIC-TA 2012). AISC, vol. 201, pp. 15–26. Springer, Heidelberg (2013)

Author Index

Amorim, Ivone 70

Bianchi, Maria Paola 84
Biegler, Franziska 98
Blanchet-Sadri, Francine 110
Brzozowski, Janusz 124

Čevorová, Kristína 136

De Biasi, Marzio 148
Demaille, Akim 162
Domaratzki, Michael 332
Dumitran, Marius 176
Duret-Lutz, Alexandre 162

Esparza, Javier 1

Fristensky, Brian 332
Fujiyoshi, Akio 188

Gil, Javier 176
Goldner, Kira 110
Gusev, Vladimir V. 200

Han, Yo-Sub 238

Ibarra, Oscar H. 211

Jirásková, Galina 136, 226

Ko, Sang-Ki 238
Krajňáková, Ivana 136
Kůrka, Petr 252

Lombardy, Sylvain 162
Luttenberger, Michael 1

Machiavelo, António 70
Maia, Eva 264
Maletti, Andreas 278
Manea, Florin 176
McQuillan, Ian 98
Mereghetti, Carlo 84
Mitrana, Victor 176
Moreira, Nelma 264

Osterholzer, Johannes 290
Otto, Friedrich 16

Palano, Beatrice 84
Palmovský, Matúš 226
Pighizzini, Giovanni 42
Pittou, Maria 304
Pribavkina, Elena V. 200

Rahonis, George 304
Rashid, Jibran 318
Reis, Rogério 70, 264
Roche-Lima, Abiel 332

Saiu, Luca 162
Sakarovitch, Jacques 162
Schlund, Maximilian 1
Šebej, Juraj 226
Shackleton, Aidan 110
Sirakoulis, Georgios Ch. 58
Szykuła, Marek 124

Vávra, Tomáš 252

Yakaryılmaz, Abuzer 148, 318